Nuclear Energy: Salvation or Suicide?

NUCLEAR ENERGY: Salvation or Suicide?

AN EDITORIALS ON FILE BOOK

Edited by Carol C. Collins

Facts On File Publications
New York, New York • Bicester, England

Nuclear Energy: Salvation or Suicide?

Published by Facts On File, Inc.
460 Park Avenue South, New York, N.Y. 10016

Library of Congress Cataloging in Publication Data
Main entry under title:

Nuclear energy, salvation or suicide?

Includes index.
1. Atomic power industry—United States—Addresses, essays, lectures. 2. Atomic power-plants—United States—Addresses, essays, lectures. 3. American newspapers—Sections, columns, etc.—Editorials. I. Collins, Carol Chambers.
HD9698.U52N818 1984 333.79'24'0973 84-5943
ISBN 0-87196-817-7

International Standard Book Number: 0-87196-817-7
Library of Congress Catalog Card Number: 84-5943
9 8 7 6 5 4 3 2
PRINTED IN THE UNITED STATES OF AMERICA

Contents

Preface

In December 1982, the American nuclear power industry turned twenty-five years old. Commercial nuclear energy was born in Shippingport, Pennsylvania, where the first commercial reactor was put on line by Duquesne Light Co. engineers in 1957. Ironically, it was also in Pennsylvania that the industry suffered its most severe setback, the 1979 accident at Three Mile Island's Unit 2 reactor.

It now appears doubtful whether the nuclear energy program will ever recover. No new domestic plants have been ordered since 1978, and every reactor ordered since 1974 has been abandoned. The future of nuclear power in the U.S. is described as "bleak" in a study issued in February 1984 by the Congressional Office of Technology Assessment. One hundred plants have been cancelled in the past decade, the OTA report noted, predicting that the industry would have to undergo a radical transformation to survive even into the 1990's. The problems cited in the report did not originate with the Harrisburg incident—high capital costs, operating difficulties, competition from alternative energy sources—but the erosion of public confidence in the nuclear power program since the mishap at Three Mile Island has been massive.

Rarely has an issue of such national economic importance involved such a range of emotionally-charged issues. Every report intended to clarify some area of dispute, whether it be the radiation hazards posed by nuclear wastes or the fear that exports of reactor fuel and technology could speed the proliferation of nuclear weapons, has only served to further divide industry adherents and critics. The opinion at one end of the scale is that nuclear energy is the most important gift we can give to future generations, supplying them not only with an inexhaustible energy source but with the means to remain economically prominent as a nation. At the other extreme is the belief that the continued use of nuclear energy is dangerous not only to those who now live near power plants but also to future generations whose health may be affected by exposure to radioactive wastes, and that its use also threatens America's economic stature by encouraging the neglect of cheaper alternative energy sources.

In the pages that follow, leading American newspaper editorialists discuss the basic issues of safety and economics involved in the nuclear power debate. Their opinions, molded by all imaginable combinations of geographical and political perspective, provide a valuable and lively treatment of the subject. No attempt has been made to favor one viewpoint over another.

March, 1984 — Carol C. Collins

Part I: Nuclear Plant Safety

The safe operation of nuclear power plants is of paramount concern both to those who vehemently oppose their use and to those who believe they are essential to fulfill this nation's future energy needs and economic potential. The legacy of fear from Three Mile Island cannot be overestimated; its impact, upon a public who had been told that such an occurrence would be absurdly unlikely, was tremendous. The distrust created not only by the very fact of the accident but by its handling has scarcely diminished during the last five years, and has had a strong effect on nearly every aspect of plant development, regulation and licensing. Adherents of nuclear energy fear that another such incident could well toll the death knell of the industry.

There are of course some popular misconceptions about nuclear power plants, as there are bound to be about any complex technology in so esoteric a domain as nuclear physics. This is particularly true because the awesome power of nuclear fission is associated indelibly with another of its applications, the creation of atomic weapons. The nuclear energy industry in addition has earned a good deal of public mistrust through its past tendency to assume a 'father-knows-best' attitude, treating consumers as children incapable of reaching a sound judgment. The resulting credibility gap, exacerbated in the early days of nuclear power by the federal desire to see the industry succeed, has caused a backlash that may be impossible to combat. It is unfortunate that most of the literature available about the difficult topic of nuclear plant safety is heavily biased by the predispositions of its authors to either favor or oppose the industry.

Beyond public relations and perception problems, however, the industry continues to discover new problems with nuclear plants that may endanger public safety. Not all of the 'bugs' have been worked out in the application of this comparatively youthful technology. Even many industry officials agree that the American nuclear energy program expanded too quickly for its own good, progressing to the building of large reactors before sufficient research had been completed. In 1983, the Nuclear Regulatory Commission temporarily closed down five nuclear plants because of uncertainties about the safety of their cooling pipes, which were suspected to be prone to cracking; hazardous cracks had been found at 13 of 17 previously inspected plants of the same type. There are also continuing complaints about quality assurance procedures. Most recently, the NRC denied an operating license for the nearly completed Byron Nuclear Power Station in Illinois, arguing that Commonwealth Edison Co., its owner, had not adequately monitored the work of contractors on the site. Paradoxically, it is in part the more stringent evaluations and safety standards required for new plants that have alerted the public to new problems and heightened fears about safety.

In the wake of Three Mile Island, the nuclear industry and the public turned their newly galvanized attention to many areas of plant construction, operation and regulation, using the knowledge gained in Pennsylvania to enhance future safety precautions. Both the original reaction to the 1979 accident and the most controversial topics to grow out of subsequent investigations—operator training, evacuation plans, reactor siting and licensing, etc.—are weighed in the following editorials.

The Three Mile Island Accident: How Close to Meltdown?

A series of breakdowns in the cooling system at Pennsylvania's Three Mile Island plant March 28, 1979 resulted in the shutdown of the facility's No. 2 reactor and the release of radiation into the atmosphere. The facility, owned by the Metropolitan Edison Co. and two other utilities, was located 10 miles south of Harrisburg, in the Susquehanna River Valley. By Friday, March 30, the Nuclear Regulatory Commission was warning of the possibility of a core meltdown of the superheated plant, with the release of lethal radioactivity over a wide area. (Ironically, such a catastrophe had been the basis for the plot of a commercial film, "The China Syndrome," released March 16.) The NRC also raised the threat that a hydrogen gas bubble that had formed in the overheated reactor vessel of the crippled plant could explode without warning. It was feared that the force of such an explosion could destroy the walls of the containment building, thus releasing all the radioactivity inside. The containment building had the highest level of contamination recorded in commercial nuclear operations history. Gov. Thornburgh March 30 urged pregnant women and preschool children within a five-mile radius of the plant to leave the area at once.

The crisis eased on April 2, the sixth day of the near-disaster, when nuclear experts announced that the gas bubble had shown a "dramatic drop" in size, and fuel temperatures in the core continued to drop significantly. It was another week, however, before Pennsylvania's governor rescinded his warning to pregnant women, and stated that there was no danger to public health from milk or drinking water in the area.

Conflicting reports about the cause of the accident were offered by officials from the NRC and from Met Ed, the plant's operator. The general outline of the accident that emerged, however, was that in the early morning hours of March 28, a valve had failed in a pump in the primary core cooling system, leading to a halt in the flow of water used to take heat away from the reactor. As a result of the interrupted water flow, the emergency core cooling system was automatically set off, to deliver adequate cooling water to the reactor's uranium fuel rods. At a critical moment, the NRC stated, a plant operator had mistakenly turned off the emergency system, and, after a period of time, turned it back on. During that period, it was believed, the core was uncovered and damaged, as some of the pellets of enriched uranium fuel became so hot that they either melted through or ruptured the zirconium-clad tubes that held them. Also, some of the water used to cool the core spilled onto the floor of the reactor building. When some of the radioactive water became steam, it was vented into the atmosphere above the plant to relieve pressure. These and subsequent ventings resulted in the release over a four-county area of small amounts of radioactive iodine, krypton and xenon. Met Ed angered Thornburgh and other Pennsylvania officials with these gas ventings, which were not announced, and also by dumping approximately 400,000 gallons of slightly radioactive waste water into the Susquehanna River March 29.

The gas bubble inside the reactor, it was believed, was the result of the unexpected starts and stops in the primary and emergency cooling systems. Technical experts theorized that when the intense pressure in the reactor vessel dropped (during failure of the cooling systems) hydrogen and oxygen gas, which was formed during fission as water molecules were torn into their constituent elements, remained in the reactor. It was revealed March 31 that a small hydrogen explosion had occurred inside the reactor March 28.

On the thirteenth day after the No. 2 reactor core was damaged, Harold Denton, the federal official in charge of the plant, announced that the crisis was over. Two and a half weeks later, the NRC said that the cooling process to bring the crippled reactor to a shutdown condition was completed. The NRC reported April 30 that it believed the uranium core of Three Mile Island No. 2, normally bathed in pressurized coolant, had been left uncovered for as long as 50 minutes. In testimony before a Senate subcommittee, Harold Denton said: "It's still too early to say how close we were to a meltdown, but the core clearly reached very high temperatures, having come uncovered several times."

The Boston Globe

Boston, Mass., March 30, 1979

It will take years to evaluate the effect on the health of persons living near the Three Mile Island nuclear power plant of the accident there Wednesday during which radioactive steam and radiation escaped into the open environment.

It took just an instant, however, to assess the accident's impact on the nuclear power industry. It is a damaging blow to an industry already buffeted by economic and political storms. And it will surely complicate efforts to fashion a coherent energy policy in the nation for the years immediately ahead.

The nuclear industry has repeatedly insisted that its plants were virtually fail-safe. That claim is simply no longer credible. Mistakes are made. Accidents can happen, and they do. The real issue is the risks those accidents pose compared with the risks posed by alternative energy sources, most notably coal mining and burning.

Two months ago, portions of the study of reactor safety on which the nuclear industry and the government has relied for several years were disavowed by the Nuclear Regulatory Commission (NRC). Then earlier this month the NRC ordered the closing of five nuclear plants because an error was discovered in the computer model on which their ability to withstand an earthquake had been predicated. And now, the Three Mile Island accident, possibly the worst accident the domestic nuclear industry has experienced. And all of this has coincided with the release of the movie "The China Syndrome," a suspense thriller centering on both technical and human failings in the operation and management of a nuclear plant.

The immediate response to the Three Mile Island accident must and will be to identify precisely the cause of the accident, either human or technological. Systematic efforts to monitor the health of those who may have been contaminated by the relatively low-level radiation that escaped from the plant will have to be inaugurated.

Beyond these steps, however, a broader initiative must be undertaken and it will almost surely have to come from the White House. Sections of the nation, most notably New England, are already heavily dependent upon nuclear power. That situation cannot be changed overnight. Further, nuclear power, despite its rising costs, still remains an economically attractive, domestically produced alternative to oil-dependent electricity production.

But a hefty segment of the American public — one surely augmented by the Three Mile Island accident — is understandably worried about more than costs, more even than foreign dependency. They are worried about nuclear safety — in the plants, in the transporation of nuclear wastes, in the storage of nuclear wastes. They are concerns worrisome enough to compel Jimmy Carter, nuclear engineer turned President, to pronounce nuclear energy our "last resort." Yet they are concerns not so overwhelming as to exclude from the Carter energy program a substantial increase in the number of nuclear plants operating in this country by the year 2000.

The nation needs from the President an honest and open assessment of nuclear power. The assessment must include a credible evaluation of its risks. It must include a credible evaluation of the costs of stemming our reliance on nuclear power. It must include a credible evaluation of the nation's ability to meet the issues of nuclear waste transportation and storage. And if, in the President's view, we must rely increasingly on nuclear power, he must tell us how the utilities, and ultimately their consumers, are to pay for it.

All of this information is now lacking. The Three Mile Island accident exposed not so much the possibilities of a "China Syndrome" as the reality of a longstanding political syndrome — the failure of the nation's leadership to articulate a coherent and forthright position on the proper place of nuclear power in our energy future.

RAPID CITY JOURNAL—

Rapid City, S.D., April 4, 1979

The accident in the nuclear plant at Three Mile Island should place the question of nuclear power squarely before the public whose future will be shaped by the decisions that are made.

Americans may be induced to look more closely at those who are anxious to commit them to a future in which their diminished standard of living hopefully will be compensated by a better quality of life.

It is being predicted that the events on the Susquehanna River will lead to the demise of the nuclear power industry. The fact is, however, that political leadership on the issue has already allowed anti-nuclear activists to bring the industry almost to its knees.

It is ironic that the Pennsylvania accident comes at a time when there seemed to be some movement to rally the country to confront the degree to which its economic future rests on the nuclear option. That movement may now be lost in the hearings and handwringings with which this nation is inclined to prolong its crises.

At least six years were lost in the development of offshore oil resources because of the 1969 oil spill near Santa Barbara. Even longer delay could be the aftermath of the Three Mile Island incident because politicians will be encouraged to join in fighting nuclear power.

In looking more closely at the nuclear energy alternative, it should be recognized that there is always risk in life, in invention and in discovery. During the 18 years in which nuclear plants have operated without inflicting serious injuries, there have been nearly 850,000 deaths caused by motor vehicles in the United States. Had those who now oppose nuclear plants been around for the launching of the gasoline buggy or the airplane they would have had reason to oppose those developments.

The hard question remains: Is the risk of nuclear power greater than the benefit? Nuclear energy provided some 13 percent of the nation's electrical needs last year. If those plants are closed down, where do the millions of barrels of oil needed to replace nuclear fuels come from?

Are Americans ready to go back to being one-car families, give up their recreational vehicles, pleasure boats and reduce drastically the fuel used for air conditioning and home heating?

We cannot have both a lifestyle based on the high usage of energy and an energy-production environment that is free from the risk of nuclear contamination and even smoke from coal.

Because there is no mechanism for a national referendum to decide the philosophic and economic issues and determine the will of the majority, the political arena is where decisions must be made. That process will require reasoning that is more balanced than what we have seen to date.

THE ARIZONA REPUBLIC

Phoenix, Ariz., April 2, 1979

THE MAN generally conceded to be the "father" of the hydrogen bomb, Dr. Edward Teller, has a simple explanation of why nuclear power has been embattled since its birth.

It was born as a weapon of destruction, not good. If electricity had first been demonstrated in a prison electric chair, Dr. Teller said, it also might never have been fully accepted, either.

The fact that nuclear power has established a safety and efficiency record far better than any other energy source has made no difference. Nuclear's critics would like to sharpshoot it into extinction.

Anti-nuclear groups will have a field day, of course, with the Three Mile Island steam radiation accident in Pennsylvania. It has all the makings of prophesies fulfilled.

But, while the Pennsylvania episode cannot and should not be dismissed, its actual proportions are not as disastrous as some would have it.

Radiation being emitted from the plant is far less than natural irradiation absorbed every year by millions of persons—from the sun, from flying in commercial aircraft, from medical X-rays, from natural geological surroundings in certain parts of the country.

But, even if nuclear power has certain dangers, the broader question is whether the American people are unwilling to accept risks in energy production as a price of technology and advancement.

One wonders that if today's nuclear critics had been alive since the beginning of man's existence would they have:

✔Banned the caveman's first use of fire because of the destruction it unleashed on forests and animals?

✔Outlawed the digging of coal in shafts because of the hundreds of thousands of lives lost to caveins and gas asphyxiation?

✔Prohibited the drilling and refining of petroleum because of the flaming disasters at wells that have taken untold lives?

✔Prevented the evolution of commercial aviation because of the thousands of lives lost in the last half-century in crashes?

✔Stopped space exploration because of the horrible burning deaths of three astronauts—more than 10 percent of the astronaut corps at the time—in their Gemini capsule at Cape Kennedy?

It's unfortunate that the events at the Three Mile Island nuclear plant will overshadow the unpublicized, uninterrupted safety of the nation's other 71 nuclear plants, and more than 100 overseas plants. In the Chicago area, for example, some 45 percent of all electrical power is from accident-free nuclear power.

The nuclear power industry, and government regulators, have an urgent responsibility to quickly ascertain the cause and consequences of the Pennsylvania accident, and put it into the proper perspective before naysayers create hysteria.

THE PLAIN DEALER

Cleveland, Ohio, March 31, 1979

As the story of a disabled nuclear plant continued to unfold yesterday near Harrisburg, Pa., only two points were clear — officials did not seem to know what was going on inside the Three Mile Island plant and they did not seem to know what they ought to do outside the plant.

Near the plant, in Bainbridge, an angry father walked into an elementary school to take his children away. "It's disgusting," he said. "They're not organized. They're not ready for anything like this."

And the shame of it is the man was right; authorities in Pennsylvania were not ready to deal with the results of radiation leaks from a nuclear power plant. Even the governor told a reporter that he was not sure of the credibility of the information he was being given. And the governor was the one who had to decide what to do with the people who live near the plant.

As more radiation leaked into the atmosphere from the disabled plant, officials could not agree whether it was an accident or on purpose, whether people should be evacuated or simply told to stay indoors. Finally, they did both, telling most people to stay inside while evacuating pregnant women and small children — those most likely to be in danger from small amounts of radiation — from a five-mile radius of the plant.

Officials described most of the half million people who live near the plant as being calm. There was little panic. But it is a wonder why, particularly when in the middle of all reports an air raid siren was triggered in downtown Harrisburg. It turned out this was a warning to stay indoors. It is surprising that thousands of people did not head for the hills.

Perhaps there is a good lesson to be learned from all this. James R. Schlesinger, U.S. energy secretary, said the accident and subsequent events "will allow us to better understand the problems of nuclear power."

It is hoped that includes both better safety measures and a plan for dealing with the results of such an accident as the one at Three Mile Island.

According to the director of the Ohio nuclear preparedness office, this state has a set of procedures to be used if radiation reaches certain levels, including the evacuation of pregnant women and small children at one level and mass evacuation at a higher level.

Still, these decisions are based on the kinds of facts available in Pennsylvania. The information there changed almost by the hour. And the residents had to accept the information on faith. It is doubtful that the residents near Three Mile Island will ever again have as much faith as they had before.

The Washington Post

Washington, D.C., March 30, 1979

THE ACCIDENT at the Three Mile Island nuclear reactor is an event that, by every prior calculation, should have been exceedingly improbable. Some of the questions are pretty obvious: Are those safety calculations reliable? Do they require reconsideration? Was it mechanical failure, or human error? If an operator made a mistake, it is not going to be corrected by redesigning pumps and pipes.

The reactor has vented some radioactive gas into the atmosphere of central Pennsylvania, and yesterday it was still emitting radiation through its concrete shield walls. In both cases, the levels of radiation were low. Federal officials said that they were of the same magnitude as background radiation from natural sources. The exposure to people outside the plant was apparently similar to the exposure that an airline passenger might receive in a flight at high altitude. But, as the Department of Health, Education and Welfare warned earlier this month, current research indicates that any increase in radiation is harmful, and low-level radiation from natural sources may be an important cause of cancer. The health danger to the general public in the accident at Three Mile Island appears to be small. But it is not insignificant.

There is no way, unfortunately, to generate electricity without risk. Good engineering and wise regulation can reduce that risk, but never to zero. Hydroelectric plants are generally the safest of all the possibilities, but even dams have been known to burst. Solar technology is available for heating bath water. But generating power from sunshine, on a commercial scale, is decades in the future. Nearly one half of the nation's electricity comes from coal-fired generators, and the process of burning coal inevitably puts highly toxic gases into the air that we breathe. A coal furnace represents a substantially greater threat to public health than a uranium reactor of the same size. It would be sadly ironic if the only effect of the accident at Three Mile Island were to shift American utilities away from uranium and more heavily than ever onto coal.

Since there are no risk-free solutions, the most sensible course is to continue to rely on a variety of different sources of power, including nuclear—and to keep developing them slowly and cautiously. But if the development of nuclear power is to proceed slowly and cautiously, two things must happen. One is that the need for power must be limited, since the faster Americans increase their demand for additional electricity, the harder it will be to remain prudent. The other thing that must happen if nuclear power is to be developed at all is that the public alarm verging on hysteria concerning the dangers be taken account of and honestly answered. That is why the explanations of the Three Mile Island accident will have to be exceedingly thorough, persuasive and public.

The Pittsburgh Press

Pittsburgh, Pa., April 1, 1979

It will be days, maybe even a week, before anyone can tell just how the nuclear drama at Three Mile Island will end.

But it's a foregone conclusion that the cause of nuclear energy has suffered a major setback.

The accident at the atomic power plant south of Harrisburg has so far harmed nobody. But it has inconvenienced many and alarmed many more.

And even if the smouldering furnace in the atomic powerhouse is finally brought under full control without further threat to the public health, proponents of nuclear energy will hereafter find theirs an even tougher row to hoe.

★ ★ ★

Indeed, President Carter already has voiced the view that this incident "will probably lead inexorably toward even more stringent safety design mechanisms and standards" in the nuclear industry.

That foreshadows ever-longer delays in the building of new nuclear power plants at a most inopportune time — for the United States is facing a growing energy crunch as a result of continuous price-jacking by foreign oil producers and the disruption of oil supplies from strife-torn Iran.

That, however, is a concern for another day. At the moment, the chief concern is extinguishing the damaged atomic furnace — the nuclear reactor — at Three Mile Island.

In this respect, the next 72 hours may well be — to use the nuclear vernacular — critical. For within this period the scientists and engineers and technicians who are striving to bring the reactor to a "cold shutdown" should be able to gauge how well the operation is progressing.

★ ★ ★

The ultimate peril, of course, is a complete "meltdown" of the heart of the reactor.

Although the structure housing the reactor is designed to contain the horrendous radiation such a meltdown would produce, nobody can guarantee this for sure — because nothing like this has ever occurred. And even now officials view the chance of such a happening as "remote."

But if a meltdown should crop up as a distinct possibility rather than just a remote risk, thousands — perhaps millions — of persons would have to be evacuated from their homes as a precaution.

The hope is that no such drastic step will become necessary. But only time will tell the tale.

Meanwhile, as the countdown at Three Mile Island continues toward a safe reactor shutdown, we will all be going through what amounts to a nuclear sweatdown — hoping, and praying, for a happy ending.

The San Diego Union

San Diego, Calif., March 31, 1979

There is confusion and alarm over the crisis at the Three Mile Island nuclear power plant in Pennsylvania. It may be several days before engineers can promise an end to the escape of radiation from the stricken plant.

Only guarded conclusions can be drawn at this point about the cause of the accident or its significance as a measure of the safety of other nuclear plants of similar design. Even presuming the worst is over at the Pennsylvania site, the accident could be a staggering blow to the future of nuclear power at a time when the country needs it most as an alternative to fossil fuels.

From what is known of the chain of events in the Three Mile Island plant, the most important safety system worked exactly as it was supposed to when trouble first occurred Wednesday morning. It shut down the reactor and triggered the emergency core-cooling system. But what should have followed — the controlled and safe cooling down of the reactor core — went awry through an incredible sequence of failures in equipment and human judgment.

This is the kind of failure of multiple safety systems that is supposed to be at the far end of the range of probabilities. Why did it happen? The answer will have to come from the Nuclear Regulatory Commission and from the consortium of utilities operating the Three Mile Island plant, which had been closed down on several occasions last year to replace or redesign some of its equipment.

Critics of nuclear power had pointed out in the past that there had been too little testing of the emergency core cooling systems that stand as a barrier to catastrophic accidents. One reason was the absence of serious accidents during the 25 years that commercial nuclear plants have been on the scene. If there is a positive side to this accident, it lies in the experience and warning that can make other plants safer.

There can never be 100 per cent assurance against accidents in any industrial activity, but generating electricity with nuclear fission demands that public risk be reduced to the remotest possibility. In the end, the impact of the Three Mile Island accident on the nation's nuclear program will depend on whether it demands any change in the calculation of risk that has brought nuclear power this far.

A study published in Science magazine last month found that the statistical risk to human life in producing electricity from coal — mining it, transporting it and burning it — is six times greater than the risk entailed in producing power from uranium in nuclear reactors. That's the other side of the coin. The nuclear accident in Pennsylvania has not made it any safer to mine coal and burn it, nor has it added one iota to our reserves of oil and natural gas.

Nuclear plants are generating about 13 percent of the nation's electricity today and could be producing twice as much by the end of the next decade. Whether they do will depend on what we learn from the grim drama unfolding in Pennsylvania.

THE SACRAMENTO BEE

Sacramento, Calif., March 31, 1979

Six days ago, we published an editorial about nuclear safety and the fear of nuclear accidents called "'The China Syndrome' Syndrome" which concluded that in situations "where both the fears and the assurances are considered excessive, it's almost inevitable that the counsel of the fearful will be regarded as the more reasonable. In a situation where the public has learned to distrust promises of safety even the guarantees of real safety will no longer be trusted."

We had no idea, of course, that those words would be so prophetic. Yet this week's nuclear accident in Harrisburg is, both in its general outlines and in the public declarations of the officials involved, a near replica of the fictional accident depicted in the Mike Douglas film. At this point, we still don't know how serious that real accident is — indeed, it may take generations before that is really known — or how much worse it may yet become. The real horror of such accidents, after all, is that they can spread their poisonous effects over an enormous area and over countless generations.

Some things, however, are already known. A few hours after officials of the federal Nuclear Regulatory Commission sought to reassure Congress that the radiation leak in Harrisburg was nothing to worry about, the governor of Pennsylvania announced that he was seriously considering the evacuation of 950,000 people in four counties. Thus the country is now almost precisely where it was before — though now even more uncertain about nuclear safety and even more suspicious of those who try to reassure the public about it.

Obviously the prime issue at this moment is the health and safety of those who work in the Three Mile Island plant and those who live in the areas affected by radiation. Beyond that, however, the false assurances of nuclear safety in the past make it plain that the basic need is not only for a reliable technology of safety but for the creation of institutions and review bodies which the public can trust. Perhaps the awesome power of nuclear energy will always make it impossible to assure complete safety. Still it ought to be possible to create regulatory bodies and safety commissions whose record for candor and accuracy will make them institutions on which the public can rely.

A week ago, we wrote that what's been most significant about the nuclear accidents of the past, both those on the screen and those taking place in real life, "is that they are concerned as much with the average person's inability to trust the assurances of the industry and the government that is supposed to regulate it as they are with the dangers of nuclear power and radiation." After Harrisburg, that statement seems more true than ever.

The Boston Herald American

Boston, Mass., March 31, 1979

It will be days, weeks, maybe years before the fallout from the accident at the nuclear power plant near Harrisburg, Pa., can be fully assessed. But it is probably the worst in U.S. history.

That fact must not be minimized, nor should it be cause for panic. What we need now are answers to a lot of questions: How and why did it happen? Can the same thing or worse occur again somewhere else? If so what, if anything, can be done to prevent it? What, precisely, are the risks inherent in nuclear power? How do they compare with the dangers posed by other forms of energy? And are they worth it?

All of those questions can and must be answered as quickly as possible. And the truth, we suspect, will lie somewhere between the emotional extremes now vying for attention.

Yesterday, Energy Secretary James Schlesinger said that Wednesday's accident was serious, but he contended that "it underscores how safe nuclear power has been in the past." His knee-jerk reaction is no more helpful than the predictable response from the Clamshell types who are sure the accident proves the fictional thesis of the movie "The China Syndrome," and want to use it as an excuse to shut down all atomic energy projects no later than last week.

Nor is the vital search for answers made any easier when industry and government officials insist upon misleading us and lying to the public. Over a nine-hour period on Thursday, the "experts" kept changing their story, escalating their assessment of the danger from "no radiation leaks" to levels "1000 times normal."

Cover-ups of that kind are absolutely inexcusable. We've got to get the facts and get them straight before a complete assessment of the accident — and the answer to all our questions — is possible.

THE SUN

Baltimore, Md., March 30, 1979

The political fallout from the malfunctioning of the Three Mile Island nuclear power plant near Harrisburg promises to be far greater than its radioactive fallout. President Carter's hopes that Congress would pass his bill to expedite the licensing of nuclear plants faded the moment radioactive steam started leaking from the nuclear power installation 55 miles from Baltimore. Not only will anti-nuclear activists be in position to intensify their protests against virtually any plant siting proposal that comes along; they also can be counted upon to take action against existing power stations where defects have been detected.

What this means is that the nuclear component that represents 13 per cent of the nation's electricity supply is under seige. Just a fortnight ago the Nuclear Regulatory Commission ordered the closing of five nuclear plants until it was determined they were earthquake-resistant. Three Mile Island makes it six plants. And the Union of Concerned Scientists has questioned the safety record of a dozen more installations. Were all these to be taken off line, the nation would suffer a power loss greater than that provided by all the oil Iran sold us in the best of times.

It is idle to suppose the electric power industry or such friendly Senate chairmen as Frank Church and Henry Jackson can quickly overcome the political catastrophe that has occurred just over the Pennsylvania line. It may be years before the likes of the Baltimore Gas and Electric Company will be able to build another nuclear power plant, because of financial as well as regulatory constraints. So the power industry will perforce have to turn to coal, perhaps even high-sulphur coal, despite the fact that the consequent air pollution will far exceed any environmental damage that yet can be traced to nuclear power.

If anyone can avert a panicky reaction against nuclear power, it is the President. Mr. Carter has consistently supported expansion of the faltering industry. But like so many of his energy initiatives, his nuclear programs have been hampered by inconsistency and contradiction. His opposition to the development of a fast-breeder reactor, for example, has complicated the search for means to dispose of nuclear wastes. And his plans to speed nuclear plant construction have been so slow in coming that they have missed the train, if the Harrisburg accident means what we think it does.

Yet the President dare not duck this issue. Even at the risk of criticism from such anti-nuclear Democrats as Ted Kennedy and Jerry Brown, Mr. Carter should boldly come to the defense of nuclear power development. Obviously, he can do so—and should do so—only by announcing a sweeping program to review and improve safety at all present and future plants. But unless Mr. Carter makes nuclear power a key component of the energy program he is about to unveil, this important industry will be dead in the water.

THE MILWAUKEE JOURNAL

Milwaukee, Wisc., March 31, 1979

The jury is still out on the accident at the Three Mile Island nuclear power station in Pennsylvania. The full ramifications of the incident — the worst in the history of nuclear power in America — remain to be determined, but an exhaustive examination and explanation of the plant's problems are imperative.

A pump broke down; the reactor was shut off; excess heat was "bled" out of the system, but in the process radioactive steam was released and workers and surroundings were exposed to potentially dangerous radioactivity. This has been followed by other discharges, a limited, precautionary evacuation of nearby areas and some concern about the remote possibility of a catastrophic meltdown.

The public urgently needs answers to such questions as: Did all the safety systems work correctly? If so, was the release of radioactive contaminants acceptable?

It is important to distinguish fact from emotion. There is much at stake in such incidents. The nuclear power industry out of self-interest will want to play down what occurred. Opponents of nuclear power will underscore the dangers, perhaps to the point of hysterics. A thorough study, open to public scrutiny, can level these biases and place the accident in proper perspective.

So far, the consequences of this plant failure seem mercifully limited. However, the accident *did* happen and potentially poisoning and lethal radiation *did* escape. Dangers do exist with nuclear power. Neither the machines nor the safety systems that surround them are perfect. If small mistakes can occur, large, very dangerous ones can also. That is why we need to know all we can about what transpired at Three Mile Island, and why large scale reliance on nuclear power should be approached with utmost caution.

The Dallas Morning News

Dallas, Texas, April 8, 1979

IF IT WERE just a question of efficiency and cost, there would be little difficulty selling nuclear power to the American people. In these two vital departments, the atom beats the competition hollow.

But of course, as Three Mile Island reminds us, it is not these things that bulk in the public imagination, when nuclear energy is considered; it is the recollection of a mushroom cloud rising high in the Japanese sky.

We are transfixed, as it were, by Hiroshima and Nagasaki, and by the legend of death that surrounds and accompanies the phenomenon of the split atom.

This is without precedent in human affairs. Fire has killed hundreds of thousands, and yet, for all our healthy fear of it, we gladly admit it to our kitchens and living rooms.

The disaster at New London, Texas, in 1937 — where a school blew up, killing 296 students and faculty. This was related to the petroleum industry or its byproducts. And yet no general fear of oil and gas resulted.

With nuclear energy, the matter is different, doubtless because of its introduction to the world via the nose cone of a bomb. Its impressionistic connection with the taking of human life will endure for a long time to come. The arcane nature of atomic energy — it is something only scientists fully understand — also intimidates.

Such considerations — combined with the doomsday reporting from Middletown, Pa. — cause us to lose sight of some facts worth noting about nuclear energy and its safety record.

The first fact is that no member of the public ever has been injured by radiation in consequence of a nuclear accident. (How different from New London!) Of course one can't say with divine certainty that no harm will flow from the Three Mile Island accident. What is plain, as of the present moment, is that provable harm has yet to flow.

A second fact is that the baneful effects of radiation are often overstated. As Professor Bernard L. Cohen, a former president of the nuclear division of the American Physical Society, expresses it: "If all our power were nuclear, and if you considered the whole nuclear fuel cycle — making the fuel, using it and reprocessing it — then the total effect of all such radiation would be equal to the risk of smoking one cigarette every 20 years, or of an overweight person's adding 1/100th of an ounce to his weight."

What about accidents, like Three Mile Island, though? It isn't plain, first of all, what caused the accident. That's under study. Whatever it was, it can be fixed — and will be fixed, because the nuclear industry is no more desirous of accidents than is Ralph Nader. Its safeguards against mishap are already vast and impressive.

Now and again there are failures. So, for technical reasons, are there airplane crashes, with attendant human casualties. But what then? Do we ground all airplanes? Or do we just strive harder to see that such accidents are kept to a minimum?

The cocoon-like world that the anti-nuclearists would oblige us to live in, free of risk and of danger, might be a cozy place. But it would be utterly unlike the real world, where hazard is as much a part of the human condition as is Sunday afternoon football.

The States-Item

New Orleans, La., March 31, 1979

Wednesday's accident which shut down the Three Mile Island nuclear power plant 16 miles from Harrisburg, the capital of Pennsylvania, dealt a severe blow to the credibility of nuclear power advocates.

Industry and government advocates repeatedly have assured the public that nuclear-powered, electric-generating plants are safe and clean. But opponents have insisted technology and human capabilities are not to be trusted when dealing with something so potentially catastrophic as nuclear power. The accident at Three Mile Island, while far from a major catastrophy, strengthens their case.

There was no "China syndrome," theoretically the ultimate accident, in which the reactor core melts through its containment and heads, beyond control, emitting deadly radiation all the while, "toward China" through the earth.

But the precise cause of the Three Mile Island mishap remains in dispute, underscoring the nagging concerns over human accountability and technical safety. Executives of Metropolitan Edison Co., operators of the plant, repeatedly cited the failure of a valve in a pump that circulated water around the reactor core. But engineers at Babcock & Wilcox Company in Lynchburg, Va., insisted there had been no valve failure or failure of a pump or pipe.

A dispute also arose over the source of radiation. Company spokesmen originally said it came from steam from the cooling system, but a Nuclear Regulatory Commission official later said it involved "direct radiation coming from radioactive material within the reactor containment."

The safety issue has dogged the domestic nuclear industry from its beginning precisely because it was not resolved at the beginning. In their eagerness to license, build and operate, nuclear power backers have tried to smooth over safety questions. Backed by government, the industry has invested billions of dollars and much of the nation's energy future in nuclear power. Now the unresolved safety issue is coming home to haunt the industry at a critical time for the nation. After so much investment, the overriding question of how to safely store vast amounts of radioactive waste is still unanswered. The suspicion grows that there might be no acceptable answer.

Encountering delays, mounting safety opposition, and dangerous inflation, the industry faces a dim future. Utility companies are hesitating to invest further in nuclear power.

As is the Three Mile Island plant, most nuclear generating facilities are located near population centers of necessity. (Louisiana Power & Light Company's Waterford Unit III at Taft is 25 miles from downtown New Orleans). Inevitably, what might have happened at Three Mile Island will be extrapolated to other localities, and public concern over nuclear power will grow.

With nuclear power, haste has been wasteful. Eager as the nation is for additional energy sources, the Three Mile Island accident flashes a distinct caution light on nuclear power development and reliance.

Newsday

Long Island, N.Y., March 30, 1979

We're still torn between relief and apprehension after Wednesday's accident at the Three Mile Island nuclear power plant: relief because the reactor's safety systems worked well enough to forestall a possible catastrophe; apprehension that they failed to prevent the escape of radiation into the countryside around Harrisburg, Pa.

While radiation continues to seep out of the plant, there has been no definitive explanation of the accident's cause. Equipment failure, by some accounts. Human error, according to others. Very likely it was a bit of both. Human beings are liable to react erratically to stress, and it's hard to imagine any more stressful situation for workers in a nuclear power plant than a serious malfunction involving its reactor. The potential risks contained in that combination are almost impossible to assess, as the Nuclear Regulatory Commission conceded recently when it disavowed the estimates of the formerly authoritative Rasmussen report on reactor safety.

The accident at Three Mile Island is only the latest in a series of close calls in the nuclear power industry—including a partial meltdown at the Enrico Fermi plant near Detroit in 1966 and a fire at Browns Ferry, Ala., in 1975. In none of these cases did the authorities responsible for evacuating area residents receive prompt notification. Yet a few hours could be crucial to the success of such an undertaking, which is difficult at best.

Utilities and government authorities have already been severely criticized for their lack of attention to evacuation planning. If nothing else, the accident at Three Mile Island is a warning that a workable evacuation plan is a must for any area in the immediate neighborhood of a nuclear power plant. And that includes eastern Long Island, where LILCO expects to open its Shoreham plant next year and plans two more reactors at Jamesport.

At Three Mile Island, as it turned out, only the plant itself was deemed to require evacuation. But it's still too early to tell how widespread or severe the contamination of the surrounding environment will be, or how long it will take for the reactor core to cool off to the point where workers can be brought in for a cleanup. Indeed, no one knows whether repairs and continued operation of the reactor will be possible at all. The Fermi plant had to be shut down permanently.

There's some comfort to be had, however, in the NRC's recent emphasis on safety. It shut down five East Coast reactors two weeks ago, despite a sharp run-up in oil prices, because of doubts that the plants' piping met earthquake standards. Those who joined Senator Gary Hart of Colorado in lashing out at the NRC for being overly cautious then ought to be eating their words now.

Still, the Three Mile Island incident is bound to diminish the public's already shaky confidence in nuclear power. The financial implications for the industry are difficult to predict at this point, but they could be considerable. And right now the long-range effects on the health of workers and residents are literally incalculable.

So even if there's no reason for panic, there's plenty of cause for concern as a result of the accident. And there's certainly a powerful argument for re-examination and possible retrofitting of existing nuclear plants that use similar cooling systems. Considering the many unknowns of nuclear energy, especially the behavior of a reactor when something goes wrong, the public has a right to insist on the best design and construction money can buy. And it should accept nothing less.

The Houston Post

Houston, Texas, April 5, 1979

The accident at the Three Mile Island nuclear plant in Pennsylvania is irrefutable evidence that we have not made nuclear power safe. This is true though most of the nation's 70 nuclear plants continue to generate electricity without posing an imminent threat to the people in the areas surrounding them. Yet, despite the U.S. nuclear power industry's excellent safety record to date, the Pennsylvania accident will have an immediate negative impact on the public's attitude toward such plants. There has been speculation, chiefly among opponents of nuclear power, that this could be the beginning of the end for this source of energy, or at least a de-emphasis of it.

Rejecting nuclear power, however, would mean drastically altering our economic growth plans and our entire way of life. But its continued development now demands a scrupulous re-examination of our assumptions about its safety. The formation of the hydrogen bubble in the Three Mile Island reactor apparently surprised experts, raising the threat of an explosion or — however remote — a meltdown of the nuclear fuel in the reactor's core. More attention must obviously be focused on the design and construction of cooling systems that will reduce the risk of similar conditions developing in other plants.

Apart from improving the integral operation of plants, more study should be given to their location with regard to population centers and environmental considerations. Should nuclear plants continue to be built at individual sites scattered around the nation or should they perhaps be grouped in nuclear parks? It has also been suggested that they be buried underground to enhance their safety, but this would add greatly to their already high cost.

Three Mile Island also showed that we are ill prepared to cope with a nuclear accident in all its ramifications—from keeping the public informed to protecting those in the vicinity and coordinating emergency efforts. The delay in reporting the leakage of radiation from the plant and the rash of conflicting statements in the first hours shook public confidence in later official statements concerning the incident. The initial confusion can be largely explained by the uncertainties facing the experts charged with handling the situation.

The nation has experienced its first nuclear accident with potentially severe consequences. It is, at the very least, a grim warning of the dangers inherent in the incredible force we have harnessed and a reminder of human fallibility. We must try to draw the proper lessons from it — including a greater respect for the awful power of the atom.

The Star-Ledger

Newark, N.J., March 30, 1979

The mishap at a Pennsylvania nuclear power plant that spewed radioactive steam into the atmosphere is a harrowing reminder of the deadly potential of these generating facilities, an important but highly controversial aspect of the nation's energy program.

Fortunately, the accident at the Three Mile Island plant near Harrisburg did not injure any worker, but the radiation emissions from nuclear-charged water could be measured as much as 16 miles from the site.

Again, there was no immediate contamination danger to residents in the area, but the mishap underscores the need for continuing vigilance to avert the ever-present possibility of nuclear accidents of catastrophic proportions.

The technical complexity of nuclear fission generating facilities makes it even more imperative that the utmost safety precautions are exercised. Despite the emotional controversy that has attended this highly hazardous method of supplying a significant portion of America's energy needs, the safety record of these plants has been highly commendable.

Until the malfunction of the nuclear power plant in Pennsylvania, there has not been a major accident in the more than two decades of operation.

It is a safety experience rating that confirms the presence of exacting technical and manufacturing safeguards that limit the possibility of mishaps . . . as close to a fail-safe priority as humanly possible.

This does not completely rule out nuclear plant mishaps. The breakdown at the Pennsylvania generating facility is an unsettling example of the lethal implications of mechanical malfunctions or human error where radioactive materials are involved.

But this is no time to press the panic button: There must be a rational balance between this country's enormous energy needs and the generating capacity to meet them. Nuclear power plants have a vital role in the energy field, even more so with our growing dependence on foreign energy resources.

This should not, per se, lessen the need for the most stringent precautions. If anything, the Harrisburg nuclear accident suggests an even higher priority for intensive safety standards that would lower the accident level to an irreducible minimum.

DAYTON DAILY NEWS

Dayton, Ohio, April 5, 1979

Only one thing is sure as the rogue nuclear reactor cools on Three Mile Island: Despite the repeated assurances of the industry, despite the pipe-puffing condescension of Energy Secretary James Schlesinger, despite supposedly onerous government regulations, nuclear power plants are not as safe as their supporters would like us to believe.

The technology is not uniquely foolproof and the engineers are not uniquely incapable of error. The Three Mile Island reactor nearly went out of control, leaving experts baffled both as to what was happening and what to do about it. We do not know whether the relatively small amounts of radiation that spewed from the plant will in coming years cause sickness, birth defects or death in some of those exposed to them.

Nor is the Three Mile Island reactor special. There are seven more in operation that were built by its maker, Babcock & Wilcox. Records at the Nuclear Regulatory Commission indicate they have been trouble prone. One of them, Davis-Besse, is between Toledo and Cleveland on Lake Erie. It has experienced some of the same problems that led to near-disaster in Pennsylvania.

Now that the "impossible" has happened, the criticisms of nuclear power's opponents are getting a respectful new hearing. There will be changes in the way we deal with nuclear power.

The decisions that face us are only partly technological; they are mostly decisions of common sense and philosophical outlook. They are decisions that will affect our way of life, our prosperity, our health, our very lives.

The short run choices, though unpleasant, are easiest to make.

Can we close down the 72 reactors that are supplying 13 percent of our electricity? It's hard to see how. But we can try to make federal regulation tighter and more effective, making reactor design changes where those seem prudent.

We should close down the Babcock and Wilcox reactors until we are reasonably sure none has dangerous problems. The NRC's post-accident statement that the reactor are safe is not persuasive. Its own documents suggest otherwise.

But what about the 80 reactors in various stages of licensing and construction? Do we let those at an advanced stage proceed to completion, while halting construction and planning for others? Probably.

Even if the reactors themselves could be assumed to be perfectly safe, their wastes still present a hellish dilemma. Until that is resolved, it is hard to see how we can responsibly continue to build the plants that produce such dangerous wastes.

But those who would ban or restrict the number of nuclear plants must realize that there would be difficult energy choices still to be made.

Are they willing to put up with possible shortages of electricity at times — say, on the hot summer days when we want to run our air conditioners? Are they willing to pay the cost of increased coal mining and the cleaning up of the poisons its burning puts into our lungs?

Some will answer no; it is easier to feel the pain in one's pocketbook than the slow deterioration of lung tissue or the slow growth of a tumor or the subtle changes of chromosomes under the impacts of radioactivity.

Some of us will answer yes; that while we live in a world filled with risks, the risks of nuclear power plants pose dangers of such scale and of such unknowable long-range peril that their number must be limited.

But no one should delude themselves that the question is the easy one of nukes, yes or no. The questions, rather, are ones of what kind of future we want, what we are willing to pay for it and what value we place, ultimately, on the safety of the millions of our fellow citizens whom we would ask to live in sight of those ominous cooling towers.

The Evening Bulletin

Philadelphia, Pa., April 3, 1979

After six harrowing days, the engineers and scientists working to clear a hydrogen bubble from the Three Mile Island nuclear reactor succeeded in bringing the plant closer to a final, safe "cool down." And a tremendous sense of relief spread from central Pennsylvania across our part of the world.

Those anxious days at Three Mile Island won't be forgotten — not by the tens of thousands of neighboring residents who left their homes, schools and businesses as a prudent precaution, or by the rest of us who tried to judge the dangerous of a nuclear accident from a welter of sometimes confusing reports.

We have all been through an ordeal that has left the credibility of the nuclear power shaken. Simple assurances that nuclear power is "super safe" won't do any longer. Despite all the engineering that went into the Three Mile Island plant, it turns out that a series of failures and the build-up of a hydrogen gas bubble hadn't been properly anticipated.

We hope we will not again be in a position in which nuclear safety procedures have to be improvised.

The wisest way to proceed from this point is for the nation, collectively, to keep *its* cool over nuclear power — to shun calls for a general shutdown of nuclear reactors, but to insist that their design and operation be more tightly controlled by the Nuclear Regulatory Commission.

Before long, peak summer energy demands will be hitting us. We need the nation's 70 existing nuclear power plants to avoid brownouts or power failures; but we need to have them run reliably, even if it means added costs in new safety features or increased government surveillance. All forms of energy are now at a premium, and safety precautions are the premium we must expect to pay for the nuclear energy we need so badly.

As the Nuclear Regulatory Commission struggled with the bubble at Three Mile Island, it ordered reports on what's being done to assure against another failure at seven similarly designed plants in other locations. We trust that's simply the first indication that the commission intends to take a more active role in supervising the operation all nuclear reactors.

There is nothing to be gained from any arbitrary moratorium on the opening of new nuclear plants — one action that's now being urged on Congress. But there's a lot to be gained in terms of rebuilding the credibility of nuclear energy by giving the NRC a more active presence in nuclear control rooms around the country — and in their design and construction.

Nevada State Journal

Reno, Nev., April 6, 1979

The accident at Pennsylvania's Three Mile Island nuclear power plant shouldn't come as much of a surprise to Nevadans.

For months, and in some cases years, Nevadans have been concerned with the tough questions Pennsylvanians now are asking themselves about nuclear energy.

But the Pennsylvania accident came at a curious time for Nevada.

Our state is at a crossroads concerning nuclear energy.

Decisions will be made in forthcoming months about the future of atomic energy in Nevada. The Three Mile Island accident has generated a consciousness-raising, if you will, in this state which will have a healthy effect.

Nevadans are looking at what has been happening in Pennsylvania, and comparing it to what they have observed in their own state.

Contrast the following:

— Despite early attempts by the power company officials to minimize the seriousness of the accident, Pennsylvania residents were warned of potential danger. Pregnant women and young children were encouraged to evacuate the area.

In Nevada, residents were never warned of possible health hazards either from the above-ground or below-ground nuclear tests, from which radiation leaked. They were not told to evacuate. Yet our residents have been consistently exposed for nearly 20 years to significantly higher levels of radiation than emitted at Three Mile Island.

— Sen. Edward Kennedy, D-Mass., chairman of the Senate Human Resources health subcommittee, urged the government to provide free medical examinations for Pennsylvania residents exposed to radiation.

As of this writing, a similar plan has not been urged for Nevadans. This is the case despite requests from citizens concerned for their health, and several months of media coverage about Nevada radiation hazards at both ends of the state.

At the same time, some disturbing parallels exist between the Pennsylvania situation and our own.

The Atomic Energy Commission, which has been the overseer for both the growth of nuclear power plants and the conducting of atomic tests in Nevada, assured everybody that both situations were safe.

Yet evidence continues to mount that the AEC (since 1971, the Department of Energy — DOE) did not adequately address the tough safety questions involved, while consistently down-playing possible safety hazards.

We are left to wonder what we can believe in information released about nuclear energy in Nevada.

Can we believe DOE assurances that underground tests are not leaking radiation, when it took the Baneberry Trial in Las Vegas to reveal that they ever leaked at all?

Can we believe President Carter has the best interests of U.S. residents — including Nevadans — at heart when he wants to pour nearly $13 million into an accelerated atomic weapons testing program in Nevada before the SALT-II treaty is signed?

Can we believe the citizens of this state will truly have the final say on whether Nevada will become a nuclear waste graveyard, when we find that nuclear wastes already have been quietly making their way to the Nevada Test Site for several years?

We have known for some time that the government has looked covetously at Nevada's vast rangeland as a likely place to bury nuclear wastes.

Now, we have just learned that a group headed by our own Department of Energy director, Noel Clark, had two years ago pinpointed the Southern Nevada desert as an ideal spot to build the nation's largest nuclear power complex.

Nuclear energy has been a fact of life with Nevadans for nearly 30 years.

And we have already paid the price for it.

We have stood by as our pristine desert environment has been pock-marked and contaminated by years of nuclear blasting.

Some of our citizens already have buried one generation of cancer victims, and believe they are still coping with the effects of radiation.

And we have become increasingly disillusioned about the credibility of our governmental leaders in the area of nuclear power.

We are tired of being misled.

We applaud the recent demands by Gov. Robert List and freshman Assembly Tod Bedrosian that the federal government should not be allowed to put nuclear power plants on Nevada soil until it can adequately answer our safety concerns.

We would like to see that demand expanded to include all nuclear projects in Nevada.

The difference between nuclear accidents in Pennsylvania and nuclear accidents in Nevada boils down to a question of magnitude. A nuclear accident in densely-populated Pennsylvania has the potential of harming many more people than in sparsely populated Nevada.

However, we believe in the rights of individuals — wherever they live — to protection of life. An individual person's life and health in Nevada is as precious as that of an individual in Pennsylvania.

We believe we have a right to know what our leaders intend to do with the Nevada Test Site, and any other nuclear energy proposals which could come to fruition within our boundaries.

We believe we have a right to decide for ourselves whether to allow any more nuclear projects in our state.

THE INDIANAPOLIS STAR

Indianapolis, Ind., April 1, 1979

It is no surprise that opponents of nuclear power development have come swarming out to exploit the attention attracted by the nuclear electric generating plant accident last week at Harrisburg, Pa.

Naturally they see this accident as a golden opportunity to impress upon the public the dangers of nuclear power plants.

Maybe it is and maybe it isn't. Contrary to some wild early statements by supposed nuclear experts, a few days after the accident it appeared that no serious general hazard to the public had resulted from it, even though Gov. Dick Thornburgh advised evacuation of preschool children and pregnant women within five miles of the mishap site, in what he called "an excess of caution."

What seems to us a calm appraisal came from Paul S. Lykoudis, a nuclear engineer for more than 20 years and now head of the Purdue University School of Nuclear Engineering.

"I have not seen enough technical information to make a full assessment," Lykoudis said. "Nevertheless I can say that every nuclear reactor has so many backup systems that an accident like this reconfirms the extreme safety of their operation."

Note well his remark that he had not yet seen enough technical information for a full assessment. Neither has anyone else. It probably will take weeks to get conditions inside the plant under control so that a full appraisal of what happened can be made.

Nevertheless the prophets of doom among the opponents of nuclear power are in full cry. The accident has already been labeled "the most serious" involving nuclear power generation ever to occur in the United States.

That will be a good statement to remember if it turns out that the accident posed very little if any danger to people.

The debate over official truthfulness on this incident is likely to continue for a long time.

We hope that as investigation proceeds there will be careful avoidance of premature comments on findings but that the full truth will be told as soon as it is determined. Nothing should be held back. Everyone involved should make a scrupulous effort to see that reports reaching the public contain nothing but the truth, but also the whole truth.

This may be hoping for too much. The controversy over atomic energy in America is deeply tainted by ideology, emotionalism, radical political dogmatism and superstition comparable to belief in witchcraft.

If the final report sustains the fears about dangers of nuclear power plants, so be it.

But if on the other hand the report sustains those who have argued that nuclear power generation is safe when plants are properly designed and built, then theirs will be the golden opportunity presented by this accident.

Meanwhile sensible people will withhold judgment.

Rocky Mountain News

Denver, Colo., April 5, 1979

It is still not known precisely how the nuclear drama at Three Mile Island will end. But it is a foregone conclusion that the cause of nuclear energy has suffered a major setback.

The accident at the atomic power plant near Harrisburg, Pa., has so far harmed nobody. But it has inconvenienced many and alarmed many more. And even if the smoldering furnace in the atomic power house is finally brought under full control without further threat to the public health, proponents of nuclear energy will hereafter find theirs an even tougher row to hoe.

At the moment, the chief concern is extinguishing the damaged atomic furnace — the nuclear reactor — at Three Mile Island. Wednesday, officials reported that a troublesome gas bubble no longer poses the threat of explosion. The most serious dangers appear to be over. But scientists, engineers and technicians continue to strive to bring the reactor to a "cold shutdown." The ultimate peril, of course, is a complete "meltdown" of the heart of the reactor.

As the countdown at Three Mile Island continues toward a safe reactor shutdown, we will all be going through what amounts to a nuclear sweatdown — hoping, and praying, for a happy ending.

President Carter says it's too early to make judgments on "the lessons to be learned" from the accident, but he acknowledges that the incident "will probably lead inexorably toward even more stringent safety design mechanisms and standards" in the nuclear power industry.

Already throughout the country, officials are taking a closer look at nuclear generating facilities in their own backyards, attempting to assess what perils, if any, they may have to cope with. One flaw in emergency procedures in Colorado has already been discovered. An emergency response telephone number at the headquarters of the state's Office of Disaster Emergency Services doesn't answer at night and is often busy during the day. Reaching this number would be crucial in the event of an accident at the state's Fort St. Vrain nuclear generating facility, since the office would serve as a center for information and coordination of emergency operations.

The Evening Gazette

Worcester, Mass., April 11, 1979

The accident at Three Mile Island has been seized by the wilder anti-nuke crowd as proof that all fission power plants should be shut down. But it also has led to thoughtful re-examinations of the future of nuclear energy by some of its more open-minded advocates.

One question being debated anew is whether nuclear energy is indeed cheaper than energy from coal. Until relatively recently, nuclear seemed to have a clear edge over either oil or coal power. Now the gap is narrowing. New safety requirements surely will increase the cost of nuclear power. How much, no one knows.

A long-time supporter of nuclear power, Lewis J. Perl, a vice president with the utility consulting firm of National Economic Research Associates, told The New York Times, "The economics of coal and nuclear are close enough, that if the safety factors alone now shifted by an order of magnitude, we'd all certainly change our view that nuclear is cheaper."

In fact, in Dr. Perl's cost extrapolations for plants that come on line in 1990, coal power is estimated to be an average of 11 percent cheaper than nuclear power. However, these costs will vary from region to region. In Western states where plants can be located close to coal mines, coal will be about 20 percent cheaper. In New England, nuclear power will be about 30 percent cheaper than coal, said Dr. Perl.

There is some doubt, however, given all the variables and unknowns in the nuclear-coal cost equations, whether any price comparisons are valid. In a Harvard Business School report on "Energy Future," to be published in July, nuclear economics expert Irwin C. Bupp found that it is almost impossible to say whether nuclear power is cheaper than coal or vice versa. The so-called experts on nuclear-coal cost analysis get radically different answers based on "choice of data and assumptions" that suit "one's interests and hunches," according to Bupp.

The cost comparisons — hard to figure as they may be — are important because one of nuclear's selling points was its relative cheapness. To discover that nuclear may not be so cheap after all makes the energy policy decisions of the country, and the region, more difficult.

The energy choices are dwindling down. Do we want to continue our dependence upon foreign oil along with the trade imbalance and price instability? Do we want to move to coal-generated plants and an increase in air pollutants, especially sulfur? Do we want to lift the current nuclear moratorium and stifle fears about its potential dangers? Is oil decontrol with its inflationary consequences a good idea?

Choosing a rational energy policy these days is a no-win situation. It is a matter of examining all of the price tags — environmental, health, safety, dollar, trade and otherwise — and picking the least objectionable way to produce the energy needed by this country.

Democrat and Chronicle

Rochester, N.Y., April 4, 1979

IT WASN'T supposed to happen but it did, and that's the single most disturbing thing about the accident at the nuclear plant on Three Mile Island in the Susquehanna River just south of Harrisburg, Pa.

In the words of Rep. Jonathan Bingham of New York, "Something is happening in Pennsylvania which the nuclear industry has told us was impossible."

The danger is not over yet. But as it's turning out, this has not been a serious accident in purely human terms. No one was killed. No one was injured.

But the sequential failure of several safeguards that were said to be foolproof has set alarm bells ringing all around the country.

The release into the atmosphere of radioactive gases demonstrated that nuclear safety has been greatly oversold. A frightening potential for harm was revealed when a "melt-down" became a possibility, with the threat of contamination of a large geographic area.

Directly as a result of the accident, the future of nuclear power has become much cloudier. It's not doubted that the industry has been sharply set back as a result of these recent anxious days.

That new, cheap, endless source of electricity that was to have revolutionized society turns out to be seriously flawed.

THE ACCIDENT occurred at a time when the nuclear industry was already being hammered.

Technically safe, long-term storage of radioactive waste has recently been shown to be a far tougher problem than had been expected.

The Nuclear Regulatory Commission not long ago questioned the once-in-a-million-years estimate of the probability of a serious accident developed by a special government committee headed by Dr. Norman C. Rasmussen of M.I.T.

Evidence has been growing that even low levels of radiation may pose hazards.

Construction costs have been soaring, and orders for new plants have dropped off severely.

Coming on top of all this, the Pennsylvania accident must strengthen doubt and opposition.

But wholesale abandonment of nuclear power doesn't seem warranted at this point.

The alternatives, after all, are pretty bleak.

Oil is getting costlier and harder to find. Solar energy is a long way from large-scale development.

And when the hazards of nuclear power are talked about, they have to be set against the deaths of hundreds of coalminers.

SOME 70 nuclear reactors today provide about 13 percent of the country's electrical power. In some areas, as in Chicago, New England and Rochester, they account for about half the power used.

So a withdrawal of this source of power would cause tremendous dislocations and hardships.

Nor is this being seriously proposed. The outlook is one of less reliance on nuclear power for the future.

Rep. Morris K. Udall, who heads the House committee mostly responsible for nuclear energy, has been quoted as saying:

"I don't think we'll close down the 70 reactors we now have operating, but I personally believe it is very unlikely that we will build anywhere near the number of new reactors projected for the year 2,000."

Even a strong supporter of nuclear energy such as Sen. Henry Jackson says the Harrisburg accident places nuclear power in "a semi-limbo."

A slowdown in development seems inevitable.

And this period of questioning and review is going to have to be well used by the experts whose job it is to see to nuclear safety.

The lessons of Harrisburg are going to have to be learned and applied before public confidence is restored.

Los Angeles, Calif., April 5, 1979

You know what troubles us most about the Three Mile Island incident?

It's not even the likely loss of a billion-dollar facility, though no one should find much comfort in that.

And while the possibility of a full-scale meltdown (though remote) was always real, even that (at least right now) doesn't disturb us as much as something else.

And that something else is the conduct of the nuclear power industry, utility company, and Nuclear Regulatory Commission officials involved in this incident. To put matters bluntly, their behavior has, on the whole, been exceedingly high-handed, condescending, and sometimes disturbingly patronizing.

They kept telling us that there was no cause for alarm, and yet with every silken reassurance there seemed even more reason for alarm than before. "The plant is in a safe condition," Jack Herbein, a Metropolitan Edison vice president said last Thursday. "We have a very minor fuel failure."

That's not the way things turned out.

That same afternoon, Three Mile Island plant official Blaine Fabian sought to reassure the public with these soothing words: "There is absolutely no danger of a meltdown. We are not in a 'China Syndrome' type situation."

And that didn't turn out to be entirely accurate, either.

One must, of course, allow for the fact that the authorities did not wish to contribute to anxiety by exaggerating the danger. But this isn't what we're talking about. Responsible officials have every right to act responsibly. What they do not have the right to do is patronize the public.

In a way, the Three Mile Island incident was instructive beyond what it taught us about the safety hazards of nuclear power. It also taught us about the attitudes of those who run this industry. Even now, that attitude seems to be that nothing has really changed. A public utilities commissioner in Connecticut sought to place the situation in perspective. While agreeing that the prospect of a nuclear power plant meltdown was a "horrendous thing," he argued: "I think the risk is worth it. Good God, 1,200 people get killed each year from electricity and there are 7,500 deaths by fire. There are risks every day." A spokesman for the Nuclear Regulatory Commission put the matter even more plainly: "Did anyone die?" Energy Secretary Schlesinger, as ever the stubborn, bull-headed and non-public "public" servant, continues to call for a speedup in the licensing of nuclear power plants.

You see, it is precisely the rising chorus of reassurances that causes one to pause, to wonder what is being pulled over one's eyes, and why. And we're not the only ones who are worried. Dr. Irwin Bross, of the famed Roswell Park cancer research institute in Buffalo, was reacting to belated news of a spillage of radioactive water from a nuclear power plant in New York that the plant chose not to report. Said Dr. Bross, "I am becoming concerned about representatives of government and the nuclear industry who really know nothing, but who are telling us not to worry. It is clearly irresponsible to tell us not to worry about these nuclear accidents."

Indeed. We have been told not to worry so often, by so many spokesmen for the nuclear power industry, from local officials to Dr. Schlesinger, that now they have us really worried. "We have every confidence that the Rancho Seco plant is safe," insists William Walbridge, general manager of the Sacramento Municipal Utility District. Well, do *you* feel reassured?

But wait a minute. Perhaps the Rancho Seco plant is safe. Perhaps all the others are relatively safe, too. And perhaps the Three Mile Island incident should not stunt the growth of the nuclear power option.

Maybe. But unless these officials start leveling with the public, they're going to give the impression they have something to hide.

Houston Chronicle

Houston, Texas, April 3, 1979

Until last Wednesday, the nuclear power industry defended itself against those predicting disaster by saying that nuclear power plants had accumulated more than 1,300 reactor-years of operation without an accident involving a significant amount of radiation.

That record no longer stands. With the release of radiation from the Three Mile Island plant in Pennsylvania, nuclear power is facing its severest test. The best interests of the nation would be served if the industry is able to meet that test, because nuclear power is badly needed now and will be even more vital in the future.

Where the industry has been confident of its safety precautions, there is now proof that there are both technical and procedural problems. Even more stringent safety regulations and standards are bound to result from the Three Mile Island plant accident. Attention up to this point has been focused on solving the emergency created by the accident. Later, there will be the necessary and exhaustive investigation of how the accident occurred and what should be done to prevent any similar event at any other plant. And even after that, undoubtedly all other safety standards will be getting a thorough review. Congressional inquiries have already begun.

Opposition to nuclear power has come from a vocal and emotional minority. Some are sincere scientists; others simply consider anything nuclear demonic. The escape of radiation at the Three Mile reactor will increase the clamor for a ban on nuclear power and a public opinion battle is to be expected.

In this connection, the Pennsylvania accident revealed another area that needs attention, and that is the way information about nuclear mishaps is handled. Confusing and contradictory statements were made and the flow of basic information to civil authorities was inadequate, leaving them unable to make proper decisions. Nevertheless, the people in the area of the plant have been commendably calm and cooperative.

There is no way to create energy without risk. Coal mines collapse. Pipelines explode. Risks are measured against gain, as in driving a car despite the growing annual traffic toll. Nuclear power poses both risks and gains. The Three Mile plant accident demonstrates the risk; the fact that nuclear power already supplies more than 10 percent of the nation's electricity shows the potential gain. The need is for the nuclear power industry to maximize its safety precautions so that the margin of risk is an acceptable one.

'I REPEAT—THERE IS NO REAL CAUSE FOR ALARM...'

The Washington Star

Washington, D.C., April 3, 1979

As the crisis at Three Mile Island entered its seventh day, the possibility seemed to brighten that the rogue reactor can be cooled, and a menacing buildup of gas pressures abated, without loss of life or dangerous radiation exposure.

But even if all now goes well, the incident will have dealt a shattering blow — a fatal one, some insist — to nuclear power prospects. Certainly, the Pennsylvania incident offers a humbling lesson in the vulnerability of sophisticated, high-technology societies to the unforeseen.

The buildup of a hydrogen "bubble" in the reactor vessel, dragging out the process of cooling the fuel core and threatening a calamitous explosion, seems not to have been among the problems foreseen by nuclear engineers; and there was no ready plan for dealing with it.

Along with this came a tangle of human and political problems: the disconcerting possibility of human error (the shutting off of an emergency cooling pump at an early stage); the confusing clash between federal authorities and plant managers about the course and dangers of the crisis. Who was in charge? Who should be? Sen. Gary Hart plans to introduce legislation providing for unchallenged federal authority in dealing with such accidents in the future: a measure that seems to us eminently reasonable.

When the crisis ends, and its causes and course are fully evaluated, we shall have a firmer grasp of the implications for nuclear technology and regulatory policy.

Meanwhile, it is not too soon to begin thinking about the major choices that were already hotly debated and will be the more so after Three Mile Island. Until 4 a. m. last Wednesday morning, nuclear power was about to be proclaimed *the answer* to the oil crisis. Now that prospect has been knocked into a cocked hat, or so it seems.

The Pennsylvania accident underscores that nuclear power-generation technology, however sophisticated, remains in relative infancy: not much more than two decades old. You might say that's old enough to be safe, and no one in his right mind will be disposed to quarrel with that. But even allowing for the acceleration of technological change, the nuclear revolution has been astoundingly rapid. Consider equivalent revolutions — steam locomotion, flight, the development of the internal combustion engine — and you must pronounce its development *comparatively* untroubled. And while it obviously has far to go, and that far more cautiously than in the past, these two decades of progress will not be tossed lightly aside.

For what, after all, are the alternatives? Assuming that electricity is not dispensable, indeed that demand will grow and cost along with it, the alternatives to nuclear generation are twofold: oil and coal.

Further dependency on oil carries risks so clear as to need little emphasis here — above all, the risk of intensifying an already dangerous dependency on foreign supply, and of continued unsettling impact on the stability of the economic system.

Coal is plentiful, but has its drawbacks too. Coal of high sulphur content, the most plentiful, cannot be burned without severe environmental hazard and without statistical risks to public health far higher than even the riskiest nuclear technologies. Technologies that reduce sulphur content in coal are costly, and bear the same marginal economic cost as foreign oil.

The United States can get along without nuclear-generated electric power, or at least rub along with the current proportion of electricity it produces: some 13 per cent. But with the normal increment of demand, any "moratorium" on nuclear development, or political obstacles and licensing complexities that make it impractical, will entail one or both of these consequences: greater mortgaging of the country's future to the oil sheikhs and further environmental degradation by air pollution or by mining techniques that limit the safety risks of miners.

In short, when the reactor is cooled and the dangerous gases dispersed at Three Mile Island, and when the immediate questions have been explored, the real debate over the future of nuclear power will begin. And real, not illusory, choices will bear upon us. It is the luxury of childhood to think that mutually exclusive wants may be satisfied, and that choices may be made without consequences. But the discipline of maturity is harsher.

AKRON BEACON JOURNAL

Akron, Ohio, April 3, 1979

UNTIL MORE is known about what caused the breakdown at the Three Mile Island nuclear generating plant than the public has so far been told, it is premature to draw sweeping conclusions about the implications for the rest of the nation's nuclear plants.

The accident there clearly adds new political difficulties to the growing problems — economic, political and technical — that the industry has been facing through recent years.

But if the root cause lies in some easily correctable design fault in the Three Mile Island plant and perhaps in the seven similar plants across the nation built by Babcock & Wilcox, the logic of the case against nuclear power should be no stronger because of the accident.

For non-technical outsiders watching, it is disconcerting to hear that a human error in valving of the coolant system at the Pennsylvania plant may have either caused or exacerbated the destructive problem there.

This report implies a more fundamental human error in the design stage, in apparent failure to provide some form of mechanical "fail-safe" control to override an operating error capable of destroying the reactor. The non-technical public had been led to believe that safety factors of this kind had been built into every commercial reactor — in so many different ways, in fact, as to make the possibility of a malfunction like this one a statistical near-impossibility.

If the cause is found to be of this kind and the likelihood of similar difficulty is appreciable in other B&W plants, they should be modified promptly to overcome the fault.

This done, the arguments against continued use of the other 71 nuclear plants in the country or construction of more like them should be no stronger than it was — except that a lot of people who previously paid little attention to arguments either way now have a new basis for concern.

THE ANN ARBOR NEWS

Ann Arbor, Mich., April 2, 1979

"TODAY, only a handful of people know what 'the China syndrome' means.

"Soon, you will know."

As it turns out, no one has to go to the film linked with that advertising pitch to find out what it means.

The loss of control over radioactivity, and risk of core melt, at a nuclear plant in southeastern Pennsylvania, has introduced everyone to the concept that a wholly uncooled nuclear plant core might burn downward toward China.

That concept, like the film portraying the possibility it might happen, is, so far as we can determine, fictitious. A totally uncontrolled nuclear plant core would more likely burn downwards 20 feet or so, creating serious radioactivity problems which would require evacuation of an affected area for a year at the very least.

* * *

IN THE REAL WORLD, no one can say with certainty exactly what will be learned from the series of accidents at the Three Mile Island nuclear plant near Harrisburg, Pa.

The easy assumption is that these mishaps prove nuclear power plants simply cannot be built with an acceptably low risk for those who work in them or live nearby.

That assumption hasn't been proven any more than coal mine disasters and oil field fires prove those energy sources to be intolerably dangerous.

However, the Three Mile Island accidents do prove that nuclear power plants can, given the wrong combination of design and/or operating errors, pour out dangers more terrifying to the general public than any oil field or coal mine disaster. Radioactivity can't be seen. Residents near Three Mile Island are getting widely varying advice from individuals who profess to know whether they should evacuate or consult doctors.

* * *

SCARCELY MENTIONED so far, in discussions of the Three Mile Island accidents, is long-range uncertainty as to whether a safe, economically acceptable means of getting rid of used nuclear power plant fuel elements is going to be found.

Michigan residents, with nuclear power plants on the shores of Lake Michigan and Lake Erie, have made it clear they will not easily be persuaded that the Alpena area's deep salt beds are an acceptable burial place for the wastes. The U.S. Energy Department is tentatively planning to spend $400 million to prepare 2,500 feet-deep burial places in southern New Mexico's salt beds, but is also experimenting to learn if burials in granite at the nuclear weapons test site in Nevada might be suitable as a national disposal site. The message to the public is that a crucial link in the attempt to substitute nuclear power for petroleum remains a puzzle.

It isn't necessary to leap to emotional final conclusions to realize that nuclear power has not been shown to be a safe means of maintaining the energy-consumption levels Americans have grown accustomed to in this century.

The Birmingham News

Birmingham, Ala., April 3, 1979

So far, no one has been killed and no one fatally exposed to radiation in the accident at the Three Mile Island nuclear power facility at Harrisburg, Pa. On the basis of present information, it appears that both mechanical failure and human error contributed to a malfunction in the nuclear-powered system that set off a series of accidents.

But regardless of whether anyone is hurt or dies as a result of exposure to radiation, many questions about nuclear energy will have to be answered before Americans — indeed the Western world — will feel comfortable again with reliance on nuclear energy to power the U.S. economy. The big question, of course, is safety.

Statistically, a nuclear system is still the safest form of energy. Hundreds are killed and thousands are maimed yearly by fires, gas, falling debris and explosions in putting coal, gas and oil to work to generate heat and electric power so essential to U.S. industry, to housing, agriculture and a thousand kinds of service.

We tolerate the dangers from these sources of energy because they have been with us for millennia and are part of our history and because little or no mystery surrounds them. On the other hand, the nuclear process is still very new. While our experience with it has been mostly positive up until the present crisis at Harrisburg, we still have a great deal to learn about what happens in an uncontrolled reaction as well as the consequence of a core meltdown.

The great fear is radiation, not only from a reactor out of control, but from radioactive pollutants accidentally emitted which may have the capacity to damage life and the environment for a long time. The accident at Harrisburg has done nothing to lessen those fears. On the contrary, it has greatly exacerbated fears of the unknowns about nuclear power.

The real knowledge most Americans have regarding nuclear power is more than offset by the fictions — both positive and negative — which have grown up around it. Apparently debate by knowledgable scientists and engineers has largely been ignored by the public. Doubtless the opinions of most Americans today are based more on pre-nuclear and anti-nuclear propaganda than upon clearly established factual data.

But regardless of cost in terms of both dollars and time, the legitimate questions of the public must be answered, in so far as answers exist, regarding safety — even to what is known and what is not known about ensuring a safe system. While the nation — indeed the world — must have increasing amounts of energy, if our standard of living is to be maintained and those on the lower rungs of the economic ladder provided a larger share of goods and services, the risk factor must be given sufficientexploration for the public to make a reasonable judgment.

It well could be that the cost of providing a fail-safe nuclear system with next to zero risk will be prohibitive, and nuclear power will no longer be an option. But it is too soon to speculate along such lines. The answers, at least those that are known, should be made as plain as possible. The Harrisburg accident has made that need clear.

TULSA WORLD

Tulsa, Okla., April 9, 1979

THE badly-damaged reactor at Three Mile Island, Pa., is being safely shut down. The story of the worst nuclear power accident in U.S. history now enters what might be called the post-hysteria period.

The important fact is that no one was killed or injured.

But Three Mile Island may yet produce a "nuclear disaster" of a different kind if it leads to a halt of nuclear power development. The threat of such a disaster could not come at a worse time.

The recent Iranian oil cutoff was another shocking reminder that the United States depends on half a dozen of the world's least stable countries for nearly 50 percent of its petroleum.

And yet we hear frightened people calling for a shutdown of all nuclear plants or a moratorium on new facilities. The plants provide more than 13 percent of the U.S. electric power supply — an energy equivalent of more than the total Iranian oil contribution.

Yes, nuclear power — like all other forms of power and all other major industrial undertakings — poses risk to life and safety. We do not have and will not have — even if nuclear plants are closed — a risk-free, perfectly safe society.

Abundant energy is in itself a major contributor to health and safety. Look at the thousands of deaths from industrial accidents that occurred routinely before the most onerous and dangerous work was taken over by machines. During a single heat wave of the 1930s, thousands died of sunstroke and related ailments. Many were laborers doing the kind of work that is now done safely with machines powered by electricity or some other form of energy.

Few who lived in those good, old days of drudge, heat and poverty really want to go back.

The future of nuclear power is of special interest in Eastern Oklahoma. In the years ahead, this area will depend on the Black Fox nuclear plant for a major part of its energy. Those who understand the need for that power plant and who appreciate the true relationship between an adequate power supply and a safe, healthy, prosperous society should let themselves be heard.

THE EMPORIA GAZETTE

Emporia, Kans., April 3, 1979

THOSE of us who have stood up steadfastly for nuclear power are beginning to waver in our tracks. The nightmare at Three Mile Island plant in Pennsylvania is enough to shake anyone's faith.

From the beginning, the public has been fed conflicting reports about the accident at the Pennsylvania plant. At one extreme were those who said the situation was under control. At the other were those who predicted a nuclear catastrophe.

As usual, the truth was somewhere between the two extremes.

—

The accident in Pennsylvania is of special interest here because of the Wolf Creek Nuclear Plant being built near Burlington. If a similar disaster happens there, will Emporians have to throw some things together and flee for their lives?

We have been assured by the Wolf Creek developers time and again that there is no danger. All is well. There is no way that radiation could escape from the plant.

Yet questions have been raised about the quality of concrete used in the base for the nuclear reactor at Wolf Creek. In another incident, a gaping hole was left in a wall of the plant because concrete was not properly poured.

Later tests have shown that the concrete in the base mat meets Federal standards, although no final clearance has been received from the Nuclear Regulatory Commission. And the hole in the wall was easily repaired.

Like the Hindenburg disaster, the Pennsylvania nuclear accident has shocked many Americans who were inclined to believe the reassurances of press agents.

—

Those in both government and industry who defend nuclear power will have to change their ways if they want continued support. They must abandon hard-sell techniques and stick to plain, hard facts. If there is a danger, however remote, they must let the people know.

After they have accurate information, Americans may be willing to accept the hazards of nuclear power in order to head off the certain disaster that will come if this nation runs out of energy. — R.C.

The Seattle Times

Seattle, Wash., April 3, 1979

THE situation at the Three Mile Island nuclear power plant in Pennsylvania reportedly has stabilized and the outlook is favorable — which is more than can be said for the future of nuclear power in this country.

Even if the hydrogen bubble inside the reactor dissipates completely, no more radiation is released into the atmosphere, and no further evacuation of residents is required, a thorough reappraisal of the nation's commercial nuclear program is now in order.

It may be an agonizing reappraisal. The answers and conclusions that can be drawn from Three Mile Island are by no means easy or clear-cut.

— Yes, this was the most serious accident in the history of U.S. nuclear power.

— Yes, more radioactivity was inadvertently released than ever before from a commercial reactor.

— Yes, the sequence of events that caused the accident — an apparent combination of mechanical failure and human error — was not foreseen despite years of tests of nuclear-safety systems.

— Yes, the hydrogen bubble that formed in the reactor was not "part of our standard assumptions" but was "a new twist," according to Harold Denton of the Nuclear Regulatory Commission.

Despite all that, a precipitous backlash against nuclear power — such as a nation-wide shutdown of all nuclear plants — would be a serious mistake.

What is needed in the aftermath of Three Mile Island is a comprehensive review of the safety systems at all 72 of the nation's operating nuclear power plants, and a skeptical preview of the systems at plants now under construction. This should be done with an eye to tightening safety standards, as President Carter has suggested.

This effort should begin at the eight other reactors with cooling systems built by the Babcock & Wilcox Co., manufacturer of the system at Three Mile Island. If firm assurances of their safety cannot be provided, they should be temporarily shut down.

Also needed is a critical survey of evacuation plans for residents around all nuclear plants. This incident proved that such plans are haphazard and inadequate. Serious consideration should be given to the suggestion that the federal government take over a reactor site in times of crisis — even though the N.R.C.'s record is not altogether reassuring.

One disturbing revelation of Three Mile Island was that the commission was aware last November of cooling-system problems at the plant, but took no action. Another troubling aspect was the steady stream of conflicting and misleading information put out by Metropolitan Edison and the N.R.C.

Finally, this accident should stimulate a profound reassessment of our national energy policy and, we hope, a greater emphasis on conservation and the earlier development of alternative energy sources that will wean us more rapidly away from conventional sources such as oil, coal and nuclear fission.

The choices we face are limited and difficult, especially in the next 20 years. But hard choices will be the inevitable and unpleasant legacies of Three Mile Island.

Oregon Journal

Portland, Ore., March 31, 1979

Harrisburg, Pa., offers the latest evidence that nuclear power should be as limited as possible and viewed as a temporary stop-gap until safer forms of energy can be developed.

The radiation leak at an atomic power plant there indicates that nuclear development is not as safe as its advocates would like to believe.

It has the potential to be very dangerous. With all the safeguards atomic scientists and engineers talk about, high-risk accidents do occur.

Pro-nuclear scientists explain away each of the mishaps or near accidents as though they are flukes that cannot happen in the future.

Yet they keep happening, not in large numbers, but with so hazardous a product as nuclear power any accident is one too many.

When he was campaigning for president in 1976, Rep. Morris Udall often made the point that the question of nuclear power could not be settled by science. One roomful of scientists would tell you that it was the ultimate power source and safe to develop, he would note, and the next day another roomful of scientists would warn that it spelled the doom of civilization. His observation was borne out in Oregon during the campaign on a ballot measure to ban nuclear power in this state.

The Journal has never taken the position that all nuclear power forever be banned. If there is no other way to meet the power needs, it may have to be used. For the time being, the output from the Trojan plant makes a substantial contribution to the Pacific Northwest's energy supplies.

But the Harrisburg incident further indicates that nuclear power should be relied upon as little as possible and that other, safer sources — such as solar — be developed and promoted as quickly as possible.

Safe disposal of wastes is not the only concern. All the safety systems notwithstanding, the release of radiation poses a genuine threat.

The Dispatch

Columbus, Ohio, April 3, 1979

A SERIOUS ACCIDENT at a nuclear power plant near Harrisburg, Pa., has temporarily shattered public confidence and produced some impassioned cries for a moratorium on nuclear power.

This type of reaction is entirely understandable, but it should not block future development of this much-needed energy source. It is evident from what is known about the technical malfunction at the Three Mile Island facility, that safety considerations need added attention.

But the nation cannot afford to abandon nuclear power because of a serious accident. Energy Secretary James Schlesinger puts the issue directly: "We will have to have nuclear power" to meet future energy needs. At the moment 72 nuclear power plants across the country provide about 12 percent of America's electricity needs.

An extended shutdown of facilities that have been operating satisfactorily would cripple the U.S. economy this summer, stifle economic growth, and force severe conservation on consumers. Over the long run nuclear power seems the only feasible option to avoid undue dependence on unstable Mideast oil exporting nations.

The Harrisburg accident, as regrettable as it is, will no doubt offer government and industry experts many important lessons to improve safety and reliability. But, America cannot afford to throw out the baby of nuclear power with the bath water. That would be sheer folly.

The Providence Journal

Providence, R.I., April 3, 1979

Unquestionably, the acceptability of nuclear power plants to the public has been severely damaged by the accident at Three Mile Island in Pennsylvania. President Carter and his advisers are now faced with the need to draft an energy policy that pushes alternative sources of energy.

Nuclear power has always been dogged by serious questions. Now that Three Mile Island has revealed problems never anticipated by the industry, public resistance can be expected to increase. That does not mean closing down of all the 72 existing plants. Probably most of them — and also those that are already under construction — should be continued. But the Nuclear Regulatory Commission undoubtedly will — and should — be giving them the tightest inspection that it can field.

Oil — much of it imported — will have to fill the gap left by unbuilt nuclear plants for the immediate future. As a a result, the conservation program that the President has been contemplating becomes even more urgent. Some consolation may be found in the newest report by the CIA that shows an actual increase in world oil production over a year ago. But the urgent need to reduce U. S. reliance on imported oil is still obvious, as is emphasized by the shutoff of Iranian oil. If this country is to reduce its vulnerability to foreign political upheavals and price boosts, it must push with renewed determination all the other means of producing electric power for industry and homes and oil for heating and transportation.

The alternatives are all well known. Domestic oil exploration and production can be stimulated through decontrol of prices as fast as feasible. But this must be coupled with an excess profits tax, to force the oil companies to reinvest the extra revenue, not simply pocket it. Conservation, to be effective, will require some mandatory features, such as the 55 mile-per-hour speed limit, lowering thermostats in winter and raising them in summer, discouraging pleasure driving, encouraging or forcing industry to convert to coal — or gas, if there is enough of the latter.

Stepped up research and pilot plants for liquefaction of coal is needed as rising oil and gas prices bring that option into the range of economic feasibility. Such plants may be able to utilize the products of strip mines in the West, where strict controls can assure restoration of land after mining and the worst health hazards of underground mining can be avoided. Similar research and plants for gasohol — alcohol that can be burned in motor vehicles — could utilize excess sugar cane or beets, perhaps even grain not needed for food.

The government ought to expedite negotiations with Mexico for import of its natural gas, which is now being flared, and as much oil as Mexico can provide. Canada, too, should be pressed for all the oil and gas it can spare. And the new pipeline, to tap the natural gas of Alaska and Canada's Far North, should be advanced out of the planning stage.

Negotiations with Hydro Quebec for import of excess electricity from its huge plants on James Bay, which are now being conducted on a low-key basis by some New England governors, ought to be moved to the federal stage and intensified. Similarly, the Department of Energy ought to be persuading Canada to get on with tidal power in the Bay of Fundy, guaranteeing a market for the power if necessary, and maybe even offering to provide some of the capital. Short of that, DOE ought to reopen consideration of the tidal project in Passamaquoddy Bay at Eastport, Maine.

Depending on the need, some relaxation of environmental restrictions such as emissions standards for motor vehicles or the prohibition against high sulfur oil in heating plants, might be warranted. The country certainly ought not to roll back existing standards, which would swamp some congested areas in health-threatening air pollution; but it can and should stretch out the schedule for implementing higher pollution-control standards.

Nuclear power has suffered a body blow. It may not be out, but it is down and undoubtedly its development will proceed much more slowly than in its first 20 years. The United States is not lacking in substitutes. The drawbacks to each must be overcome. The future of the country and its economy need not be dimmed by the nuclear setback any more than by the threats to oil imports, if the same ingenuity that fashioned nuclear power is focused on the alternatives.

The Chattanooga Times

Chattanooga, Tenn., April 3, 1979

Even if technicians at the Three Mile Nuclear Power Plant near Harrisburg, Pa. are able to stabilize the reactor by alleviating the hydrogen bubble that has complicated seriously the situation that resulted from what must be described as the most serious accident in the history of the nation's nuclear power program, there can be little doubt that procedures for dealing with such incidents must be improved.

Initial reports say the accident began with a stuck valve — an eerie reminder of the situation in the current movie, "the China Syndrome" — that triggered a sequence of increasingly serious events: a pump failed, automatically shutting off the turbine and sending the control rods back into the reactor's fuel to shut it down. But one of the two major cooling systems shut down after a leak occurred, thus increasing the pressure that blew out some safety valves. An emergency core-cooling system activated, thus preventing a complete meltdown of the reactor, the accident that nuclear experts consider to be the ultimate nuclear disaster.

The accident forced the evacuation of hundreds of persons in the area around Harrisburg. Already, President Carter has created a special task force under the National Security Council to monitor the incident and the Defense Department has ordered a complete review of the nation's nuclear policy.

These actions are the least that can be done for now but they are merely the preliminaries to what must be done to insure the safety of persons living near the plant. So far, according to officals of the Nuclear Regulatory Commission, the radioactive steam released from the power plant poses no danger to public health, presumably because it has been dissipated by the wind. There is no guarantee, however, that the situation won't become worse, thus posing a more serious threat to the countryside.

If the Three Mile Island plant accident was caused primarily by design failures — the NCR has recorded several malfunctions since the plant went "on line" last April — it need not mean that the U.S. should abandon its reliance on nuclear power. At the very least, however, Sen. Edward Kennedy, chairman of a subcommittee that oversees energy matters, is correct in asking the Department of Energy to abandon its plan, at least for now, to devise ways to speed up the licensing of such plants.

THE CHRISTIAN SCIENCE MONITOR

Boston, Mass., March 30, 1979

There appears to be every reason to be thankful and relieved that the accident at the Three Mile Island Nuclear Power Plant was not more damaging. It was no "China syndrome" type of situation, as one plant spokesman put it. There were no fatalities, no serious personal injuries. A handful of workers were exposed to radiation but none was seriously contaminated or hospitalized, according to wire service reports.

However, the full public-health impact of the accident in the three-month-old nuclear plant in central Pennsylvania is still being evaluated. State officials continue to monitor surrounding areas for signs of radioactive iodine, which could be ingested through breathing or drinking milk.

At this writing the cause of the accident, described as the worst ever in a nuclear generating plant, is still not known. Officials at the US Nuclear Regulatory Commission say it will probably be some time before an investigation discloses what actually happened. Senator Hart of Colorado, chairman of the subcommittee on nuclear radiation, says part of the radiation problems was caused by human error – an atomic reactor operator apparently turned off a cooling system too soon, which led to the release of radioactive steam and radiation that was detected at low levels as far as 16 miles away. Plant officials, however, initially put the blame on the failure of a valve in a pump in the cooling system.

This is just one of several questions that will need to be explored. Another is why there was a three-hour delay in reporting the mishap to state civil defense officials. NRC officials say the problem may not have been detected within the plant until 7 a.m. – three hours after the leak started. If so, this obviously calls for closer monitoring and far more stringent safety precautions within the plant itself.

It is just such apparent inattentiveness to potential safety problems that has helped feed the fears of nuclear critics. Those familiar with the entire nuclear field complain of too often finding a relaxed "make do" attitude among engineers and technicians – a willingness to cut corners rather than abide strictly by regulations. This does not stir confidence.

Certainly there was and is no cause for public panic over the Three Mile Island mishap. The mere fact that radiation and damage were contained as well as they were says something for current nuclear safeguards. But where the potential for destruction is so vast, there can be no room for possible slip-ups, human or otherwise. And that should be the lesson gleaned from the Pennsylvania accident.

Sentinel Star

Orlando, Fla., March 30, 1979

UNTIL A great deal more is known about Wednesday's accident at Three Mile Island's nuclear reactor, the issue is certain to be clouded by emotion.

Nuclear energy opponents will call for the shutdown of existing U.S. plants while proponents of atomic reactors minimize the danger. Each side has valid points, but condemning an entire industry because of an isolated malfunction would be as irresponsible as failing to heed its warning.

The ultimate disaster, it must be remembered, didn't happen. The plant's safety system prevented meltdown of the radioactive nuclear core. Indeed, the history of nuclear generation in the United States is remarkably free of radiation-related hazards, a health and safety record that hasn't been equalled by coal-burning plants.

The subject of nuclear catastrophe has been explored by qualified physicists and the consensus supports continued development of the program. A typical nuclear plant, said MIT's Norman C. Rasmussen, an international expert, poses a risk of one in 20,000 for five deaths to occur; the typical coal plant takes about that toll each year.

Yet we are dependent on coal to replace oil and gas and to take up the slack brought on by delays in nuclear plant construction.

Unfortunately, Three Mile Island's accident will be another deterrent to progress. But it also could alert industry officials, lulled by 25 years of relative safety, to risks that require closer monitoring.

For the remainder of this century nuclear energy is the only available alternative to the problems of producing, transporting and burning coal. It makes fewer demands on resources, causes fewer occupational deaths and injuries and is much less hazardous to the public.

Let's judge it by its achievements, not its few **failures.**

The Salt Lake Tribune

Salt Lake City, Utah, April 1, 1979

Malfunction of the Three Mile Island nuclear power plant near Harrisburg, Pa., could become a milestone in the future of this country's nuclear power industry.

Compared to what might have happened, and indeed, still could happen, the incident is a benign one. Yet it has provided a textbook case of what is wrong with nuclear power regulation and ways of dealing with emergencies.

On the basis of what is known so far about the Three Mile Island plant there are insufficient grounds for giving up on nuclear power which is vitally needed in the energy-hungry United States. But it is abundantly clear that new safeguards and new perceptions of possible danger must be fashioned.

Events at Three Miles Island chillingly demonstrate that plant designers and government regulators did not, as was widely claimed, anticipate all possible consequences of a breakdown and figure out ways to deal with them. For example, the potentially dangerous "gas bubble" that hindered cooling of the reactor core was described by Dudley Thompson, executive officer for operations and support at the Nuclear Regulatory Commission, as a "new problem," and he added in the strange language of the bureaucrat, "We are in a situation not comparable to previous conditions."

Another glaring inadequacy exposed by the Pennsylvania incident is a paucity of plans and personnel to deal with even the thus far low-level emergency at hand, much less a full-blown radiation catastrophe. Confusion surrounding events near Harriburg is prima facie evidence that no reliable central information agency existed and conflicting reports of what happened and what would happen next added to the rumor-charged uncertainty.

Further, the General Accounting Office released a report which concluded that existing evacuation plans for areas surrounding nuclear facilities are inadequate to protect the public in event of a serious accident.

Much of the adverse reaction the nuclear power industry will receive in the aftermath of the Three Mile Island incident is a direct result of that industry's over-the-years emphasis on the fail-proof safety of nuclear reactors. And the government did little to discourage this effort.

On the record, nuclear reactors have proved safe and can continue to be so in the future. But they apparently have never been and cannot be as safe as the industry has consistently pictured them. Had the public been prepared for the possibility of an accident instead of the virtual impossibility of one, backlash now building might have been averted or greatly reduced.

In the wake of Three Mile Island, benefits and drawbacks of nuclear power generation will be reevaluated. As things now stand there is insufficient reason to reject this timely power source. But there is a plain and urgent need to reassess the chance of such failures elsewhere and to prepare realistic contingency plans for coping with future accidents — or even sabotage. This must now be recognized as part of the overall price of America's hefty — and growing — energy appetite. It is a price that will have to be paid until consumption is cut back substantially or less threatening energy sources are more fully developed.

BUFFALO EVENING NEWS

Buffalo, N.Y., April 5, 1979

The worst of the crisis at Pennsylvania's crippled nuclear plant has apparently eased, but it may be days or even weeks before President Carter can fulfill his personal promise to tell the American people in plain language what went wrong.

Certainly nothing short of candor will do if nuclear power is to regain credibility on the part of a gravely worried public as an important part of the country's future energy "mix."

By virtue of his early nuclear experience, as well as his executive responsibility for the government's regulation of the nuclear industry, Mr. Carter is plainly the appropriate source of whatever reassurances it may take to ease the credibility problems intensified by the nightmarish confusion of official statements and scientific jargon at Three Mile Island.

And in fact, Mr. Carter already has made a promising beginning. He calmed fears by his pledge of a full report on the mechanical and human errors during his first-hand visit to the facility. He has followed this up with orders for a federal review of the reactor mishaps. And now, as a prudent precautionary measure, the government is moving to make sure that there are no design defects that could trigger similar crises in any of the other eight plants with reactors made by the same firm.

* * *

But all of this is still only a beginning in addressing a national crisis of confidence in claims of nuclear safety. The depth of this is indicated by Gov. Carey's hasty judgment that the Harrisburg accident "spells the end of the future for nuclear power now" in New York. Yet when thousands are warned of a possible evacuation in the event of a melt-down or explosion, when the government's foremost nuclear authority admits scientists were caught off guard by unforeseen incidents, when citizens confess they don't know what to believe — something much more convincing than bland assurances by nuclear-power defenders is needed to fill the credibility gap.

What is needed, before an anti-nuclear "critical mass" sets off a public opinion chain reaction that precludes any rational decisions, is a searching review of all the relevant questions — about both the accident and the future role of nuclear energy — by a blue-ribbon presidential commission, along the lines urged by Pennsylvania's Senator Richard S. Schweiker.

* * *

Let's have full and frank answers to such key questions as these: What defensive safeguards must be strengthened at existing and future plants to avert overlooked hazards or the margin for human error in theoretical calculations? How does the government propose to dispose of nuclear wastes and the future decommissioning of obsolescent plants? Must plants in earthquake zones be redesigned to avert potential calamity? Are the risks in man's taming of the nuclear genii simply too great for an expansion of nuclear power's present role in our energy supply — and if so, what are we prepared to give up in a trade-off?

Such questions, to be sure, deal only with safety considerations, but it is these which account heavily for an escalation in construction costs and delays in licensing which have dimmed the optimistic expectations for nuclear power as a cost-saving substitute for oil and environmentally-hazardous coal.

The nation cannot afford to permanently write off nuclear power in an age of exorbitantly expensive and depleting fossil fuels. But neither can it afford the risk of any more nuclear emergencies and the write-off of billion-dollar plants because of either safety-system failures, or leaving to chance the hope that later scientific "breakthroughs" will solve certain critical waste-disposal or other problems that are still beyond the reach of present technology.

San Jose, Calif., April 3, 1979

UNTIL now, we were convinced that the burden of proof rested with the critics of nuclear power. The preponderance of respected scientific opinion was that nuclear energy is dependable and safe. But then came the malfunction at the Three Mile Island nuclear power plant near Harrisburg, Pa.

What nuclear scientists assured us was virtually impossible — a radiation leak spreading fear and unknown dangers to hundreds of thousands of nearby residents — proved all too possible.

And with that accident, the burden of proof shifted. Advocates of nuclear energy must now prove with greater certainty the safety of nuclear power plants and the reliability of their safety systems.

In California, the U.S. Nuclear Regulatory Commission should give immediate consideration to Gov. Brown's request for a "precautionary and temporary" closure of the Rancho Seco nuclear plant near Sacramento. Rancho Seco, a near twin, was built by the same firm as the Three Mile Island plant. Fear that the two nuclear facilities may share a "serious generic defect" is not confined to the governor. This fear will persist and fester in the absence of thorough safety investigations.

Above all, as the governor emphasized at a news conference Sunday, "It is appropriate to err on the side of caution."

In a sense, the widespread calls for reexamination of the nation's nuclear energy policies are superfluous. Yesterday's assurances about nuclear safety perished at Three Mile Island. All the old assumptions are open to new challenge.

Nuclear development, dealt its worse blow yet, has a bleak outlook in the United States. Despite this, all the nuclear plants are not about to be switched off tomorrow, not when they generate 10 percent of the nation's electrical energy. A precipitous shutdown would mean energy turmoil.

A priority review of the safety systems of nuclear plants similar to the Three Mile Island facility is the more realistic alternative. This, however, should quickly expand into extraordinarily intensive safety reviews of all nuclear reactors. The guiding principle should be that set forth by Gov. Brown: " . . . to err on the side of caution."

We know nuclear plants are not as safe as we were formerly told. Now we are left to determine if the added risks outweigh the apparent benefits.

The News and Courier

Charleston, S.C., April 3, 1979

On the defensive, proponents of nuclear power are saying press and public have over-reacted to the accident at the Three Mile Island plant in Pennsylvania, and have blown the danger potential out of all proportion. People have been hitting the panic button, the proponents say, because they don't understand nuclear power principles, and do not know the first thing about radiation.

There is partial truth in that contention. The American layman has little or no grasp of the nuclear process, of how nuclear reactors at electric generating stations work. What concerns the average newspaper reader or TV watcher is that the experts — the nuclear scientists, the operators of the Three Mile Island plant, and Nuclear Regulatory Commission officials — don't agree on how critical is the situation. The apparent reason they cannot agree is they have not before faced a comparable situation.

"...we suffered damage to the reactor core (where the uranium fuel is), which is significant," Joseph M. Hendrie, chairman of the Nuclear Regulatory Commission, told a House subcommittee a day after the accident. "This is the first time we've had this level of core damage at all in the history of the civilian program." Moreover, there was disagreement on how best to cope with the potentially dangerous gas bubble in the reactor at the Three Mile Island plant. Asked at one point if he had ever heard of a bubble in a reactor before, the operations director for the NRC answered: "We've not had a situation like this...until today."

Man always has feared the unknown. His fear has not diminished in the nuclear age. On the contrary, it has heightened in some regards. Citizen knowledge of radiation may be sketchy at best, but millions of people have enough common sense to recognize that the risks involved in nuclear power plant operations can't be equated with the risks involved in producing and utilizing other energy sources. A mine explosion and cave-in can be a tragedy, but its effects are localized and limited in time frame.

In a nuclear accident, radiation effects can be extensive, and persist for no one knows how many years. That realization is what makes many people uneasy about the Three Mile Island situation—plus the fact that those supposed to know apparently don't have the answers.

The Kansas City Times

Kansas City, Mo., April 3, 1979

Once the crisis is completely past at the Three Mile Island nuclear power plant near Harrisburg, Pa., the nation is entitled to a definitive answer on what went wrong — and a quick answer, not one that comes months later through the customary snail's pace of a congressional investigation. Human survival without fear is the first consideration and then back to the gnawing question of sufficient energy and its relation to the entire economy.

President Carter promised over the weekend to be "personally responsible" for thoroughly informing the American people of the status of nuclear energy. That's a promise that must be kept and, again, with maximum speed. In the wake of the nation's worst nuclear accident, the future of nuclear power as a bridge over the energy shortage is in serious doubt.

Reaction was swift in Congress where a request was made that the Carter administration pull back from its support for legislation that would speed the licensing of nuclear power plants. Energy Secretary James Schlesinger has not withdrawn his contention that nuclear power is still essential. Temptation may be strong in some quarters to pull the rug out from under Schlesinger by giving in to legitimate public concern that should be addressed rather than exploited.

The Pennsylvania disaster points up how far down the road this country — to say nothing of Western Europe — has gone in relying on nuclear power. Seemingly overnight, 14 percent of the nation's electricity during peak periods of power usage is generated by 72 nuclear reactors. Within the next decade, until its future was clouded by events in Pennsylvania, nuclear plants were expected to provide more than one-fifth of all electricity.

If not nuclear power, the question comes back to the hard choices of more and more expensive oil if it is available, the stripping of land and pollution that accompany coal usage, drastic conservation that alters the economy and lifestyles while awaiting the possible development of renewable energy sources. The energy crisis already had become nearly unmanageable because of the temptation for public officials to blink in the face of political pressures. Difficulties have been aggravated considerably by the Pennsylvania disaster.

Look at the possible scenario. Three years ago the Seabrook nuclear plant in New Hampshire, which is still under construction, sent ripples through the presidential primary in that state's overblown election. Now the nuclear debate and Seabrook in particular may distort the entire political process and the energy debate if Gov. Jerry Brown of California, a critic of nuclear projects, elects to challenge Carter administration policies in presidential primaries.

The first responsibility of government and nuclear power experts has been to bring the Pennsylvania crisis to a safe ending. What happened at the Three Mile Island plant was an accident that nuclear facilities are designed to prevent — a temporary loss of coolant. The chain of events had not been anticipated by nuclear experts, which is a reminder that the most brilliant of nuclear scientists are pioneers on a hazardous journey.

It may be too late or even completely unnecessary to turn away from nuclear power, but the public needs new and convincing assurances that health and safety have not been sacrificed in a head-on plunge into a brave new age. As a starting point, guarantees must be forthcoming that what happened at Three Mile Island cannot happen again.

The Philadelphia Inquirer

Philadelphia, Pa., April 4, 1979

On the record of the last seven days, it would be unwise to place much confidence in statements coming from the operators of the Three Mile Island nuclear generating plant or from the federal officials who are, and always were, responsible for its safety. "They are way out in an unknown land with a reactor whose instruments and controls were never designed to cope with this situation," Prof. Henry Kendall of the Massachusetts Institute of Technology said, entirely accurately, to Newsweek, *"They are like children playing in the woods."*

Nonetheless, it is profoundly hoped, of course, that the crisis in the Susquehanna is passing.

In the enthusiasm all the world must feel for such optimism, it is important to remember that the putative gas bubble in the reactor, now ostensibly being reduced or eliminated, was never even thought of until Saturday, the fourth day of the crisis.

The root problems were never foreseen and to date have been only abstractly theorized. Today, the residual problems are only being guessed at. That guessing has responsible members of the Congress estimating the clean-up may cost $1 billion, or that it may be impossible, leaving a virtually eternal residue of deadly radioactivity entombed 10 miles south of Harrisburg.

Assuming the plant, in one more eruption of invincible ignorance or arrogant irresponsibility, does not explode, whether it is torn down or left as a mausoleum for radioactivity 100 times the levels lethal to humans, Three Mile Island is an eloquent monument to one truth: The U.S. nuclear power industry and government failed.

Three Mile Island must be a turning point in American public policy.

It is premature today to conclude that the only prudent policy for the future is to turn immediately and forever away from nuclear power generation in any form.

It is equally premature to conclude that the U.S. *shouldn't* turn away from nuclear power, forever.

What Three Mile Island does demand, unequivocally, is that a national debate begin to examine which of those alternatives is prudent, or if there are others.

There is a great deal of talk in Washington about the forum for that debate. The White House press secretary Monday spoke of an interagency study in the Executive Branch. President Jimmy Carter has pledged he will personally report to the people.

That is not enough. It alone would be negligent. An exhaustive investigation is needed, held by the Congress in open forum, in which no secret is kept from the public and no witness is allowed to go unheard or to evade thorough examination.

If there were need to argue that the Executive Branch is incapable of managing the debate responsibly, it was filled Monday by James R. Schlesinger, who as Secretary of Energy is primarily responsible for the bureaucracy and the decisions which affect America's power policies. "It is my judgment," he told a House subcommittee, "that the nuclear licensing bill indeed will be resubmitted."

What is the bill? A proposal, pressed forward last year and left in limbo, that licensing standards and systems required before a nuclear plant can be put into operation be loosened and eased, reducing the present 10-to-12-year lag by half.

"It's a question of trading Iran off against Three Mile Island," he said.

Mr. Schlesinger's position, judged in human terms, is outrageously indifferent to the dreadful — and entirely realistic — fears which millions of Americans have lived through for the last week. Beyond its arrogance and insensitivity, however, it is a vital reminder that Mr. Schlesinger and his colleagues in the government are so imprisoned by their conviction that nuclear development is desirable — come hell or high water, come Three Mile Island or worse — that their judgment cannot be allowed to dominate the debate if America is going to make prudent assessment of the future of nuclear power.

The Oregonian

Portland, Ore., April 3, 1979

The dramatic events at the Three Mile Island nuclear plants have not shaken The Oregonian's confidence in nuclear power. We continue to believe that alternative fossil fuel sources of energy are far greater long-term risks to the nation's health and will exact a higher economic and social price on consumers and workers than will nuclear power.

The nation does not have any short-term acceptable option to nuclear power if it is to rid itself of foreign oil blackmailers, price gouging and outrageous political demands that have characterized the oil cartel since 1973. Inflation would be worse without the atom's energy.

The nation cannot permanently close all 72 U.S. operating reactors, particularly in regions where some utilities last winter were generating 40 percent of their electricity from the atom. While inflation is taking a heavy toll all along the energy front, Commonwealth Edison of Chicago ordered two more reactors late last year, bringing its total to 15. These were the only new orders in 1978, but they demonstrated the company's belief that nuclear power would prove 10 percent cheaper than coal, and faith that the reactor climate would improve.

Events in Pennsylvania, while unlikely to cost a single life, have shaken public confidence in nuclear power, despite its 20-year safety record that is unmatched by any other energy source. But it is the unknown, the obscurity of the technology, the linkage with holocaustic weapons and the confused reporting of complex events that have fanned fears since the accident in Pennsylvania emerged as a worldwide news event.

Unfortunately, spokesmen at the scene did not always inspire belief they understood or were in control of events taking place inside the reactor. Only a complete investigation can hope to reveal the extent of the threat. We do not know whether causes were mechanical, human or both. That investigation must be extensive and public if confidence is to be quickly restored in nuclear power safety.

The accident, by focusing the public's attention briefly on nuclear power, can be hoped to help the public understand that nuclear plants are not all alike, any more than are all airplanes. Different companies build them to different designs, although all are supposed to meet uniform safety tests. The Trojan plant in Oregon has a Westinghouse reactor system, while the proposed Pebble Springs plant would be built by Babcock and Wilcox, the company that built the Three Mile Island plants.

Before the situation worsened last week, and while the problem was a minor event, The Oregonian saw some early reports flowing from the scene as more a "media meltdown" than any actual destruction of the uranium core. Even when unexpected problems began to occur, belying earlier reports, some newsmen were frankly admitting they could not understand the technical language they were getting at the scene. "Meltdown" is one of those "buzz" words, having no precise technical meaning and being better suited for movies and headlines. It did not accurately describe what our editorial intended to convey, making us guilty of some of the rhetoric we were deploring.

Politicians, of course, will now begin to take cover, adding to nuclear's damaged reputation. The administration had hoped to see some 380 more reactors built by the end of this century. This goal, while well under what the industry believes is needed even with conservation and other new fuel sources, was not going to be met before Three Mile Island, and certainly will not now.

That the nuclear industry is in grave danger is the same as saying that the nation's energy program is also in critical distress, because they are locked together.

The accident near Harrisburg may well scare off investors, concerned about federal insurance and the long and expensive lead time needed to put a plant on the line. Early this year, there was some good nuclear news on the horizon. Congress is working on an accelerated licensing plan; progress was being made on the waste disposal problem; uranium mining expanded 30 percent; and some 100 organizations, ranging from scientists to housewives, have been formed to support the power resources of the atom.

All these efforts have been damaged, although the accident at Three Mile Island, unlike fires in nursing homes and a boarding house that took at least 35 lives this week, has, so far, not produced a single fatality.

Richmond Times-Dispatch

Richmond, Va., April 3, 1979

The Three Mile Island nuclear accident apparently did not unleash enough radiation to endanger human life, but it spewed out enough fears and doubts to cripple the nuclear power industry severely. Some pessimists are predicting that its wounds will prove to be mortal.

Even the staunchest advocates of nuclear energy must admit that the accident has evoked frightening concerns about the safety of nuclear power plants everywhere. It is soothing, but not totally reassuring, to say that other plants are constructed differently and that the Harrisburg accident was the result of a local peculiarity. There must be a general review and strengthening of safety plans and procedures throughout the industry.

Foes of nuclear energy would not be satisfied with this, of course. Nothing would please them except the total abolition of nuclear power, and they will gleefully cite the Three Mile Island accident as proof positive that nuclear power should be banned as a completely unacceptable threat to mankind.

Existing evidence does not justify such an extreme verdict. So far, the horrors of the Three Mile Island accident have remained only *potential* horrors. The fact that they have been contained strengthens the nuclear industry's safety record, which is the best of all power industries. It would be foolish to ban nuclear power not because of a disaster that occurred but because of a disaster that was *prevented.*

There is, moreover, no such thing as a totally safe source of energy.

Coal? Each year scores of men are killed and maimed and thousands more are permanently sickened by the hazards they encounter in the dark holes in which they work. And the public is threatened, environmentalists insist, by the pollutants that burning coal emits.

Oil certainly is not a totally safe source of energy. Nor is hydroelectric power, the source of which is "clean" water, for dams have been known to burst and send forth torrents of water to sweep people to their deaths.

The risks of any source of energy must be weighed against its benefits to mankind, and at this point the scales remain tipped in favor of nuclear power. A ban on nuclear energy would endanger the nation's economy, for alternative fuels could not meet the nation's growing energy needs. And a ban on nuclear power would require personal sacrifices that most Americans, in the long run, would be reluctant to make. Restrictions on the use of air conditioners on the hottest summer days and on the use of household appliances are among the strong possibilities, for most power companies could not meet their peak summer demands for electricity without their nuclear generators.

We do not seek to minimize the implications of the Three Mile Island accident. Obviously nuclear plant safety systems are not so reliable as the American people have been led to believe. But the accident has not confirmed the arguments of nuclear opponents, either. The United States would make a tragic mistake if, as a result of the Three Mile Island mishap alone, it abolished the industry upon which its energy hopes rests.

The Idaho STATESMAN

Boise, Idaho, April 3, 1979

One of the anti-nuclear theories has it that we're on borrowed time — that Americans are fortunate not to have experienced a nuclear disaster. The accident at Harrisburg appears almost tailor-made by proponents of that argument.

The accident at the Three Mile Island nuclear power plant was blamed for damaging the reactor core, sending radioactive material into the atmosphere and creating a giant, potentially explosive gas bubble. Radiation levels inside the plant's reactor building registered 1,000 times the normal amount.

It was fortunate that no one was killed. It also was fortunate, in a sense, that the accident happened. Better to have some radiation emitted into the atmosphere and some relatively harmless contaminants released into a river than to experience the stuff nightmares are made of.

It is ironic, or perhaps providential, that the Three Mile Island accident occurred at a time when the national nuclear consciousness has been raised by, of all things, a movie. Those who saw *The China Syndrome* noticed similarities between the fiction of the film and the reality of Harrisburg. Perhaps the result will be heightened concern by the public and officials over the potential hazards of nuclear power.

So far, it appears we have got off easy at Harrisburg. But if the federal people in charge of licensing nuclear power plants are wise enough to prevent a real disaster, they will treat the incident as if it had been precisely that. Harrisburg refutes the nuclear industry's statements about safeguards and theoretical probability of accidents. The reality is that something went wrong, and it was serious. Who can guarantee that worse accidents aren't in store?

Nuclear power is such a new endeavor, and the risks are so awesome. Even if you disregard the potential for plant accidents, the question of nuclear waste disposal alone is sufficient cause to re-examine the nuclear issue. We need the power, but we can't afford unnecessary risks.

The ancient Greeks and Romans, faced with fuel shortages, turned to the sun. They built entire cities based on using its life-giving energy. About the worst consequence of this policy was a case of the goose bumps. In turning to nuclear power, about which so much still is uncertain, we risk our own extermination.

Unless we slow down and re-examine every aspect of nuclear safety, we may be on borrowed time. And who knows how much is left?

Beyond TMI: Frequency of Nuclear Plant Mishaps Alarms Public

Although no nuclear plant incident has had as large an impact on public perception of the industry as the 1979 Three Mile Island episode, minor mishaps at nuclear reactors are by no means infrequent. Less than a month after the 1979 accident, it was reported that the files of the Nuclear Regulatory Commission for 1978 showed 2,835 "reportable occurrences" at nuclear power plants in the U.S. Such incidents, which violated NRC rules or in some measure threatened public safety, took place at dozens of the nation's nuclear reactors. The NRC files showed that every plant in operation in 1978 had had at least one unscheduled shutdown. Reports in succeeding years by Critical Mass, an antinuclear group affiliated with Ralph Nader, put the number of incidents of malfunction at 2,300 in 1979, 3,804 in 1980 and 4,000 in 1981. Industry and NRC officials, while not disputing the actual figures in the reports, which were taken from NRC records, protested that they grossly distorted the true safety record because the great majority of the reported incidents were minor ones.

Some of the more notable incidents which have occurred since the accident at Three Mile Island are discussed in the editorials that follow. In January 1982, a steam generator tube ruptured at the Ginna nuclear plant in Ontario, N.Y., resulting in the automatic venting of radioactive steam into the air. Although the reactor's core was not damaged, about 11,000 gallons of radioactive water were dumped onto the floor of the containment building. This was considered the most serious incident since Three Mile Island until the failure of New Jersey's Salem I reactor to shut down automatically in February 1983. The failure occurred twice when the plant was restarted following shutdown for fueling. Upon restarting, the reacter experienced drops of water level in the core; the automatic shutdown system to prevent overheating should have been tripped off but wasn't. According to the NRC, plant operators did not notice the failure on the first occasion because they opted for a manual shutdown when the water-level problem was recognized. The NRC blamed the mishaps on improper equipment maintenance and on inadequate training and supervision of the plant's operators.

DAYTON DAILY NEWS

Dayton, Ohio, October 13, 1979

A thousand or so folks got their jollies trying to occupy the Seabrook, N.H., nuclear power plant site the other day, but it's doubtful that they did their cause much good. That is, if their cause is stopping nuclear power, and not in trying to use the issue to radicalize the populace.

The anti-nuke coalition has a sound and reasonable basis, but it has attracted its share of fruitcakes. And there is a distinct aroma of brandy and candied cherries wafting from Seabrook. That is about the only thing the industry's public relations folks have had to get cheerful about in a long time. The industry's got *real* problems, as a glance at the headlines of recent days shows:

—A leak of radioactive steam led to the closing of a nuclear generating plant at Red Wing, Minn.

—Someone, for unknown reasons, used pencil and paper clips to prop open a switch at a nuclear power plant near Richmond, Va., and radioactive gas "burped" out.

—A reactor at UCLA leaked radioactive gas — nobody knows how much — into a building on the college campus.

—Arizona National Guardsmen seized a plant making luminous clock and watch dials because the plant had been leaking radioactive tritium for months.

—The Nuclear Regulatory Commission closed a uranium plant in Tennessee that apparently "lost" at least 20 pounds of highly-enriched uranium — the kind needed to make a bomb.

—Perhaps most damning of all, Washington Gov. Dixie Lee Ray, former head of the Atomic Energy Commission and a proponent of nuclear power, closed a nuclear disposal site because wastes being sent to the dump site were improperly packaged.

These incidents, along with scores of previous ones and the near-thing at Three Mile Island, are enough to persuade many Americans that the folks running the nuclear industry are neither as competent nor as careful — perhaps as honest — as they would like us to believe. And that considering the great capacity that industry has for doing great harm to great numbers of people, it ought to be reined in and watched carefully.

Couple that reasonable public unease with the horrendous costs of building nuclear power plants and the uncertain economic future they face, and for the next few years at least, it is not likely that any utilities will order new nuclear plants.

The real questions are what we do about the nuclear plants that are being constructed now, and how we meet the energy demands they are designed to fill. Those questions ought to be settled by rationality, not by counting who marched for a plant and who marched against it.

Long Island, N.Y., October 15, 1979

First there was the candle that started the fire at the Browns Ferry nuclear power plant in Alabama five years ago. Then there was the maintenance crew at Three Mile Island that closed off the emergency core-ooling system and forgot to open it up again. And now it seems a paper clip and a pencil were the tools that brought about the accidental shutdown and release of radioactive gases at the North Anna nuclear plant in Virginia, 70 miles south of Washington.

It's hard to escape the suspicion that anything everyone thinks has been made safe can be rendered unsafe by one human being. And testimony last month before a Senate subcommittee on the bumbling of the Nuclear Regulatory Commission staff during the accident at Three Mile Island cast further doubt on the knowledge and competence of the agency that's supposed to police the nuclear industry.

Officials at Virginia Electric and Power Co., which owns the North Anna plant, have informed the NRC that they believe workers in the emergency control room had used a pencil and a paper clip to wedge an important water control switch into a closed position. A series of events resulted, culminating in the release of radioactive xenon gas. The plant will remain closed for at least 12 weeks while the NRC completes its investigation.

There's nothing new about human error, but it can foul up the most sophisticated technology. And that means no amount of safety engineering is enough unless the people who operate something as potentially dangerous as a nuclear reactor are trained to follow the rules.

That *doesn't* mean the country wants robots operating the plants; it needs people who can think and react properly in a crisis. But the technocrats and probability experts who design the plants must take the human factor into account at every turn. Even though no electrician may ever again use a candle to check for air leaks in a reactor cable system, or paper clips to fix a stubborn valve, it's human nature to take shortcuts—and to forget.

In most cases, of course, the effects are minor and short-lived. But in the case of nuclear reactors, they could be disastrous. That's why Congress, the NRC, state and local governments, utilities and the general public must make preparations for worst-case accidents as long as nuclear plants remain in operation.

The Dispatch

Columbus, Ohio, October 9, 1979

THAT IMPORTANT report from a blue ribbon commission investigating the Three Mile Island nuclear power plant incident still is awaited, but a significant dividend already is evident.

Last week, at the Prairie Island nuclear power plant near Minneapolis, engineers detected the emission of radioactive steam. Unlike engineers at Three Mile Island, they acted promptly, including notification to authorities and the public.

A federal Nuclear Regulatory Commission report says the leak was detected at 2:14 p.m. in one of two generators. The faulty generator was shut down at 2:24 and the alarm sounded at 2:30. By 2:41, a rupture in one of 3,300 tubes in the generator was isolated. Controlled cooling was started. A NRC officer reports conditions now are normal.

In a period of 27 minutes, a nuclear power plant malfunction was detected and corrections placed in motion. It is convincing testimony to what can and must be done to assure that nuclear power can be harnessed wisely and calmly.

The Virginian-Pilot

Norfolk, Va., October 3, 1979

Three questions from the latest accident at a Virginia Electric and Power Company nuclear plant demand honest answers. They go to the very heart of Vepco's competence to operate nuclear reactors safely.

First, there was the bizarre tampering with a switch—its automatic operation blocked by a paper clip and a pencil, Vepco said Monday—by unidentified control-room personnel at the North Anna power plant. Consequently, a valve controlling the flow of reactor coolant failed to function properly, contributing to the accidental release of radioactive gas. But yesterday, surprisingly, Vepco denied its own paper clip story.

While Vepco questions its employes about what really happened, the public—represented by Nuclear Regulatory Commission inspectors—must question the company's overall standards and commitment to safety. Has Vepco's management encouraged shortcuts or makeshift rigging in hopes of avoiding costly shutdowns? The NRC should question Vepco's workers about their perception of the utility's priorities and what is expected of them.

It is one thing for a small business to take modest risks with jerry-built machinery to avoid costly production delays. But if that mentality dominates a nuclear utility, the risk to public health demands a change of thinking.

Second, there was the long delay following last week's accident before Louisa County officials were notified. The NRC, as required by law, was told promptly, but Vepco failed to tell the county's director of emergency services until 11 hours later. Nor was the public alerted through the news media until 12 hours later—and then evidently only because the NRC held a public briefing in Washington.

Did Vepco believe it could keep the accident hushed up to avoid further damage to its public image? Doesn't Vepco understand that candor is the surest route to public confidence and credibility, that long delays in releasing information breed suspicion?

Third, a vent pipe designed to dilute and filter radioactive gas discharges was not connected, and so gas was released to the atmosphere and into an auxiliary building occupied by two workers. Fortunately, neither was injured.

But why was the vent not connected? Why didn't the resident NRC inspector spot that deficiency before the accident? One company official said it may not have been hooked up during the 15-months that the North Anna reactor had been operating.

Would Admiral Rickover have let a nuclear ship leave on patrol without all its pipes connected, its valves working as intended, its switches unjammed by paper clips? Not likely. And what do you suppose the admiral would do to a skipper who tolerated such sloppy, reckless operation?

When the safety of the DC-10 aircraft was questioned last spring, the Federal Aviation Administration grounded all the planes—at great expense to the airlines that fly them—until inspectors cleared each one as airworthy.

The same searching look at Vepco's nuclear power plants should be conducted before any of its reactors at North Anna or Surry are started up again. Another fine, readily paid indirectly by the ratepayers, would be an inconsequential penalty.

We want Vepco to operate its nuclear plants, but only if it observes the highest safety standards. Its record, however, is that of repeated violations for which the NRC has fined it $127,400, more than any other utility in the nation. This record, plus last week's accident, should impel the NRC to consider whether the utility is capable of operating its reactors without endangering public health, or whether it must undergo internal reforms before having its operating license renewed.

The Topeka Capital Journal

Topeka, Kans., June 27, 1979

A suspended Virginia nuclear power plant employee told the Newport News Daily Press that he sabotaged fuel rods at the plant in order to force the company into making changes in safety and security operations at the station.

Such action can never be condoned or championed as in the public interest. The man, if found guilty of his admitted crime, should be prosecuted.

William E. Kuykendall told the newspaper he and another man poured sodium hydroxide on the rods in a storehouse at the plant. The man said he attacked the most vulnerable area of the plant in a manner that showed the potential for disaster and yet did not create any life or health-threatening hazards.

He said he and his accomplice entered the fuel storehouse at the Virginia Electric and Power Company's Surry nuclear power plant with a container of the caustic, devised a cup from material in the building, and poured the substance over the separately stored fuel rods.

"You can hang me by my thumbs in Richmond Square as long as you put a sign in front of me saying what I did and why I did it," he said.

If the man wanted to call attention to disaster potential or plant vulnerability he could have gone to company officials. Or, he could have gone to the newspaper with his story of how it allegedly could be accomplished. He did not need to resort to actual sabotage, and no one should be allowed to escape punishment for carrying out such deeds.

The plant discovered the sabotage 10 days later. Of 64 unused fuel elements, 62 were damaged. They are valued at $30 million. Cleaning them cost the plant, its stockholders and its customers $6 million.

Richmond Times-Dispatch

Richmond, Va., September 29, 1979

A Richmonder said a "cold chill of fear" gripped him when he heard Tuesday a broadcast report that there had been "an accident" at the North Anna nuclear power plant and that details would be given in a later news broadcast.

On the basis of the facts now known, and pending the official Nuclear Regulatory Commission report, the accident was an extremely minor one, and the radiation leaked was insignificant. Stories in this newspaper have cited the fact that the amount of radiation to which a few workers were exposed was about one-seventh the amount those persons would have received from chest X-rays, and that it exceeded by only a minuscule degree the amount of radiation an individual is exposed to, because of the high elevation, on a coast-to-coast air flight.

The Nuclear Regulatory Commission was informed of the accident immediately after it occurred, but there have been suggestions from some quarters that the Virginia Electric and Power Co., which operates the nuclear plant, was remiss in not also notifying state and Louisa County officials.

Emergency procedures, approved by the Nuclear Regulatory Commission, provide for immediate notification of state and county authorities when it appears that radiation of certain specified amounts will be released into the air. The amount of radiation leaked inside the nuclear plant was a tiny fraction of the amount which calls for notification, and the amount released into the air outside was too little to measure. So Vepco was under no obligation to notify state and county authorities.

Aside from the legal obligation, should Vepco immediately notify state and county authorities when there is a minor accident such as occurred Tuesday? We believe the answer is no. There apparently was not the remotest threat of any damage to people outside the plant Tuesday, and official notification of state and county authorities, with word spreading through the community, might have resulted in totally unjustified alarm and even panic.

Even a minor mishap in a nuclear plant is not to be taken lightly, obviously. But such an occurrence should be kept in perspective. Unless the Nuclear Regulatory Commission finds otherwise, it appears that what happened at the Louisa plant Tuesday was inconsequential and that its inclusion, for example, in the national television news broadcasts Tuesday evening gave emphasis to the incident far beyond what was merited.

SAN JOSE NEWS

San Jose, Calif., October 18, 1979

ONCE again civil disobedience of a sort is in the news, this time in Surry, Va., where two men were convicted Tuesday of sabotaging fuel rod assemblies at a nuclear power plant. The case is believed to be the first of its kind in the nation.

Surry Circuit Judge Ligon Jones accepted the jury's recommendation and sentenced the defendants to two years in prison, the minimum allowed.

William E. Kuykendall, 26, and James Merrill, Jr., 24, former employees of the Virginia Electric & Power Co., could have been imprisoned for a maximum of 10 years. They are free on $5,000 bail pending appeal.

The two admitted pouring a caustic chemical, sodium hydroxide, over 62 nuclear fuel rod assemblies stored at VEPCO's Surry generating plant last April 27. They did it, they said, to dramatize lax security and unsafe working conditions at the plant.

They sought initially to offer "moral necessity" as a defense. By sabotaging the plant, they explained, they were protecting themselves and others from potential nuclear catastrophe. Last week Judge Jones refused to entertain that line of defense. The decision was proper.

Kuykendall and Merrill may have been correct in their assessment of VEPCO's shortcomings. The lengthy list of particulars they forwarded to the Nuclear Regulatory Commission may have been accurate in every detail, though the utility denies it.

They had every right to be concerned when neither the utility nor the NRC reacted as promptly as the conditions they described seemed to warrant. Indeed, over the years VEPCO has been fined $125,000, more than any other utility, for violation of federal nuclear regulations. The NRC did not respond to Kuykendall's and Merrill's charges until Oct. 10, the day their trial began.

But none of that gave them a basis, in logic or in morals, to conclude that sabotaging VEPCO was a necessary recourse, or even their only one.

On April 27 the plant was shut down for tests; it still is. Nobody was, or is, in imminent danger of anything.

If neither VEPCO nor the NRC would listen to their warnings, they could have taken their charges to the many environmental and anti-nuclear organizations which abound in the nation — and to the press. In today's atmosphere of skepticism regarding nuclear energy, there is no question Kuykendall and Merrill would have received a wide, and respectful, hearing.

They chose instead to attack VEPCO property. The company says the damage will amount to $1 million. Kuykendall and Merrill admit breaking the law; they say their consciences made them do it.

That being so, they should accept the responsibility that attaches to every uneasy conscience. The history of this country is replete with men and women who have gone to jail to challenge what they considered unjust laws or to preserve the integrity of their consciences. Refusing to be drafted comes to mind as an example.

Kuykendall and Merrill may have been right about VEPCO, but they were wrong to seek martyrdom on the cheap. We're glad Judge Jones and the Surry jury saw it that way, too.

OKLAHOMA CITY TIMES

Oklahoma City, Okla., October 18, 1979

A JURY in Virginia has refused to buy the weird argument of a couple of nuclear power plant employees that the way to make the facility "safe" was deliberately to make it unsafe.

They were convicted of single felony counts of damaging a public utility. The two men had admitted pouring a caustic chemical on fuel rods at the Virginia Electric & Power Co.'s Surry nuclear plant. Their act of sabotage caused an estimated $1 million in damage.

The men claimed they did it to dramatize what they regarded as lax security and unsafe working conditions at the plant. But that is specious reasoning. If they felt the plant was unsafe, how could they possibly improve the situation by doing those things that would make it even less safe, and, indeed, dangerous for themselves and their fellow workers?

The act was stupid and proved nothing except that virtually any equipment can be fouled up if the employees working with it have the wrong motivation. An employer has to assume a certain degree of responsibility and good will on the part of his employees. As the prosecuting attorney pointed out, the two men — control room operator trainees — had been given positions of trust, and they betrayed it.

Upon their conviction the pair may have achieved what they wanted: the status of heroes to the anti-nuclear-energy crowd. But to the general public they are simply common criminals and they will carry that stigma the rest of their lives, whether or not they serve the full minimum two-year sentence recommended by the jury.

There is a dedicated effort in this country to destroy the nuclear energy industry out of an unreasoning fear of alleged potential danger to the populace. That fear results from a lack of knowledge and understanding about what is essentially a foreign and hostile technology, in the words of Dr. Ralph Lapp, a leading nuclear energy consultant. But he points out, too, that nuclear power is the only form of energy that has an unparalleled safety record.

Denver, Colo., October 18, 1979

Colorado Department of Health officials and law enforcement officers were angry that it took Public Service Company representatives eight hours to notify anyone after a leak occurred in the coolant system of the Fort St. Vrain nuclear power plant near Platteville. They should be. Now it turns out that a department emergency line went unanswered when attempts were made to report the radiation leak.

Whatever the cause of the mixup, it was bad news.

Plant service manager H. W. Hillyard said, "It just wasn't that big a deal, although we can certainly understand the public's concern, what with the Three Mile Island incident and all."

Well, it is "that big a deal," and Public Service Company officials should realize that it is. This is not to dispute utility officials' contention that the amount of radiation released into the plant was so tiny that probably no harm will be done.

But if Public Service Company officials really understood the public's concern — and there is no doubt large segments of the public are edgy since the Three Mile Island incident — then they should also understand that complacency about the delay is the very thing that fuels that concern. If PSC wants the public to tolerate continued operation of the Fort St. Vrain plant, the lessons of the Three Mile Island incident should not be dismissed so lightly. Already, because of frequent shutdowns during the past two years, some people have questioned the plant's practicality, including Chris Zafiratos, a physicist at the University of Colorado's Boulder campus.

Lt. Gov. Nancy Dick, acting governor during Gov. Richard Lamm's trip to China, said she learned of the accident 13 hours after it happened — from a reporter who called her for a comment. She rightly observed, "One of the causes of over-reaction is a delay in reporting. It really heightens people's anxiety when it (an accident) is heard of several hours later."

The Hartford Courant

Hartford, Conn., December 20, 1979

The amount of radioactive gas that escaped from the Connecticut Yankee plant in Haddam last Sunday exceeded federal limits. State and local officials should have been notified of the accident at once. They were not.

Plant staff at the Northeast Utilities' facility did not relay the information until at least six hours after the fact.

The company concedes the error — a frustrating public relations setback during the industry's campaign to calm fears about the Three Mile Island accident.

The explanation for the delay is intended to be reassuring. Northeast says the Connecticut Yankee staff knew the radioactive release was minor, and chose to calculate exactly how much gas had escaped before reporting the incident.

The Sunday tardiness was not the first time Northeast has failed to report accidents promptly to state and local officials. Gov. Grasso grew quite testy at one point, when she realized that she and other state and local officials were not at the top of the list of persons who were notified. Northeast promised to do better. And, for the most part, it has.

But one such episode is one too many. Executives at Northeast have to make it crystal clear to employes at the nuclear plants that instant notification is vital. Delay creates serious misunderstandings and, quite possibly, panic. State and local authorities should always be informed within minutes, not hours, of an accident.

A company spokesman says investigations are going on to prevent similar radioactive discharges. Solutions being considered include venting the degassifier into a tank rather than the stack, improving the reliability of the water level controller and installing automatic shutdown devices which would stop the degassifier when pressure got too high.

That is well and good, but that is not enough. The President's Commission on the Accident at Three Mile Island concluded that equipment can and should be improved to add further safety to nuclear power plants. "But as the evidence accumulated, it became clear that the fundamental problems are people-related problems and not equipment problems," the commission noted. "In the testimony we received, one word occurred over and over again. That word is 'mindset,' " the commission continued. "After many years of operation of nuclear power plants, with no evidence that any member of the general public has been hurt, the belief that nuclear power plants are sufficiently safe grew into a conviction . . . The commission is convinced that this attitude must be changed to one that says nuclear power is by its very nature potentially dangerous . . ."

The commission said a comprehensive system is required in which equipment and human beings are treated with equal importance. It pointed out that there is a lack of communication among key individuals and groups.

Northeast insists that it already has the system suggested by the commission, and it is constantly improving this system.

Obviously, it needs to do better. The problem last Sunday was "people-related." The experts must not only study changes needed for venting in the degassifier and improving the water level controller. The experts must learn a skill more basic than nuclear engineering — reporting the accidents quickly, and all the time.

The Cincinnati Post

Cincinnati, Ohio, October 24, 1979

Two employees who sabotaged a Virginia nuclear power plant have been convicted by a jury. Any other outcome would have been an invitation to open season on nuclear power installations.

The pair actually got off lightly. The jury recommended only a two-year prison sentence, the minimum for the crime of which they were charged. They could have received up to 10 years.

The defendants had intended to plead "moral necessity" on the charge under state law of damaging a public utility, but the judge refused to allow it.

During the trial the employees admitted damaging the plant but said they did it to call attention to lax security and unsafe working conditions.

That is so much malarkey. They were in fact trusted employees who had security clearances. They used the plastic coded-card "key" of one of them to enter a restricted area and pour a caustic substance on 62 fuel rods, which caused an estimated $1 million in damage.

That isn't calling attention to lax security and unsafe conditions; it's a deliberate abuse of trust in order to sabotage a plant in using a power-producing process that has become controversial.

It's high time to get tough with such violators. Law breaking of any sort under the guise of "moral necessity" cannot be tolerated. Perhaps the verdict in the Virginia case will discourage others who have been encouraged by well-engineered and well-publicized protests to believe that nuclear power plants are fair game for saboteurs.

The Washington Star

Washington, D.C., October 21, 1979

The trial in Surry County, Va., was *not* about how safely Vepco ran or did not run its nuclear generator. The trial was about sabotage at a nuclear power plant.

Circuit Court Judge Ligon L. Jones correctly refused to permit defense attorneys to try the Virginia Electric & Power Company on allegations of unsafe operation. By not allowing the trial to veer in that direction, the judge maintained a focus appropriate to the case.

William E. Kuykendall and James A. Merrill Jr., both former control-room trainees at the nuclear plant on the James River, were convicted by a Surry County jury Tuesday on one felony count of damaging a public utility and were given the minimum sentence, two years in prison. The case is said to involve the first known act of sabotage at a nuclear facility in this country and, given the inflamed tempers over fission power, the guilty verdict is notable.

The two men, 26 and 24, attempted to plead "moral necessity" — contending that pouring a caustic chemical on fuel rods at the plant April 27 was an act of conscience committed because, they said, their complaints to Vepco about supposedly unsafe and unhealthy conditions had not been heeded. Judge Jones would not allow the plea.

That was proper — quite apart from Vepco's plant procedures. (The Nuclear Regulatory Commission will investigate the charges by the defendants.)

What the community faced in Surry was an instance in which two persons disdained — for whatever motives and with whatever possible result — more pedestrian avenues of remedy, elevating their personal sense of rectitude above all else. Mr. Kuykendall, a veteran of the Navy's nuclear reactor program, testified that he targeted the spare fuel rod assemblies because they would not affect the plant's safe operation and because they would not cause a health hazard.

But that is not the point. Suppose — and it is not too great a supposition — that his calculations had gone awry and that his act of conscience had seriously injured others or put thousands of persons in jeopardy from a nuclear accident?

The defendants claimed that Vepco repeatedly ignored their complaints about the plant. Was sabotage *the only* course available to them? Is it likely that the two men would have been ignored, in the post-Three Mile Island atmosphere, had they gone to state or federal officials with their complaint? Had they called a press conference on the subject of an unsafe nuclear power plant, it would have required a caravan of buses to carry reporters to Surry County.

"Moral necessity" is a profound claim, one to be invoked *in extremis.* That hardly appears to have been the situation in Surry. The young defendants were, in short, both righteous and reckless.

THE TENNESSEAN

Nashville, Tenn., October 1, 1981

WITH the nuclear-power industry already facing ruinous interest rates, other escalating costs and decreasing power demand, it has just been handed another worry: many of its reactors may already be wearing out after only a few years of service.

Safety officials of the Nuclear Regulatory Commission say the thick steel shells of 13 reactors in nine states — none in the TVA area — are being turned brittle by radiation so fast that some may become unsafe to operate by the end of next year.

Officials of all the utilities concerned disputed the NRC's assessmemt, some nonchalantly declaring they "are not worried at all." However, the NRC inspectors weren't speaking out just to hear themselves talk. Commission staff members are still trying to determine the seriousness of the problem. But what they have to say is cause for worry.

Dr. Thomas Murley, director of safety technology, said new information could show the problem to be more distant, but that on the basis of information now available, "I would start to say we'd get very nervous after another year or so."

The problem is that radiation affects the metal of the reactor pressure vessel, a 40-foot-high steel cylinder that surrounds the uranium core, where the nuclear chain reaction takes place.

The vessel, which cannot be replaced after a plant is built around it, is designed to last for the 40-year life of the plant. But recent tests have shown that the brittleness is developing far faster than expected. The brittleness reduces the ability of the steel to withstand temperature and pressure changes and could cause cracks and water leaks in the vessel. Some officials said that with a loss of coolant the reactor core could melt down and an accident result.

"I don't think anyone would feel confident that a vessel would hold water after a crack," Dr. Murley said. "You'd lose cooling, and eventually you'd have a meltdown."

Thus, it seems the industry technologists may have made another miscalculation — this time on the lifetime of the reactors — which could slow down the industry's development and run up the long-range costs of generating nuclear power.

The utilities who own these deteriorating plants may whistle past the graveyard, but the truth is the radiation-caused weakening of the reactor walls is likely to become a major issue in the nuclear industry's future. The problem should begin now to get the industry's serious attention rather than being played down as if it didn't exist.

THE LOUISVILLE TIMES

Louisville, Ky., July 16, 1980

If there was any remaining doubt that nuclear reactors are subject to Murphy's Law, it was surely dispelled last week by the news that 2,300 "incidents" occurred at America's 68 atomic power plants last year.

This information was gleaned from Nuclear Regulatory Commission files by Critical Mass, a group associated with Ralph Nader. "Incident" is an inoffensive word for mechanical failure or human error.

Industry spokesmen of course dismiss the number of foul-ups as irrelevant. Detailed reporting requirements do no doubt tend to make safety problems appear worse than they are.

For years, however, the public was assured that nuclear plants were for all practical purposes absolutely safe. The principle that has always been the bane of engineers — "if anything can go wrong, it will" — presumably didn't apply to reactors.

Almost no one believes that now. Even if many of the "incidents" counted by Critical Mass were minor, the public is painfully aware that small failures can develop into major disasters.

Since the Nader group opposes nuclear power, skeptics may question its conclusion that another serious mishap like the one at Three Mile Island is likely.

A new book on atomic energy by David Lilienthal will be far more difficult to dismiss.

Mr. Lilienthal, a former chairman of the Atomic Energy Commission, argues that the design of the "light water" reactors generally used in the United States is so complicated as to be inherently unsafe. He suggests that problems are inevitable when error-prone human beings operate extremely intricate machinery.

Mr. Lilienthal, however, is not against nuclear energy. He says we need more of it. But he also contends that the possible safety advantages of other reactor types, including the "heavy water" model used in Europe and Canada, have not been adequately studied. He calls for a radical new design that is safer and simpler.

Mr. Lilienthal and the Naderites make essentially the same point: Things often go wrong in nuclear plants and in the long run that means trouble. Indeed, unless the improvements Mr. Lilienthal hopes for are some day realized, nuclear energy will be of limited usefulness until the fusion reaction is harnessed some time in the distant future.

ST. LOUIS POST-DISPATCH

St. Louis, Mo., June 23, 1981

Not everyone is "cool," nor is the quality easy to define. Paul Newman, Willie Nelson, Miles Davis, the Fonz and O.J. Simpson are all cool, but in different ways. Cool is not only found among entertainers. Cops can be cool. But so can robbers, because cool is neither good nor bad. It is a quality that implies understated braggadocio and imperturbable self-confidence that is, at root, audacious.

Consider, for example, those who have "nuclear cool." Herman Kahn may not have started it all by "thinking the unthinkable," but he is one of a type given to such cool phrases as "An effective deterrent requires mutually assured destruction" or "No one was killed at Three Mile Island."

A recent example comes from a South Carolina nuclear facility that produces the raw materials for atomic weapons — an operation that has been plagued by over 200 incidents with "nuclear hazard potential" in the past two years. That such incidents are occuring at triple the rate of five years earlier, is "not considered alarming" by the facility's nuclear cool manager.

Then there is the government-owned fuel enrichment center at Oak Ridge, Tenn. An independent consultant says there is "no doubt" that accidental leaks of more than 11,270 pounds of radioactive uranium from the plant have posed health hazards, even though all 121 releases were within "allowable limits." Energy Department bureaucrats doubt that any "measurable" damage was done — which is not the same as saying no damage was done. And they insist that the plant has always operated with a concern for its employees — 53 of whom were exposed in an accident last month. As for the plant's loss of an average of over 300 pounds of radioactive uranium a year: It is not "unreasonable," they say. Only the uncool would think otherwise.

Rockford Register Star

Rockford, Ill., October 2, 1981

In a highly-technical dispute, the Nuclear Regulatory Commission (NRC) claims thick steel shells of some of the nation's atomic generating plants are being turned brittle by radiation. This makes them subject to cracks and the traumatic results of water leaking out of nuclear core cooling chambers, NRC officials say.

Although the brittleness was expected, the NRC says, the condition is developing much faster than anticipated and is threatening to seriously shorten the life-expectancy of the plants. If the condition continues, NRC officials warn, the plants will have to either shut down years earlier than planned or risk the catastrophic results of a "melt down." The condition has been discovered in 13 plants (none in Illinois, but two in Wisconsin).

Spokemen for the utilities involved dispute the NRC assessment and say, "We are not worried at all." They claim only a very unusual combination of circumstances could create any problems.

This dispute, centering on technical knowledge in several fields, obviously is beyond the ability of any layman to evaluate. But it is one more reminder that in dealing with nuclear power we are involved in exploring new areas where "knowledge" is often theory, where "facts" are often untested, where "experts" can be wrong.

It's also a reminder that, although we in Illinois right now have no option to accepting nuclear-produced energy, it still is not an industry which has won the right to unquestioning acceptance as the only choice for the future.

The Evening Bulletin

Philadelphia, Pa., October 5, 1981

If you thought your only worry about nuclear power accidents was from someone forgetting to turn a valve, think again. Constant radiation makes metal brittle, even the 9-inch-thick walls of a reactor vessel.

It's similar to metal fatigue in old planes. High pressure and a dramatic drop in temperature, which could occur in an accident, could produce cracks.

The nuclear power industry has been aware of the problem for years but it has resurfaced lately. A Nuclear Regulatory Commission official says some reactors may have to be fixed or shut down by the end of next year.

Does that mean we're in danger, especially in this area?

Not at present, it appears. But a lot of information about the reactors must be developed. The NRC should have firm criteria for judging when a reactor is in danger.

One important measure of a reactor's viability is its "nil ductility transition temperature" — reference temperature, for short. That's the point at which the reactor loses nearly all its ability to withstand the 2,200-pound reactor pressure.

The lower that temperature, the better. Over 300 degrees Farenheit spells real trouble. Around 200 or over brings special NRC scrutiny.

Oddly, there is no figure at which the commission blows a whistle. One official's explanation is that too many other factors in reactor safety preclude setting a shutdown figure. Until the commission develops a clear threshold, however, the public has a right to be nervous.

Three Mile Island Unit 1 (Unit 2 was the one crippled by the accident) has a reference temperature of 180. It is one of eight reactors in the nation about which NRC has asked operators for extensive information.

Salem No. 1 in New Jersey is a cool 90 degrees. Philadelphia Electric's Peach Bottom and Limerick plants, both boiling-water reactors, have no pressure problem.

Brittle reactors can be fixed: By annealing — which means superheating the reactors to eliminate the effects of radiation — or by removing the outer row of fuel assemblies. Neither is easy but it can save the reactor.

NRC has asked for more information. That's what the public needs, too. Not necessarily the kind of highly technical stuff the power companies will send to the commission. But an accurate lowdown on when a reactor near them is getting to the brittle-breaking point and what is to be done about it.

The Charlotte Observer

Charlotte, N.C., January 27, 1982

A striking juxtaposition of news about nuclear power on The Observer's front page Tuesday reminded Piedmont Carolinians of the relative benefits and risks of a technology that supplies 30% of their electricity. There was a report about an accident Monday at a nuclear power plant in New York; adjacent to that story was one about a celebration at the McGuire nuclear plant 17 miles from downtown Charlotte.

Business and political leaders from the Piedmont had gathered with Duke Power officials to mark the opening of Duke's newest nuclear generating plant. They listened as U.S. Energy Secretary Jim Edwards proclaimed nuclear power the "cheapest, safest, cleanest" method of generating electricity.

Indeed, the uranium that fuels McGuire costs much less than the coal that would be required to generate a similar amount of power; and nuclear power does not as a matter of course spew pollutants into the air.

But the accident several hundred miles north was a reminder of nuclear power's troubling side. At the Ginna nuclear plant in New York, experts worked to control a leak in the plant's complex steam generation system.

By late afternoon, puffs of radioactive steam had vented into the air outside the Ginna plant several times. Experts said the releases posed no danger to area residents' health, but the plant was shut down and will not generate power for many weeks, forcing ratepayers in that part of New York to pay for higher-cost replacement power from some other source.

Proponents are correct in saying nuclear plants have accumulated impressive safety records. On the other hand, there is the possibility of a catastrophic accident, and the experts still haven't put into place a permanent method of containing the wastes — some of them lethal for hundreds of years — that the plants produce.

Considering such risks — as well as nuclear power's benefits — will become increasingly important for Piedmont Carolinians. By 1985, Duke Power expects to generate more than 40% of our electricity at nuclear plants, compounding this region's exposure to both the risks and the benefits of nuclear technology. Right now, no good alternative is apparent, but Carolinians ought to be aware of the technology's impact for good and ill.

The News American

Baltimore, Md., January 27, 1982

One of the weaknesses of nuclear power plants — and perhaps the only baffling weakness — is found in the system of tubes in the steam generator component. The tubes are prone to ruptures and cracks and dents, and nobody understands exactly why; two reasons may be a lack of complete knowledge about water chemistry and hydraulic forces.

It was a steam tube that "broke dramatically" at the reactor near Rochsester, N.Y., releasing a small and apparently not dangerous amount of radioactive steam into the air and flooding part of the plant with thousands of gallons of radioactive water. Shades of Three Mile Island? Yes, but the engineers at the plant operated by the Rochester Gas & Electric Corporation, took emergency measures two minutes after instruments showed that a relief valve on an emergency tank failed to close; at the Pennsylvania plant, the identical valve stayed open for an hour before anybody noticed.

Indeed the accident at Three Mile Island, in 1979, may very well be the reason the "site emergency" at Rochester apparently will not have as dire results. The Nuclear Regulatory Commission has insisted on emergency drills at the nation's atomic plants, and recently there was such a drill at Rochester. No matter what nuclear power proponents say, every plant has the potential for disaster, the reason why constant vigilance and painstking safety procedures are necessary. Scary as it was, we feel encouraged by the quick response at Rochester. But with a nuclear plant right here in Maryland, we'd feel a lot more comfortable when the mystery of those tubes is solved — if it can be.

THE SUN

Baltimore, Md., January 27, 1982

A loss-of-coolant accident at a nuclear power plant in upstate New York Monday apparently never endangered the public health or safety. The accident was of an anticipated kind, and operators therefore did exactly what they were supposed to do in response. Emergency equipment brought the plant to a "cold shutdown"—a safe condition in which the dreaded meltdown of the radioactive reactor core is not possible—with only a minuscule release of radioactivity to the environment.

It would be reassuring to be able to say the handling of the New York accident shows that safety at nuclear power plants has improved so much since the frightening partial meltdown at Three Mile Island in 1979 that the public may rest easy now. That's only partly true. Although the NRC has tightened safety procedures in plant construction, maintenance and operation, too many factors that could lead to accidents remain partly uncontrolled. The New York accident is a case in point; although operators handled it well once it occurred, that still leaves one very large question unanswered: Why did it occur at all?

Physically, the New York events began with the rupture of pipes carrying water through a heat transfer system. Apparently, impurities in the water caused sludge buildups in the pipes—which had not been fabricated from the correct alloys for resisting sludge-caused ruptures. The NRC and the industry had known of this problem for years, but believed it was not especially threatening—not threatening enough to justify the high costs of frequent plant shutdowns to replace pipes showing signs of leaks.

The leaks in the New York plant were major ones and occurred, as one official put it, with "dramatic" suddenness. But NRC officials insist that they had anticipated even such dramatic events. That the New York operators had been drilled in handling just such an emergency seems to substantiate this claim.

The fact the New York operators behaved with far greater professionalism than their counterparts at TMI indeed is a commendable sign of heightened awareness of the risks of loss-of-coolant accidents. Yet the NRC's revelation that it is not always economical to have nuclear power plants in tiptop operating condition—and that plant safety sometimes depends, as it did Monday, on hair-trigger reactions by plant employees—surprises and worries us. The question inevitably arises: How reliable is a technology if it is economically prohibitive to keep its components in safe operating condition?

Roanoke Times & World-News

Roanoke, Va., January 27, 1982

AS NUCLEAR accidents go, the one Monday at an Ontario, N.Y., electrical generating plant was not a biggie. Radioactive steam escaped into the atmosphere when a cooling-system tube broke, but the plant was shut down quickly and the steam probably posed negligible threat to public health.

The incident, however, could put another nail in the coffin of nuclear energy in this country.

That's not so much because of the scare factor. This was not another Three Mile Island. Even there, the lasting legacy of Three Mile Island could well be its crushing expense: hundreds of millions of dollars to clean up and rehabilitate a multibillion-dollar plant that has been idle since the March 1979 accident.

Since this was not a serious incident, the Ginna plant of Rochester Gas & Electric Corp. will not be down very long. Still, it could be several weeks before it is back in service. Cleaning up 11,000 gallons of radioactive water and doing other maintenance will be expensive.

While the plant's shut down, the utility's 285,000 customers, residential and commercial, will have to pay the New York state power pool $250,000 to $400,000 every day for supplemental electricity. For the average residential customer, the added expense will be $5 to $8 a month. On top of $38 for the normal 500 kilowatts a month, that's a big jump.

This illustrates what many have come to call the chief hazard of nuclear power: finances. These plants carry astronomical price tags. The 12-year-old Ginna facility is relatively small, with an output of 470 megawatts. But it cost $88 million to build, and since it was completed RG&E has spent another $150 million on it.

Ginna has looked like a good investment: It has had few maintenance problems and its operating efficiency has been 74.9 percent, excellent for a nuke. When such a plant is on line, the figures look very good. When it's not, the figures can look extremely bad. RG&E depends on Ginna for nearly half its generating capacity, and for the next few weeks the plant will be generating nothing but expense.

The Ginna accident comes at a time when another nuclear plant in that part of New York state is mired in controversy — not over environmental factors but over money. The Nine Mile Point No. 2 plant has been under construction since the mid-1970s. Prospective cost has ballooned from $380 million at the start to $4.9 billion now, and some estimates run as high as $6 billion.

About $1.3 billion has been put into the plant, and some critics want the state's public service commission to order construction stopped rather than put more funds into it. If that happens, a fresh controversy will erupt over who should pay for writing it off, customers or stockholders.

Nuclear plants have been steadily improving their operating efficiency. But they depend on a technology that is not only terribly complex and expensive, but also unforgiving. Even minor accidents and upkeep problems can bring things to a crashing halt. Long shutdowns often follow while technicians do trouble-shooting and utility officials explain things to the public and to the Nuclear Regulatory Commission.

Increasingly, utilities are deciding this is too risky and expensive a game to play. The nuclear industry is running on momentum alone; it's had no new reactor orders for years now. The fact is that the peaceful atom's potential was grossly oversold. Most of its salesmen have passed quietly from the scene, leaving their customers holding the bills. Gigantic ones.

The State

Columbia, S.C., January 30, 1982

THE INCIDENT at the Rochester Gas & Electric Co.'s Ginna reactor Monday at Ontario, N.Y., argues well for the safety and safeguards of America's nuclear power plants.

There was no damage beyond the reactor site and minimal radiation was released in the steam after a pipe burst. The Associated Press reported the amount of radiation at the boundary of the plant was 3 millirems. That compares with an average 20 millirems from a chest X-ray. A lethal dose of radiation is considered to be 600,000 millirems.

Warning systems and emergency procedures by the power reactor staff worked smoothly and effectively. And, in a day, the site emergency was formally called off.

Nuclear plant accidents are nothing to be celebrated, but they have proved to be instructive. The nuclear power industry can be proud of its safety record, which is greater than that of any other industry.

Nothing that happened at the Ginna reactor should undermine the public's confidence in nuclear power.

The Providence Journal

Providence, R.I., February 1, 1982

Compared with the bumbling performance, confusion and recriminations after the Three Mile Island nuclear power plant accident, the relatively smooth shutdown of a potentially dangerous plant in upstate New York was a credit to the industry.

All is far from rosy, however. The trouble at the Ginna plant at Ontario, N.Y., on Jan. 25 started with a bursting tube or tubes in the steam generator. The thousands of small tubes in these particular units, used in 49 U.S. plants, have become a major nuisance. Corrosion has caused pinhole leaks, small cracks and four outright ruptures.

In fact, just such tiny leaks have been detected in the steam unit of the undamaged No. 1 reactor at the Three Mile Island plant in Pennsylvania (the big 1979 mess was in No. 2), delaying its resumption of power production for another six months.

The bad accident three years ago at Three Mile Island was initiated by a pump failure and compounded by malfunctioning valves and an appalling series of missteps on the part of the plant operators. The nation chewed its fingernails while the utility company achieved an agonizingly slow control of the situation.

Fortunately, the sequence of events at the Ginna plant proceeded rapidly to a "cold shutdown" of the reactor 31 hours after the tube rupture. The clothing of a dozen workers became contaminated, some radioactive steam was vented to the open air and more than 1,000 gallons of "mildly radioactive" water were released into a sump pit under the reactor. The utility people said the public was at no time in real danger.

The performance by employees of Rochester Gas & Electric Co. and by the equipment at this 12-year-old plant was, in the main, praiseworthy. A few dubious moves were made and at least one valve misbehaved. The worst news is that nobody can guarantee that the same sequence will not be triggered again by a defective tube — at the Ginna plant or any place using this particular kind of steam generator.

The Ginna accident is now conceded to have been serious in its implications, although relatively harmless in its real consequences. Its lesson is simple: There must be extreme care in operation, play-it-safe replacement of steam generator tubes without trying cost-cutting remedies and a restudy of the emergency evacuation strategy in the plant's neighborhood.

The Knickerbocker News

Albany, N.Y., March 22, 1982

Three Mile Island alerted the nation to the need for stringent safety requirements at nuclear power plants — and evacuation of workers last January at the R. E. Ginna plant near Rochester underscored that need. The Nuclear Regulatory Commission, in fact, has linked the two incidents: The formation of a potentially dangerous steam bubble in the Ginna reactor head was "the worst nuclear accident since Three Mile Island."

Yet despite these warnings, the state Legislature has not approved a package of bills designed to protect workers and customers from danger and ratepayers' wallets from unnecessary costs. The Assembly last week approved the bills, but the Senate has yet to do so. It's a scenario similar to 1980 and 1981, when assemblymen gave wide bipartisan support to the measures, only to watch them die of neglect in the upper house.

The aims of these bills, which were drawn during the last three years by the Special Committee on Nuclear Power Safety chaired by Assembly Speaker Stanley Fink, are eminently worthy of support. They would:

- **Require that a resident inspector — paid by the state, which would be reimbursed by the owner — be present at all nuclear power facilities. Trained back-up personnel would also be required to assure constant supervision.**

- **Establish guidelines for the transportation of radioactive materials in the state and for training public safety personnel in responding to accidents involving these materials.**

- **Require utilities to set aside a portion of their income in a sinking fund to pay for decontamination and decommissioning of nuclear plants.**

Mr. Fink says the first two proposals are "timely and vital," while the third would address accounting procedures "currently employed by the Public Service Commission (that) deceive ratepayers into thinking that funds for decommissioning nuclear reactors are available."

The speaker has common sense and history on his side. He notes that last year "in conjunction with the strong support of Gov. Carey, legislation requiring evacuation planning in the event of nuclear plant accidents was enacted. The accident in January at Ginna shows the necessity for and value of this evacuation planning law, and that incident and the one at Indian Point in 1980 reinforce the need for additional laws to provide nuclear safety."

In most cases, a nuclear accident is minor in origin but can become gravely serious if neglected. The same can be said about neglect at the Capitol.

FORT WORTH STAR-TELEGRAM

Fort Worth, Texas, January 30, 1982

There is something to feed the arguments of both proponents and opponents of nuclear power after the recent accident at the Robert E. Ginna (ji-NAY) plant in upstate New York last week. But of greater concern to all should be the implication from this latest plant accident that nuclear power may not be the great supplemental energy source that utilities have hoped it would be.

The accident at Ginna near Rochester on the shore of Lake Ontario was handled effectively and with a minimum release of radioactivity — less than one would receive in a cross-continent airplane trip. The accident, a ruptured tube in the cooling system, was similar to the one that occurred at Three Mile Island nearly three years ago but the procedure to correct it went smoothly at Ginna and the damage caused was far less.

There were some minor problems in communication in the early stages of the accident but for the most part the emergency procedure — largely because of what was learned from the Three Mile Island accident — went smoothly.

The effective handling of the emergency is comforting. Not all nuclear plant accidents have to be as potentially dangerous and as costly at the one at Three Mile Island. But it is also foreboding. There are 39 nuclear plants in this country with similar steam generators. The Ginna plant will be closed for several weeks, perhaps months, while repairs are being made. Repairs in nuclear plants are costly and that cost will be passed on to utility customers.

Nuclear power plants are costly to construct because of the extreme care that must be taken. They are costly to repair for the same reason. Add to that cost the time a plant must be out of service for repairs and there appears to be an extremely unreliable and costly supplemental power source in nuclear plants.

It behooves us then to not put to much reliance in nuclear power as a large portion of this nation's energy supply unless major improvements can be made in the facilities. And it behooves us to seek out other alternative power sources to supplement the basic fossil fuel generation of power.

We've learned a lot from the Three Mile Island lesson and we should be learning something from the Ginna experience. Nuclear power isn't a Damoclean sword but neither is it the answer to our energy problems.

The Seattle Times

Seattle, Wash., November 15, 1982

PEOPLE love to be scared — witness the enduring popularity of horror movies, scary novels, ghost stories, and other "what if" tales.

This month, most of the nation's newspapers and other media gave prominent play to one of the scariest "what if" scenarios of modern times: The prospect of a major accident at a nuclear-power plant, complete with core meltdown, massive failure of all safety systems, breach of the reactor containment vessel, release of a huge radioactive cloud, the deaths of thousands of residents in surrounding communities, cancer cases for decades to come, etc.

Making it all the scarier, the stories were based on documents leaked to The Washington Post that were part of a study done by Sandia National Laboratories for the Nuclear Regulatory Commission, giving them an air of respectable credence.

Using a sophisticated computer model and the most comprehensive meteorological, population and economic data available, the study calculated the consequences of a "worst case" accident at 80 nuclear plants now operating in the United States. And no doubt about it, the results were scary: More than 100,000 deaths and $300 billion in damages at certain reactors near heavily populated urban areas. What's more, The Post's story said there was a 2 percent chance of such an accident before the year 2000.

However, the stories omitted at least one pertinent fact: The scientists concluded that the odds of such an accident were so remote that the scenarios were discounted and weren't even included in the final report Sandia sent to the NRC. And the NRC later challenged The Post's 2 percent figure as an incorrect interpretation of the data.

The head of the NRC's risk-analysis division said the chances of such a catastrophic accident were about one in a billion — or comparable to the likelihood that a fully loaded 747 would crash into the stadium on Super Bowl Sunday.

As the Three Mile Island accident proved, there is reason to be concerned about nuclear-power-plant safety. But safety standards have been tightened considerably since TMI, and nuclear power still has a far better safety record than any other major source of energy.

The main worries of the nuclear industry today have less to do with safety than with financing — and that's something worth getting scared about.

WORCESTER TELEGRAM

Worcester, Mass., March 19, 1983

Although no one planned it that way, the movie "The China Syndrome" was released just days before the accident at the Three Mile Island nuclear power plant in Pennsylvania. As a result, the movie drew even better at the box office than it might have otherwise. The film dramatically summed up many of the charges against the nuclear power industry — reliance on automatic controls that may not be foolproof, lack of adequate training by plant operators and faulty construction covered up by unscrupulous inspectors and businessmen.

Events last month at a nuclear power plant in Salem, N.J. read like the script from "The China Syndrome." Relays stuck open, cooling water in the reactor core dipped to dangerously low levels. Control rods that should have automatically cooled down the reactor did not function. Attempts to repair the situation were delayed when a control handle snapped off and had to be fixed. Most alarming was an apparent human error of potentially disastrous proportions. The crew either did not understand what had happened or was reluctant to pass on the information so that the defect could be corrected. Officials of the Nuclear Regulatory Commission say the situation was the worst since Three Mile Island.

On Feb. 25, it took a 32-year-old control-room operator 23.8 seconds to realize there had been a failure in the automatic controls and to move toward shutting down the plant.

At the time, the plant was operating at 14 percent of capacity. The Nuclear Regulatory Commission says the operator's action might have been futile had the reactor been operating at full power. The reactor core might have been exposed, releasing enormous radiation.

The commission has cited the Salem power plant operators before for "below average performance." Salem has been fined for security breaches and for insufficient protection of workers from radiation. The pair of 1,100 megawatt power plants is 20 miles south of Wilmington, Delaware, in a populous region of the Delaware Bay.

While the events of Feb. 22 and 25 posed no immediate threat to public safety, they "raise serious safety questions" and "could lead to a serious accident" if not resolved. The Salem facility has also been plagued by a strike. The F.B.I. has checked five separate incidents at the plant for evidence of sabotage without a finding.

For the time being, the Salem reactor remains closed. The regulators say it will remain closed until they are satisfied that the operators have made the required corrections and have adopted proper maintenance procedures.

Operating a nuclear power plant is a complicated and potentially dangerous business. There is no room in the nuclear power industry for mistakes of the sort that have been adding up to trouble for Salem I and its owner, the Public Service Electric and Gas Co. They have to be examined, understood and corrected before that plant goes on line again.

Sunday News Journal

Wilmington, Del., March 6, 1983

THERE WAS another potential disaster at the nuclear generating plant across the river at Salem, N.J., 11 days ago: "double failure of the . . . station to shut down automatically."

If one of the circuit breakers involved had failed to break the circuit involved, it would have been front page news. In fact, both circuit breakers involved failed to break the circuit and it was front page news for two days.

As far as we can tell no one died prematurely in the potential disaster; no one was injured.

If Friday, the 25th of February, the day the Salem plant's "scram" system failed, was a typical day in the coal industry, between 100 and 200 people suffered early deaths in this country related to mining and burning coal. (Related is another one of those words like *potential*, but the deaths were real.) Another 100 to 200 Americans stepped into automobiles that day, if it was a typical day, and didn't come out alive. Others started their car engines and, emission controls notwithstanding, began pouring pollution into the atmosphere that will contribute to other early deaths.

Some people persisted in smoking, to what we are told was the detriment to their own health and the health of others.

The temporary circumstance of abundantly available crude oil contributes to gas guzzling that adds both to carcinogenic material in the air and highway crashes that turn people into corpses.

Salem's most recent accident was the first in the history of nuclear power generation in which all of the *automatic* redundancies built into the system installed to ensure *automatic* shutting down of the plant's reactor failed.

When the automatic shutdown devices failed, an alarm automatically went off and a mere person had to shut down the plant manually. The highest priority, following this accident, should be for the federal regulators of nuclear energy to pursue an unflagging training program to make those mere humans, given the possible failure of redundant safety devices piled upon redundant safety devices, more and more capable of reacting to such alarms and responding properly.

It was indisputably a serious matter and it was reassuring to hear from the Nuclear Regulatory Commission that Salem's No. 1. generating unit was not to be restarted pending assurance of progress in avoiding repetition of such an incident.

It was reassuring that reports already were being written for the NRC on how the facility plans to avoid similar incidents.

It is reassuring also that publicity's full glare is turned on to every such incident.

A real added potential danger is that we might become so inured, so habituated to such reports that diligence in supervision of our power generating stations might eventually be relaxed.

But we live in a world of both risk and potential.

The Salem plant and others like it, if they can clean up their acts, have the potential for eventually producing energy too cheap to meter. We will need such energy, if for nothing else to forestall potentially catastrophic war over ever dwindling supplies of fossil fuels.

Should we turn our backs on that potential, and stand by in masterly watchfulness for the dreamed of breakthrough that will give us nuclear fusion, and unlimited power from sea water?

As to potential risk, there is no use being lighthearted about it, but it is everywhere.

If we eat a black jellybean, elect a legislator, split an infinitive or engage in some industrial activity that generates chemical wastes along with jobs, we are asking for trouble.

The watchwords must be both Eternal Vigilance and Never Despair.

The Philadelphia Inquirer

Philadelphia, Pa., March 21, 1983

The frightening 1979 film *The China Syndrome* was on television last week. And what is doubly frightening is how little things have changed since its release four years ago — in Philadelphia the same week that the nuclear plant at Three Mile Island came close to causing a catastrophe of untold proportion.

The movie is about how greed and duplicity and shortcuts nearly resulted in the meltdown of a fictional atomic plant in California. If that scenario — far more simplistic than the TMI reality — might have seemed unreal before the near-tragedy of errors of March 1979, the real-life episode in central Pennsylvania yanked the rug from under decades of reassuring safety promises by the nuclear industry. In a matter of days, public confidence in nuclear power's fail-safe reputation was completely rattled.

The industry — and, to some extent, the Nuclear Regulatory Commission — also got the scare of its young life. At TMI, a plant had been losing reactor cooling water for more than two hours and no one had seemed to realize it or, when they did, know what to do about it.

There were promises all around of "never again." The public could rest assured that management — in this case General Public Utilities (GPU) — would respond, that operators would be upgraded and that everything would be checked and rechecked to guard against any recurrence.

But just as it seemed on the cautious mend, public trust has again been shaken in the last few months, first by an eye-opening lawsuit that GPU filed against its supplier, Babcock & Wilcox, and second by a startling safety failure at the Salem nuclear reactor No. 1 in Lower Alloways Creek, N.J., 30 miles from Philadelphia.

During the unseemly courtroom bickering between GPU and its supplier, a picture every bit as horrifying as *The China Syndrome* began to emerge: There was evidence that TMI operators faked earlier reports on coolant leaks to the NRC, that the quality of operator personnel (according to GPU itself) was going down because of poor training and that maintenance funds had been slashed drastically.

U.S. District Judge Richard Owen, who presided until a hasty out-of-court settlement ended the case in late January, opined from the bench at one point that, instead of everything being "hunky-dory," as GPU executives portrayed it, "the operator training [at TMI] was pretty atrocious."

Enter the events of Feb. 25 at Salem 1. In what NRC officials called "the most significant accident precursor since Three Mile Island," two relays that were supposed to automatically shut down the overheating reactor core both stuck, forcing a last-minute manual shutdown of the plant. The event, said Harold Denton, director of the Office of Nuclear Reactor Regulation, "raises serious safety questions regarding the safe operation of the Salem facility."

Like a lost echo from 1979, the NRC staff reported to the full commission the problems it found with Public Service Electric and Gas Co.'s operation of the Salem plant: There was an absence of proper maintenance on the relays that failed, control-room operators seemed unaware of certain warning signals and how to cope with them, and management didn't have its act anywhere close to together.

Those were the lessons of Three Mile Island. Yet despite promises by the nuclear industry and the NRC that such shortcomings would be resolved, they have flared up again at Salem. If straightening up its safety record was to be the basis for renewed public trust in nuclear power, that trust unfortunately is yet to be earned.

As the March 28 anniversary of the accident at TMI approaches — and as GPU assures one and all that it can safely restart its undamaged Unit 1 at TMI — it is unsettling to reflect on how vulnerable the industry and the public still remain to a replay, or worse, of the events of four years ago.

Fears Mount over Operator Error, Sabotage

As with many public apprehensions about the safety at nuclear power plants, the concern over the training and reliability of plant operators was increased by the 1979 accident at Three Mile Island. In the report prepared by the Kemeny Commission, it was suggested that government-accredited schools for operator training be established, and that licensees be periodically reviewed to be sure they were complying with government standards. (See pp. 48-55.) The new regulations instituted by the Nuclear Regulatory Commission in the aftermath of the accident included many that were concerned with more stringent qualifications for operators and their supervisors. There is, of course, always a risk of error when a human being is operating any kind of equipment, and it was acknowledged that plant operators could not reasonably be expected to react correctly in every potential circumstance. After the Three Mile Island accident, however, it was widely recognized that even a small margin of error when combined with certain types of plant malfunctions could create a disproportionately large danger, as it did in Pennsylvania.

A new type of anxiety over plant operators began to grow after the first instances of sabotage by nuclear reactor employees. (See pp. 20-29.) In May 1983 the NRC sent warnings to plant managers about the increasing incidence of these acts by plant insiders. A study conducted by the commission found "32 possibly deliberate damaging acts at 24 operating reactors and reactor construction sites" from 1974 to 1982, including 11 since 1980, "directed against plant equipment in vital areas at operating stations."

The Washington Post

Washington, D.C., April 19, 1979

THERE IS AN ANALOGY worth pursuing between running a nuclear reactor and flying a commercial airliner. An airline pilot, like a reactor operator, holds the safety of a great many people in his hands. The job is generally pretty routine, but there is always the possibility of a sudden desperate emergency. Pilots are trained—and paid—for that moment. But there the analogy breaks down: Power station operators are not.

Among all the safety improvements that will follow the accident at Three Mile Island, reexamination of the operators' training will be crucial. In the sequence of things that went wrong, someone mistakenly turned off the primary cooling pumps. The Advisory Committee on Reactor Safeguards recommends substantial improvements in instrumentation to get more and better information to the control room. Doubtless that can help to prevent reactor accidents in the future. But the value of the information will always depend on the speed and skill with which technicians react to it.

Which brings us back to the airlines analogy. Senior pilots flying the big jets for the airlines can earn upwards of $80,000 a year. Reactor operators get perhaps one third as much. Airlines require their pilots to be college graduates. Reactor operators are typically high school graduates who learned the job in the Navy or, more likely, by running conventional power plants fueled by coal or oil. Both pilots and reactor operators must hold federal licenses. But the pilot's license comes up for renewal every six months. Every airline carries on a program of continuous training, under the eye of an inspector from the Federal Aviation Administration. There is constant drill in responses to the wildest imaginable range of malfunctions.

Like the airlines, utilities and reactor manufacturers have simulators on which their personnel are trained. But it appears that the simulators stay within a narrow range of predictable troubles. The various inquiries into the Three Mile Island accident will presumably see whether the men on duty there had ever faced, on a simulator, anything like the original pump failure. Certainly they had never confronted the situation that rapidly developed after the wrong switches were subsequently pulled.

The point is that in all respects—education, training and pay—standards for airline pilots are far higher than for reactor technicians. The jobs are similar only in that a serious mistake, in either of them, can be tragic. There is a clear need for the Nuclear Regulatory Commission to begin designing a licensing and training system on the same scale as the FAA's.

The Boston Globe

Boston, Mass., April 2, 1979

A preliminary review of the accident at Three Mile Island found that human error figured in at least six different steps that ultimately cascaded into a breakdown of the system. In light of this, the investigatory commission called for by President Carter should consider not only the specific technical causes of the accident but the broad personnel issues as well.

The problem may be compounded by the fact that the 72 power reactors operating in the United States were constructed by four different companies. Each has its own idiosyncracies. A trend toward uniformity is far from realization and this places additional burdens on operating personnel, who must be highly knowledgeable about the ways reactors function. They cannot depend on standardized components to alert them to malfunctions.

Three Mile Island was a warning that even the newest reactors are vulnerable to human lapses. It is not yet clear whether those lapses could have been prevented by better training. Nor is it clear whether the three-hour failure to notify the Nuclear Regulatory Commission (NRC) contributed to the deepening crisis, delaying intervention by outside experts better equipped to deal with the problem. The President's commission should address these issues in detail.

The commission should also look into the practicality of establishing a special trouble-shooting team to be on call for such emergencies. There is already a precedent for such teams in the petroleum industry. A strike team located in the center of the country and equipped with jet transportation could reach any reactor in less than three hours.

By good luck and hard work by those at the Three Mile Island reactor, no one was injured. The accident might not be totally in vain if it leads to more rigorous standards for the training and performance of those who work on nuclear reactors.

THE PLAIN DEALER

Cleveland, Ohio, April 20, 1979

Nuclear Regulatory Commission officials rate operation of the Davis-Besse nuclear plant so low as to be barely acceptable. If it were any worse, ". . . we would have justification to shut it down," said Norman C. Mosely, NRC director of reactor operations inspections.

For residents of northern Ohio that is a chilling appraisal of the nuclear plant co-owned by Toledo Edison Co. and Cleveland Electric Illuminating Co. near Port Clinton, about 70 miles west of here.

Responding to those and other comments by NRC officials, John P. Williamson, chairman of Toledo Edison, also expressed dissatisfaction with plant operations. He added, "Our first priority is safety."

Yet average citizens must wonder at his statement when they hear federal inspectors criticize plant operations as being at the lowest possible level to justify keeping the facility open.

The regional NRC director, James G. Keppler, told The Plain Dealer that Davis-Besse has "had more human errors than any plant in my region," which includes more than a dozen operating reactors.

He previously had reported there was such a large number of personnel errors that one NRC inspector said the plant should be shut down because the same mistakes kept being repeated.

In addition, in January there were two "abnormal events," as they are called in the nuclear industry, that Moseley said "you could liken somewhat, I guess, to the situation at Three Mile Island."

The "situation" at Three Mile Island, near Harrisburg, Pa., occurred March 28 and continues today, more than three weeks later. The Davis-Besse plant is of similar design.

Keppler said that even with its deficiencies Davis-Besse should not be closed. However, he also said he believed the plant, which has been out of commercial operation since March 30, should not be reactivated "until these problems are satisfactorily resolved."

Williamson promised a thorough safety review before restarting the plant. But indications are that it might be back on line in the next several days. The out-of-service period seems hardly to be adequate time to correct the shortcomings identified by the NRC.

Williamson said, "I can replace dollars, but I can't replace lives." We trust that was more than public relations talk. If he sincerely believes everything that statement implies, Davis-Besse will not be restarted until its personnel and equipment can be counted upon to be as fail-safe as possible.

THE ATLANTA CONSTITUTION
Atlanta, Ga., August 6, 1979

Memories of the Three Mile Island nuclear "accident" have faded somewhat, but that episode did a lot to alert people to both the dangers and the uses of nuclear power.

Last week the Nuclear Regulatory Commission reported that an investigation shows the accident might have been prevented if plant operators had allowed safety equipment to function as it was supposed to do.

While nuclear power remains a controversial subject, most people, especially as the energy crunch worsens, would probably agree that it is going to be part of our future. And everybody would agree that if it is, safety is an absolute must.

The NRC report implies that, because humans are human and because humans err, safety may be more of an ideal than a practical possibility. We would have thought that anyone operating a nuclear plant would be super stringent, but apparently that wasn't so. All our technological knowledge and skill is, in the end, at the mercy of human error.

Like fire for primitive man, nuclear power is both a great danger and a great boon. And it is up to us which it will be.

The Salt Lake Tribune
Salt Lake City, Utah, April 22, 1979

The evidence is still being gathered and the "full accounting" President Carter promised the American public on April 5 about what went wrong at Three Mile Island is still some distance in the future. Still, indications in the press suggest a considerable amount of human error was involved.

Operators of the nuclear power plant apparently shut down auxiliary cooling pumps for maintenance two weeks prior to the March 28 accident. This was one of several mechanical and human failures that might have contributed to the atmospheric release of worrisome levels of radioactive vapor near Harrisburg, Pa.

In the wake of the Three Mile Island accident much has been printed and broadcast about the adequacy of the federal government's nuclear power plant licensing procedures. Critics lament that they aren't stringent enough and industry representatives are working feverishly to reduce the 12 years it presently takes to get a nuclear power plant "on line."

Noticeably lacking in all this is any reference to the licensing of the people, the individuals, operating them. The discussion always centers around the plants and whether this collection of pipes, pumps, concrete, nuclear fuels, control rods, instruments and so on are adequate. Virtually nothing is ever mentioned about the qualifications of the people who are pushing the buttons and read the dials in these plants.

It is thus legitimate to ask whether these people have been certified as competent to run these plants. Are they, individually, sufficiently knowledgeable about the plants to cope with the emergencies like that at Three Mile Island? Will they respond positively to alarm signals rather than ignore them or "override them" as was reportedly the case at Three Mile Island?

The Federal Aviation Administration licenses both planes and pilots. Thus, the flying public is doubly assured that the Boeing 727 it is riding in is certifiably airworthy and that the pilots on the flight deck are clearly competent to fly the plane.

In addition to the licensing or certification of individual nuclear power reactor operators, perhaps it would be in the public's best interest to begin a national training program, using reactor simulators like those the National Aeronautics and Space Administration uses for training astronauts.

For eight hours a day at the Johnson Space Center astronauts training for Space Shuttle missions sit in the cockpit of a simulator, going through flight procedures. Teams of instructors, using computers, can feed up to 4,500 malfunctions into a training run. The astronauts respond to each malfunction, carrying out procedures that will keep their "pretend" space orbiter alive and well.

If NASA finds it necessary to go to these lengths to assure the success of a space mission and protect the lives of a relatively few crew members, it would seem several times more urgent that the Nuclear Regulatory Commission launch its own reactor malfunction training program, via simulators. After all, there are the lives of hundreds of thousands people involved in the safe operation of a nuclear power plant.

The Washington Star
Washington, D.C., July 15, 1979

A subsidiary question to emerge from the accident at Three Mile Island concerns the competence of those who operate nuclear power plants. Who are these people, and may they be counted upon to respond properly in case of an emergency? It is much the sort of question one asks upon boarding an airplane.

A study just completed for the Tennessee Valley Authority attempts to deal with these questions. Indeed, this "task force on nuclear safety" recommends significant revisions in the selection and training of licensed plant operators.

What is novel in the TVA study is the emphasis on intelligence-testing for nuclear plant operators — including those currently employed. The report urges not only that, but also giving operators "the equivalent of a college education . . . and paying salaries that will attract the best people into the program." Salaries for senior licensed operators now average between $19,000 and $40,000 nationwide, which makes one wonder if there are parallels between salary and performance.

The emphasis on intelligence comes with this useful working definition: "Intelligence distinguishes those who have merely memorized a series of discrete manual operations from those who can think through a problem and conceptualize solutions based on a fundamental understanding of possible contingencies."

Now, testing intelligence is a strange and controversial business, and so-called IQ tests can't accurately gauge all the complexities and varieties of human intelligence. But such tests can be a reasonable guide as to how someone will react intellectually to certain environments; and few environments have quite the combination of logic and complexity of nuclear power plants.

The report simply proposes that TVA "pursue a long-range goal of having the operator training program accredited as a program culminating in a recognized academic degree or certification"; and it recommends a training and retesting program that goes beyond psychological profiles, simulator training, and written and oral examinations of mathematical and mechanical aptitude. All this, of course, the better to predict an operator's on-the-job responses.

TVA has a special interest here. It not only operates three nuclear plants at Brown's Ferry; it has 14 more reactors either planned or under construction. But the standard proposed seems to us sensible for any utility or, for that matter, any agency charged with regulating nuclear utilities.

The Nuclear Regulatory Commission, which licenses operators and senior operators, is aware of the TVA study, but its own recommendations for change remain very much in the planning stage. The nuclear power industry, meanwhile, is planning a "national institute" to "establish industry-wide benchmarks for excellence in the management and operation of nuclear power programs," which seems very promising but also a bit vague.

If nuclear power is going to be with us for a while, and if human reaction remains a significant factor in the safe operation of nuclear plants, it is only sensible to ask that those who operate plants be screened as carefully as possible, trained in the most rigorous fashion and licensed only when these procedures set the highest standard.

The recommendations to TVA suggest ways in which the NRC, too, could broaden its examinations for licensing operators — and strengthen its criteria for approving industry training programs.

The nuclear power industry and the NRC ought to take a close look at the TVA report. If nuclear power is to continue to win public support, the industry must be able to guarantee that those in the control room meet the highest standards of intelligence and training.

THE SACRAMENTO BEE

Sacramento, Calif., June 6, 1979

The presidential commission investigating the near-disaster at the Three Mile Island nuclear reactor in Pennsylvania is just getting under way, but preliminary reviews already have located a critical area of neglect: the training programs and licensing requirements for reactor operators.

The Nuclear Regulatory Commission says it had harbored doubts about the adequacy of the training programs conducted by the firms that build the reactors even before the March 28 accident. Now Harold Denton, the NRC's chief regulator, says training deficiencies represent the biggest lesson from Three Mile Island. "We did not spend enough time on the human side, as opposed to the hardware side, to prevent accidents like this one," Denton says. "TMI presented the operators with a combination of events they weren't trained to cope with."

The NRC is now in the process of re-evaluating operator training programs and licensing requirements, but the need to broaden the training and tighten the licensing requirements is already apparent.

Government and industry rules, for instance, require licensed airline pilots to be better educated and better trained than reactor operators. Yet, as Three Mile Island demonstrated, many thousands of lives could depend on the judgment and decisions of those who man the control rooms of the country's 72 commercial nuclear power plants. At present they are trained and licensed in about half the time it takes a pilot to qualify as a licensed airline pilot.

The most damning criticism of reactor training came from one of Three Mile Island's operators. He said the program concentrated heavily on "teaching the test" rather than on careful and thorough instruction in the processes involved. Trainees were drilled on likely test questions and required to memorize answers whether or not they fully understood the significance of the question or the answer. The emphasis was on getting trainees to pass the licensing examination.

By contrast, one important reason for the thorough airline pilot training and follow-up is because the industry insists on it.

In tightening up reactor operator training and instituting retraining programs, the NRC should demand no less from the nuclear power industry. Failing that, the commission should take over direct supervision of operator training.

Arkansas Gazette.

Little Rock, Ark., May 9, 1979

Whither nuclear power?

A question not easily answered, we would say, but one that all Americans are going to have to be thinking a great deal about in the coming months.

Some points about nuclear power, in our own observation, already are beginning to come into focus, as more is learned about the accident on March 28 at the Three Mile Island plant in Pennsylvania.

Certainly there was little to cheer from the visit made to the plant Monday by the House Energy and Environment subcommittee, headed by Representative Morris Udall of Arizona, who said, later, that the nuclear industry's future "hangs in doubt."

Although the feeling on the subcommittee is divided, Udall is trying not to judge the incident prematurely. After viewing the "immensity and complexity" of the reactor control room panels, however, he said that "I think we have got to focus more on the human element and less on all of the backup systems and computers and all of the rest" under investigation.

There are recurring indications, from several sources, that fundamental error may have been made throughout the brief history of nuclear power in not setting high enough standards for reactor operators. This is the generality. It may or may not have valid application in specific cases, even at Three Mile Island. It is, by and large, a boring job, watching for long hours all those dials and gauges, and simulated training exercises sometimes do not assume that many things can go wrong at once.

The House subcommittee's visit did shed some light on events at Three Mile Island on March 28. A control room supervisor told the congressmen, for example, that Nuclear Regulatory Commission inspectors stood by with plant operators when that unexpected hydrogen explosion occurred 10 hours after the start of the accident. The NRC, however, says it was not aware of the explosion until two days later. Officials of Metropolitan Edison, the operating company, say the significance of the indication — a "pressure spike" on a monitoring gauge — was not known until March 30.

This is an unsettling revelation, for it indicates that no one in charge at Three Mile Island — government or private — could interpret correctly what was clearly showing on a gauge. The information suggests as well that this one mistake was not an isolated incident. Could this pattern not easily be repeated elsewhere with results too horrible to contemplate?

It is the posture of the various components of the nuclear power industry — including the electric utilities operating the reactors — that the industry is right on top of everything, that if the rest of us will just trust them everything will be all right. It is a familiar position to take in the public relations arts, but it is working less and less frequently these days, as more Americans become aware of the dangers even if they do not possess all the technical knowledge that those with vested interest might have.

As the Udall subcommittee was touring the Three Mile Island plant, President Carter was visiting at the White House with six co-ordinators of Sunday's anti-nuclear demonstration at the Capitol. The co-ordinators did most of the talking during the meeting with Mr. Carter, and their presentation could have impressed upon him the deep concerns of those who oppose all nuclear power.

President Carter, a former nuclear engineer with Navy training, clearly has a bias in favor of nuclear power, as does his Energy secretary, James Schlesinger. Mr. Carter, however, would seem to be more open to doubt than is Schlesinger even though the President told his anti-nuclear callers that shutting down all of the nation's nuclear power plants was "out of the question."

What we suspect is going to happen, as the ground swell of opposition to nuclear power continues to grow, is that national policy is going to move toward moderation.

Mr. Carter himself, while speaking of the need for nuclear power in some specific areas — Chicago, for example — also says that America must "shift toward alternate sources of energy supplies." This should mean solar power on a scale heretofore dismissed at the urging of those entangled with competing power sources. It should mean some more use of coal. In some areas of the country, other means of power generation might well be appropriate. Certainly a more intensive effort must be made to conserve.

The nation can either decide to promote more use of nuclear power or less, and it would appear in light of all the doubt raised by Three Mile Island and subsequent events that less is better. Just three months ago, for instance, the Council on Environmental Quality, a White House advisory group, said the United States could achieve, "a healthy, expanding economy" by the year 2000 while increasing its energy consumption by only 10 to 15 per cent. This requirement could be met by nuclear plants already built or licensed for construction. It is reasonable to suggest further, in our view, that if strong efforts toward the energy alternatives Mr. Carter speaks of were combined with aggressive conservation, construction on some of the plants already licensed could be stopped without severe penalties.

Those who wish to shut down all nuclear power plants now could well be correct. They are not kooks, but instead are persons sincerely concerned about this tiger the country has by the tail. They are more nearly right, certainly, than are those who would proceed ahead, full speed, to promote and expand the use of nuclear power.

WORCESTER TELEGRAM.

Worcester, Mass., April 6, 1979

Federal safety investigators say that the dangerous malfunction at the Three Mile Island nuclear plant was caused by three equipment failures and three human errors.

The equipment failures do not alarm us as much as the human misjudgments. We assume that the hardware can be redesigned to operate properly in case of any similar emergency.

But what can be done about the so-called "idiot factor"? How can we rest assured that valves supposed to be open are not left closed for weeks before an emergency takes place? How can we make sure that people under stress will not empty the coolant to the point where the nuclear fuel rods are exposed?

As our readers know, we have favored nuclear power as dependable, inexpensive and safe. It is still a fact that nuclear power plants have, over the past 20 years, compiled a safety record that is the envy of all other power sources — coal, oil and even hydroelectric.

But the accident at the Pennsylvania plant — an accident that was never supposed to happen — raises questions that must be answered. The defenders of nuclear power are making a bad mistake if they think the Three Mile Island accident can be dismissed by the reassurance that "the system worked."

For example, The Wall Street Journal, normally a source of perceptive thinking, says that "the salient fact is that despite the high drama no one was hurt." As we see it, the salient fact is that the psychological damage may be beyond measurement. People whose lives have been filled with fear overnight have not got through unscathed and it is idle to pretend otherwise.

Another salient fact, as we see it, is that the government and the nuclear industry have an enormous task in restoring confidence. The contradictory stream of information and misinformation that poured out of Pennsylvania last week badly eroded the faith of a lot of people in the ability or willingness of those in charge to tell the truth.

We continue to believe that nuclear power can be operated safely and will continue to be an essential part of our energy supply in the years ahead, although we doubt that it will grow very much. But, in order to even hold its own, those in charge must devise some way to convince the people that there is no China Syndrome of inutterable disaster in the offing.

The Birmingham News

Birmingham, Ala., May 22, 1979

Investigations to untangle what went wrong at Pennsylvania's Three Mile Island nuclear power plant — and the state of atomic energy generally — have only begun. But aspects of the Pennsylvania accident already point to the need for precision and uniformity in devices which monitor nuclear safety conditions and for upgraded training for nuclear employees..

Congressional probers say Three Mile Island, a brand new facility, got into trouble largely because inaccurate gauges misread vital data. And certain emergency controls, supposed to keep the reactor bathed in cooling water, also malfunctioned. While the sequence of events tends almost to rule out "human error" as such, it is possible someone present in the control room with a more extensive engineering background might have more readily sensed something amiss and taken corrective action sooner.

The various inquiries do not need to run their course without reaching some strong assumptions now: A key one is that nuclear reactors must have multiple sensing devices, in depth and as precise as monitoring equipment routinely used in the space program. It appears that despite their high cost, nuclear reactors do not now all have such capability and, moreover, rely on devices provided by a variety of manufacturers. Absolute standardization and calibration should be the goal.

Yet to be answered is the question of training of nuclear personnel—who are actually employees of the utility owning the power plant—in terms of proficiency and standard operating procedures. A possible answer is that on-site safety controllers should be highly-trained federal technicians.

Such an arrangement would be analogous to federal air controllers in charge of safety and plane traffic control at airports. While they do not supplant local decision-making in airport operations, they do have the last word on safety, and an enviable record due to rigorous and continuing training. A good case could be made that nuclear controllers should be no less proficient.

Even so, changes in key equipment and training standards do not amount to an indictment of atomic power, but the demonstrated need for improvements. In theory and in practice, atomic energy has a good safety record and is an essential component of the nation's energy supply. Unplugging the reactors makes as much sense as prohibiting flights of jumbo jets over metropolitan areas in the belief something could happen. Aircraft moving at high speeds through congested airlanes could be banned as unsafe, too; but it is high technology and highly-developed human skills which makes air travel work. The same extra ounce of prevention can be applied wherever nuclear power is employed.

The record of the atomic energy plants shows that they have done a commendable job, in what is still an infant industry. The lesson of Three Mile Island is that more can and should be done to prevent a recurrence. Harnessing atomic energy has proved scientifically possible; the remaining issue is to build in all safeguards for the unexpected, the malfunction of a small part or the tiniest error or lack of judgement, to make it fully practical.

Post-Tribune

Guarding Your Interests Daily

Gary, Ind., May 24, 1979

There is little comfort for anyone in the latest "official" report on the Three Mile Island nuclear accident.

A congressional task force says "design and equipment error" was more to blame than human error.

Several days ago, some Nuclear Regulatory Commission people said operator error played the major role.

The NRC's decision to impose a three-month delay on issuing licenses for nuclear plants was prudent. Some time is needed to assimilate the lessons learned from the Three Mile accident.

Still to be heard from is a presidential commission appointed to conduct hearings and issue a report on the Three-Mile Island scare.

What is the public supposed to believe? The conflicting reports, released separately, are adding fog when sunlight is essential. About all that is certain now is that the mishap was not an act of God.

The congressional report, did, however, include two points that should be kept in the open.

One is that the control room operators "...weren't highly trained engineers. They weren't the most profound people in the world. And they weren't prepared for this type of thing." But the report blamed the design more than the men.

The other is that "Such an accident not only could happen again but it is likely to at any time."

That first point surely emphasizes the need to have highly trained people running such complex equipment. It needs no elaboration.

The other point, well, it probably is speculative, but to dismiss it as politics or as scare tactics would be folly. Yet, what is the public supposed to do now?

Mere humans are groping for explanations, sincerely so. But the cloud of uncertainty and unanswered questions isn't likely to dissipate soon.

This paper will continue to publish reports of the studies, conflicting though they may be. Obviously there are many more questions than answers in this trying period as we back off and look at nuclear power. But history tells us that unless hard questions are asked, few satisfactory answers can be found.

The three hard questions still before the house are: What happened? Why? Where do we go from here?

The Post-Tribune has long held that nuclear power is essential to an efficient economy, even more so in these days of spiralling oil prices. But the air must be cleared on the Three-Mile Island accident and the public needs to be assured that the heretofore excellent safety record of nuclear power plants will be maintained.

The Evening Bulletin

Philadelphia, Pa., August 6, 1979

An alarm bell is still faintly clanging in the sealed reactor building at the Three Mile Island nuclear power plant. The alarm can't be shut off until the containment building is opened for what will be a long and costly cleanup of half a million gallons of radioactive water on the building's floor.

There's a kind of symbolism in that ghostly alarm bell. It's reminding us that we can't rest easier about the safety of nuclear power plants until President Carter's Commission on the Accident at Three Mile Island gets to the bottom of why the operators at Unit No. 2 either failed, or were unable, to keep the overheated reactor supplied with cooling water.

A report by the Nuclear Regulatory Commission blames the Three Mile Island accident mainly on human errors, mistakes by the operators who were running the reactor last March 28. If the operators had paid closer attention to their controls, the NRC alleges, the plant's emergency equipment would have kept the reactor's core from being damaged and emitting radiation.

Despite ingenious backup systems, the "human element" will always be crucial to the safety of nuclear power plants. An operator's grasp of what's actually happening when alarm bells go off can make the difference between a "routine" malfunction and a near-disaster. The nature and extent of the training he receives beforehand is critically important.

What troubles us most about the NRC's appraisal of the Three Mile Island breakdown is its assertion that the plant's operators were working under a "mind set" that may have put the uninterrupted generation of power ahead of anything else. Rather than risk supplying Unit No. 2 with too much water, a situation in which the reactor would have had to be taken "off line" until the water level could be lowered, they fed it too little water.

At least that's how the NRC sees it. If that finding holds up, it suggests that the training TMI's operators received led them to be more concerned with producing power than exercising extreme caution in the interests of safety.

As much as anything, President Carter's TMI commission needs to scrutinize the training nuclear plant operators receive, and the presumptions behind it. Until then, the alarm bell will keep ringing.

Los Angeles, Calif., May 11, 1979

"Human factors engineering" is a technical way of describing what housewives do when they hang the spice rack so it's handy to the stove; it's what Detroit does when it places the fuel gauge on the dash next to the speedometer. All it means is that the person designing a machine takes into account the ability of the operator to efficiently absorb the information he or she needs to run the machine.

"Human factors engineering" has been with us since World War II, when the military started designing machines too complex for its operators to handle. Operators were getting more information than they could use, process, or act on; subsequently,the machines failed. Whole systems went down, and the military went back to the drawing boards.

It began collecting information "regarding the abilities and limitations of the human operator...so that mutually supportive man-machine interfaces (i.e., jargon for today's systematic encounters between technology and human) could be designed." In the years since then, "human factors engineering" has become an indispensable design component in fields as diverse as auto manufacturing and the aerospace industry.

But to everyone's growing dismay, it appears that the advances made in human factors engineering were all but ignored by an important sector of our society: the nuclear power industry. And now an interoffice report on that subject from Governor Brown's office, obtained by The Herald Examiner, charges that "modern nuclear technology is harnessed by an antiquated control technology which represents a clear and present hazard."

This frightening charge appears not to be unfounded. A host of industry and government studies are cited in this report — all focusing on shabby control-room designs in nuclear reactors as a potential stumbling block in controlling "in-plant accidents." Even the now-in-dispute Rasmussen report, the "Reactor Safety Study," which said it was more likely we'd be killed by lightening than by a nuclear meltdown, states that nuclear control-room designs deviate "from human engineering standards generally accepted in other industries."

This is a troubling allegation, coming as it does on the heels of the Three Mile Island accident. After the accident, you remember, we were informed that the incident was the result of "operator error." But what the governor's report makes clear is that the "operator error" cited at Three Mile Island was inevitable, given the preposterous design of nuclear control panels.

Reactor control rooms, after all, are formidable. They typically contain more than 7,000 controls set on a panel which stretches more than 30 yards. Crucial indicators and switches are not separated from more mundane controls, and switches linked to one another may be yards apart. According to the Rasmussen report, "the volume of raw information (to be mentally processed) **exceeds the saturation point** of the operator" on the typical control panel. In another report, investigators asked operators what might help them handle crises in the control room; "statements frequently alluded to the need for roller skates."

In an NRC report submitted to Congress in 1978 — precisely one year before the Three Mile Island accident — more than 200 reactor design factors were reviewed in an attempt to upgrade nuclear plant safety. Of the 200, five were given top priority attention — and one of the five was better "human factor engineering" in control rooms. The study noted that control rooms could be made more manageable for a measly $50 million per plant, but the changes were not made. Instead the "design improvements" were relegated by the NRC to "further research," which presumably was being done when Three Mile Island hit the news.

The governor's report also makes it clear that nuclear power plants "affecting the public health and safety are poorly instrumented and designed for adequate reactor operator control." It also makes it clear that the nuclear industry and the NRC knew about the shortcoming for years — and did nothing to rectify the problem.

So we are confronted with the incomprehensible situation wherein "human factors engineering is utilized fully in developing safe ski bindings, but is not utilized fully in nuclear power plants."

This is folly of the worst sort, and raises nagging and ugly doubts about the nuclear industry's integrity, and the competence of the NRC to regulate that industry. Accidents, we all know, will happen. But we must do everything in our power to ensure that nuclear accidents *don't* happen; and the most obvious place to start is in the control rooms of nuclear reactors. Why control rooms have still not been upgraded is a question the NRC must answer — and hopefully, before we experience another Three Mile Island. Because from now on, no one is going to buy the story that the accident was the result of "operator error." If such an accident occurs, it may be the result of design deficiencies bordering on the criminally negligent.

Roanoke Times & World-News

Roanoke, Va., June 17, 1979

Advancing technology has brought us planes that can cross the country or the ocean; railroad engines that can pull, and push, up to a hundred cars; productive nuclear power plants; and automobiles that have cut our country down to size. But planes still crash, trains derail, and we need not be reminded of what can happen at nuclear power plants or behind the wheel of an out-of-control car.

Human error: Neither computers, special radar, automatic transmission, nor electronic "transition" can make up for the engineer, driver or pilot who's trying to swat a fly or keep little Johnny from throwing the dog out the window; trying to watch several hundred dials and gauges at the same time or trying to remember every possible configuration of events that might lead to — or avoid — nuclear meltdown. Imperfect humans, subject to distractions and panic, are at the controls of all our marvelous — though fallible — machines.

The National Transportation Safety Board and the Nuclear Regulatory Commission have, out of necessity, been paying more attention lately to human error. Distractions, or other imperative responsibilities, have been blamed in the wreck of a United Air Lines DC-8 that crashed near Portland, Ore., last December; in the Atlanta-to-Washington Southern Crescent derailment in Nelson County; in the Three Mile Island incident; and in the mid-air collision over San Diego last fall.

It is unfair, although unavoidable, to lay blame on specific individuals; anyone can make a mistake. But when a pilot makes a mistake, hundreds are likely to die; and the loss of life and property is little less when an engineer or driver makes a mistake.

Perhaps, in thinking about the development of more and more complicated machinery, we need to consider the competency of single individuals. Planes and trains should not, like the Ship of State, become too big and complicated for one person to handle.

The Philadelphia Inquirer

Philadelphia, Pa., July 28, 1979

There is a saying among engineers and technicians: "The reliability of a complex system is no greater than that of the nut holding the switch." It is an old saying, and it can be found pasted up on walls, bulletin boards and control panels in just about any technical establishment in the United States.

It is the technologists' wry assent of the fact that no matter how good a machine is, its designers, its operators and the managers who make decisions about it are human after all, and can and do make mistakes.

Testimony given at hearings of the presidential commission investigating the nuclear accident at Three Mile Island indicates that the managers of the nuclear power industry have forgotten that old saying. Discussing the attitudes and beliefs of people working in the industry, officials of Babcock & Wilcox, the firm that built the reactors at Three Mile Island, said their firm's safety programs had been "focused on the machine, not the man."

One executive admitted that the industry had begun to believe its own claims so completely that a kind of a "mind set" had developed about the infallibility of the equipment. This acccounts for the way B & W officials ignored warnings that reactor operators needed better training on how to handle the kind of problems that occurred in the failure at Three Mile Island. Over the quarter-century history of nuclear power, this mind set apparerntly has become so pervasive that it is difficult for reasoned, technically accurate warnings or even real breakdowns to rattle the industry's placid, we-have-it-all-under-control superiority.

Item: In 1974, the Rasmussen Report, a definitive study of the probability of various kinds of nuclear accidents, predicted the type of failure that did occur at Three Mile Island. The report, commissioned by the Atomic Energy Commission, predecessor of the Nuclear Regulatory Commission, has been criticized both for underestimating and overestimating dangers. What is pertinent, however, is that the report predicted the failure and assigned a specific probability and nothing was done about it.

Item: In 1977, more than a year before TMI, a similar accident occurred at the Davis-Besse plant near Toledo, Ohio. This time, safety systems had not been blocked by maintenance crews, but the operators did prematurely shut off a high-pressure cooling system, as their counterparts did at TMI. It was that accident that prompted Babcock & Wilcox engineers to ask that new instructions be given to plant operators.

But the operators never got the new instructions, and the simulator used by the company to train reactor operators at its Lynchburg, Va., facility was not programmed to teach them to cope with that kind of failure.

Item: The questions now being asked by the presidential commission about the training of plant personnel are the sort of questions that have been raised all along by opponents of nuclear power: How good are the men and women who run nuclear reactors? How much do they understand about the stupendously powerful processes they are trying to control? How well are they trained to recognize dangerous trends, and do they know what to do to correct them?

From the testimony, it appears that many in the industry have not addressed these serious questions fully. The radiant glory of technological success has kept them blind to the certainty that human fallibility could affect the performance of such magnificent machines.

For that kind of thinking, another saying, from the world of the computer specialists, seems to provide appropriate definition: Garbage in, garbage out.

The Wichita Eagle-Beacon

Wichita, Kans., August 4, 1979

The Nuclear Regulatory Commission's finding that the Three Mile Island nuclear accident was preventable doesn't relieve our uneasiness over the inherent dangers of nuclear power generation. In fact, it reinforces our fear that what apparently was merely an accident caused by human error could have been a disaster.

If it can happen once, it can happen again. And unfortunately, human error is among the more difficult problems to correct.

Yet it must be corrected, because the nation already depends on nuclear plants for 12.5 percent of all its electricity. And as President Carter has pointed out, to close down the existing plants, at least, "is out of the question."

So we are stuck with nuclear power — at least for several years to come. That doesn't mean we should ignore its dangers. What has happened at Three Mile Island and at such sites as the Marble Hill nuclear power plant in southern Indiana, where utility officials are accused of covering up construction defects in the plant's reactor, should convince any prudent person that caution is needed.

As we have said before, not a single new reactor ought to be started in the United States until the government can assure the people that every step is being taken that can be taken to protect the population from a nuclear accident. Further, no more should be started until there is reasonable assurance that nuclear wastes can be stored permanently and safely.

To accomplish this, tough new standards should be established for the certification of plants, for the training of operators and for federal inspection. Rep. Dan Glickman's amendment that would provide money to station federal inspectors at nuclear plants has passed the House and Senate and is before a conference committee. Inspectors should be on duty permanently at nuclear sites, and should have the authority to shut down a reactor in an emergency. The Glickman plan provides the money, but the NRC will have to see that these inspectors have the authority they need.

An even touchier problem is the disposal of nuclear waste, and no really suitable answers to that problem have been provided by the NRC or by the president.

Presently, spent fuel rods are stored under several feet of water at reactor sites. This clearly is not a permanent solution, nor is it even a satisfactory temporary one. A crash program to find a way to dispose of long-lived, highly toxic wastes is needed.

If no suitable plan is found in the next five years or so, the future of the entire nuclear industry well may have to be reassessed.

ST. LOUIS POST-DISPATCH

St. Louis, Mo., August 5, 1979

In the early hours of the nuclear accident at Three Mile Island, Gary Miller, station manager of the facility, was getting confusing readings of the temperature within the reactor core. Some of the 52 measuring devices were registering 700 — 200 degrees above normal — and as high as the computer was programmed to report. Then he sent a technician to get readings from the devices before they were fed into the computer. Some registered zero, others 2,400 degrees, which is near the melting point of the rods that contain the radioactive fuel. In the tense and uncertain atmosphere that prevailed, he chose to disregard the high readings. "I wanted something I could believe in," he later explained. Similarly, when top utility executives were informed of the crisis situation they refused to believe it was happening. Such things weren't supposed to happen. But they did.

The Nuclear Regulatory Commission's office of inspection and enforcement has released the first of what promises to be a series of reports on what went wrong at Three Mile Island Unit 2. Other reports will be forthcoming from both houses of Congress and a commission appointed by President Carter. Stripped of bureaucratic double talk, the NRC staff report boils down to a technological tautology : If things had been different, it says, the nation's most serious accident at a commercial nuclear power plant "could have been prevented."

Its summary says that *if* the plant operators had allowed the emergency core cooling system to do its job, and *if* they had been better trained, and *if* the instruments had worked as they were supposed to and *if* Babcock & Wilcox Co. had designed the facility differently, and *if* there had not been so many apparent violations of the NRC operating guidelines (including release of radiation 11 times the permissible levels) — things would have been different.

If wishes were horses, beggars would ride. We'll have to side with the spokesman from General Public Utilities Corp., the holding company that owns the stricken facility: the NRC report "isn't a comprehensive analysis of the contributing causes of the accident." It is wishful thinking.

THE BLADE

Toledo, Ohio, June 10, 1979

THE more the Three Mile Island nuclear power-plant accident is studied, the stronger the case becomes for entrusting the operation of all such facilities only to specially skilled personnel. That does not mean just better-trained technicians but graduate engineers and nuclear experts thoroughly indoctrinated in the intricacies of atomic energy and the potential perils involved in its use.

Four operators of the Three Mile Island reactor testified before the presidential commission investigating the mishap there that they were simply not equipped by experience or schooling to handle the situation that confronted them. And there is mounting evidence that inadequate preparation of operational personnel is by no means a problem unique to the Pennsylvania plant.

According to one report, the average incoming class of 30 at a major training center for reactor operators includes only five members with engineering degrees. The instruction that most operators get anywhere is apparently aimed primarily at giving them the minimum information necessary to pass the written tests for licensing by the Nuclear Regulatory Commission. Many get no simulator training at all, an NRC survey discovered, and such simulators as exist are "primitive" devices incapable of setting up emergency problems for trainees to deal with.

Even if such gross inadequacies were corrected and the training of the customary technicians were vastly improved, however, the question remains whether that alone would provide the level of competence that is needed among nuclear power-plant operators. The answer was suggested, probably unwittingly, by a member of a congressional task force which is also looking into the ramifications of the Three Mile Island accident. Commenting on the misleading or conflicting information provided by control-room equipment at the plant, he said that a sophisticated nuclear engineer might well have been able to appropriately interpret other conditions and diagnose the true situation despite the inaccurate readings.

It is well recognized by this time that improvements are necessary in certain design features in nuclear power plants; some changes, in fact, already are being made at Davis-Besse and similar facilities. It is also acknowledged that the NRC itself must upgrade not only many of its own standards for the design, construction, licensing, and operation of the plants but also its performance in seeing that the standards are met and maintained.

But there is absolutely no way to remove all risk from nuclear power-plant operations, any more than every hazard can be prevented at coal or oil-fired generating facilities or any other human activity. What is needed at nuclear plants is special understanding of the unique characteristics of atomic energy and the expertise to deal with both ordinary operations and abnormal circumstances — including some that simply may not be predictable despite the most careful contingency planning. This kind of responsibility calls for persons with engineering and/or scientific credentials well beyond those of even the most conscientious and well-trained technician.

AKRON BEACON JOURNAL

Akron, Ohio, August 6, 1979

A REPORT made last Thursday by the Nuclear Regulatory Commission probably did not change by one whit the convictions of anti-nuclear activists who rallied at the Davis-Besse plant near Port Clinton Sunday.

But it must be somewhat reassuring to those whose electricity comes from nuclear sources and to many others across the country who believe nuclear power must play a continuing, even expanding, role in the struggle for energy independence.

The report, in brief, said the safety system of the Three Mile Island nuclear power plant was adequate — if emergency procedures had been carried out as designers intended.

It did not say the design of the equipment could not be improved, but that the accident might have been prevented if plant operators had not taken inappropriate actions.

The resulting radiation risk was termed "minimal."

The obvious response of nuclear opponents is, "But there *is* a risk."

Of course there is, but it must be seen in perspective. When it is, rational arguments can be made for and against nuclear energy.

Almost simultaneously with the NRC report, the Wall Street Journal cited a study put out by the Health and Safety Commission in Britain, where debate about the future of nuclear power is also going on.

Comparing accident deaths that can be expected per unit of electrical energy produced from coal, oil and gas, and nuclear sources, the report rated nuclear safest and coal, by a considerable margin, most dangerous.

Besides the accident potential in coal mining and in oil and gas production, all these ways of producing energy produce pollution health hazards for the public, the report said.

That is not an endorsement of nuclear power. It is simply an attempt to be fair in assessing its danger and its potential.

The Dispatch

Columbus, Ohio, February 17, 1982

RECENT EVENTS are causing heightened concern over the safety of nuclear power plants and over the credibility of power plant operators. They place in some doubt the future of nuclear energy in this country.

Many of the 72 power plants in the nation seem to be falling apart. At the beginning of this month, 24 of the plants were closed because of various problems.

In addition, the Nuclear Regulatory Commission (NRC) recently announced that thousands of pipes that conduct steam and water within reactor and coolant systems are corroding in many plants and have to be repaired or replaced.

The credibility of power plant operators and owners is also deteriorating. It started with conflicting, confusing, less-than-forthcoming and sometimes false information that was issued by utility spokesmen during the accident at Three Mile Island. Various government reports since have indicated that misinformation from the utility hampered efforts to control the problem and served to incite widespread anxiety.

The performance of utility officials at the Diablo Canyon nuclear power plant in California has also been damaging to the industry. They long maintained that the plant was safe to begin operations and ridiculed critics of the plant.

Then the NRC discovered that diagrams used in the design of earthquake supports for piping in the reactor had been reversed, calling into question the plant's ability to withstand shocks from a nearby offshore earthquake fault. Subsequent investigations found that the safety data compiled by the utility could not be verified. The NRC, noting that it could not certify the plant's safety, revoked a low-power testing license.

Late last month, a pipe burst at the Ginna nuclear power plant outside of Rochester, N.Y., causing a brief release of radioactive steam into the atmosphere. The cause of the rupture has not been determined.

At a three-hour NRC meeting last Wednesday, Ginna' owner, the Rochester Gas and Electric Corp., argued for permission to resume operations. "We're 99.99 percent sure — nothing is for absolutely sure — that we can run safely," a utility official told the NRC. The request, fortunately, was rejected.

A few hours after the meeting, an inspection of pipes in the Ginna plant revealed new and "very dramatic" defects in water tubes inside the same steam reactor that failed last month. "It looks like somebody went in with a hacksaw," an official stated. "Some of the tubes show severe denting and external degradation." He said he mentioned the hacksaw only to provide a graphic description, adding he did not know what caused the damage. A piece of one pipe is missing, and officials fear it may be in the steam generator, posing a threat to one or more of the generator's 3,259 other pipes. At the NRC hearing earlier in the day, the utility assured the commission that pipes in the reactor were "healthy" and incapable of causing problems.

These kinds of developments are bound to have a negative effect on a public already skeptical of nuclear power. A recent *Associated Press-NBC News* poll found that 56 percent of those polled oppose further nuclear plant construction. This represents a dramatic turnaround from 1977, the last time *NBC News* sought opinions on further construction. In 1977, 63 percent said more plants should be built.

Utilities and consumers are also finding that nuclear plant construction can be risky and expensive. When the Washington Public Power Supply System, which serves 115 utilities in eight western states, decided in January to scrap further construction of two plants, it faced debts amounting to $343 million in termination costs plus money needed to retire $2.25 billion in bonds for work to date. It informed customers they faced rate increases ranging from 50 percent to 300 percent over the next three years.

No industry can long endure such questions of safety, credibility and expense. The NRC and the nuclear industry must find ways to address these serious problems. Nuclear energy is a vital component of the nation's energy future. Its continued acceptance and dependability must be assured.

The Times-Picayune
The States-Item

New Orleans, La., August 26, 1981

When the Advisory Committee on Reactor Safeguards advised the Nuclear Regulatory Commission recently that "The Louisiana Power & Light management has not yet been successful in putting together the team of experienced and qualified personnel which we believe will be necessary to successfully operate the (Waterford 3 nuclear) plant," thus putting another potential hold on that facility's operating license, the ACRS was putting its finger not so much on LP&L as on a nationwide industry problem.

The industry says there are not now enough trained licensed operators to fill the expanded staffs now required for 70 operating plants and 75 under construction. One professional estimate is that 3,000 more operators will be needed over the next five years, double the present force.

This would seem to open up to the nation's present and potential labor force a happy hunting ground for jobs. "No doubt the market is tight," comments a Middle South Services Inc. manager. "Everyone with a nuclear power plant is advertising for a nuclear power plant operator." Companies are stepping up recruitment efforts, consultant firms are hiring out to hire on, and some utilities, according to an NRC project director, "are picking up additional people with experience from the nuclear navy and raiding other utilities."

But the hunting grounds are not that happy. One cannot go to school and graduate as a nuclear power plant operator. Indeed, there are few formal requirements for beginners — not even a college degree. One must get the job to get the company training. Experience gained outside such training programs tends to come from time in the military, at fossil fuel plants and work as students with small university reactors. And beginning salaries are not attractively competitive.

The whole matter is in a state of flux, however, and all parties seem to be feeling their way to a better situation. Utilities and universities are looking into special programs, and the NRC's authority as final arbiter of standards and qualifications gives it considerable powers of creative command. But until the pay and prestige of the operators themselves are attractive enough — a point made in the report following the Three Mile Island incident — finding personnel may be a major restraint on the continued safe development of the nuclear power industry.

NRC Evacuation Rules Threaten Licensing of Some Plants

At the time of the Three Mile Island incident in 1979, there was no approved evacuation plan for Pennsylvania. This was part of the impetus for a provision included by Congress in the June 1980 authorization bill for the Nuclear Regulatory Commission. The bill required the NRC to deny operating licenses to future nuclear plants not covered by standby state and local plans for evacuation in the event of a nuclear accident. The NRC, in enacting its own regulation, went beyond the congressional instructions. The commission decided that after April 1, 1981, it could order an existing plant closed if the commission decided the utility's own emergency plan was inadequate. They could also close one down if informed that the emergency plans of state and local governments would not safeguard the local population.

Two of the most heated battles over evacuation plans have concerned nuclear plants located in New York. In May 1983, the NRC voted to shut down the Indian Point nuclear power plant in June unless an adequate evacuation plan was developed by that time. Indian Point's twin reactors, No. 2 and No. 3, are located near Buchanan, N.Y., 30 miles north of New York City. Opponents of the plant contended that it posed a hazard to the health and safety of some 290,000 persons living in the most densely populated area surrounding any U.S. nuclear plant. The NRC ruling was the first that threatened to close a plant for its lack of effective evaluation procedures. Protection strategies, according to NRC emergency procedures, had to serve all residents living within a 10-mile radius of the plant. There were two main deficiencies in Indian Point's procedures: Rockland County refused to join in evacuation exercises, and bus drivers in Westchester County had failed to aid evacuation efforts in earlier drills. After a "highly efficient and well-coordinated" drill was conducted in August 1983, the evacuation procedures for Indian Point were finally approved by the NRC. (State and utility officials filled in for the Rockland County officials.)

The evacuation plan for New York's Shoreham plant, in Suffolk County, has also run into opposition from local officials. A major reservation of the county is that the only escape route from Shoreham, on Long Island, would be through New York City. The problem at Shoreham is still unresolved, and its future may rest with the financial health of its owner, the Long Island Lighting Co. In February 1984, LILCO defaulted on its partnership with four other utilities in the construction of Nine Mile Point 2, another New York nuclear plant, citing financial difficulties caused by the delayed Shoreham reactor.

Herald News

Fall River, Mass., May 17, 1979

The director of the state's Civil Defense Agency says that plans are being drawn up for possible evacuation of a 10 mile area around nuclear power plants, including the one in Plymouth.

Evacuation plans had been limited to a five mile area until now, but the public's concern since the Three Mile Island incident has prompted state officials to expand them to cover a wider area.

It should be stressed that the change is not motivated by any specific fear of an incident either at Plymouth or at Rowe, where the state's two nuclear power plants are located. It stems from the general uneasiness following the Three Mile Island incident and a fear that nuclear authorities may not know themselves just what the possible dangers may be.

The President and Governor King have both reiterated their confidence in nuclear power since the Pennsylvania incident. They are probably right, but it remains true that the public wishes as much reassurance as possible about precautionary measures such as the expanded evacuation plan.

The Civil Defense Agency is doing precisely the right thing. It is drawing up plans which, in all likelihood, will never be used, but which are essential to public confidence right now.

Meanwhile the Pilgrim I station at Plymouth is out of commission so that an exhaustive survey can be made of its pipes. The station will be shut down for at least two weeks. This painstaking survey is evidently the result of the Three Mile Island incident.

In a quiet way the federal and state governments are doing everything possible to allay public fears and to forestall any possible repetition of what happened in Pennsylvania.

While their steps will obviously not satisfy those who wish to eliminate the use of nuclear energy altogether, they are at least steps in the right direction.

They should prove reassuring to everyone living in Massachusetts.

The Burlington Free Press

Burlington, Vt., May 22, 1979

The lesson of Pennsylvania's Three Mile Island nuclear power plant accident as it has been learned in Vermont is that we do not have a workable plan to protect the people of the southeastern part of the state and neighboring areas in New Hampshire should an accident occur at the Vermont Yankee nuclear power plant in Vernon.

State officials have relied on a plan developed in 1971 that calls for the evacuation of residents in a five-mile circle from the plant. Ever since Three Mile Island, that limit has been deemed inadequate and does not take into consideration prevailing winds.

Other flaws include the problem of the best way to notify area residents, many of whom cannot get television reception or radio signals from Brattleboro stations which presumably would broadcast an emergency evacuation. The plan calls for Brattleboro to receive evacuees, but the southern half of the town is within the five-mile limit. There are an inadequate number of shelters in the area and those that do exist are inadequately stocked with food and medicines. One evacuation center is Windham College, which is defunct and auctioning off everything — including beds and other furnishings necessary to take care of large numbers of people.

In short, the state plan is no plan at all and Gov. Richard A. Snelling has recognized this fact by appointing William H. Baumann of the Criminal Justice Training Council to draft a new evacuation plan. His task is a monumental one. Not only must he decide what areas should be evacuated, but where the evacuees would go and how they would be cared for. His job also is complicated by the large influx of tourists during the summer and winter months who also must be protected. Baumann also must have the cooperation of local officials of the sending and receiving towns which is vital in any evacuation plan.

And there is the broader question of dovetailing Vermont's plans with those of nearby states which would be affected by any accident at the Vernon plant, as well as the possibility of an accident at the Rowe, Mass., nuclear plant, just south of the Vermont border.

Although Vermont Yankee officials are confident that a Three Mile Island accident can't happen in Vernon, state officials are clearly concerned, and they should be. They must devise a more realistic plan than the one that now exists on paper and they must waste no time doing it. There are potentially a great number of lives at stake.

Arkansas Gazette.

Little Rock, Ark., May 10, 1979

When the Three Mile Island nuclear power plant accident happened, it suddenly dawned on authorities in many states housing nuclear plants that they did not have adequate plans for coping with similar situations. Arkansas, alas, was among them.

In the weeks following, however, a substantial effort has been undertaken in Arkansas, and when the state emergency services officials met this week to evaluate their progress it was obvious that large gains had been made. A final, comprehensive plan is expected for presentation to Governor Clinton within three weeks.

A part of the plan dealing with public health, it seems, has already been approved by the federal Nuclear Regulatory Commission, thus making Arkansas the 12th state to reach this goal. This part was drawn up by the state Health Department's Bureau of Environmental Health, which has created four classifications of nuclear accidents based on radiation dosages.

Evacuation measures clearly are going to make up the principal part of the public safety considerations in the state's final plan. Most of the evacuation would be by highway, using ambulances, trucks and buses but some helicopters would be available as well

It is essential, in any case, that Arkansas settle on a workable plan as quickly as possible, for the probability is that both units at Nuclear One near Russellville will be allowed to resume operation in a few weeks, when requirements of the NRC staff are satisfied. Accidents don't follow schedules and they often give no warning. Three Mile Island holds many lessons for improved safety in dealing with nuclear power plants. If the accident had happened in Arkansas instead of Pennsylvania, our own state would have been sorely pressed to cope with the possibility of mass evacuation or rapid monitoring of escaping radiation in the plant's general vicinity.

Pittsburgh Post-Gazette

Pittsburgh, Pa., April 6, 1979

Among the numerous unpleasant surprises of the nuclear plant accident near Harrisburg was the revelation that Pennsylvania does not have a federally approved plan for dealing with such on-site emergencies.

That fact was brought out by the General Accounting Office, Congress' watchdog agency. And apparently Pennsylvania is not alone; most states with nuclear facilities also lack such contingency plans.

The GAO's report suggests that the nuclear industry's insistence that the plants could never suffer a calamitous accident persuaded officials in Pennsylvania and other states not to "alarm" the public with discussions of on-site emergency plans. That's rather like taking down fire exit signs in a building lest people worry that it might be a firetrap.

The GAO also noted there was no single federal agency that could direct a mass evacuation that might be prompted by a major disaster. So much for all of the money the taxpayers have pumped into civil defense since World War II.

Gov. Thornburgh is worried — and rightly — about refurbishing the image of the Harrisburg area, and Pennsylvania as a whole, in the wake of the Three Mile Island disaster. A first step should be to appoint two task forces, one to draw up on-site emergency plans to submit for federal approval, the other to draft a mass evacuation plan for every area of the state which has a nuclear power plant — including the Beaver County area to the northwest of Pittsburgh surrounding the nuclear power plants on the Ohio River.

The utilities and the nuclear plant manufacturers who serve them should be the first to support these efforts in every way. The best reassurance to the public about the future of nuclear power would be the adoption of specific plans for coping with the emergencies that we now know can happen, regardless of the best intentions and precautions.

DAYTON DAILY NEWS

Dayton, Ohio, May 14, 1979

One of the major failings of the government's regulation of nuclear power plants is that it has allowed those plants to be placed within killing distance of thousands of persons.

The governmnet had assumed that this to be safe, because it was considered next to impossible for a nuclear plant to go out of control and blast radiation into the air. We are all wiser now. And once we have acknowledged that a serious accident — though unlikely — is possible, it follows that people should not live so close to nuclear reactors that they cannot get away if it blows.

Two professors at the Massachusetts Institute of Technology say that a catastrophic release of radiation could spread 18 to 29 miles down wind from a reactor. The distance might be greater, depending on conditions.

Within that area, say the professors, "the possible radiation levels could mean immediate death for half the exposed population." Immediate death for half the exposed population. And more deaths later, perhaps decades later, for the lucky ones.

If their studies are reasonably accurate, it seems only prudent to locate any new reactors — if they are ever built — at least 30 miles from any population concentrations and to prohibit development within that radius. It also seems only prudent to clamp tighter standards of design and operator training on existing plants that are near population centers and ultimately, to phase them out.

John F. O'Leary, who once headed the Atomic Energy Commission and is now Deputy Secretary of Energy, warned candidate Jimmy Carter two and a half years ago to keep nuclear plants away from population centers. He called the chances of an accident "probable," and he also recommended slowing down existing reactors in populated areas and developing evacuation plans.

Those were strong words coming from a man who was a member of the atomic club. Now that he is an assistant to Energy Pooh-Bah James Schlesinger, however, Mr. O'Leary seems to have moderated his view.

But it doesn't take long experience in the industry or a Ph.D. from M.I.T. to know that nuclear plants and people don't mix. If we ultimately decide that more such plants will have to be built, we at least ought to have the brains to put them in the wide open spaces, from which farmers and small towners can easily and swiftly flee. Built near cities, they are mass death biding its time.

Lincoln Journal

Lincoln, Neb., May 14, 1979

A Senate committee's recommendation for switching off all nuclear power plants where geographical emergency evacuation plans haven't yet been formally approved by Washington is absurd.

One of the lessons of the Three Mile Island plant accident is that such evacuation planning ought to be highly developed and periodically tested by those who will be in charge. That's little different, really, from Civil Defense evacuation plans associated with nuclear attack.

But surely it is possible to perfect these plans without requiring nuclear plants to stop producing electricity.

Not quite as silly in the other direction, although still questionable, was the action the other day of Rep. Douglas Bereuter and associates on the House Interior Committee. They endorsed a six-month moratorium on new nuclear plant construction.

Bereuter says the nation "should have the results of what went wrong at Three Mile Island" before proceeding with additional plant construction.

All right, that's prudent. What's out of phase is the six-month moratorium.

Technicians say it's not likely anybody will be able to physically check the interior of the damaged reactor container vessel at Three Mile Island for maybe a full year, not six months. That's how intense the radioactivity is.

Surely a complete understanding of what happened to the reactor core is needed before experts can satisfactorily tell Bereuter "what went wrong," let alone postulate systems for preventing the same kind of future accident.

Congress better be ready to make that freeze on new construction a year or more. But closing down functioning plants to await bureaucratic paper work on emergency planning options is nonsense.

THE SACRAMENTO BEE

Sacramento, Calif., July 1, 1979

Until the Three Mile Island events proved their folly, government agencies had no plans for coping with a major nuclear power plant failure, like a core meltdown or a radioactive leak, because such an accident was considered too unlikely to bother preparing for. Now, however, the federal Nuclear Regulatory Commission and many states have started to revise their emergency planning. In California, the governor's Nuclear Power Plant Emergency Review Panel, which studied this state's ability to handle a Three Mile Island-type accident, came up with a list of the new safety measures that are needed, and Sen. John Garamendi introduced three bills in the state Legislature to implement the panel's recommendations.

The proposals are simple, cautious and not overly dramatic. Garamendi's bills call for the preparation of evacuation and emergency plans for an area around each nuclear power plant in the state larger than the three- to six-mile radius covered by existing plans; for automatic notification of the appropriate public agencies in case of specific equipment failures; for continuous radiation monitoring that is made available to government agencies as well as to plant operators, and for better public education about emergency procedures. These are all measures that, had they been taken before the Three Mile Island accident, would have greatly reduced the confusion that followed it.

Garamendi has tried to avoid stirring up the usual nuclear power debate and to keep these safety measures from becoming identified with anti-nuclear politics. His proposals assign responsibility for overseeing local emergency planning to the state Office of Emergency Services rather than to the Energy Commission, which critics consider too anti-nuclear to be neutral. And they require legislative approval of the new emergency plans before they are adopted in order to assuage the fears of pro-nuclear legislators that planners might come up with emergency safety requirements so extensive and expensive that no power plant or community could afford them.

Despite these efforts at non-partisanship, however, and despite the fact that all the state's utility companies were consulted in drafting the bills and the major utilities have withdrawn their original opposition to the emergency planning, two of Garamendi's bills have not made it out of committee in the state Senate and the third is expected to have trouble on the Senate floor. The senators cannot seem to get beyond the political split that usually dominates nuclear power legislation.

The bills are vague about how much the new precautions will cost, which understandably worries the utilities that would have to pay for them. But the financing mechanism itself — a user's fee — is a sensible one that would make the cost of nuclear power reflect the cost of power plant safety. And if the legislators are worried about the cost of the program, it would be more appropriate to impose a spending limit on the emergency planners than to forego emergency planning altogether.

The precautions Garamendi has proposed are too vital to the health and safety of the community to hold them hostage to a different dispute. All three bills should be passed.

The Des Moines Register

Des Moines, Iowa, June 21, 1979

One of the most disturbing lessons from the accident at Three Mile Island was that officials had not prepared an adequate plan for evacuating the thousands of persons living near the nuclear power plant. A mass evacuation did not become necessary, but if it had, the lack of an adequate evacuation plan could have led to mass confusion and loss of life.

It is surprising, therefore, that the House of Representatives this week rejected an amendment to require every new nuclear power plant to have an approved evacuation plan before going into operation.

A recent investigation by the General Accounting Office found that, of the 43 states in which nuclear plants are operating or are under construction or consideration, only 12 have evacuation plans approved by the Nuclear Regulatory Commission. Not all of these 12 states have adequate plans.

Some opponents of the House amendment argued that it didn't go far enough. Others said that the country couldn't afford the amendment since it might delay the licensing of nuclear plants.

Fortunately, Congress is not finished with this issue. The Senate Public Works Commitee has approved a bill to require that new nuclear plants have approved evacuation plans. The measure would give operating plants six months to develop adequate plans. The least Congress can do is to insist that every nuclear power plant have an adequate plan to evacuate the public in event of emergency.

The Birmingham News

Birmingham, Ala., June 21, 1979

The House of Representatives is to be commended for looking beyond the current tension over the problems at Three Mile Island nuclear generating plant to the larger issue of the nation's energy needs. In so doing the House also moved to ensure greater protection against accidents in nuclear facilities.

On a 350-10 vote, House members approved $5 million to hire 100 new inspectors for the Nuclear Regulatory Commission to strengthen surveillance of safety measures in nuclear generating plants. At the same time they voted 235-147 to reject a proposal that would have prohibited for one year federal licensing of nuclear plants in states that have failed to submit emergency evacuation plans to the NRC. Both actions are in the national interest.

Houston Chronicle

Houston, Texas, June 20, 1979

It is encouraging to see a vote by the House of Representatives that seems to indicate the congressmen are not being carried away at the moment by the fit of hysterics some people have thrown in the aftermath of the Three Mile Island nuclear power plant accident in Pennsylvania.

The House decisively (235-147) rejected a move that would have prohibited for a year issuing new licenses to nuclear plants in states that have not submitted plans for evacuating residents in case of accidents. We don't think anyone has a quarrel with such contingency planning but we also don't think there is valid argument that things have to come to a halt because of it. Feasible contingency planning can be done while the process of bringing new energy supply into use continues.

This proposal has all the earmarks of a disguised delaying tactic against nuclear power in general, something at which its more rabid opponents are quite adept.

The House also seems to have acted responsibly in requiring full-time federal safety inspectors at all nuclear power plants and providing the money to hire them. It is part of the same overall legislation. Only 22 of the 70 operating plants are said to have such resident inspectors. Extending this further precaution to all plants appears appropriate.

The Salt Lake Tribune

Salt Lake City, Utah, July 18, 1979

A Senate bill that would shut down in six months all nuclear poers plants operating in states without federally approved evacuation plans could be a boon to anti-nuclear power forces.

Although the proposal is well intentioned, the practical effect could be to withdraw a nuclear power plant's operating authority because officials over which the operators had no control failed to comply with the evacuation plan mandate.

Thus a reasonably safe nuclear power plant supplying badly needed electricity could be shut down because state agencies stalled.

Presumably local pressure from electricity consumers would force prompt state action. But the reverse could be true. Anti-nuclear forces might be able to mount even greater influence and by frustrating formulation of an evacuation plan shut down the nuclear plants in their state.

A better approach would be to give states the opportunity to devise their own evacuation program. Those states which would not or could not come up with an acceptable plan within a stated time would be obligated to accept one made in Washington.

The goal should be to see that each state is covered by a good evacuation procedure. It makes no sense to hold the nuclear industry hostage when the goal can be accomplished without that potentially damaging side effect.

THE KANSAS CITY STAR

Kansas City, Mo., July 20, 1979

Ever since the Three Mile Island accident in late March, Congress has been eager to demonstrate its dedication to making nuclear power safer for the public, especially those persons living near a nuclear generating station. So there were sundry amendments waiting when the bill to approve $373 million this next fiscal year for the Nuclear Regulatory Commission came up in the Senate.

The most stringent one approved would shut down, as of next June 1, all nuclear plants in states which do not have a federally approved emergency evacuation plan. The lack of any such plan, prepared in advance, was one of the more worrisome aspects at Harrisburg, Pa., in those first confused hours when authorities debated whether a major evacuation was necessary or desirable.

At present 39 of the nation's 70 licensed nuclear plants are located in 16 states which do not have such plans. Ten other plants are nearing completion in states which have no plans. Twelve states in which plants are sited do have the required emergency plans. At all plants the operating utilities have their own evacuation plan for the immediate vicinity, but the question of moving threatened populations beyond that area is left up to local or state authorities.

The Senate also acted to assure that the NRC will have enough inspectors to maintain one in residence at each operating nuclear facility, but the Senate refused to be stampeded into crippling the nuclear program. It rejected a six-month freeze on new construction permits as an arbitrary delay in an energy program which has been busy since the Three Mile Island incident trying to apply the lessons learned in plant redesign, equipment inspection and better operating procedures. And it also refused to give each state veto power over federal efforts to dispose of nuclear wastes within its borders, once a satisfactory disposal method has been developed.

The issue of safe storage of radioactive wastes must be resolved before the nuclear power program can move ahead permanently, and the federal government must have the final say in such decisions. Even so, the requirement that the states act to develop emergency evacuation plans for their respective nuclear installations could severely damage nuclear energy in such places as California, whose current governor considers himself committed to an antinuclear stance. If the state did nothing, under the Senate amendment its nuclear reactors would have to shut down next year.

The need for new and improved safety rules for the nuclear power industry was clearly shown last March, but it would be regrettable if any such could be used by all-out nuclear opponents to arbitrarily block further licensing.

Oregon Journal

Portland, Ore., September 28, 1979

Portland General Electric and state and local government officials are piecing together a new evacuation plan around PGE's Trojan nuclear plant. The stakes are high: The plan must be approved by the Nuclear Regulatory Commission (NRC) by June 1980 or the plant must be shut down.

PGE has drafted a plan for evacuating residents within 2.5 miles of Trojan, which involves removing a few hundred people in case of a serious nuclear accident.

A legislative mandate and anticipated NRC action have pushed PGE toward a 10-mile plan which could involve 60,000 people.

PGE sees its responsibility as mainly within the plant site. Utility officials would order an evacuation, should an accident ever occur, but it would be up to state officials in Oregon and Washington to move people, re-route traffic and control food supplies.

Columbia County has been working closely with PGE and state officials, and Cowlitz County, Wash., is becoming increasingly involved as the plan spreads to 10 miles.

If such a plan were activated, a governor's decision center, probably in Salem, would become the headquarters for emergency planning. Two Portland hospitals, Good Samaritan and St. Vincent, and Columbia County in St. Helens, have been designated for interim treatment. Specialized treatment would be available at the Hanford Environmental Health Foundation, 300 miles away.

A prodigious amount of planning is occurring for a nuclear accident which everybody hopes never will occur. Not even the Three Mile Island accident involved such large-scale evacuation as could occur around Trojan.

One of the considerations the Pennsylvania governor's office had to weigh at the time of the Three Mile Island incident was the fear that more people might be hurt in an evacuation, through such things as heart attacks and traffic accidents, than were at risk from a possible escape of radiation from the nuclear plant if they stayed out.

It's wise to spend time and money to prepare for such an eventuality, however. The NRC is expected to provide $100,000 for planning. If that money doesn't appear, the legislative Emergency Board should provide needed funds. So should Washington state, which happens to find itself a short distance from the plant.

More than 30 governmental agencies — including the U. S. Coast Guard — which would handle river traffic — have been involved in previous exercises to test response to an emergency. A major exercise is scheduled next month.

Such planning is the price we are paying for nuclear power. It's well that PGE and a host of government agencies take such planning seriously. Too much is at stake to view it any other way.

The Birmingham News

Birmingham, Ala., July 20, 1979

The U.S. Senate did a good job of separating the chaff from the wheat in its actions on two nuclear power plant bills that would have placed a blanket moratorium on the licensing of any new nuclear power plants for six months. Later, it voted, however, to shut down by June 1, 1980, nuclear power plants in any states that do not have emergency evacuation plans developed for use in the unlikely event of a nuclear accident.

Nuclear power, of course, is not the best of all possible worlds. But it is a form of energy that produces 12 percent of this nation's electricty and which will have to be relied upon even more heavily in the future if America's dependence on foreign energy sources is to be reduced.

The blanket moratorium was called for, according to its sponsers, in order that we might learn "the lessons of Three Mile Island." It is doubtful, however, that a six-month moratorium would have provided many more safety checks than are in the voluminous regulations that already control nuclear plant construction and operation. What such a moratorium would have actually done is make those planning nuclear plants in the future a lot more hesitant, *vis a vis* an unpredictable Congress, to consider investing the millions of dollars it would take to bring them on line.

The second, less severe, moratorium that the Senate OKed does seem reasonable, however. Though the risks of an actual nuclear mishap appear infinitesimally small, the potential destruction of such an accident could be horrific. It only makes good sense to be prepared when the stakes are so high. Alabama already has such an evacuation plan drawn up and other states which do not should follow suit.

St. Louis Globe-Democrat

St. Louis, Mo., July 18, 1979

The U.S. Senate should be commended for demanding that all states have a federally-approved emergency evacuation plan in the event of a nuclear accident.

While the chances of a nuclear mishap are very slim, it is only good safety practice to have a statewide plan ready in the event that something does happen.

If this becomes law, it could shut down, as of June 1, 1980, all nuclear power plants for which a state has no federally-approved emergency evacuation program.

Theoretically 39 nuclear plants could be closed if all the states which still don't have such plans failed to comply. But, realistically, states and the government are not going to allow even a small percentage of the 39 to be closed. If ordered to meet this deadline, states would meet it and the federal governmemt would help them do it because the need for the power these plants provide is too essential to be cut off.

Some areas of the country get as much as 40 to 50 percent of their power from nuclear generating plants. Chicago, for example, gets about 50 percent of its power from atomic plants. It would be a catastrophe of first order to impose massive shutdowns of nuclear plants due to the heavy dependence on their electricity output.

Requiring states to have these evacuation plans makes sense. It is one of the beneficial results to come from the Three Mile Island accident. Another plus is the upgrading of safety equipment and procedures to prevent future accidents at nuclear plants.

These precautions, however, should not be used as an argument against expansion of nuclear power. They simply have made an industry that has achieved a phenomenal safety record over three decades even safer.

The Boston Globe

Boston, Mass., October 3, 1979

The Nuclear Regulatory Commission showed a welcome, if unsettling, concern about the evacuation plan that would be used in the event of a nuclear accident at the Pilgrim I plant in Plymouth. The plan devised by Boston Edison Co., operators of the Pilgrim plant, and by Plymouth officials had been considered a reasonably good one, and was one of the few in New England that had been tested to see if it would work at all.

But the hearings revealed that the evacuation plan relied far too heavily on a chain-of-command communications network which might not be fully activated before a dangerous release of radioactivity could have occurred. And the NRC seemed to find it hard to take seriously a plan which depended on evacuating 600 Plymouth residents to a high school in Plympton, which has not existed in some 60 years, and depended on the warning being given by a dozen police cruisers going up and down the streets with bullhorns.

The NRC, which made Plymouth its first stop in a cross-country investigation of evacuation plans, decided that the Pilgrim plan must be expanded to cover the 50,000 persons living within a 10-mile radius of the plant, not merely the 15,000 persons living within a five-mile radius. Moreover, a tightly scheduled deadline for giving a warning was set, with Edison now given 15 minutes to notify local public safety officials — significantly, at the first sign of trouble, not after trying unsuccessfully to control the problem themselves — and the local officials, in turn, have another 15 minutes to complete the warning of residents of their communities.

But beyond the specific improvements recommended at Plymouth — improvements that will be expensive whether local communities or Edison pick up the cost — the hearings provided a forum for the NRC to make it clear that it now takes seriously the risk of nuclear accidents and the need for realistic public safety precautions. "The accident at Three Mile Island was a slowly developing emergency," a member of the NRC task force said at Plymouth. "We can't predict how the next accident might come. It could happen very quickly." That concern must now be taken just as seriously by local public safety officials.

Long Island, N.Y., September 23, 1979

Thirty years into the atomic age, the Nuclear Regulatory Commission finally seems ready to consider the possibility of a major accident at a nuclear power plant. Its staff is now saying what many experts have proclaimed for years: that the potential for disaster exists, and it can't be ignored when reactor sites are chosen.

Until now the NRC and its predecessor, the Atomic Energy Commission, have always insisted that the possibility of a "China Syndrome"—a reactor-core meltdown combined with a failure of the containment vessel—was so remote it didn't even need to be taken into account in siting or emergency planning. As a result, many reactors in this country are located close to population centers or in areas that would be difficult to evacuate.

The areas around nuclear power plants where the NRC requires emergency and evacuation planning are called Low Population Zones. These LPZs are often quite small. At Indian Point north of New York City, for instance, the radius is only 0.6 miles, even though 4.3 million people live within 30 miles of the plant. The LPZ for the reactor at Shoreham is two miles, but LILCO is in fact planning for a 10-mile radius. In the past, some utilities have been allowed to shrink their LPZs if they added safety features to their reactors.

The underlying assumption was that the most serious accident possible would involve a loss of coolant but no dangerous radioactivity escaping beyond the LPZ. At Three Mile Island, where the LPZ has a two-mile radius, that appears to have been true: Although radioactive gases were carried far beyond the LPZ, apparently no one outside it was hurt.

Still, Three Mile Island shook the industry and the public, as well it should have. Nothing out of the ordinary—a plane crash, earthquake, sabotage—triggered the accident there. Sloppy design and operation were apparently to blame, and both will presumably be upgraded as a result. Emergency preparedness must be improved as well, and the Senate has passed a bill directing the NRC to prepare new rules covering population density and the siting of new reactors.

The NRC should adopt the siting recommendations of its staff, on the theory that late is better than never. But those recommendations are restricted to future sites; the status of reactors now operating and those still under construction is left up in the air. A report by a House Government Operations subcommittee released last summer went much further and merits the most careful consideration by the NRC and Congress.

The nation has been lucky so far, but luck is no substitute for meticulous planning to reduce the risks of nuclear power. With a de facto moratorium on new reactors, improving the safety of existing ones should be the NRC's top priority, even if that could mean a shutdown for those that threaten the greatest number of people.

The Oregonian

Portland, Ore., October 2, 1979

One of the lessons learned from the nuclear accident at Three Mile Island in Pennsylvania is that emergency response plans, no matter how precise and definitive, tend to unravel during a real crisis. That does not mean emergency plans are useless. But to be effective, the plans must be tested. Often. More full-blown, nuclear accident simulations should be staged at the 72 nuclear plants throughout the nation. But at what cost and who pays?

No one really knows how much it will cost a utility, state, county or city to test its emergency response plan to a nuclear accident. Taxpayers, not a utility's ratepayers, are the likely ones to foot most of the bill, either through funding for local government emergency services or through federal grants. The Price-Anderson Act, which insures a nuclear accident up to $560 million, likely will pay for the estimated $18.2 million it cost residents of Pennsylvania in evacuation expenses and lost wages. But no nuclear insurance covers a prescription for preventive medicine.

A major test of Oregon's emergency response plan for the Trojan nuclear plant near Rainier has been scheduled for October. PGE officials feel the cost of the exercise to the utility will be nominal since most of PGE's personnel will be diverted from normal working pursuits to participate without incurring overtime expense. The cost of equipment and one-day deployment of personnel by state and county agencies will not be known until the exercise is complete, if then.

The cost of testing a community's response capability to a potential nuclear accident is justifiable no matter who pays. The expenditure purchases protection of public health and welfare, not to mention peace of mind.

But the cost should be quantified, not hidden, and added as an item, however small, to the true cost of building future plants. Even if 15 or more simulated exercises were scheduled in the life of a 30-year plant, the cost of staging them probably would be small compared to the expensive evacuation foul-ups that occurred at Three Mile Island.

Nuclear power, produced at approximately 30 mills per kilowatt-hour, can withstand the added preventive medicine costs and remain attractive relative to other conventional thermal plant sources, which range from 45 kwh for coal to 80 kwh for oil-fired plants. The power produced at Trojan — currently at 19 mills per kilowatt-hour — would be a bargain even if emergency response plan exercises were financed out of power sale receipts.

The Providence Journal

Providence, R.I., November 21, 1979

In the continuing scrutiny of nuclear power plants that is being carried out by the Nuclear Regulatory Commission in the aftermath of the accident at Three Mile Island, one of the stickiest problems is going to be a decision on permitting continued operation of plants close to population centers.

This problem has been highlighted in controversy over the Indian Point reactors of Consolidated Edison Co. only 40 miles from Times Square in New York City. Two reactors there are in operation, a third awaits licensing. In the event of a serious accident, hundreds of thousands of persons might be exposed to radioactive fallout. But that is only part of the problem.

Even if the area of serious danger were confined to a 10-mile radius, a total population of 330,000 would be involved. And any attempt to evacuate that many people under emergency conditions would be subject to all kinds of foul-ups. It would be almost impossible to plan adequately for such an evacuation in advance.

The Presidential Commission on the Three Mile Island Accident recommended, among other things, that no new plants be licensed until the states where they were situated had drafted satisfactory plans for evacuation. That proviso, if followed to the letter by the NRC, will improve safety for new plants. But the really tough question is whether safety requires the shutdown of Indian Point and other sites with similar population density.

The NRC has identified seven other sites where the population within a 10-mile radius is more than 100,000. These sites must be reviewed on a case-by-case basis. To the task the NRC — or its successor agency — must bring a much tougher attitude than it has used in the past. Such a change in attitude was urged by the presidential commission. It was stressed again in a talk here last week by Victor Gilinsky, one of the five NRC members.

Commissioner Gilinsky conceded the difficulty in deciding on guidelines that will assure enough safety for all nuclear plants. He indicated that the NRC is struggling toward answers to the question, "How much safety is enough?" He insisted that the NRC itself and the whole nuclear utility industry must abide by much higher standards than in the past.

Whether enough safety features and backups can be provided at plants like those at Indian Point to offset the risk of accidents affecting not only the immediate area but the whole New York City area is still a matter for study and debate. The industry takes a more optimistic view. Consolidated Edison points to 17 years of operation at Indian Point without an accident affecting the public. Critics argue that where human operators are concerned there will always be accidents.

Closing of existing plants obviously would entail severe losses for the utilities. Since monopoly utilities cannot be allowed simply to go broke and liquidate the cost could be heavy for consumers in their areas, too. In short, the drastic action of declaring a plant unsafe, at least in its location, could work a substantial hardship on the very people it was intended to serve.

Thus the pressure is on to find ways to make these plants safe — that is, safe enough. There will always be some risk of a serious accident. But the extra burden that has been placed on both the utilities and the regulatory agency by the Three Mile Island accident should force everyone concerned to conduct his business with a whole new dimension of care.

There may be some utilities that still do not measure up to the standards that are now being developed for safe construction and operation of nuclear plants. Previously, the NRC has accepted the policy that any utility was entitled to build a nuclear plant, just as it had been entitled to build conventional plants. That policy will have to change. As Commissioner Gilinsky pointed out in his talk here, a certain size may be necessary for a utility to meet all the requirements of maintenance and operation and trained staff.

As Gov. Bruce Babbitt of Arizona said, in his personal addendum to the presidential commission's report, operation of nuclear plants may have to be concentrated in the hands of a few highly skilled and carefully monitored agencies, private or public, from which utilities could buy the electricity. This is a version of the suggestion put forth by Alvin Weinberg, former director of the Oak Ridge Laboratory, for nuclear "parks." But that concept, if it should be adopted, can apply only to future development. The emphasis in the next few months must be focused on tightening up the safety and the training of all personnel, from the bottom to the top, at the 72 existing plants.

Rocky Mountain News

Denver, Colo., July 10, 1979

Compared to Three Mile Island, Rocky Flats has a good plan in the event of a nuclear accident, according to Phil Stern, an engineer and investigator with the Boulder District Attorney's office. Stern, while working with a commission appointed by President Carter, helped analyze the emergency plan put into effect after the Three Mile Island nuclear power plant accident in Pennsylvania.

But Stern says he is skeptical of the Rocky Flats emergency plan all the same, and for a number of reasons. To mention just one, he says the plan recommends residents of Boulder County call 911 if they need transportation should an evacuation be ordered.

"What are we going to have, 100,000 calls pouring into the switchboard for people looking for a ride?" he asked.

That sounds like cause for skepticism.

Minneapolis Star and Tribune

Minneapolis, Minn., October 6, 1979

One lesson from the leak this week at NSP's Prairie Island nuclear power plant is clear: Minnesota's emergency plan is just that. It does not provide adequate public notification in a non-emergency. And a non-emergency is what happened when a tube ruptured in a steam generator and radioactive gas escaped into the atmosphere.

The emergency plan directs NSP to notify a small number of federal, state and local officials immediately after an accident at a nuclear plant and then to concentrate on the plant problem. Public agencies take over the remainder of the situation, with the Emergency Services division of the Minnesota Department of Public Safety in charge of communications.

The plan also sets out four categories of situations, from a full emergency requiring evacuation down to a "controlled situation" with no potential for danger. But only in the worst-situation category does the plan direct Emergency Services to alert the public, using the state emergency broadcasting system. The plan contains no specific guidelines or procedures for public notice in the three lesser categories.

After the accident, one NSP official said that "there was no reason to inform the public ... inform them of what?" And Gov. Al Quie's emergency staff coordinator said the governor's office knew within a half-hour of the radioactive release that it was not serious, but added that notification of residents near the plant "could create a panic."

Events show both men are wrong. Emergency Services or the governor's office should have told the public promptly that the leak was not serious. The lack of any clear, official information for about three hours after the rupture raised understandable and unnecessary fears in the immediate Prairie Island community, which knew from activity at the plant only that *something* was wrong. And the delay allowed time for incomplete information to spread to the general public on radio and press association wires.

Other evaluations are more thoughtful. Some NSP officials say that the plan clearly should be revised to include communication with the public in non-emergencies. And an Emergency Services employee suggests that special provisions be made for notifying Prairie Island residents, including the Prairie Island Sioux community, because of their proximity to the plant. Those are good suggestions.

Rockford Register Star

Rockford, Ill., July 27, 1981

Even when unwelcome, the predictable often can be comforting.

Thus there is something almost comforting in learning that Commonwealth Edison has missed the federal deadline for installing warning sirens at its Illinois nuclear power stations. Did anyone really expect the company to conform?

And there is something almost comforting in Commonwealth Edison's explanation: "We didn't think it was practical and waited" hoping the rule would be eased.

Now don't you find it comforting that Commonwealth Edison still knows how best to protect the country?

Of course, following the company's reasoning, a driver approaching a red light could decide it wasn't practical to stop and hope, instead, that the truck speeding through on the green would give way.

Rockford, Ill., December 22, 1982

Looking for safety assurance in an era when more and more residents must live close to nuclear power plants, we got instead a disquieting story of a new enterprise in the east.

Out in Harrisburg, Pa., they're selling pills that could prevent cancer of the thyroid among residents absorbing large amounts of radioactivity in a nuclear plant accident.

"In case of nuclear accident, listen to the radio and get your potassium iodide tablets ready," are the directions that go along with the pills.

It's an Rx we may not be ready for, but it signals we live in a nuclear-powered era and people are worried about it — no matter what the Nuclear Regulatory Commission (NRC) says or fails to say.

Folks in the Harrisburg area catch our attention. They're the same people who brought us the Three Mile Island nuclear plant accident story in 1979. We now live in the shadow of Commonwealth Edison's Byron Station cooling towers.

The idea of "nuke pills" is not new, nor are we endorsing it. But it's a concept worth noting in case those who build nuclear reactors and our NRC protectors begin thinking the public is indifferent to the possible dangers.

In the fall of 1981, Tennessee state officials distributed potassium iodide tablets to 6,000 families living within five miles of the twin-reactor Sequoyah plant near Chattanooga. The recipients were told to swallow the tablets only at the direction of the governor and state public health commissioner.

This was not reassuring to residents nor pleasing to officials in Alabama where a similar power plant is on line. Critics point out that a tablet is only good for one radioactive release. Another tablet would have to be taken for each subsequent release.

The Harrisburg entrepreneur apparently has tired of waiting for other states to join Tennessee in public purchase and distribution of the tablets. They will be sold like aspirin with child-proof caps.

We will take a pass on the pills, but we're ready to buy all the real nuclear plant safety assurance anybody has got to sell.

SYRACUSE HERALD-JOURNAL

Syracuse, N.Y., October 15, 1981

During August, the Federal Nuclear Regulatory Commission found that "the present state of planning" in nuclear plants in the state to be "adequate to carry out the responsibilities of the state and local government in the case of an accident."

Then, another outfit, the Federal Emergency Management Agency, tested state and county plans with a mock emergency exercise a month later.

The result?

So poor that the agency's regional chief announced the Nine Mile Point plant near Oswego, operated by Niagara Mohawk Power Corp., might be shut down.

If the county and state emergency services failed to minister to the ill and injured, failed to evacuate those exposed to a nuclear threat, contain the damage, even failed to threat, even failed to warn downwind neighborhoods, we must ask first:

Did those in charge know their assignments, the layout of the plant, the expected demand for transport and medical care? Had they been briefed? Had they been exposed to a scenario that duplicated on paper what happened at Three Mile Point in Pennsylvania?

The local and state disaster teams failed 55 points out of 110, according to the evaluating service.

Consequently, the FEMA regional director warned the plant may be closed.

We suggest examining the entire chain of command from the federal agencies, the nuclear regulatory commission and emergency management office, to the Oswego County ambulance service and ask:

Where did the chain of communication break down?

Perhaps the Federal Emergency Management Agency hasn't lived up to the assignment of "management" that graces its title.

If it hasn't, should we dispose of the entire agency?

Neither the shutdown threat nor its parallel makes sense.

Syracuse, N.Y., January 27, 1982

If one of the nuclear power plants nearby blew a gasket freeing radioactive particles into the atmosphere, we wouldn't know in this part of the state for lack of a warning system keyed to atmosphere measurements.

That's what the experts tell us.

Apparently Onondaga County is dependent on instruments in place at the nuclear plant sites and on the measuring-alerting equipment operated by Oswego County 10 miles downwind from the Nine Mile Point and Fizpatrick nuclear plants.

Shouldn't Syracuse and Onondaga County tie into those detecting systems?

As we recall the aftermath at Three Mile Island in Pennsylvania, the public wasn't given a true radiation picture for days.

The Hartford Courant

Hartford, Conn., November 8, 1981

The Tennessee Valley Authority is distributing a peculiar kind of morning-after pill to about 7,000 families living near the Sequoyah Nuclear Plant in Soddy-Daisy, Tenn.

The pill consists of potassium iodide and is intended to prevent thyroid cancer in those residents who might be irradiated by a nuclear accident before they have time to flee the area.

The TVA is acting quite responsibly in handing out the pills, but it provides a rather morbid reminder of Nevil Shute's novel from the 1950s Cold War era, "On the Beach."

In that book, the Australian government distributes suicide pills to its populace, which is slowly dying of radiation spreading to those parts of the globe not already blown away by a nuclear holocaust.

In this new Cold War-like era, with all the talk going around about the possibilities of nuclear conflicts, perhaps the governments of the world ought to take a cue from the TVA.

Perhaps they should distribute a morning-after pill to improve the odds of survival for those who wouldn't die outright in a nuclear war. The chances of such a war are probably higher than of an accident at the Sequoyah Nuclear Plant.

BUFFALO EVENING NEWS

Buffalo, N.Y., May 12, 1983

The federal Nuclear Regulatory Commission has ordered the closing of the Indian Point nuclear power plants, north of New York City, unless improvements are made in emergency evacuation procedures.

Evacuation plans are obviously important in the case of a nuclear complex that is only 35 miles from New York City. There are 288,000 people in a 10-mile zone surrounding the plants. On the other hand, while present evacuation plans have been found inadequate, there is no reason to believe that, with concentrated effort, the deficiencies in the plans cannot be eliminated. If they are not, then the NRC would have no alternative but to close the plants.

The evacuation regulations were formulated in the wake of the Three Mile Island nuclear accident in 1979. While there turned out to be no actual danger to residents, the accident dramatized the possible dangers in the unlikely event of a nuclear catastrophe. There was a feeling among the NRC commissioners that the new regulations must be enforced to prevent a general downgrading of emergency planning across the nation. Evacuation would be especially important in the case of Indian Point, since it is surrounded by a denser population than any other U.S. nuclear plant.

The problems, however, do not seem insuperable. Two criticisms made by a federal monitoring group were a lack of bus drivers and the failure of neighboring Rockland County to take part in an evacuation drill. One of the NRC commissioners said he believed it would be possible to develop a satisfactory evacuation plan. The owners of the plants, Consolidated Edison and the New York State Power Authority, plan to make improvements to avert a scheduled June 9 shutdown.

In commenting on a similar critical report last year, a Con Edison spokesman called it a "real Catch-22," saying that state and local governments are responsible for emergency planning. "We have no authority over these bodies," he said, "yet if their plans don't meet federal requirements, the continued operation of our plant could be jeopardized."

It would be absurd if the Indian Point plants were closed down simply because a nearby county refused to cooperate. Rockland County leaders are in favor of closing the plants. If they persist in their foot-dragging, then state officials could help to fill the gap in emergency planning.

While there must never be any compromise with safety standards there is cause for concern about the impact on available power supply and downstate utility bills if the plants are closed down. Indian Point is Con Ed's most efficient power plant and produces 9 percent of its total output. A State Power Authority spokesman said that the permanent shutdown of the two plants would mean added electrical costs of $18 billion in the period from 1984 to 1999.

For Western New York, the loss of this big block of power could mean increased demands from downstate areas to share our Niagara power or to "average" power rates across the state.

With the deadline now set before them by the NRC, the utilities and state and local officials must work together to devise a plan that would assure safe operation of the plant throughout its useful life.

THE ARIZONA REPUBLIC

Phoenix, Ariz., April 14, 1983

DEVOUT opponents of commercial nuclear power plants continue to distort facts of the accident at Three Mile Island.

While the Three Mile Island accident may well have frayed the nerves of residents near the plant and revealed incompetence on the part of plant operators, the event came nowhere near the disaster suggested by nuclear foes.

However, the accident did have beneficial results on the nuclear industry and the public at large.

From it has come a new awareness by the nuclear power industry of the responsibility for training plant operators, planning to cope with accidents that might occur, and developing a more sophisticated and comprehensive system for alerting government and community in the event of an accident.

Nowhere have these improvements been put into place more conspicuously than in Arizona, where a new $700,000 alert center will be completed this week in Phoenix.

The alert center will monitor the operations of the Palo Verde Nuclear Generating Plant 50 miles west of Phoenix.

Were an accident of some sort to occur at Palo Verde, the alert center, located at the Papago Buttes, would be responsible for notifying government agencies and the media, and for implementing disaster precautions.

Its existence undoubtedly is comforting to those who live in constant fear of ubiquitous contamination from nuclear power.

However, much more comforting is the indisputable fact that commercial nuclear power plants — operating in the United States for some three decades — have been safe, dependable sources of electricity, and not the fearsome technological Frankensteins that nuclear opponents have been predicting they would be.

Los Angeles Times

Los Angeles, Calif., June 12, 1983

Members of the federal Nuclear Regulatory Commission often wonder aloud why so many Americans worry that they are soft on safety. Last week they voted themselves an answer.

By a margin of one "reluctant" vote, the five-member commission agreed to allow a nuclear power plant north of New York City to keep operating without an approved plan for evacuating nearby residents in an emergency. No other nuclear plant operates in such a densely populated area.

Only a month ago the commissioners warned the operators of the plant that if a proper evacuation plan were not in place by June 9 they would order the plant closed until the operators produced a system for moving residents out of harm's way.

Nothing significant changed after the warning. Regulations that clearly demand such planning were not amended. New York state government officials talk about taking a more active role in shaping a plan, but that is not the same as having a plan. Shutting down the plant would be very costly to its operators and to consumers, but that had not changed, either. The commission knew the potential cost when it issued the ultimatum.

There is only one way to look at the action. In nuclear parlance, it is a regulatory equivalent of the most serious failure that a power plant can suffer—a spinal meltdown. Are there any further questions?

Post-Tribune

Guarding Your Interests Daily

Gary, Ind., May 16, 1983

A more immediate evacuation question threatens the future of nuclear power across the country. The federal requirement of an emergency plan for each nuclear plant area is turning into an anti-nuclear weapon.

Many local governments won't develop or participate in emergency preparedness plans. The price of that is clear in the case of New York's Indian Point power facility, which could be shut down unless the local government officials come up with an adequate plan.

At 37 of the country's 53 sites for nuclear power plants, local governments have not met the federal requirements. Obviously, a plan to evacuate people in case of a nuclear attack is not the same as moving them away when there is a serious accident at a nuclear power plant. Secretary of Energy Donald Hodel is right, the breakdown of federal, state and local governments and private utility cooperation "has become a serious national issue."

Until the 1973 Three Mile Island accident, little attention was given to emergency planning. That incident revealed an abysmal ignorance not only about nuclear plants but about moving people, and about telling them the facts.

The reaction to that scare brought federal orders to create plans on evacuation, communications, warning systems, and several other concerns. The Nuclear Regulatory Commission relies on the Federal Emergency Management Agency's findings in deciding whether to license a plant. Many utilities are caught in a squeeze not of their own making.

Some members of Congress have urged the regulatory commission to find ways to issue an operating license regardless of local participation. Don't do it. The rules are being used in some cases to obstruct plant operation, but the basic fact remains that without reasonably good emergency plans, the power plants should not be operating.

The Philadelphia Inquirer

Philadelphia, Pa., May 13, 1983

Last week the Nuclear Regulatory Commission said the atomic plants at Indian Point north of New York City might have to be shut down. There's no plan to get people living nearby out in case of emergency, the NRC said. Close to 300,000 people live within 10 miles of the plants. Back in 1979, one of the NRC's own, the director of state programs, Robert Ryan, said it was "insane" to build the things there in the first place. But there they are.

The folks living around Indian Point are not exactly cooperating with the directive to get an evacuation plan approved. They take something of the attitude that many cities have taken on the evacuation plans for nuclear attack: If something goes haywire, the rulebook is out the window. If the NRC takes away the plant's operating license, the Indian Point neighborhood — Westchester and Rockland and adjoining counties — is not exactly going to go into mourning.

In Lower Alloways Creek, N.J., scene of the infamous Salem nuclear plant, residents are being cooperative, but the results leave a lot to be desired. The Federal Emergency Management Agency (FEMA) said the disaster drill around the plant exposed a general sense of confusion and lack of training. That's unfortunate because Salem was just hit with the largest fine in NRC history for not paying attention to safety.

It also led the list of increasing examples of "insider sabotage" that the NRC issued the other day. One year ago this month, the NRC report said, a valve feeding coolant to the plant was "deliberately mispositioned," apparently during a labor dispute.

Absent risks of accident or sabotage, evacuation plans might be academic. But as long as those risks remain, the question is whether evacuation plans can be made to measure up.

ST. LOUIS POST-DISPATCH

St. Louis, Mo., June 17, 1983

The federal Nuclear Regulatory Commission has made a sham of its own rules by its decision the other day to allow the Indian Point nuclear power plants in New York to continue operating despite the absence of emergency plans to protect area residents in case of an accident. After the 1979 accident in a nuclear power plant at Three Mile Island in Pennsylvania, the NRC adopted a rule stipulating that it would not allow reactors to continue operating without "reasonable assurance that adequate protective measures can and will be taken" in case of an accident. Nearly every plant in the country missed the April 1, 1981, deadline for compliance with the rule. And scores of plants, including Indian Point, were threatened with fines or shutdowns.

As the first nuclear plant site in the country to be given two four-month deadlines for compliance with the NRC rule, Indian Point became a symbol for whether the regulatory agency meant business. When a third deadline passed earlier this month and the NRC voted 3-2 to ignore its own requirements, the agency became in effect an apologist for inaction on safety. The NRC majority's excuse was that New York State and the plant operators were developing an emergency plan. But the government of Rockland County, where Indian Point is located, is not satisfied.

The Three Mile Island accident, called the worst in the history of the nuclear power industry, was supposed to have galvanized the country to action on evacuation and other plans to protect the public. If the NRC, after more than two years of delay, is still unwilling to enforce its own shutdown rule against a delinquent nuclear installation in a heavily populated area (only 35 miles north of midtown Manhattan), that suggests that the lesson supposedly taught by Three Mile Island meant nothing.

THE LOUISVILLE TIMES

Louisville, Ky., May 12, 1983

"Hiroshima-on-the-Hudson" was the name that government engineers once playfully conferred on the town of Buchanan, N. Y., home of the celebrated, and possibly doomed, Indian Point nuclear power station.

This was not a reference to the power plant's explosive potential, but rather to its unfortunate, some say "insane," location 35 miles from Times Square in New York City. A serious accident could endanger millions of people in the surrounding metropolitan area.

However, the old Atomic Energy Commission, which approved the site, was never known for its sentimentality. Those were the days when the federal government was going all out to make nuclear power appear competitive with other energy sources. The regulators talked about safety, but apparently gave greater weight to the argument that power stations in built-up areas would avoid high transmission costs.

Public relations also figured in the decision to build near population centers. Nuclear advocates were determined to sustain the fiction that reactors were safe in any location. And no one was eager to suggest that rural folks should accept the risk of radiation exposure so their city cousins could have lots of electricity.

Congress did its part by passing a law limiting the accident liability of nuclear utilities to $560 million. Urban sites would have been unthinkable if power plant owners had to buy adequate insurance coverage.

The upshot is that 18 years later the owners of two operating nuclear power plants at Indian Point are beating their brains out trying to find a way to evacuate the nearly 300,000 people who live within a 10-mile radius. The Nuclear Regulatory Commission says it will shut down the plants unless an approved plan is ready by June 9.

The action was a surprisingly gutsy one for an agency whose members are inclined to think evacuation plans deserve a low priority. The commission correctly realized its credibility would plummet if it failed to enforce its own rule, which went into effect after the Three Mile Island fiasco.

Nuclear promoters have little to rejoice about. For one thing, the industry will be more vulnerable to local officials who, as in the Indian Point case, wage guerrilla warfare by refusing to take part in emergency planning. The next crisis looms on Long Island, where a completed power plant at Shoreham could be denied a license because of the near impossibility of evacuating thousands of residents.

New York state, possibly with federal help, will no doubt come to the rescue rather than allow the abandonment of reactors designed to operate into the next century. But public policy makers and nuclear entrepreneurs should not miss a crucial point: The industry's dim prospects today are in part a result of government decisions to shield it for so long from economic, scientific and political realities.

While the "Hiroshima" tag overstates New York's dilemma, residents of the area have ample reason to take the risks seriously. The first Indian Point plant was closed permanently years ago because of defective plumbing. The second has had various difficulties, including a major leak that forced a six-month shutdown. The third has been out of operation for more than a year because of corroded steam generator tubes.

And just last week, the NRC fined a New Jersey utility $850,000 because sloppy management led to the failure of the automatic shutdown system at a plant 20 miles from Wilmington, Del.

The nuclear wizards have, in short, been their own worst enemies. Their bumbling makes the NRC's decision to protect us rather than them all the more welcome — and necessary.

The Star-Ledger

Newark, N.J., June 17, 1983

The federal Nuclear Regulatory Commission has been facing some hard choices of late. On the one hand, it is under considerable pressure to enforce regulations making sure that proper safety precautions are taken at all nuclear plants. On the other hand, there is the practical need of drafting regulations that can reasonably be complied with.

Of all these regulations, the most difficult to enforce are those needed for emergency planning. Ever since the Three Mile Island accident, there has been concern that in the event of a more serious accident in which it was necessary to move large numbers of people from the area, poor planning would make this impossible.

And so a rule was adopted requiring emergency evacuation plans. Despite the rule, these emergency plans have been slow in developing. A major problem is that local governments in the area of nuclear plants have often failed to cooperate.

The matter came to a head in New York State when the federal agency gave the operators of two nuclear plants in suburban New York counties—Con Edison and the New York Power Co.—a deadline to produce a satisfactory emergency plan—or face a shutdown of the facilities.

The deadline was not met, but the Regulatory Commission still decided, by a 3-2 vote, to allow the plants to continue to operate.

The split on the commission was deep and the commentary bitter. The majority contended that it would be a "capricious violation of our regulatory responsibilities" to close the plants down under these circumstances. The minority contended the majority had made "a mockery of our emergency planning regulations."

Both positions are reasonable and the matter is a serious one, for it is likely to occur again in other areas. Perhaps the paramount issue should be one of public health and safety. In this case, there was no question of any health or safety defects at the nuclear plants. Under those circumstances, the majority decision was a pragmatic one in view of the extremely difficult circumstances.

AKRON BEACON JOURNAL

Akron, Ohio, May 23, 1983

NUCLEAR POWER has been taking it on the chin of late.

In April, it was the Supreme Court decision upholding a California ban on future nuclear power plants until a safe method of nuclear waste disposal is in place.

This month, the Nuclear Regulatory Commission — in surprisingly tough talk considering its past record — said it would shut down two New York plants unless an acceptable emergency evacuation plan is developed by June 9.

The two plants in question, just north of New York City, are operating even though their owners failed twice to meet previous deadlines for formulating workable plans.

The chief stumbling block, in New York's case, has been one county's refusal to participate in evacuation drills. County officials maintain there is no quick way to evacuate thousands of people living in the shadow of the reactors.

This is not just New York's problem. Thirty-seven other nuclear power plants now operating or soon to start lack evacuation plans approved by the federal government. That fact has not been ignored by nuclear power opponents, who are using the federal regulation to challenge the role of nuclear power in this country.

Nuclear plant operators were ordered to develop emergency plans for safely removing people after the 1979 accident at Three Mile Island. In some cases, local governments consistently refuse to cooperate.

Examples reported by the New York Times include a county in Kentucky that insisted the state build a sought-after road before it would participate. More typical demands involve money to pay for evacuation, medical and communications equipment.

There is a breakdown in communications, a troublesome one, between power plant operators and government officials responsible for the safety of citizens. It threatens to turn billion-dollar plants into useless relics, in some cases before the "on" switch is even flipped.

Energy Secretary Donald Hodel has appointed a committee to recommend federal action to clear up this mess. Since states don't seem to be successful at solving the spreading dilemma, it will probably take the heavy hand of Washington to knock some heads together.

Emergency evacuation plans are vital, as Three Mile Island grimly proved. They must be realistic and effective, not some impressive scheme on paper that is next to impossible to implement.

Nuclear power is a complex issue. Every level of government should have some say about it. But local officials, by themselves, must not be allowed to drown out all others.

Transcripts, Kemeny Panel Report Spur Criticism of NRC

In the aftermath of the Three Mile Island incident, two occurrences served to further undermine public confidence in the Nuclear Regulatory Commission and in the safety procedures at nuclear power plants. Transcripts of the NRC's closed-door meetings held during the Three Mile Island crisis were made public by a House committee April 12. The records of a fairly continuous meeting of the commissioners from Friday morning, March 30 through Wednesday, April 4, showed that the commissioners and the NRC's technical staff were at times highly disturbed by receiving incomplete information from the accident site. NRC Chairman Joseph M. Hendrie said March 30 of the crisis facing both him and Gov. Richard Thornburgh: "We are operating almost totally in the blind; his information is ambiguous, mine is nonexistent and—I don't know—it's like a couple of blind men staggering around making decisions." The officials were unsure about whether to recommend the evacuation of people living around the Three Mile Island plant.

Later in 1979, the President's Commission on the Accident at Three Mile Island urged President Carter to abolish the Nuclear Regulatory Commission altogether, and create a new executive agency charged with policing the nuclear power industry. The 12-member commission, headed by Dartmouth College President John G. Kemeny, presented Carter Oct. 30 with a 179-page report, which said that the NRC was so preoccupied with the licensing of atomic energy plants that it had not given "primary consideration to overall safety issues." Kemeny stated: "The NRC is beset by an overconcentration on regulations, an attitude that tends to equate the meeting of regulations with safety." The Kemeny Commission did not recommend any delay or halt to nuclear construction, but did say that fundamental changes had to occur in the way nuclear reactors were built, operated and regulated if the risks of nuclear power were "to be kept within tolerable limits."

The Washington Star

Washington, D.C., April 15, 1979

There has been a heavy drain, in recent years, on public confidence in the expertise and credibility of public officials; and the doubts may be enlarged by 800 pages of transcribed Nuclear Regulatory Commission proceedings in the Three Mile Island nuclear crisis.

They show that the general impression of confusion was not misleading. The accident was, in the jargon of one NRC expert, a "failure mode the likes of which (has) never been analyzed" and its dangers were neither easily nor quickly gauged. The transcripts reveal disquieting lapses of liaison between the NRC and its agents at the scene, and the operators of the Three Mile Island plant. They show that evacuation plans were sketchy and out of phase with the needs of state officials, that dangerous events, such as Wednesday's "funny blip" (a possible hydrogen gas explosion within the reactor vessel) went unreported for days to the NRC. They show the NRC uncertain about whether to share deepening apprehensions with the public and Pennsylvania officials. Finally, for whatever it's worth, they show the White House and the NRC trying to manage the news about the crisis: often for good reason.

Is it really surprising? One might have expected more readiness to cope on the NRC's part, yet few emergencies are handled with the cool dispatch and calculating mastery of legend.

These transcripts are disturbing, but they do not warrant facile judgment on the NRC's apparent lack of foresight and mastery. "No plant," said the NRC's director of reactor safety at one point, "has ever been in this condition . . . been tested in this condition . . . been analyzed in this condition." The commissioners were dealing with unknowns, even unknowables, and that fact must temper judgment.

It might indeed be easier on our nerves *not* to obtain, so soon after a potentially lethal nuclear accident, so complete a picture of the groping "blindness" (to use Chairman Joseph Hendrie's own term) of responsible officials who are assumed to be expert.

But the NRC transcripts are valet's-eye history, and valet's-eye history is often unheroic. Anyone who ever watched people wrestling under pressure with an unanticipated crisis hardly needs the NRC transcripts to discover that these are not the best of times.

Most societies, at most times, insist for that very reason on concealing, at least for a time, the untidy inside lore of "crisis management." But this society has chosen instead to broadcast the *arcana* of government. We tend to rush, with the barest reflection, towards a tell-it-all-and-tell-it-now approach to official actions and deliberations, of which this is only one example. We deluge ourselves with information and impressions sometimes difficult to fit into proper perspective.

The resulting grind upon our nerves and our confidence in those who govern us or guard us against erratic technologies is not yet fully measured. We assume that all disclosure, however premature, is a good thing; and perhaps it is. But whether the effect is eventually helpful or immediately demoralizing always depend on the maturity with which we assess our valet's-eye history and what policy conclusions we draw.

These bold visitations to the tree of knowledge are in our tradition. But our capacity to react with discrimination to raw information, and to act on it wisely, is not always as fully developed as our capacity for getting it.

Towards the end of the Three Mile Island crisis, NRC Chairman Hendrie, musing on a White House press briefing, spoke of his problem with "creative interpretation" of the crisis. Indeed, that has been a problem for everyone. But creative interpretation — though certainly not in the negative sense Mr. Hendrie probably had in mind — is very much needed.

The Hartford Courant

Hartford, Conn., April 16, 1979

No wonder the Nuclear Regulatory Commission resisted congressional efforts to release transcripts of secret meetings on the crisis at Three Mile Island.

The regulators apparently wanted to protect themselves from embarrassment, at best, and indications of nonfeasance, at worst. They turned over the transcripts to a House subcommittee only under threat of a subpoena.

Portions of the NRC secret meetings deal with the problem of communicating with the public through the press. Joseph Hendrie comes through as an extremely careless, or amazingly uninformed, commission chairman. "Which amendment is it that guarantees the freedom of the press? Well, I'm against it," he says at one point.

At another point, Mr. Hendrie notes, "We'll probably enter — what is it? — four or five months of over-regulation of the nuclear industry." In a more serious moment, the chairman admits,"It seems to me I have got to call the governor. We are operating almost totally in the blind. His information is ambiguous. Mine is nonexistent and — I don't know. It's like a couple of blind men staggering around making decisions."

Mr. Hendrie's imagery changes when he tries to humor himself and his colleagues by wondering aloud whether they should be called "a pride of commissioners" or "a gaggle of commissioners." All this, during a time when a meltdown at the Three Mile Island nuclear power plant was considered a possibility, when radioactivity was being released in the atmosphere, when thousands of people were fleeing the area.

The most revealing aspect of the transcripts is not the contempt shown by Mr. Hendrie and his colleagues for open meetings and for the need to inform people. It is the alarming ignorance of what was happening at the plant site during the critical days. Something is terribly wrong if the Nuclear Regulatory Commission finds itself "almost totally in the blind" during the most serious nuclear accident ever reported.

Mr. Hendrie talks about entering a period of "over-regulation" of the nuclear industry. There is no need for over-regulation, whatever that means. There is need for adequate regulation, so that the regulators do not act like "blind men staggering around making decisions," in case of another accident.

THE BLADE

Toledo, Ohio, April 16, 1979

THERE is no denying the dismaying impact of transcripts depicting rank confusion among the Nuclear Regulatory Commission and its staff after the accident at the Three Mile Island power plant. But unnerving as the revelation may be for the public and embarrassing as it certainly is for the agency, release of the documents is actually a blessing in disguise.

Diehard foes of nuclear power can, of course, be expected to have a field day with the records of the NRC's closed-door sessions in the aftermath of the incident at the Pennsylvania plant. The more than 700 pages of taped discussions and informal comments will no doubt become a prime exhibit in the case the opponents try to build against the nuclear-power industry generally and the NRC in particular.

Viewed in perspective, however, the difficulties the agency had in coming to grips with the situation — determining what had happened and deciding how to deal with it — constitute a valuable lesson. It is, indeed, one of those lessons which we said earlier stand to be learned from this country's most serious accident in more than 25 years of experience with commercial nuclear power generation.

While the NRC cannot be excused for glaring flaws in its ability to mobilize its forces as quickly and to address its immediate responsibilities as effectively as it should have, there are some fundamental factors that must be underscored. The most important is that, despite the picture of total ineptitude suggested by the transcripts, there obviously were persons available who had expertise in how to proceed. That is clear from the fact that the most threatening problems within the plant were dealt with and the crisis was resolved without apparent harm to the public or the environment or even a need for mass evacuation from the area in the interim.

Another point to be borne in mind — an explanation if not an excuse — is that the NRC's failure to respond as instantly and efficiently as it should have may simply lie in the fact that it has not had to face such a situation before. The quarter-century record of public safety in U.S. nuclear power generation must surely be accepted as a sign that something is being done right in the design, construction, and regulation of such plants.

It is not necessary and it would not be sensible to shut down the nuclear-power industry or to halt its continued development. Some parts of the country already depend upon it for most of their electric power, and it will become increasingly important in the production of the energy the nation needs.

What is absolutely necessary is that the agency charged with overseeing the industry improve its preparedness for handling emergencies as well as review and greatly tighten its standards both for the plants themselves and for its own enforcement of those standards. To that end, the record of the NRC's performance in the Three Mile Island incident — and the publication of that record — should, in the long run, be highly beneficial.

The Salt Lake Tribune

Salt Lake City, Utah, April 18, 1979

Don't be too hasty in condemning members of the Nuclear Regulatory Commission on the basis of published transcripts of NRC meetings after the Three Mile Island nuclear reactor accident.

There seems little doubt that the five commissioners were pretty much in the dark about what was taking place at the scene near Harrisburg, Pa. But so were officials on the spot. The commissioners sitting in Washington had to depend on reports which were far from precise and sometimes contradictory.

Readers who recall the Watergate tape transcripts should know that the published versions were sometimes incomplete and misleading. A transcript simply cannot convey the whole picture.

Portions of transcripts carried in news stories reflect the judgment of individual newsmen making the choice. As such they can be structured to present a picture ranging from utter confusion to mild bewilderment. Most accounts, we suspect, emphasized the confusion.

It is reasonable enough to expect that top people who regulate operation of potentially dangerous nuclear reactors will take charge when something goes wrong in a big way. But that is not necessarily how such regulatory bodies are set up. Did members of the National Transportation Safety Board, for example, have any clearer picture of what caused that spectacular midair collision over San Diego last fall? A transcript of their after the crash deliberations might show similar befuddlement and want of accurate information.

The shocking thing about the NRC's fuzzy performance is that it suggests the entire nuclear regulatory system is equally ineffective. That may indeed prove to be the case. But it cannot be assumed solely on the basis of tranxcripts of NRC meetings in the first days after the Three Mile Island accident.

THE RICHMOND NEWS LEADER

Richmond, Va., May 15, 1979

The incident at the Three Mile Island nuclear reactor in Pennsylvania now rates as old news. An emergency occurred, it was brought under control, and no one died or was injured. But, in the aftermath of the incident, the public is receiving a chilling report on the extent of ineptitude and incompetence within the federal Nuclear Regulatory Commission.

The NRC, charged with regulating the nation's nuclear industry, revealed its inability to cope with a nuclear emergency in transcribed meetings that began within hours of the Three Mile Island episode. The proceedings disclose an obsession with public image, and with "modes" of handling both the emergency and the press — tasks at which the NRC was woefully inadequate.

The taped discussions show that the NRC did not understand what was happening at Three Mile Island. Commission members and their advisers seriously discussed several times ways of causing other accidents at the plant "to get into a mode for which all these systems were designed and we could cope with.... What we need at the moment is a good pipe break."

Later: "We think we've got a way we can break the control rod. . . . We've got people looking at the way to fail a control rod drive on purpose and provide a crack. Unfortunately, the only way you can do that we know is to heat it; in other words, you want to start a fire." The NRC eventually rejected that possibility and devoted much time to its relations with the press.

During the final hours of the emergency, however, the NRC reported the danger of explosion within the reactor as a result of the formation of a hydrogen bubble. That possibility dominated headlines and added immeasurably to the climate of dread and fear that gripped the Three Mile Island area. The prospects seemed to suggest the formation of a mushroom cloud at any moment.

Now, it turns out, the possibility of an explosion *never existed*. "The amount of concern was entirely undeserved," an NRC spokesman says. "There never was any danger of a hydrogen explosion in that bubble. It was a regrettable error." The NRC asked numerous nuclear experts how much oxygen would be needed to generate an explosion, but it took 36 hours for the answer to sink in: There would be no oxygen generated at the Three Mile Island plant. The error never was explained publicly: "I guess we have not been in a mode of [public relations]," an NRC official explains.

Apparently, the NRC is not in much of a mode for any useful purpose. In dealing with the Three Mile Island incident, it has left an indelible image of bumbling incompetence and inexplicable ignorance of the technicalities the Commission is supposed to understand. The emergency at Three Mile Island underscored the fact that there is no risk-free means of producing energy. Yet nuclear power is essential to meet the nation's future needs; its promise far outweighs the negligible risks incurred. But the NRC evidently is not up to its task.

The Star-Ledger

Newark, N.J., November 5, 1979

A quarter of a century after the United States began an extensive commitment to harness the power of the atom for the purpose of generating electricity there remain critical aspects of safety.

This highly negative aspect is offset by the prominent positive role that nuclear reactors now have in generating a significant share of energy required in an industrialized society.

These are immutable facts that must be responsibly acknowledged in the wake of the unsettling findings by the President's Commission on the Accident at Three Mile Island.

If the nuclear mishap at the Pennsylvania generating plant does nothing else, it will serve as an everlasting reminder of the need to exercise the greatest possible technical and human precautions in minimizing the possibility of recurrence.

Complacency no longer can be tolerated. Fortunately, the radiation that escaped at Three Mile Island will have a "negligible effect" on public health, but the catastrophic potential is sufficiently harrowing to underline the importance of wide-ranging remedial action that the executive and legislative branches must undertake after a comprehensive review of the panel's recommendations.

The grim conclusions by the commission rule out absolute guarantees — even with the most stringent preventive measures — that a failsafe level can be achieved, completely eliminating the possibility of a serious nuclear accident in the future.

Short of that full assurance, it becomes imperative that fundamental changes in the organization, procedures and practices of the nuclear industry are instituted in an effort to prevent a repetition of Three Mile Island.

What began as a minor affair as a result of equipment breakdown was transformed into a terrifying situation by human error, haphazard emergency instructions and an inadequately designed control room. These cumulative defects made an accident like Three Mile Island "eventually inevitable," the panel found.

While it raised troubling questions about the future of the nuclear industry, it was apparent that the presidential commission did not have all the answers. It did not appear to have a comprehensive grasp on whether the conditions it reviewed at various facilities were common in the nuclear generating sector.

The panel underlined the urgency for stronger regulations, an upgrading in the quality of work practices and personnel and a reorganization of the Nuclear Regulatory Commission, an agency that was found critically wanting in the Three Mile Island emergency.

The future of nuclear energy ultimately must be resolved on a delicate balance of our needs in practical economic terms and the public welfare. Essentially, this is a political decision, albeit an extremely difficult one. The country, meanwhile, has to accept the bleak realities of Three Mile Island . . . and more important, the urgency for sweeping reforms and changes to minimize risks.

Three Mile Island is an apocalyptic portent that must be fully addressed by the public and private sectors.

The Globe and Mail

Toronto, Ont., November 2, 1979

The report of President Jimmy Carter's commission of inquiry into the accident at the Three Mile Island nuclear plant was anything but a whitewash.

Among contributing causes of the nightmare it found ignorance, incompetence, undermanning, confusion, operating errors, faulty design, and blindness to what should have been recognized as warning signals. The company that ran the plant wasn't up to the job. The Nuclear Regulatory Commission didn't know what it was doing. Equipment failed. And improperly trained operators made one error after another.

It all adds up to a dismal picture and a heavy and urgent agenda for radical improvement in the design, construction, operation and supervision of nuclear power plants in the United States.

But by the same token it also adds up to a very solid and persuasive set of credentials for going ahead with the development of nuclear generating capacity, provided it is done with the necessary competence, care and concern for safety. The Three Mile Island operation made a hash of the application of nuclear energy, the commission found. And it was the hash, not the use of nuclear energy in itself, that was at fault.

So the commission's report, while calling for change at almost every stage of the development and operating process, from the drawing board to communication with the public, also recommended that the development of nuclear plants be continued on schedule.

And that should be encouraging to all who are open to reason and to factual evidence on the issue of nuclear power, to those who fear that necessary power development will be halted by opposition based on fear and superstition rather than on hard fact, and to those who have been alarmed to find that reasonable questions about safety and emergency procedures are too often brushed off with glib, unfounded assurances that nothing can possibly go wrong.

Things can go wrong in every kind of power generation. Fossil fuels can explode. Hydro power dams can collapse. Coal mines can cave in. Coal smoke can poison the atmosphere. From none of these sources of energy can we expect a guarantee against failure or error. What we can expect is that everything humanly possible be done to prevent these errors and failures and to ensure that when they do occur they do not lead on to disaster.

THE CHRISTIAN SCIENCE MONITOR

Boston, Mass., November 1, 1979

One overriding message emerges from the Kemeny commission report on the nuclear accident at Three Mile Island. The United States, if it continues to pursue the development of nuclear power, cannot do so entirely without risk and, if this risk is to be kept to "tolerable levels," there have to be major institutional changes in the nuclear energy industry. President Carter and Congress will need to follow through swiftly on this broad recommendation if they are to assure the American people that present nuclear power generation, let alone expansion of it, is desirable. Failure to do so can only fuel the efforts of those scientists, environmentalists, and others who would call a halt to all nuclear development. The economic cost of this could prove high.

It is not only the construction of new power plants which must be addressed — and here Americans will be sobered by how close the commission came to recommending an outright moratorium. The public is also bound to ask questions about the 70 some nuclear plants now in operation. Are the safety risks at these within tolerable limits? It would be overreacting to suggest that all these facilities be shut down until this can be thoroughly ascertained, a drastic move which could entail unacceptable economic dislocations and hardship. But it seems to us the US Government could institute a case-by-case review of existing plants — their design, siting, training programs, emergency procedures, management methods — and relicense them if satisfactory. In fact there perhaps ought to be periodic recertification of nuclear plants to assure fullest compliance with upgraded safety standards.

As for the Kemeny commission's broad recommendations, these ought to be carefully studied and pursued. It makes sense, for instance, to grant operating licenses for nuclear reactors only after the federal government has approved state and local emergency plans, and after the plants have been reviewed for safety improvements, competence of their training program, siting away from population centers, and so on. There also can be no quarrel with the recommendation that reactor operators be given much more rigorous training than they have to date. Perhaps the most significant finding of the commission's probe is that the major cause of the Three Mile Island accident was not the deficiencies in equipment design but the failures of the people involved — the way plant operators and later the Nuclear Regulatory Commission reacted. It is clear that, to bring about the radical change in attitudes toward nuclear safety which the commission says is needed, there will have to be a substantially higher level of education.

Major changes will also have to take place at the NRC but whether that agency should actually be abolished and replaced by an independent agency as the presidential commission recommends is open to debate. Following the NRC's poor handling of the Three Mile Island crisis, a good case can be made that new leadership is needed at the NRC. A bolstering of its inspection and enforcement branches and some structural changes also may be warranted. But the NRC itself is the product of a reorganization and, as we have seen from the establishment of the new Department of Energy, mere organizational changes do not automatically promise efficiency and effectiveness. This will have to be looked at carefully.

But looked at it must be. It will not please the President or the Congress that, amid all the other burning issues which must be dealt with, and dealt with in an increasingly politicized atmosphere, the question of nuclear safety is now added to the agenda. This is a volatile issue. But inasmuch as the United States now depends to a degree on nuclear power to meet its energy needs, it would be disserving the national interest not to give the Kemeny commission report the highest priority.

FORT WORTH STAR-TELEGRAM

Fort Worth, Texas, November 2, 1979

The president's commission on Three Mile Island did not recommend, as some had hoped, a moratorium on nuclear power plant construction but, if its recommendations are adopted, the effect may be the same as a moratorium.

Two major points made by the commission were that the Nuclear Regulatory Commission was lax on enforcement or nuclear power regulations and that the regulations were complex and confusing; and that the nuclear power industry should concentrate on the training of personnel to cope with emergencies. The commission said the industry has tended to concentrate on equipment.

To that respect, the commission said: "We are convinced that if the only problems were equipment problems, this presidential commission would never have been created. The equipment was sufficiently good that, except for human failures, the major accident at Three Mile Island would have been a minor incident.

The commission recommended drastic changes in the composition of the NCR and a concentration by the power industry on safety training for personnel.

Thus, if the federal regulations are tougher and the safety precautions and training for power industry personnel are mandated, it may become too costly to construct a nuclear power facility. The cost already is close to being prohibitive.

And, on the other side of the issue, if no attention is paid to the presidential commission's report, we have this warning from the commission: "We are convinced that, unless portions of the industry and its regulatory agency undergo fundamental changes, they will over time totally destroy public confidence and, hence, they will be responsible for the elimination of nuclear power as a viable source of energy."

Either way, the future does not look bright for the nuclear power industry. However, the nation needs alternative energy sources and nuclear power is one. At present nuclear plants produce 12.5 percent of the nation's electricity. It is projected that by 2000 nuclear power will produce 27.8 percent. If we don't have nuclear power as a supplement, we must have something else. While we debate alternative power sources, we continue to increase oil consumption and that is not economically or politically feasible.

It is well past the time for this nation to get off dead center and do something about its energy problem.

"It'sh (hic!) time t'take a s'hobering look at th' whole N-plant messh!"

RENO EVENING GAZETTE

Reno, Nev., November 19, 1979

The Kemeny Report is a scathing indictment of the nuclear power industry and its regulators. But at the same time it is a most positive report for those who believe in nuclear power.

It is positive because it finds fault mostly with men, and very little with machinery. If the hardware had been found to have serious defects, nuclear power would have been in great jeopardy. Years might have been required to restructure this complex equipment. But one can restructure human processes fairly quickly, if the desire is there.

The desire should be there, because the Iranian crisis has told us again, as the gas lines did this spring, that the United States must develop alternate sources of energy. Atomic power is one of those sources — if it can be made safe.

The Three Mile Island accident raised serious doubts about that safety. But, even at the time, it seemed that a large percentage of the problem was due to personnel. Neither utility workers nor the Nuclear Regulatory Commission (NRC) appeared to know what was going on.

The 12-person Kemeny Commisson confirmed the truth of this suspicion. It found what it called an almost total lack of emergency planning by the utility, the state, and the NRC. "Wherever we looked," said the report, "we found problems with the human beings who operate the plant, with the management that runs the key organization, and with the agency that is charged with assuring the safety of nuclear power plants."

The commission noted an appalling lack of training of control room operators. According to the report, the operators did not have the most rudimentary knowledge of atomic reactors or the hazards of radiation. As a result, they were unprepared when trouble came.

This training should have been provided by management. But when management chose to whistle in the dark, training should have been required by the NRC. But the NRC did nothing, either.

The Kemeny Commission found that the NRC concentrated on large, catostrophic, and highly unlikely accidents rather than the smaller, more probable ones such as occurred at Three Mile Island. The near-disaster at Three Mile Island was foreshadowed by several similar episodes, but no corrective steps were taken. The commission said it got the impression that all major offices of the NRC acted independently of each other, and warnings "fell between the cracks."

The failure of the NRC to act is no surprise to Nevadans. The NRC has failed time and again to do anything to improve the safety of nuclear waste shipments to diposal sites in this state and elsewhere. It was not to be expected that the agency would be any more responsible when it came to the power plants themselves.

It is true that the NRC has at last declared a moritorium on all new nuclear plants until safety measures can be made adequate. But the NRC made this move only under the considerable pressure of the Kemeny Commission. The moratorium by itself is no guarantee that the NRC is now ready to take its duties seriously.

If there is to be hope for atomic power, the NRC must be restructured, in order to provide proper leadership and follow-through. The one-man directorate proposed by the commission might be the answer. But whether the NRC is led by one man or five, the quality of the leadership must be improved. The work of the present directors has clearly been unsatisfactory.

Equally important are the other suggestions of the commission: to divest the NRC of all responsibiities not directly pertaining to nuclear safety; to dratiscally increase the inspection and enforcement function; to review nuclear plants periodicly instead of granting them licenses for life; and — perhaps most important of all — to establish an oversight committee to make periodic reports on the status of nuclear safety efforts.

Both the president and Congress should take all steps necessary to see that these recommendations are followed. They are absolutely necessary if there is to be any hope of assuring safe atomic power.

The Kansas City Times

Kansas City, Mo., November 10, 1979

There was widespread concern when the presidential commission on the Three Mile Island nuclear accident, in its final report, failed by one vote to recommend a moratorium on construction and operating licenses for new plants. After the commission's severe indictment of the nuclear power industry and the federal Nuclear Regulatory Commission which controls it, a shutdown of further nuclear growth seemed only logical while the 40 safety recommendations by the commission can be reviewed and implemented.

The concern has proved needless, however, since the NRC, no doubt stung by the criticism of its operations, has itself declared a licensing moratorium which could run from six months to as long as two years. Joseph M. Hendrie, NRC chairman, said the agency's first responsibility is "to apply those remedies and lessons" learned from the TMI incident to plants presently operating. This would include review of plant design features, operating and management procedures and retraining of the 2,500 plant control room operators.

The impact of a moratorium on the nuclear power industry will not be as great as it might seem. A licensing halt has been in effect since the Harrisburg, Pa., accident in late March. But even before that the nuclear growth rate had slowed markedly under increasing criticism and vigorous opposition from environmentalist groups. No new reactors have been ordered from manufacturers this year and only two were sold in 1978. The Department of Energy now estimates that there will be from 150 to 200 reactors in operation by the year 2000, which is 50 to 100 fewer than predicted early this year.

In addition to the 72 operating reactors, 92 others are under construction, but only eight or nine of these would have begun generating electricity within the next year under normal conditions. Four of these would probably have received operating licenses by the end of the year.

Continuing uncertainties over this country's future oil supply heighten the need, in the long range, for increased reliance on nuclear power as an alternative energy source. But for now the shortcomings revealed and the lessons learned at Three Mile Island must be applied to keep the risks of nuclear power within tolerable limits.

The Arizona Republic

Phoenix, Ariz., November 1, 1979

SEVEN MONTHS after the hysteria over Three Mile Island, a special presidential commission has issued its report on the accident at the Pennsylvania nuclear power plant.

Its findings have drawn the wrath of anti-nuclear activists, but by no means was the commission performing a whitewash job for the benefit of America's power companies and the people who regulate them.

Indeed, one of the commission's leading recommendations is that the five-member Nuclear Regulatory Commission be abolished in favor of an agency that would be part of the executive branch.

What nuclear opponents wanted — and did not get, nor should have gotten — was a pitch for a moratorium on atomic-plant construction.

Nor did they get a grisly pronouncement about the harm done to central Pennsylvania residents when the plant's water system went awry. Mental stress was the worst thing commission members, including Arizona's Gov. Bruce Babbitt, could find. "The amount of radiation released," said their report, "will have a negligible effect on the physical health of individuals."

Just the same, the commission called for fundamental changes in the way nuclear generators are built, operated and regulated.

And even though the group's report amounts only to advice which may or may not be used by President Carter or Congress, it was taken with good grace by the owners and designers of Three Mile Island.

The president of General Public Utilities says his company already is taking steps to improve its nuclear equipment and to do a better job of training the people who operate it. From Babcock & Wilcox, designer of the plant, came the reaction that the commission's work will "provide a framework for the continued acceptability of nuclear power."

The NRC, described by Babbitt as "a headless agency" that he fears "lacks the sense, direction, the vitality that is necessary to administer safety consideration on a day-to-day basis," came off badly in the commission's eyes for its failure to have recognized danger in its early stages.

The presidential commission is not alone in its criticism.

A Senate subcommittee doing its own investigation faults NRC inspectors for ignoring tell-tale gauge readings at Three Mile Island. In Washington state, Gov. Dixy Lee Ray, a proponent of nuclear power and former chairman of the Atomic Energy Commission — forerunner of the NRC — has stopped delivery of radioactive material to a nuclear-waste dump because the NRC failed to enforce rules for safe disposal.

The commission's report is being viewed by nuclear industrialists as the good thing it is: advice that can help them mend their ways — and in so doing restore the confidence of the American people in a crucial supply of power to an energy-hungry nation. □

The States-Item

New Orleans, La., November 3, 1979

The greatest mistake of the nuclear power industry — its rush to construct without adequate regard for public concern over safety — is underscored by the report of the presidential commission on Three Mile Island.

The 12-member commission, in its report to President Carter this week, said portions of the industry and the Nuclear Regulatory Commission must undergo fundamental changes or totally destroy public confidence and be responsible for the demise of the industry.

The commission was formed by President Carter after the frightening accident last March at the Three Mile Island nuclear plant in Pennsylvania. The commission condemned the NRC as incompetent to ensure public safety and urged its replacement by a single administrator under the executive branch. The commission also recommended much stricter licensing requirements for new plants, including federal approval of state and local emergency plans before licenses are granted.

From the beginning, government and industry promoted nuclear power in lockstep, blandly assuring the public of its safety while brushing aside its early opponents. Former Sen. J. William Fulbright of Arkansas spoke, when he was chairman of the Senate Foreign Relations Committee, of an "arrogance of power" in this country's policy toward Vietnam. There has been a similar arrogance of power in the attitude of industry leaders and successive federal administrations toward the development of domestic nuclear power. Now the industry faces political opposition strikingly similar to the anti-Vietnam movement, including some of the same participants.

The recommendations of his own commission presents a particularly difficult political problem for Mr. Carter, who is on record in favor of more expeditious development of nuclear power and who faces intra-party challenges from Sen. Edward M. Kennedy and Gov. Edmund G. "Jerry" Brown. Both are decidedly more sympathetic than has been Mr. Carter to anti-nuclear forces.

Neither the president nor industry leaders can afford to sluff off the commission's recommendations. As much was recognized by Floyd W. Lewis, chairman of Middle South Utilities Inc., who took a positive stance toward the commission's report. "I think the commission has done a workmanlike and balanced job," said Mr. Lewis. (The commission) has given us and the American public a simple message on nuclear power: Proceed, but proceed with caution."

If domestic nuclear power is to have a future the industry and government will have to proceed in a way that will convince the American people they are not unwilling guinea pigs in a potentially catastrophic experiment. Although the commission concluded that the long-term danger to public health from the Three Mile Island accident had been exaggerated, one more such accident and it could be lights out for the industry.

THE  SUN

Baltimore, Md., October 24, 1979

There can be no doubt now that the partial meltdown of the Three Mile Island nuclear power plant last spring can be attributed to gross negligence on the part of the federal Nuclear Regulatory Commission, the contractors who built the plant and the utility which operated it. These findings emerge clearly from a final draft of the report of the President's Commission on the Accident at Three Mile Island, the so-called Kemeny Commission.

The commission has recommended that the five-member, independent NRC be abolished and replaced with an agency in the executive branch with a single director appointed by the president and confirmed by the Senate, much as is the head of the Environmental Protection Agency. Nuclear power plant licenses would be subject to periodic renewal only after open, public hearings and training programs for nuclear power plant operators and regulators would be totally revamped in the interests of safety. That last is most important, but probably can be implemented only if the commission's other recommendations are followed as well. As it is now constituted, the NRC simply cannot be relied upon to implement the necessary reforms in plant construction, operation and inspection.

The commission also came near recommending a temporary moratorium on new nuclear plant construction. A commission majority favored the moratorium, but the proposal died because of a procedural rule requiring unanimity on such issues.

We believe this is the proper outcome in view of the tremendous costs for the nation that a moratorium might impose. The amounts, and the costs, of petroleum imported into the United States are rising so rapidly as to impose an economic threat far more serious than the physical threat posed by nuclear energy. Likewise, the environmental costs of burning coal in power plants are reaching a point where they may soon be palpably worse than the risks in nuclear energy.

This is not to suggest that the most rigorous safeguards against future accidents of the Three Mile Island type should not be put into effect quickly. It may be only sheer luck that the accident at this plant on the Susquehanna River did not result in massive radiation exposure of large numbers of people even as far away as Baltimore. The only reason safeguards—which are fully within the nation's technological capabilities—were not implemented earlier was because of the negligence of the builders, operators and regulators. When the president officially receives the Kemeny Commission report next week, he should not hesitate to begin immediately to put its recommendations into effect so that such accidents do not happen again.

Chicago Tribune

Chicago, Ill., October 25, 1979

We've learned not to expect too much from the reports of study commissions that are so often appointed by elected officials as a means of avoiding tough or unpleasant decisions themselves. Measured against this modest standard, the report of the President's Commission on the Accident at Three Mile Island comes off a bit better than average.

It is free of the emotion and demagogery that blurred the facts at the time of the accident in Pennsylvania last March. And on the most controversial question—whether the accident justifies a moratorium on all nuclear power construction—the commission came down on what strikes us as the side of common sense: no moratorium.

True, there appears to have been some controversy and confusion over this even within the 12-member commission. But to have halted work on the many plants now under construction would have been stupendously costly and short-sighted in view of today's uncertainty over future sources of energy and the speed with which other countries are pushing ahead with nuclear power.

This is not to brush aside the dangers involved in producing nuclear power—dangers that are not even yet fully identified and defined. But they can be exaggerated [as the dangers of electricity and automobiles were in years past]. The report acknowledges that despite the hullabaloo at the time, the leakage of radiation at Three Mile Island turned out to be so small that there will be "no detectable" effect on those who may have been subjected to it.

The commission found no single villain in the picture. Its report finds fault with the Metropolitan Edison Co. for operating the plant without "sufficient knowledge, expertise, or personnel" to do so properly. It criticizes the federal Nuclear Regulatory Commission for being so preoccupied with the nitty-gritty of licensing procedures that it fell short in its primary purpose, to ensure safety. Concentrating on the trees, it lost sight of the forest.

The report's chief recommendations were more or less predictable and are sound. It says that reactor licenses should be subject to periodic renewal on the basis of a utility's performance. This would give the government added power to close down a plant if there was evidence of trouble. The report also recommends stiffer government requirements for the inspection of plants and the qualifications of those who operate them.

The trouble is, of course, is that government inspectors are likely to be less sophisticated in the technology of nuclear power than the people they are judging. Perhaps the government's chief role should be to see that the people running a plant are kept on the alert and do not succumb to the carelessness that can contribute to human error when accidents are unlikely and thoroughly guarded against by presumably infallible instruments.

The report also says that the independent, five-member NRC should be abolished and replaced by a White House agency with a single accountable director. Possibly so. But if the NRC had already been a White House agency, the recommendation might well have been to replace it with an independent commission. The efficiency of an agency depends not so much on its structure, after all, as on the caliber of those who run it. If they are not alert themselves, they are not going to create an alert industry.

Certainly there are dangers in nuclear power, just as there are in so many other areas of life. Especially, we must have confidence in the methods of disposing of radioactive material. But man has gotten where he is by studying these dangers and learning to cope with them, not by running away from them. Radiation is not a new or unknown phenomenon. To forsake nuclear power until every possible doubt is resolved—or until other forms of energy are adequate to replace it—would involve an intolerable degree of dislocation.

The Virginian-Pilot

Norfolk, Va., November 1, 1979

Nuclear safety procedures of the government and industry are so deficient that half of the dozen members of the president's Commission on Three Mile Island favored a moratorium on the construction of more nuclear power plants until reforms are made. But because the commission adopted only those suggestions that its majority could support, a moratorium was not among the 44 recommendations given President Carter this week.

Fearing that a moratorium recomendation might induce Congress to halt new nuclear stations, the utility industry was greatly relieved. It interpreted the report as a warning to proceed with caution.

That's putting the best possible light on the commission's evaluation of the state of the nuclear art today. In fact, the 6-to-6 moratorium vote revealed the gravity of its concern.

After investigating the near-disaster at Three Mile Island it concluded that such an accident was "eventually inevitable," given prevailing safety deficiencies. "These shortcomings are attributable to the utility, to suppliers of equipment, and to the federal (Nuclear Regulatory) commission that regulates nuclear power."

The commission came down hard on the NRC: "With its present organization, staff, and attitudes, the NRC is unable to fulfill its responsibility for providing an acceptable level of safety for nuclear power plants."

Its recommendation—to abolish the NRC and create a single administrator to guard nuclear safety—is questionable. That was the argument five years ago for abolishing the Atomic Energy Commission and creating the NRC. A new agency with the same purpose as the old imitates its predecessor because it employs much the same staff, only with new flow charts and signs on the door.

The problem is internal, endemic to bureaucracy at large. "The NRC is beset by an overconcentration on regulation, an attitude that tends to equate the meeting of regulations with safety," says commission chairman John Kemeny, president of Dartmouth. "We find that's not the way to achieve safety. It is only the total dedication to safety that will assure the safety of nuclear power."

If the NRC has been entangled in its own red tape, no wonder the utility industry itself has not set adequate—and usually costly—safety standards. Utility companies, as Admiral Rickover noted recently, "are interested in the bottom line: How do you make more money? If you start anything that costs more money, they generally will object to it."

While 60 percent of the personnel operating nuclear power plants come out of the Navy's reactor program run by the admiral, he told Congress "they don't handle 'em the way we do. We have different systems that are far more rigorous than what the utilities do."

The Commission on Three Mile Island took note of utility personnel deficiencies and recommended changes in their recruiting and training programs. Industry would do well to emulate the Rickover approach.

The president, Congress, and industry may differ over the commission's recommendations. But they cannot ignore the report's bottom line: the course we are traveling to generate cheap nuclear energy is indefensibly dangerous. A mid-course correction is imperative.

THE DENVER POST

Denver, Colo., October 28, 1979

IF NUCLEAR power is to fulfill the important role assigned to it in the U. S. energy timetable, it must gain public confidence that its safety procedures are finally secure. To put it in a phrase: no more Three Mile Islands.

The March accident at Harrisburg, Pa., besides bringing on much acrimony about nuclear power, also prompted President Carter to appoint a commission to study what happened.

This prestigious group, headed by Dartmouth College President John G. Kemeny, now has completed its final draft of findings after six months of study. It is a most significant document.

The report disappointed some people by not calling for an end to licensing of nuclear power. It merely said that nuclear power is dangerous and must be handled accordingly.

That conclusion then leads to the meat of the report: how to handle nuclear power carefully. And as a preface, the president's commission establishes a familiar base: blame for the accident.

Metropolitan Edison Co., operator of the plant, mishandled the accident from its modest beginnings, making it, in effect, worse than it would have been if automatic controls had been permitted to function. James Schlesinger, then secretary of energy, used the phrase, "human error" within hours of the Harrisburg accident.

The commission reinforces this point specifically. Its report said: "The utility did not have sufficient knowledge, expertise and personnel to operate the plant or maintain it adequately."

That is a strong judgment but it is justified by the facts. The obvious response must be to require utilities to train staff personnel better, pay them better and build a professionalism among employees that will allow nothing short of excellence in plant procedures at all times.

The report doesn't let government off the hook, either. Pointing to confusion among federal regulators following the near-disaster, the commission recommends reorganization of the U. S. Nuclear Regulatory Commission. Instead of having a five-man board which spent hours *debating* what to do about Three Mile Island, the commission urges having a single administrator, after the fashion of the Environmental Protection Agency, where one director digests staff input and makes decisions. That makes sense in the light of NRC confusion last spring.

In the broader view, the Kemeny commission recommends that nuclear power plants be put through licensing renewal procedures. Periodically, a plant would come up for renewal at which time its safety record would be reviewed. Also, the commission recommends no nuclear plants be built in heavily populated areas.

Many of these ideas surfaced in the weeks after Three Mile Island. The Denver Post editorially has urged several of these steps. But the presidential commission adds impressive weight and direction to the effort that now must be made: Creating within the nuclear power industry expertise and dedication of sufficient magnitude to insure that another Three Mile Island — or any nuclear accident of that magnitude — never happens again in the United States.

The Detroit News

Detroit, Mich., November 2, 1979

The President's Commission on the Accident at Three Mile Island has concluded that the system humans use to control, manage, and operate nuclear power plants is as much in need of repair as the reactor that failed so dramatically last March.

The commission's findings are laden with evidence of incompetence, inadequate training, poor communication between compartmentalized units of the Nuclear Regulatory Commission (NRC) bureaucracy, poor coordination of safety programs, and a general "mind set" in the industry that is too promotional and that fails to relate people to machinery.

In spite of this remarkable compendium of faults, the commission stopped short of recommending a moratorium on licensing and construction during the long period that will be required to implement its sweeping recommendations.

The commission faced a dilemma. It uncovered troubles enough to conclude that an accident, somewhere, was inevitable. Yet, the investigators could not avoid the knowledge that the 72 operating reactors provide 13 percent of the nation's electrical generating capacity, and there is no substitute immediately available for this power. Further, four states rely on the working atom for more than half their electricity (Vermont, Maine, Connecticut, Nebraska) and these states would be all but closed down if nuclear power plants were turned off.

It is as much in the national interest today to retain nuclear power as it was to build these non-fossil sources of energy in the first place. The problem addressed by the commission was, simply, how to utilize the resource and minimize its risks.

Significantly, the investigative commission does not fault the basic technology. It severely criticizes, however, the way humans direct, manage, and operate it. And the faults, the commissioners say, begin at the top.

The President's commission recommends that the NRC be scrapped because, in the view of the panel, it has failed to do the job. Responsibility for licensing and regulating, says the panel, should be transferred to the administration, and the President should appoint one man to run the operation.

Is this the best solution?

Regulating the nuclear industry is different from any other federal enterprise because of its high technology and associated hazards. Finding one person qualified to direct this operation would be extremely difficult. Should he or she be vulnerable to a massive, politically-inspired cabinet reshuffle, like the one that followed Mr. Carter's long examination of conscience at Camp David?

Is the NRC a useless institution simply because it was populated by poor performers? Or should it be reconstituted with good performers?

A sensible approach that would deal with the basic problem and yet keep nuclear regulation relatively protected from suddenly shifting political winds, would be to reappoint the NRC, to restrict its five members to policy-making functions, and to have them appoint that one person to be a chief executive officer — to administer the agency and be directly responsible for its actions.

One fact is certain. Although nobody was hurt in the Three Mile Island accident, it provided a timely warning that much is wrong and must be corrected if the nation is to benefit fully, and without unnecessary risk, from nuclear energy.

The San Diego Union

San Diego, Calif., November 12, 1979

The Nuclear Regulatory Commission received a stinging vote of no-confidence from the president's commission investigating the accident at the Three Mile Island nuclear plant. The commission stated flatly: "With its present organization, staff and attitudes, the NRC is unable to fulfill its responsibility for providing an acceptable level of safety for nuclear power plants."

The commission recommended that the NRC be abolished and replaced with an independent agency headed by a single administrator appointed by the president. Whether such drastic reform is necessary, or even advisable, is for Congress to decide.

Meanwhile, the NRC is the only agency we have which is empowered to pass judgment on the safety of specific nuclear plants. As if in response to the barrage of criticism it has received, the NRC appears to be shifting toward a tougher stance on safety issues. Four new plants awaiting licenses apparently will have to wait longer, and a few plants already licensed and operating may face temporary shutdowns.

After Three Mile Island, it is hard to fault the NRC for wanting to double-check on its licensing decisions. But it is fair to ask whether the delays and interruptions in power production now in prospect are fully justified for technical reasons or simply must be endured because of the NRC's political problem. Are they necessary for public safety or to restore the NRC's credibility as a safety-conscious agency?

In the case of California's Diablo Canyon plant — one of the four with a new license pending — there will be a delay for another study of how the plant's emergency core-cooling system might be affected by an earthquake. The NRC only recently concluded a restudy of the earthquake hazard, however, and reported on Oct. 1 that Diablo Canyon could safely withstand a quake of 7.5 on the Richter scale, which is quite severe, on the fault closest to the plant.

Piling safety studies on top of safety studies can help soothe lingering doubts about nuclear power, and restore faith in the NRC as a tough watchdog, but the public will pay dearly for any extra measure of reassurance it gets. If Diablo Canyon cannot start up by next year, for instance, power blackouts and brownouts are almost a certainty in northern and central California.

The president's commission called for more rigorous policing of safety factors in nuclear power. Let's have it, but we can also hope that the new policies of the NRC are based solely on prudent and reasonable safety standards and not influenced by a desire to prove that the agency is on its toes.

The Des Moines Register

Des Moines, Iowa, November 1, 1979

The report of the President's Commission on Three Mile Island is a sobering warning to those who have listened to the nuclear industry's assurances that the public has nothing to fear from nuclear power.

The commission concluded that the management of nuclear power has been so riddled with problems that an accident as serious as the one at Three Mile Island was "eventually inevitable." Moreover, if the nation hopes to avoid another TMI, "fundamental changes will be necessary in the organization, procedures, practices and, above all, in the attitudes" of the nuclear industry and its government watchdog, the Nuclear Regulatory Commission.

The commission did not conclude that nuclear power is beyond salvation, or that the industry should be shut down tomorrow. Indeed, the commission gave nuclear power equipment a fairly clean bill of health. It found that the equipment at TMI was "sufficiently good" that, without the human failures, the incident would have been so minor that no presidential commission would have been created.

Although there has been much concern about the release of radiation, the commission reported, "It is entirely possible that not a single extra cancer death will result" from this release. The commission concluded that even if there had been a meltdown, a "China Syndrome" involving mass release of radiation probably would not have occurred.

"The containment building and the hard rock on which [the plant] is built would [probably] have been able to prevent the escape of a large amount of radioactivity," the report said.

The surprising conclusion of the commission was that "the fundamental problems [with nuclear power] are people-related problems and not equipment problems." In referring to people, the commission meant not just individuals, but the entire system, including the utility companies and the NRC.

Above all, the commission singled out the "mind-set" of the nuclear industry that its technology is "fail-safe." The commission believes this mind-set was one of the fundamental causes of the Three Mile Island accident.

It was the mind-set that allowed the industry and government to tolerate the fact that the people who ran TMI had "greatly deficient" training for handling nuclear accidents. Similarly, the control room at TMI was confusing: More than 100 different alarms went off within a few minutes of the accident.

The NRC had not even required states to come up with emergency plans for handling nuclear accidents. When the accident occurred at TMI, "the response to the emergency was dominated by an atmosphere of almost total confusion."

Metropolitan Edison, the utility that ran TMI, "did not have sufficient knowledge, expertise and personnel to operate the plant or maintain it adequately," the commission found.

Finally, the commission issued a scathing denunciation of the agency that is supposed to ensure the safety of nuclear power — the NRC: "With its present organization, staff and attitudes, the NRC is unable to fulfill its responsibility for providing an acceptable level of safety."

THE SACRAMENTO BEE

Sacramento, Calif., November 2, 1979

"We are convinced that an accident like Three Mile Island was ... inevitable."

No one was spared in the report just issued by the prestigious commission President Carter appointed to study the Three Mile Island nuclear power plant accident. The industry, it said, tends to "focus narrowly on the meeting of regulations rather than on a systematic concern for safety"; the Nuclear Regulatory Commission is a managerial mess and is still promoting nuclear development, even though that conflicts with its job of insuring nuclear safety; the law itself is too lenient with violators of safety regulations.

The accident at TMI, the panel concluded, was inevitable not because of faulty technical equipment, but precisely because the "organization, the procedures and, above all, the attitudes" of the entire nuclear establishment revolve around equipment. And all are wedded to the mistaken assumption that nuclear power plants can be made "people-proof."

As a result, power plant personnel have been inadequately trained and monitored. Their errors in past accidents are rarely analyzed, and the lessons of past errors rarely passed on to them. The control rooms in which they operate are not designed for efficient use and are not improved as scientists learn more about the interaction of people and machines. In the first few minutes of the TMI accidents, for example, more than 100 different alarms went off, a cacophony that surely would have strained any operator's ability to quickly sort things out.

Even more important, the panel found that the industry and the NRC have been so convinced of the effectiveness of their "people-proof" safety equipment that neither has prepared for failures. Workers in the nuclear power industry are not adequately monitored for radiation exposure. Research into the effects of radiation is not being pushed. Emergency planning for communities surrounding power plants is almost non-existent, and what there is is hopelessly uncoordinated. The plants are not even designed to accommodate the cleaning up required after an accident, which, as a result, has been unnecessarily dangerous at TMI.

The NRC's confidence in the safety of nuclear power plants, the panel found, borders on recklessness. The NRC licenses plants to operate while particular safety issues are still unresolved. It is "reluctant to apply new safety standards to previously licensed plants." Its inspection and enforcement division relies heavily on industry data rather than independent investigation, almost never uses its power to fine violators and, until TMI, did not even attempt to systematically analyze the vast amounts of information about accidents and near-misses that it acquired.

"Nuclear power," the panel warned, "is by its very nature potentially dangerous, and therefore one must continually question whether the safeguards already in place are sufficient to prevent major accidents." The panel's key finding — more important than the specific recommendations developed from it — is that the current organization of both the NRC and the nuclear industry actively inhibits this questioning.

Although the panel could have interpreted its task more broadly, it chose not to answer the most obvious safety question itself. It refused to recommend either for or against a moratorium on nuclear power development, saying that the question of "how safe is safe enough" must be resolved by "the political process." Yet by confining itself more narrowly to the lessons learned from Three Mile Island, the panel may have more impact than if it had taken sides on the overarching question.

This report should be a powerful goad to the President and the Congress to provide the kind of "fundamental changes" the panel has called for — and without which new TMI-type accidents are as inevitable as the first one.

Design Flaws Found in Diablo Canyon Plant

The Diablo Canyon power plant in California was the target of two weeks of antinuclear demonstrations in September, 1981; the protests were aimed at preventing the newly licensed Unit 1 reactor from being loaded with fuel and tested. The protest was sponsored by the Abalone Alliance, a coalition of 60 antinuclear and environmental groups. The alliance contended that the plant, a $2.3 billion project owned by Pacific Gas & Electric Co., was inherently unsafe because it was located about 2.5 miles from an undersea earthquake fault, the Hosgri offshore fault. The official total of arrests made during the protests, which ended Sept. 28, was 1,893—the highest number of arrests ever recorded in an antinuclear action.

The Nuclear Regulatory Commission granted a temporary operating license to the plant Sept. 21, after its Atomic Safety and Licensing Appeal Panel had found that Diablo Canyon could be operated safely and with a minimum risk of sabotage. On Sept. 22, however, a PG&E engineer inadvertently discovered that a blue-print used to design the earthquake supports for Unit 1's secondary cooling system had been reversed with a blueprint for Unit 2, which was still under construction. Unit 1 therefore had earthquake supports that were designed for its sister reactor. The NRC called a halt to testing on Unit 1 and began an extended probe into its key safety systems. A month later, the NRC announced that, in addition to the already-discovered flaw, PG&E engineers had apparently miscalculated the weight of Unit 1's heat removal system, which was suspended above the reactor floor by steel support braces. In the event of an earthquake, the braces could prove inadequate and buckle. An NRC spokesman called the error "potentially more significant" than the flaw found in September. The flaws revealed at Diablo Canyon raised questions about the procedures normally used by the NRC to evaluate a plant for licensing.

The Dispatch

Columbus, Ohio, November 9, 1981

ENERGY INDUSTRY officials have complained for years that unrealistic and unnecessary safety requirements were strangling the growth of nuclear power even as the nation was trying to free itself from dependence on Middle Eastern oil.

Since taking office, the Reagan administration has moved forcibly to shorten the licensing procedures for nuclear plants and to limit public debate on safety considerations — debate which, at some times in the past, amounted to little more than obstructionist tactics by anti-nuclear forces.

However, the problems being encountered and the mistakes being uncovered at the Diablo Canyon nuclear power plant in California give reason for pause in the rush to speed nuclear development. Government officials have had the needed access to information, the authority to investigate plans and the time necessary to identify problems before the plant was put into use, thereby possibly avoiding a major catastrophe sometime in the future.

The problems involve the power plant's ability to withstand earthquakes. The plant is situated two-and-a-half miles from an offshore earthquake fault. On Sept. 22, inspectors discovered that diagrams used in the design of earthquake supports for piping in the reactor secondary cooling system had been reversed for the plant's two reactors.

Now the plant's owner is unable to demonstrate that the weights it calculated for the plant's equipment are accurate. If the weights are inaccurate, the earthquake supports may have been constructed improperly and the plant could be subject to quake damage. The weight calculations could be correct, but nobody knows whether they are or not.

The lesson to be learned here is that a new abuse should not replace an old abuse. Certainly the regulatory review and licensing process can be streamlined and made more efficient, thereby saving utilities, investors and consumers money.

It would be a mistake, however, to render meaningless the review process by reducing government authority over power plants and by failing to provide adequate time spans to double-check and triple-check calculations, assumptions and plans.

Changes can and should be made in the review process governing nuclear power plants, but the changes should be made carefully, keeping ever in mind that nuclear reaction is capable of producing widescale personal injury and property destruction.

The Washington Post

Washington, D.C., October 6, 1981

WHAT CAN YOU say about the situation at the Diablo Canyon nuclear reactor? The facts aren't all in yet, but it appears, as protesters have insisted for years, that the plant is not sufficiently earthquake-proof. But that doesn't mean that the members of the Abalone Alliance—a quintessentially California happening of underworked TV actors and overgrown flower children, complete with folk songs and "affinity groups"—were right. At best they were right for completely the wrong reasons.

Opposition to Diablo Canyon stems from its proximity to an active earthquake fault that was not discovered until long after construction began. (The site, incidentally, was chosen in a joint effort that included the utility and prominent environmental groups.) In order to withstand seismic stresses, piping in a reactor has to be specially strengthened in particular places. There are actually two reactors at Diablo Canyon, and they are mirror images of each other (like your right and left hands). Apparently in making the extremely complicated seismic calculations, blueprints for the two units were switched, so that reinforcements in Unit One are where they should be in Unit Two and vice versa. So Unit One, which was just about to receive its operating permit, might not be able to withstand an earthquake.

Is this another failure of the Nuclear Regulatory Commission? Probably not. The error was at a level of detail far below what the NRC can detect in its audits. To have found it, the agency would need a staff big enough to duplicate everything a utility and its subcontractors do, and that would create an impossible mess. At some level the NRC has to be able to rely on the utilities' quality control. This mistake—at the very reactor the industry has made the symbol of the costs of unnecessary regulatory delay—seems to reinforce those who believe that the utilities simply aren't up to the task of managing nuclear plants.

On the other hand there is the argument that just as at Three Mile Island the system worked: the mistake was caught in time. True, but just barely. The mistake was found by someone reviewing plans for Unit Two, which is still under construction. It was missed in all the reviews of Unit One. One or both of the reactors could easily have gone to full power without its ever having been found.

No matter how big the problem eventually turns out to be at Diablo Canyon, the nuclear industry is likely to have lost a little more public confidence—confidence it desperately needs to retrieve. Unfortunately the mindless school of nuclear protest may gain in the process. Counting ourselves among those who want to see nuclear power prove to be safe and economically competitive, we hope the incident will send a message to the industry that it still needs to try harder.

The Providence Journal

Providence, R.I., September 2, 1981

Three Mile Island was a watershed in the nuclear power industry. The accident that almost destroyed the Pennsylvania reactor should have taught everybody that a much higher degree of vigilance is needed if further accidents are to be avoided. The final verdict is still not in; and new lapses in construction and operation dog the utilities.

The industry was shaken this week by the revelation that the wrong blueprints had been used for computations on the ability of a plant in California to withstand earthquakes. Five safety systems there have been subjected to doubts, pending complete refiguring of seismic stresses. The start of operation has been delayed.

This lapse by Pacific Gas and Electric Co. was astonishing, because the plant in question is Diablo Canyon, the very one besieged by antinuclear demonstrators the past two weeks. As Peter Bradford, member of the Nuclear Regulatory Commission, put it: "Here you have the most controversial area of discussion in what is probably the most controversial nuclear plant in the country."

He was referring to the protests

The containment shell problems might have been avoided if the nuclear power program had proceeded more slowly

that Diablo Canyon was situated only a few miles from an offshore earthquake fault. Opponents will be quick to use the lapse as ammunition against all nuclear power. More to the point, however, would be a campaign to increase the discipline exercised by the utilities in all phases of nuclear power.

Obviously, the blame must be placed on those officials of the company who goofed. They allowed blueprints for another plant under construction at Diablo Canyon to be sent, instead of the correct ones, to the consulting engineers who were designing strengthened supports to meet higher safety standards. Presumably, the human mistakes made at Diablo Canyon can be corrected and the safety of the plant there increased to a level acceptable to the Nuclear Regulatory Commission.

A more serious problem, basically, has been posed by the discovery that the steel shells of 13 nuclear power plants are turning brittle from the bombardment of neutrons in the reactors. These stainless steel shells, inside very thick concrete containment walls, are supposed to keep radiation from leaking out. But brittle spots could crack under extreme stress and result in breaches of the containment.

Again, how serious the problem is remains to be calculated. None of the reactors poses an immediate threat: when one does, it will have to be shut down. Whether remedial measures can be taken is not clear. The inside of the containment cannot be worked on until the reactor core has been removed and the vessel decontaminated, a long and costly process.

Had the nuclear power program proceeded at a slower pace, some of these problems could have been found and dealt with before a large number of plants was in operation. Today there are 69 plants producing power and 72 under construction or planned. Now the regulatory authorities must deal with the problems as best they can. The utilities have to stick with the plants, because scrapping them would be even more expensive than remedying defects.

When it comes to the danger of containment vessels, neither the industry nor the NRC can afford to be wrong.

Los Angeles Times

Los Angeles, Calif., November 2, 1981

A staff of federal nuclear regulators will meet Tuesday to discuss three questions about the Diablo Canyon nuclear power plant on the Central California coast:

Have they found all the engineering mistakes that Pacific Gas & Electric Co. and its contractors made in bracing the new plant against possible earthquakes? Should they keep looking? If they keep looking, when will it be safe to quit?

It seems to us that the Nuclear Regulatory Commission would be wise to take the advice of Gov. Edmund G. Brown Jr. and others, and let somebody else answer those questions—an outside consultant, or perhaps a panel that would include nuclear engineers along with the critics of the plant.

For a long time, the only serious question about Diablo Canyon was whether it could ride out a serious earthquake without damage so severe that there would be a danger of radiation leaks.

Now there is another problem—one involving confidence in what the experts say in answer to that question. The federal inspectors, the utility company and the contractors have all been over the blueprints and the engineerng data who knows how many times over a number of years. They missed both mistakes that have recently come to light.

In one case, they missed the fact that plans for bracing pipes and other equipment in one of the plant's two nuclear reactors were used to do the work on the other reactor.

In the other case, they missed the fact that the contractor miscalculated the weight of the equipment that some braces must support.

Those may be the only two miscalculations that have been made in the construction of the $2.3 billion power plant. But is anyone likely to believe the people who missed the mistakes in the first place if they say that there were no others? We doubt it.

Brown's proposal makes sense. And the sooner that outside help is sought, the sooner the questions can be answered.

THE TENNESSEAN

Nashville, Tenn., November 7, 1981

FREQUENT discoveries of improper design and other errors in the construction and operation of nuclear power plants are costing the industry heavily and raising serious doubts of whether the nation can ever feel comfortable in the production of nuclear energy.

After the costly accident at the Three Mile Island plant in Pennsylvania nearly three years ago, the leading proponents of nuclear energy gave assurances it would not happen again. The people operating Three Mile Island just made dumb mistakes, they said. With the exercise of a normal amount of caution, which was well within the abilities of the people in charge of nuclear energy in the U.S., the atom could be made just as safe as a dam.

Well, it is already obvious that dumb mistakes have not been eliminated in the nuclear industry, and they are not likely to be as long as error-prone human beings are putting up the plants and running them. The question is, can the mistakes be held within reasonable limits to permit the nuclear generation of electricity to go ahead?

The potential for disaster wherever a nuclear plant is located, is creating a serious drag on the development of the industry.

The Diablo Canyon plant being constructed in California just two and a half miles from a major earthquake fault is a case in point.

The Nuclear Regulatory Commission's Atomic Safety and Licensing Appeal Board said in July that the plant was "adequately designed to withstand any earthquake that can reasonably be expected."

In September it was discovered that engineers on the Diablo Canyon plant used the wrong blueprints to build a set of earthquake safety supports. Just a dumb error, but it was a costly one — both in money and public confidence — and it necessitated rebuilding the weak supports to conform to the standards which the NRC thought they already met.

Then in October new design errors "potentially more significant" than those found earlier were discovered at Diablo Canyon. These mistakes involved steel support braces that could fail in a major earthquake, allowing tons of equipment to fall to the reactor floor.

Many Americans, perhaps most, have tried to maintain a neutral position between the emotional opponents of nuclear energy on the one hand and, what some call, the cheerleaders for nuclear energy on the other. But many neutrals are finding it more and more difficult to put any faith in the assurances of the cheerleaders.

St. Petersburg Times

St. Petersburg, Fla., October 10, 1980

The discovery that the wrong blueprints were used to build earthquake safety supports for the Diablo Canyon nuclear plant was a major shock for an industry rattled by bad news. Considering the location of the California plant — just three miles from an active earthquake fault — the blunder is enormous.

"It is a first-rate screw-up," said Peter Bradford, a member of the Nuclear Regulatory Commission (NRC). "Here you have the most controversial area of . . . what is probably the most controversial nuclear plant in the country. To commit an error of that sort is almost analogous to a student copying down a wrong homework assignment: No matter how brilliant the work from then on, he's just not going to get the right answers."

The location of the controversial plant and the last-minute discovery of the blueprint blunder by a company engineer raise disturbing questions about the NRC.

WHY WAS Pacific Gas & Electric allowed to build a potentially dangerous power plant so close to an earthquake fault in the first place?

To protect the public health, the risk of radiation leaks should be minimized at all costs in every nuclear plant. Building one in an earthquake zone is a curious way to minimize the risk. Proof that low-level radiation is hazardous, in a backhanded way, comes from the Reagan administration. It has opposed giving medical care at Veterans Administration hospitals to men exposed to radiation from nuclear tests. Why? One reason, according to William Taft, general counsel of the Department of Defense, is that there would be a negative impact on the nuclear industry if the government acknowledged officially that exposure to radiation is a significant health hazard.

Why didn't NRC inspectors find the blueprint error at the Diablo Canyon plant during the numerous reviews required before an atomic power plant can obtain an operating license?

Harold Denton, director of reactor regulation, said that "such a level of detail would require thousands of manhours to find one problem. We just don't have those kinds of resources." That explanation is not very reassuring. Americans rely on the NRC to certify the safety of all nuclear plants. Is that trust ill-founded?

DOUBTS ABOUT NRC's performance are particularly disturbing at a time when the nuclear industry seems filled with doom and gloom.

Two weeks ago, for instance, NRC officials announced that the thick steel shells of 13 nuclear reactors — including Turkey Point 3 near Miami — are being turned brittle by radiation so rapidly that some of the plants may become unsafe to operate by the end of next year. Although spokesmen for all the utilities affected claim that their plants are safe, NRC officials say that some will need substantial repairs. If a brittle shell cracks and water leaks out, the core could melt down, a catastrophic accident.

In June, an Energy Department report was released that conceded the United States would be burdened with millions of tons of nuclear waste by the year 2000, but had no way to dispose of it safely. That report was followed by one from the Union of Concerned Scientists about the danger of highly toxic radioactive waste from nuclear reactors.

"Every year, a single 1,000-megawatt nuclear reactor produces about 30 tons of spent fuel. There are some 650 pounds of long-lived, highly radiocative elements distributed throughout this waste," the report said. "If just the plutonium-239 contained in this fuel were dispersed as fine particles and inhaled, it would be sufficient to cause fatal lung cancers in the entire population of the United States."

Mr. Reagan's call for the swift development of a permanent national high-level nuclear waste disposal site is the best aspect of the administration's strong pro-nuclear policy announced Thursday.

NUCLEAR POWER will continue playing an important role in the United States' transition to renewable energy sources and other technologies that do not have the risks of radiation. But, considering the shock waves that have rocked the nuclear industry in the past few months, shouldn't the administration reconsider its program of nuclear boosterism?

The administration has proposed cutting Energy Department programs in conservation, solar, renewable energy, oil, gas and coal by approximately $1-billion in 1983. Only nuclear power programs would not be bludgeoned by the budget cutters.

Even if the administration believes that the nuclear industry ought to be boosted to meet the nation's short-term energy needs, it still does not make sense to cut back research on alternative energy sources that future generations will need.

And since Mr. Reagan is encouraging greater reliance on nuclear energy at a time when the industry seems to have more problems than solutions, he has a responsibility to tell Americans what he plans to do about the poor performance of NRC.

Oregon Journal

Portland, Ore., November 20, 1981

It wasn't the frequently expressed fears about earthquake that stopped the Diablo Canyon nuclear power plant. It wasn't the protesters who stormed over the fences and were dragged off like limp rag dolls.

It was snafu — that old gremlin of World War II. Drawings used to label pipes in Unit One as safe from earthquakes were really drawings for Unit Two. The Nuclear Regulatory Commission has voted to hold up Diablo's permit until remedial action has been taken and is verified by an independent commission.

It does seem incomprehensible that such an error could have been made, but it must be investigated thoroughly.

While the investigation is going on, it might be prudent to consult a memo written Sept. 30 by two U.S. Geological Survey scientists, Thomas Heaton and Carl Johnson. They are quoted in Discover magazine in an article by Dennis Overbye: "If a large earthquake were to occur in the near future, many would claim that there were abundant examples of precursory phenomena."

The phenomena include dramatic increases in radon, a naturally occurring radioactive gas; renewals of water in wells which had been dry, or the decreasing of wells; and new steam vents in a hot springs area.

The changes seem to indicate new underground activity along the San Andreas Fault, the boundary between two of the tectonic plates that make up the earth's crust. Los Angeles and much of the California coast rides the Pacific plate while the eastern part of the state is on the North American plate.

Such changes as have been observed in California are accepted in China and Russia as forerunners of major earthquakes. All three of the unusual indicators are present not far from Diablo Canyon. During this enforced wait, take a second look at the ground around the plant, while the NRC looks at the pipes.

The San Diego Union

San Diego, Calif., November 23, 1981

The low-power operating license for the first unit of the Diablo Canyon nuclear power plant is being suspended by the same Nuclear Regulatory Commission that granted it in September. The discovery of errors in the construction of safety systems in the plant should be as sobering to the NRC as to the Pacific Gas & Electric Co., which is responsible for the mistakes.

The NRC is demanding a new, independent review of the earthquake safety features of the $2.3 billion plant built by PG&E on the coast near San Luis Obispo. The emphasis will be on the objectivity of the review, for the Diablo Canyon case has shown that the procedure allowing builders of nuclear plants to submit their own evidence of compliance with construction standards is not reliable.

The willingness of the NRC to acknowledge this defect in its own regulatory process can help restore public confidence in nuclear power. While scientists and engineers assure us that nuclear plants can be built to operate safely even in earthquake country, those assurances mean nothing if there is no guarantee that the safety standards are being followed.

The nuclear industry and electric utilities have taken heart from President Reagan's support of nuclear power and his promise to seek regulatory reforms that will reduce the time it takes to get a nuclear plant into operation. The new chairman of the NRC, Nunzio Palladino, a Reagan appointee, seems to be emphasizing in the Diablo Canyon decision that there will be no compromise with safety in the pursuit of that goal.

That the start-up of the Diablo Canyon plant will be delayed even longer by this latest development is bad news for PG&E and its customers. But there is good news in the evidence that the NRC is going to be more exacting in what it accepts as evidence that a nuclear plant is safe enough to operate. The public's faith in nuclear power can be no greater than its faith in the agency responsible for policing its safety.

The Oregonian

Portland, Ore., November 27, 1981

Just as the Three Mile Island nuclear plant accident provided a valuable learning experience for the relatively young nuclear industry, so should the quality assurance problems discovered at California's Diablo Canyon nuclear plant motivate the industry and the federal government to do a better job in making future plants safer.

Concern expressed by federal Nuclear Regulatory Commission Chairman Nunzio J. Palladino that the industry and the NRC have not been "as effective as they should have been" in developing guarantees of quality assurance is, at first glance, more troubling than the Three Mile example because this shortcoming is fundamental to the safety of all nuclear plants, not merely a problem related to a specific nuclear design.

That a relatively large number of construction-related deficiencies at Diablo Canyon and four other nuclear projects in the nation have come to light under recent NRC scrutiny should build confidence that the licensing agency, despite the Reagan administration's passionate support of nuclear power, is committed to learning lessons from past mistakes and will not plow ahead in approving the operation of a nuclear plant until the mechanism by which a safe operation can be predicted is improved.

The Des Moines Register

Des Moines, Iowa, December 1, 1981

In 1975, when construction of the Diablo Canyon nuclear power plant on the California coast was three-fourths completed, the U.S. Geological Survey reported that the Hosgri fault, just 2½ miles away, could experience a major earthquake measuring 7.5 on the Richter Scale. Pacific Gas & Electric Co., builder of the Diablo Canyon facility, said the plant was designed to withstand the shock of a major quake.

Last summer, as the plant was being prepared for operation, a young engineer discovered that some design charts had been mixed up so that supports for the plant had been put in the wrong places. PG&E said the error could be corrected, and everything else was fine. The U.S. Nuclear Regulatory Commission was less optimistic.

Two more errors involving structural safety were quickly uncovered, neither related to the first. So the NRC voted the other day to suspend the operating license for the $2.3 billion plant pending further seismic tests for earthquake potential. It was the first such suspension ever ordered.

It is encouraging that a watchdog agency is protecting the public. It is discouraging that, the further the expansion into nuclear power, the greater the number of potential dangers uncovered. Yet the Reagan administration has ordered a 12-percent cut in the NRC's budget, most of it in safety research, while pushing the agency to speed up its licensing procedures for new reactors.

In the American Association for the Advancement of Science's current issue of Science, Eliot Marshall notes: "Three Mile Island showed that the small and far more likely slipups may be just as disastrous as the big ones. Thus, the NRC has opened up a whole new category of worry. In the process, it has produced an enormous list of safety problems that must be investigated and remedied quickly."

The operating reactors around the country are showing signs of age and wear that could lead to breakdowns with frightening consequences. Yet the NRC, crippled by limited finances, must pick and choose among the problems needing study.

The Philadelphia Inquirer

Philadelphia, Pa., November 25, 1981

During the 4½ months that he has served as chairman of the Nuclear Regulatory Commission, Dr. Nunzio J. Palladino has seen first-hand the inner workings of the nuclear utility industry. That inside view, he said last week, "sort of clouds the high degree of confidence" that he once had in nuclear power. He is on sound ground.

Last Thursday, in his first appearance on Capitol Hill, Dr. Palladino, former dean of the college of engineering at Pennsylvania State University, told the House Interior Committee that the nuclear industry "has to reorient its thinking" if it is to restore public confidence in the atom as a safe and acceptable source of energy.

"After reviewing both industry and NRC past performances in quality assurance, I readily acknowledge that neither have been as effective as they should have been in view of the relatively large number of construction-related deficiencies that have come to light," Dr. Palladino testified. Then, Dr. Palladino and his colleagues on the NRC put some force behind those words and suspended the operating license for the Diablo Canyon nuclear power plant, a "strong sign that the commission doesn't like what it's seen," in the words of the chairman.

What the NRC saw was a series of gross design and construction errors at a plant that has become a symbol in the public debate over nuclear power. The twin reactors, located on the California coast 2½ miles from an earthquake fault, were first proposed 18 years ago.

Critics of the plants argued that they were unsafe and should never be licensed to operate. Supporters of nuclear energy countered by saying that the reactors had been built with sufficient safeguards to protect against earthquakes, and the NRC apparently bought those assurances at face value.

A few days after receiving authorization from the NRC to start up the reactors, Diablo Canyon officials discovered that the blueprints for structural safety equipment had been reversed, invalidating their certifications of strength. They then discovered other major construction errors that raise serious questions about the plant's ability to withstand any major seismic activity.

The events leading to last Thursday's action by the NRC reflect poorly on Diablo Canyon's owner, Pacific Gas & Electric Co., and the NRC.

The NRC is charged with overseeing all aspects of a plant's construction, from the earliest design to the completed project, independently confirming a utility's claims that the plant is safe. That obviously did not occur. Had the utility not reported its findings, quite probably Diablo Canyon would have been approved for full power operation within a few months.

"We barely caught these things in time," Dr. Palladino told Interior Committee members.

That's hardly the kind of disclosure designed to reassure a worried public. But words, no matter how candid, are meaningless. The fact that the discovery was followed up by swift and decisive action is a welcome change. It is to be hoped it is also a harbinger of new thinking at the NRC.

The Chattanooga Times

Chattanooga, Tenn., November 30, 1981

When the Nuclear Regulatory Commission gave Pacific Gas & Electric Co. the authority two months ago to begin low-power testing at its Diablo Canyon nuclear power plant, the decision was marked by vociferous protests by persons opposed to nuclear power. Now, however, the NRC has reversed its decision, voting 4-1 to suspend the license. It is a remarkable turn of events which has pleased, obviously, the plant's opponents.

Not so officials of PG&E, which believes, according to its general counsel, that it has become the victim of what he calls a "concerted attack" by those who have demonstrated against the Diablo Canyon project. But the NRC's decision was not influenced by the actions of a few hundred "anti-nukers." Rather, the suspension stemmed from concern over evidence of several design and construction errors.

Those errors — one of them involved pipes which were installed backwards — did, however, lend substance to at least one complaint voiced by the protesters. Noting that the Diablo Canyon plant is built less than three miles from an earthquake fault, the protesters alleged that the plant was not strong enough to withstand the shock of an earthquake.

NRC Chairman Nunzio Palladino, a Reagan appointee, said the commission's action was most symbolic, and explained, "It's a strong indication the commission does not like what it is seeing in this case." The symbolism will be expensive. The suspension means that the plant's design and construction data must be re-reviewed, and PG&E says that the delay in putting the plant into operation will cost more than $1 million daily in added interest charges.

Looked at another way, that is less expensive than if the plant failed during an earthquake, releasing radiation into the atmosphere.

Post-Tribune

Gary, Ind., December 11, 1981

The impact always seems stronger when an influential person changes his approach on a major issue.

Take Nuclear Regulatory Commission Chairman Nunzio Palladino, for example. He once seemed eager to help promote the Reagan administration's desire for revitalizing the nuclear power industry.

Now the former dean of engineering at Penn State, who spent years designing nuclear submarine reactors, has seen the civilian nuclear power industry close-up. And what he saw obviously surprised him; his eagarness has turned to caution.

In a speech to the industry's main trade association, he said there are deficiencies of construction and preparation at some plants "which show a surprising lack of professionalism ... The responsibility for such deficiencies rests squarely on the shoulders of management."

Then last week he joined other commissioners in telling Congress that the NRC has significant reservations about the effectiveness of the International Atomic Energy Agency's system of safeguards, a position that may complicate nuclear exports.

Palladino is still a proponent of nuclear energy, but his frankness about what he's discovered and his new cautious approach is reassuring to a public gravely concerned after the Three Mile Island accident and the fiasco of Diablo Canyon.

His attitude, if it continues, might just increase credibility for his commission. In the long run, that could help an industry which, more than most, must learn to look thoroughly before it leaps.

Democrat and Chronicle

Rochester, N.Y., December 4, 1981

THE damaging criticism of the nuclear power industry by federal Nuclear Regulatory Commission Chairman Nunzio J. Palladino is dismaying. The March 1979 accident at Three Mile Island in Pennsylvania should have etched an indelible lesson about nuclear safety.

Palladino says his first five months in office have brought to his attention deficiencies at some plants "which show a surprising lack of professionalism in the construction and preparation for operation of nuclear facilities.

"There have been lapses of many kinds — in design analyses resulting in built-in design errors, in poor construction practices, in falsified documents, in harassment of quality control personnel and in inadequate training of reactor operators," he said. He further suggests that "just as utilities have certified independent financial audits of their fiscal activities," there should be similar audits of quality control measures.

We'll let the experts debate whether Palladino's criticisms are valid. But the fact that the NRC head would feel compelled to make them suggests that the lessons of Three Mile Island may not yet have been heeded.

Nuclear power is a valuable tool worthy of further development. But the public's need to be assured of its safety is both great and understandable. Any signs that utilities and builders of nuclear power plants are tolerating slipshod work are, to borrow Palladino's word, "inexcusable."

ST. LOUIS POST-DISPATCH

St. Louis, Mo., December 7, 1981

Nunzio J. Palladino, the chairman of the Nuclear Regulatory Commission, did not mince words the other day when he addressed the annual convention of the Atomic Industrial Forum. The industry, he said, was guilty of a "surprising lack of professionalism in the construction and preparation for operation of nuclear power facilities."

Among the "lapses" he cited were poor construction practices, built-in design errors, falsified documents, harassment of quality control personnel and inadequate training of reactor operators. "The responsiblity for such deficiencies rests squarely on management," he said. Given that even the most fervent advocates of nuclear power admit that it is "inherently dangerous" technology, Mr. Palladino was nicer than he might have been.

The NRC chairman's finger-pointing exercise can be interpreted as a response to the assertion by General Public Utilities, owner of Three Mile Island, that NRC regulatory failures are responsible for the accident that crippled the Pennsylvania power plant and threatens the utility holding company with bankruptcy.

It is true that the industry did not get where it is today all by itself. Nuclear power plants are a product of massive federal subsidies and of government regulation that has been as deficient, in its own way, as the private builders and operators of nuclear power plants Mr. Pallidino criticized. But as Mr. Pallidino told the AIF, "quality cannot be inspected in. It must be built in."

As far as TMI is concerned, the debate over whether government or industry is to blame is essentially an oblique way of arguing the really difficult question that has yet to be resolved. Who should pay to clean up the mess the accident has created — and for others that may yet come along?

The Hartford Courant

Hartford, Conn., December 11, 1981

The Nuclear Regulatory Commission, under chairman Nunzio J. Palladino, is hitting the nuclear industry like a cold shower. The effect may be a bit shocking, but it is salubrious.

The commission has recently grown more aggressive in trying to ensure the safety of nuclear power stations, something it should have done long ago. Mr. Palladino, who has been on the commission only five months, has come down hard on utilities for their part in diminishing public confidence in nuclear power.

Both the industry and the NRC have been reeling since the nuclear accident in 1979 at Pennsylvania's Three Mile Island plant, which is still closed and is still not cleaned up. More recently, there have been damaging reports of safety problems at the Diablo Canyon plant in California before it has even begun to generate electricity, as well as problems in steel pipes caused by radioactivity at operating plants.

Mr. Palladino, who designed reactor cores before becoming dean of the college of engineering at Pennsylvania State University, has shown a commendable determination not to become an industry apologist.

Speaking at the Atomic Industrial Forum in San Francisco, he sharply criticized the industry for building in design errors, for poor construction practices, falsifying documents, harassing quality control personnel and providing inadequate training of reactor operators.

The new emphasis on providing NRC inspectors at nuclear power plants and construction sites should help, but the industry itself must take more responsibility to prevent more safety lapses.

Mr. Palladino proposed instituting quality control audits, similar to fiscal audits, but designed to eliminate shoddy workmanship and poor practices. He went so far as to suggest that if the industry doesn't do it voluntarily, the NRC may have to require them.

The new chairman, an appointee of the Reagan administration, also has shown a willingness to risk the administration's wrath with his memo casting doubt on the effectiveness of international procedures to prevent the diversion of nuclear materials to build nuclear weapons. The letter had the approval of all five commission members.

The president wants to restore the United States as a "reliable supplier" of nuclear materials and technology, a goal that might be endangered if the commission exercises its power to reject or delay nuclear exports.

The commission should do so until it can be sure that American nuclear supplies will be used solely to generate power and will not add to the growing danger of nuclear weapons proliferation. That, in fact, is the implication of Mr. Palladino's memo, which was sent to several congressional committees.

These examples of growing assertiveness and independence at the NRC, which in the past has been criticized for its timidity, are heartening developments.

If the NRC succeeds in instilling a scrupulous regard for safety in the nuclear industry, it might help restore lost public confidence which could eventually accrue to the industry's benefit.

And by acting responsibly to help prevent the diversion of nuclear materials into weapons, it could provide world benefit

Reagan Administration Presses for Deregulation of Nuclear Industry

Vice President George Bush, head of a special task force to recommend reductions in federal regulations, announced in August 1981 that the government would begin reviewing more than 30 regulations. Included in the review were the paperwork requirements for power plants seeking nuclear licenses. President Reagan confirmed his Administration's commitment to deregulation of the nuclear power industry in a policy statement of October 1981. The statement called for a push to reduce the time it took to license a nuclear plant to between six and eight years from the current average of 10 to 14 years. The President said that nuclear power had "become entangled in a morass of regulations that do not enhance safety but that do cause extensive licensing delays and economic uncertainty." Nunzio Palladino, chairman of the Nuclear Regulatory Commission, revealed that the commission was exploring reforms that would facilitate the licensing of 33 new plants within the next two years.

Funding for the NRC in fiscal 1982 and 1983 was held up by extended debate over new "interim licensing" provisions. In fact, passage of the authorization bill took so long that fiscal 1982 and the first quarter of 1983 were over before the bill was signed. The controversial provisions, sought by the nuclear industry and bitterly opposed by environmental groups, allowed the NRC to issue temporary licenses to new nuclear plants before the public hearings on them were completed. The legislation also allowed the NRC to issue amendments to existing licenses before hearings were held on the changes. The nuclear industry contended that these provisions were necessary to avoid operating delays at about a dozen plants.

In January 1983, the NRC complied with a recommendation made by the commission which had investigated the 1979 Three Mile Island accident, by approving a set of safety standards for the operation of nuclear power plants. The commission set acceptable risk factors of 0.1% for "prompt fatalities" among persons in the vicinity of an accident, and 0.1% for "cancer fatalities" among persons within 50 miles of a damaged plant. The NRC also suggested that the likelihood of a nuclear plant undergoing a large-scale core meltdown "should normally be less than one in 10,000 per year of reactor operation." NRC Commissioner Peter A. Bradford created a stir by suggesting, in a statement attached to the proposed safety standards, that the goals contained "an implicit maximum theoretical acceptable consequence from nuclear power plant accidents of some 13,000 deaths over the life of the 150 plants now in operation or under licensing review." (Nuclear power plants were designed to last at least 30 years.)

The Louisville Times
Louisville, Ky., March 24, 1981

The atomic power industry reacted with subdued rejoicing last week when government regulators proposed a plan to speed up the licensing of nuclear electricity generating stations. The rest of us should view the proposals with skepticism and even alarm.

The pro-nuclear forces have long complained that licensing procedures take too long and cost too much. Eleven nuclear plants are complete or soon will be, according to the industry, but five at most will be cleared for full operation this year. President Reagan, who thinks the nation should have more nuclear energy, and Joseph Hendrie, chairman of the Nuclear Regulatory Commission, are sympathetic to those complaints.

But Mr. Hendrie's solution — cutting back on citizen involvement in nuclear licensing procedures — can't possibly serve the best long-term interests of either the public or the industry.

Even if the time required to get a plant on line is reduced by eight months, as Mr. Hendrie hopes, the cost to the nuclear establishment in increased opposition and heightened suspicion would be high.

Under the rules change proposed by Mr. Hendrie, regulatory commission staff members would no longer be required to produce documents or give testimony at the request of nuclear opponents or other citizens. The result would be to deny the public the detailed information needed for effective participation in hearings about a power plant or its site. However, the commission could cooperate voluntarily.

The advantage, from the industry's standpoint, is that licensing would less often become entangled in lengthy hearings and appeals. The Atomic Industrial Forum claimed recently that the cost of delays in granting operating licenses to the 11 plants could reach $2.4 billion.

There are problems with the Hendrie plan, however. For one thing, it's a sound principle that citizens and communities must have access to the information necessary to make informed judgments about projects that can have a monumental effect on their lives.

Secondly, citizens' groups do spot blunders that mysteriously escape utility engineers and federal regulators, as witness the recent events at Marble Hill.

Finally, the Hendrie plan smacks of an arrogant coverup. If the industry thinks it is the target now of protests and dark suspicions, wait until the government refuses to part with information about a controversial reactor.

The reason for the current backlog of paperwork, of course, is that few licenses have been issued since the Three Mile Island accident. But is it logical to deny citizens information because of a misfortune brought on by industry and government blunders? On the contrary, the TMI mess suggests that the industry bears more watching rather than less.

If nuclear proponents want more public confidence and fewer delays, they have to demonstrate that proper siting, design and management can make nuclear energy considerably less menacing. Shunting the public aside is hardly the way to accomplish that.

St. Louis Globe-Democrat
St. Louis, Mo., March 11, 1981

A major challenge to the Reagan administration is whether it will be able to induce the Nuclear Regulatory Commission and other agencies concerned with atomic power to speed up their now extremely slow licensing procedures.

A recent status report the NRC sent to Congress indicates the commission plans to drag its feet for a long time in making decisions on operating licenses for 11 completed, or nearly completed, nuclear plants.

After saying that the hearings on the permits take longer than expected, the NRC said that action on the 11 permits would be moved from 1981 and 1982, to 1982 and 1983. It is estimated that the new schedule will delay issuance of operating licenses to the 11 plants by an average of more than seven months.

Francis M. Staszesky, chairman of the Atomic Industrial Forum, in protesting the delays to former NRC Chairman John Ahearne, said the "11 plants will collectively stand idle for some 80 months waiting for the administrative procedures to conclude and for operating licenses to be issued." Staszesky estimates the unnecessary delays could add about $2.4 billion to the cost of these plants, costs that will have to be paid by customers.

If the Japanese can build a safe nuclear plant in three to four years, why should it take 10 to 12 in the United States? The answer appears to be that the energy bureaucracy and opponents of nuclear power now are allowed to do virtually as they please while the interest of the public in gaining the immense benefits from nuclear power is being largely ignored.

The Evening Bulletin
Philadelphia, Pa., May 22, 1981

The Nuclear Regulatory Commission, under the questionable direction of outgoing Chairman Joseph Hendrie, seems hellbent on disregarding the lessons of Three Mile Island — to date the nation's most frightening nuclear failure — and speeding the licensing of future nuclear plants.

We won't argue that the licensing process has become overly encumbered in giving all challengers a forum for protest. We do think, however, that silencing debate is equally foolish — and potentially more dangerous.

Under intense pressure from Congressional sources, the NRC this week voted 3-1 to lop two months off the time required to get a license. This means the NRC could OK a plant in 30 days. It effectively silences any appeal *before* a plant goes on line.

NRC general counsel Leonard Bickwit called the rule a compromise between divided commission members and between advocates and opponents of nuclear power. He conceded, however, that appeals probably would not be heard before a plant went into operation. And once a plant is operating, if the temporary licensing of Atlantic City casinos is any example, it will be difficult to convince the NRC and the industry to shut it down.

What the rule has done, in effect, is to shift the burden of proof from the industry to anyone who questions the safety of a plant. John Aherne, the only commissioner to vote against the rule, said he didn't think it provided adequate review. And even Bickwit admitted the nuclear industry had objected on the grounds the NRC could not reach any decision in only 30 days.

Fortunately Chairman Hendrie's tenure ends next month. He is to be replaced by Nunzio J. Palladino, a native of Allentown and dean of the college of engineering at Penn State. Dr. Palladino has an impressive background in the nuclear field. He should make the revision of the 30-day procedure a top priority.

THE WALL STREET JOURNAL

New York, N.Y., June 11, 1981

Congress appears ready to cast off the "no-nukes" monkey that has weighed on the back of the nuclear power industry since the Three Mile Island accident more than two years ago. A House committee approved a measure last week that would allow the Nuclear Regulatory Commission to grant interim operating licenses to new nuclear plants, even when the full hearings procedures have not been completed. A similar bill is awaiting floor action in the Senate, and supporters predict passage of a nuclear speedup bill by the end of this session. This would mark a welcome return to sanity in our handling of nuclear power development.

The prerequisite hearings procedures for plant licensing have become ever-lengthening and are assuredly driving nuclear power development into the ground. What started out as a good-faith federal effort to give state and local officials a say at hearings into new plant construction has turned into an obstructionist weapon for those who categorically oppose nuclear energy. It used to take six years from conception to final operation of a nuclear power plant; this time span has now stretched out to 12 years, in part due to construction delays but mainly because of the growth in cumbersome and unnecessary procedures.

These startup delays have become even more exaggerated since the Three Mile Island incident. No-nuke forces have become more vehement in their opposition to nuclear power, and the NRC's staff has been preoccupied by a thorough review of operations at all existing plants, each of which has undergone some modifications due to TMI.

According to a recent survey by the Atomic Industrial Forum, the 13 units nearest completion face a cumulative delay in operation of more than 90 months for lack of licenses. (The NRC and Department of Energy dispute this figure, saying utilities hardly ever meet their planned construction completion dates.) In any case, the costs to the economy of keeping a $2 billion nuclear plant idle are enormous. The industry estimates that it costs an average of $1 million per day in extra capital carrying costs and additional conventional-fuel costs that would have been displaced by the nuclear plant's operation.

Other than placating a few local officials and nuclear power opponents, the licensing hearings have rendered little, if any, substantive benefit in improved plant operation. The House bill would require the commission to draw up proposals to cut by half the time needed to procure an operating license.

The new congressional licensing measures would go far in precluding the procedural sabotage which is now being conducted by the no-nukes lobby. At the same time, the NRC—and ultimately the President—would remain responsible for ensuring the safe operation of nuclear plants, with or without full licensing. Interim nuclear licensing is a safe, economical and sensible alternative to the current procedural circus.

The San Diego Union

San Diego, Calif., March 31, 1981

The Nuclear Regulatory Commission has taken a first step toward streamlining the process for licensing nuclear power plants. A change in the rules would trim eight months off of what is now an 18-month procedure for granting an operating license to a new plant.

As expected, the proposed change in regulations is under attack by anti-nuclear organizations on grounds that it would reduce the role of the public in the licensing of plants. To the extent that protesters would have less time to build a case and fewer opportunities to raise objections, this is true.

But the problem with the existing procedure is that it can be exploited by people who object to nuclear power in principle and raise technical questions that are not relevant or have been examined and settled earlier. The new rules would appear to treat this problem without curtailing the review of legitimate safety issues.

The economic problems facing the utility industry are being worsened by regulatory delays in the start-up of nuclear plants which have been under construction for many years and represent a considerable investment of capital. Nuclear units at Diablo Canyon and San Onofre in California are among those with licensing dates slipping into next year because of prolonged safety hearings.

Delays which arise from consideration of new and bona fide safety questions are part of the price to be paid for the benefits of nuclear power. But delays which result from deliberate tactics to play on public fears should not be part of that cost.

The Boston Globe

Boston, Mass., May 19, 1981

The troubled nuclear power industry has less to fear from its antagonists, with their hostile bumper stickers and periodic protests, than from some of its misguided friends.

The latest such friend is the Subcommittee on Energy and Water Development of the House Committee on Appropriations, chaired by Rep. Tom Bevill (D-Ala.). In a rider attached to the Supplemental Appropriations Bill for 1981, now in conference committee, Bevill's subcommittee has set out to help nuclear power by punishing the Nuclear Regulatory Commission.

The NRC's sin? Two years ago, in the aftermath of the accident at Three Mile Island, the agency shifted some staff from its licensing division to strengthen a safety program focusing on the 70-odd other reactors already in operation. The aim was to learn from the accident, to remedy deficiencies that Three Mile Island revealed in systems and operating procedures so that similar accidents would not occur elsewhere.

The extra workload created a backlog in issuing licenses for new reactors ready to go "on line." A dozen new plants that are expected to reach the turn-on stage in 1981 and 1982 may have to wait an average of five or six months for their permits to operate at full power (although the NRC wants to permit "low-power testing" during these intervals to identify and work out any bugs.)

The Bevill subcommittee has been an aggressive booster of nuclear power, but not a very thoughtful one. Its consistent message to the NRC has been: You regulators are doing the wrong thing. Nuclear power works fine. Stop your nitpicking on safety, streamline that pesky licensing process and crank those licenses out.

That message is written boldly between the lines of the current budget rider which aims to cut $10 million from the agency's budget and shortcut the mandated hearing process that precedes the issuance of operating licenses. Punitive elements in the subcommittee report would also restrict the commissioners' travel, cut their staffs and deny the use of consultants. Finally, the rider demands a shift in personnel to allow full speed ahead on licensing.

This budget cut comes over and above the Reagan-Stockman trims. It's an attempt, pure and simple, to rap the NRC's knuckles, and it's dangerously wrong-headed.

Though no one was killed, Three Mile Island was a true disaster for the nuclear power industry. The spectacle of three control room operators and a two-hour "accident sequence" converting a $2 billion operating plant into a cleanup bill now estimated at $1 billion gravely shook the confidence, not only of the general public, but of the investment community.

Ever since, the task for any real friend of nuclear power has been to regain public confidence and to rebuild the political and investment climate that is essential in any industry with such huge capital requirements and long leadtimes for construction. That's a tall order, but probably not impossible, so long as the right lessons have been learned. That's where the NRC, with its make-haste-slowly review of safety issues and temporary slowdown on licensing, has been right and where Bevill's subcommittee is dead wrong.

The subcommittee's approach is a throwback to the old Atomic Energy Commission mindset that got nuclear power into trouble in the first place. All serious reviews of Three Mile Island, including the Kemeny Commission report, have noted that nuclear power's troubles began during the late '50s and early-to-mid '60s when exaggerated estimates of the need for nuclear power (1000 plants by the year 2000) and glib underestimates of potential safety problems sent the industry and the regulators tripping blithely down the primrose path that led to Three Mile Island. The basic critique is that the regulators were too much oriented in those years to getting new plants up and running, and not enough to making sure that all was well in those already in operation.

To forget the lesson of Three Mile Island is to ask the public to accept a level of jeopardy that it will not, in the long run, find acceptable. For the good of the nuclear power industry, the message that the Bevill subcommittee is trying to send to the NRC must be reversed. The cost of another serious accident – measured not only in the cleanup bill, but also in investor confidence – could leave the nuclear power industry dead.

The Register

Santa Ana, Calif., August 16, 1981

The current hearings over the license application for the San Onofre nuclear plant, and most discussions of nuclear energy, don't begin to get to the heart of the problem.

What makes the issue of nuclear power so troublesome is not that the technology is not sufficiently regulated, but that it is regulated too much, by the wrong people and for the wrong reasons. The history of nuclear energy provides virtually a textbook example of the dangers of leaving technology to the vagaries of the political arena. Decisions in that arena are too often based on emotionalism, power blocs and publicity, and seldom on a rational approach to the facts.

We're not sure whether nuclear power is the best hope for clean, safe, inexpensive energy or a hazard of incalculable proportions — and we seriously doubt whether anybody is really sure. The debate has been contaminated from the start, largely because it has taken place in the political arena rather than in a truly scientific forum.

Nuclear power was incubated with government subsidies and extraordinary government protections. The taxpayers footed the bill for most of the initial research and development. The Price-Anderson Act limits the liabilities of nuclear operators in the event of a disastrous accident. Nuclear operators claim that without this government-mandated limitation of liability the insurance companies wouldn't touch them. Combined with the immense subsidies to get the technology started, Price-Anderson constitutes an indirect subsidy that may (or may not) be the key to a viable nuclear industry.

We can't change the fact that the taxpayers paid for a lot of nuclear R & D, though it would be an interesting notion to require nuclear operators to pay off those costs through some kind of royalty or licensing fees. We could repeal Price-Anderson, and thus require anybody who wants to build a nuclear plant to assume full responsibility and full risk in the event of an accident. The result would be either abandonment of the idea or such strenuous attention to safety that insurance companies would be willing to sell insurance. We don't know which course would be followed, and don't much care. Either way the public would be safer than it is under the care of government licensing and monitoring agencies.

Recent history has demonstrated the instability inherent in entrusting a promising yet risky technology to the political process. The political climate has shifted since the heady days of the 1950s when almost everybody thought that nuclear power was the key to energy abundance. Those more concerned about the potential dangers than the potential benefits have achieved more clout. It is likely that many of their concerns should be addressed, but in the emotional climate of politics it is virtually impossible for a dispassionate observer to determine the validity of all the competing claims and counterclaims.

The key to a solution to the impasse is not to erect more government boards to license, monitor and supervise, but to deregulate nuclear power while eliminating all the direct and indirect subsidies to the industry. Then those who think it is the wave of the future would be able to proceed, while taking on themselves the full risks, responsibilities and liabilities involved.

The potential risks of nuclear power are quite high. The initial cost of building nuclear plants is also quite high — the San Onofre expansion alone has cost $3.3 billion. A private investor or company contemplating such start-up costs and risks will require a substantial payoff in terms of cheap energy down the road, and the best safety features that can be designed, now and in the future. People who understand the risks, and are willing to assume full responsibility, should be free to proceed, using their own money.

Frankly, we doubt that nuclear power without subsidies and government limitation of liabilities would seem as attractive to investors and entrepreneurs as it seems now. Though the technology holds great promise, it also holds awesome risk. But we'll never know whether nuclear power can stand on its own until we stop subsidizing it

Get the government out, and see what happens. That's a solution we could live with. And it's a solution that would enhance the possibility that we would all live to see the outcome of the debate over nuclear power — in operating plants or abandoned shells — in the real world, rather than in the hothouse atmosphere of a commission hearing or a demonstration.

Post-Tribune

Gary, Ind., October 25, 1981

A shortcut to approval of nuclear power plant operations? Grant temporary operating licenses for reactors before public hearings are concluded?

The country does not need that kind of speedup, but it is what Congress is considering, at the suggestion of President Reagan and to the delight of the nuclear industry.

There are at least two major flaws in this push:

- The talk about red tape and regulatory delays causing hardships for the nuclear industry is based more on malarkey than facts.
- Safety is essential — there should be no end run.

A member of the Nuclear Regulatory Commission says the length of nuclear construction times isn't longer than in most countries. "This much maligned licensing process," says Peter Bradford, "has yielded more nuclear power plant licenses than the rest of the free world combined."

Only the Diablo Canyon plant in California "can legitimately be said to have had its operation delayed by hearings, and it has a number of unique problems," according to Bradford. This is the plant where hundreds of protesters and demonstrators gathered, as the public was told the plant was safe. But severe deviations from the specifications were found while fuel loading was in process, and the plant is near an active earthquake fault.

Bradford suggests that a slack in demand for electricity and the difficulties in financing cause more delays than the regulations. They surely are a part of the changing face of the nuclear industry.

Indiana's only nuclear plant under construction, Marble Hill on the Ohio River at Madison, has been hurt by safety problems, and completion has been delayed by almost 10 years. A flaw in concrete brought a 20-month delay at one time. That's elementary safety, not an overload of regulations, and that delay has helped drive up the cost of the Marble Hill plant.

Delays did finally cause the proposed Bailly Plant in Porter County to be scrapped — in the end, the spiraling costs did it. Considering the site, among other factors, the area will, in the long run, be better off because of that decision.

Of course, regulations should be reasonable. But the industry ought to prove beyond reasonable doubt that it knows just where it is going and that plants will be safe.

The chairman of an environmental group opposing the Marble Hill plant said of the utility constructing it: "They have never known what the hell they were doing, and now we are all in the soup with them, as consumers and taxpayers."

He is biased and perhaps exaggerates. But it is a refreshing contrast to the continued "expert" claims, that "shortsighted, ignorant marchers are blocking the free world's best hope for cheap power." That charge is as much nonsense as claiming that excess regulation is strangling the nuclear industry.

The fact is, nuclear plants are expensive and their edge over coal plants is slipping. Department of Energy figures show that in 1980 the cost of producing electricty was $2.32 a kilowatt hour for nuclear reactors and $2.33 for coal-fired plants. So there are more than regulatory problems facing the nuclear industry.

NUCLEAR REGULATORY COMMISSION DISCOUNTS ACCIDENT REPORTS.—NEWS ITEM

Richmond Times-Dispatch

Richmond, Va., June 16, 1981

A congressional committee's criticism of the federal nuclear power plant inspection system, announced Sunday, should not obscure the fact that the present process for licensing new plants is too slow and too costly. Largely because of unnecessarily complicated licensing procedures, it now takes about 12 years for a nuclear power plant to move from conception to operation. As a result, a survey by the Atomic Industrial Forum has shown, utilities and their customers are burdened with hundreds of millions of dollars in avoidable costs.

As we understand the congressional committee's report, it does not make a case against a speedier licensing process. Based upon an investigation of a water spill at a New York nuclear plant last year, the report concluded that the Nuclear Regulatory Commission does not have enough qualified inspectors. This can be attributed to budgetary restraints and is, therefore, a situation that can—and should—be remedied.

The costly consequences of the existing nuclear licensing system fully justify the hiring of additional inspectors to help accelerate the process. For each day that an operational plant is idle because it lacks a federal license, the cost is $1 million. Most of this goes for the more expensive substitute fuel, usually oil, that the utility must use during the licensing delay, and the rest goes for higher interest charges on the construction costs. The industrial forum survey revealed that "the 13 nuclear units nearest to completion may encounter more than 90 months' delay due to licensing snarls, at a cost of $3 billion." Eventually, nearly all of the money spent as a result of prolonged licensing procedures comes from the pockets of the utilities' customers.

Of the 19 utilities that responded to the survey, 14 intend to replace oil-fired plants with nuclear units. The result would be a reduction in oil consumption of 350,000 barrels per day, enough to heat seven million American homes. In time, if not immediately, the conversion from conventional to nuclear fuel also results in significant savings to consumers. The survey showed that some utilities expected immediate savings of from $2.79 to $15.60 per month to a typical residential customer, and all the utilities predicted lower bills over the years.

From all of this it is obvious that the Nuclear Regulatory Commission's license procedure needs to be streamlined. While that agency must be extremely cautious in approving new nuclear plants, its licensing procedures should not be inordinately long and involved. Actually, many nuclear power critics now use the NRC's licensing process not to ascertain a new plant's safety but to block its operation altogether.

Fortunately, there is growing awareness in Congress that the situation needs to be improved. A bill pending in the House of Representatives would require the NRC to devise procedures that would reduce the time required for the licensing process by half. Both the House and the Senate are considering measures that would authorize the NRC to issue an interim operating license for any plant that its own experts considered to be safe even before the agency had completed hearings.

These are constructive proposals. Nuclear energy is a highly sensible alternative to fossil fuels, and the nation's welfare requires the removal of unnecessarily costly and complicated impediments to the development of nuclear plants Congress should proceed promptly to do whatever is necessary to establish a reasonable licensing process.

THE BISMARCK TRIBUNE

Bismarck, N.D., November 1, 1982

You've heard it said that some people never learn. How true, how true, how true.

And the axiom applies to multi-billion-dollar industries and to government officials, as well.

Take, for example, the nuclear power industry and the Reagan administration.

A few weeks ago, the Energy Department announced proposals that would do away with some safety improvements at existing atomic power plants and limit the role of nuclear opponents in some public hearings during the licensing process for new reactors.

Shelby T. Brewer, assistant secretary for nuclear energy, said the proposals are an outgrowth of President Reagan's goal "to restore the light water reactor industry and the utility industry to economic health."

Apparently, Mr. Reagan, Mr. Brewer and other boosters of the nuclear industry fail to realize that many of the people who oppose the expansion of nuclear power generation don't have anything against the nuclear power, per se.

They do, however, have a great deal of concern over how the concept is transferred to the drawing board and how it then takes shape in their neighborhoods.

How should the government and the industry cope with such concern?

They first should do everything possible to assure safety. They shouldn't, as Mr. Brewer's agency is proposing, eliminate safeguards.

They next should do everything possible to establish a good track record — good experience — so that people can see that everything is going well.

On that point, it is worth noting two stories that appeared in The Tribune the day before the Brewer story.

One carried the headline "$40,000 Fine Urged For Nuclear Plant," and the other carried the headline "Doctoring Of X-rays Charged."

Those stories, which are by no means isolated incidents involving the nuclear power industy, note some serious problems.

In the one case, the Nuclear Regulatory Commission found that operators of the Vermont Yankee nuclear power plant had failed to realize that the emergency core cooling system — the most extreme safety system — had been activated one day last April. The operators further failed to notify state officials of the problem.

A spokesman for Vermont Yankee, when questioned about the incident, would say only that the notice of violation was under study.

Vermont residents surely rested easier that night knowing their utility was studying the matter.

The other case involves Niagara Mohawk Power Corp., which is building a $3.7 billion plant in New York State.

A subcontractor, under orders from the NRC, inspected X-rays of piping welds and found that 73 X-rays had been doctored. All of the X-rays are of welds involving pipes within safety-related systems.

The examination of the X-rays found mostly "lead-pencil shadings," according to a Niagara spokesman.

He said, "I think we have to say that just because they have been doctored, that is no indication that the welds themselves are in any way defective."

The people near the reactor surely sleep better these nights knowing that.

But of course some may not. And some Vermont residents may lie awake nights wondering if their nuclear power plant operators will foul up again, and, if they do, what the consequences will be.

The industry continues to make mistakes, continues to stone-wall the public and continues to foster distrust among the general population.

And the Reagan administration, with the proposed changes in safety rules, apparently does not intend to do much to help foster trust in the industry.

As we say, some people never learn. How true, how true, how true.

And how sad.

WORCESTER TELEGRAM

Worcester, Mass., October 18, 1982

The Reagan administration has unveiled a set of plans to speed the licensing and reduce the cost of nuclear power plants.

In proposals that drew immediate comment and criticism from nuclear power opponents, the administration proposed five major changes in present licensing requirements.

If the new plan is adopted, licensing hearings will be limited to factual issues. Engineering improvements to existing plants will be required only if an existing plant is found unsafe. The current two-step process of obtaining both construction permits and operating licenses will be converted to a one-step procedure. The Nuclear Regulatory Commission will be allowed to approve sites as being suitable for nuclear plants in advance of a utility's decision to build a power plant. And the NRC will be allowed to approve, in advance, major power plant subsystems.

Shelby Brewer, assistant energy secretary for nuclear energy, said the new regulations are being suggested because nuclear power plant construction is practically at a standstill. He largely blamed the present rulebook, which often adds years of delay to the licensing process and causes huge cost run-ups for power plant projects. He says the present licensing procedure means that it takes a utility 12 to 14 years to build a plant from start to finish.

Unless the regulations are eased, there will be no significant number of new nuclear power plants built in this country, he said.

Those opposed to nuclear power, including the Union of Concerned Scientists, object to the proposed rules changes. Eric Van Loom, executive director of UCS, characterised the proposals as ". . . an attempt to move away from the strong feeling of the American people that there should be close scrutiny in the licensing process."

And U.S. Rep. Edward Markey, D-Mass., charged that the administration is afraid that the nuclear power industry will "meet its maker" in the marketplace without some major invervention of this type by the federal government.

The rule changes are clearly on the table for discussion. Industry supporters blame the present rules for major parts of the high cost of building nuclear power plants. Opponents will charge that the changes represent a weakening of the commitment to nuclear power plant safety.

But the arguments to come seem academic.

The escalating cost of nuclear power plants now under construction has pretty much killed new projects. Several proposed plants — including Pilgrim II in Plymouth — have been scrapped. Several others — including Millstone III in Connecticut and Seabrook II in New Hampshire — may never be completed. Safety requirements contribute to those costs, it is true but inflation has contributed to them as well.

If the administration can show that the proposed changes will cut down the cost of construction without sacrificing safety, then let it proceed.

But it's not at all certain that changes in the licensing procedure can revive the almost moribund nuclear power industry, as far as new plants are concerned. When a nuclear plant could be started with the assurance that it would produce electricity for four cents a kilowatt-hour when completed, that was one thing. But the electricity produced at Seabrook and Millstone, if those plants are completed, will cost up to 20 cents a kilowatt-hour — far more than coal power costs. That's the factor that is going to decide the future of nuclear power.

The News American

Baltimore, Md., November 30, 1982

There is not much question that the process for licensing nuclear plants needs overhauling. The way it works now, the process can take 12 to 14 years to complete, and is incredibly expensive. But the comprehensive package of reforms being recommended by a special task force of the Nuclear Regulatory Commission needs to approached very cautiously indeed.

The three major recommendations are in these areas:

- Backfitting: The NRC now orders utilities to "backfit" newly developed engineering improvements on existing plants to improve safety. These improvements cost millions and take months to implement.
- Hearings: NRC licensing hearings are conducted like court proceedings, with lengthy cross-examination of witnesses. Anti-nuclear groups contend that being able to question industry and government intensively is a key tool in their licensing battles with better-funded utilities.
- One step licensing: Utilities must now apply for separate construction and operating licenses in a two-step process. The task force recommends a legislative proposal to consolidate the process.

The intent, again, is to save time and cut costs. Both ends are laudable. But already the proposals are triggering controversy. For example Rep. Edward Markey (D-Mass) charges that the hearings changes would "dramatically curtail citizen participation in the licensing process and undermine the federal government's ability to ensure that nuclear plants are safely designed and operated."

It remains to be seen whether such specific criticisms are sound. But we share the underlying concerns of Rep. Markey and other complainants. The road to reform has to be trod cautiously. Whatever their other merits, reforms that undermine established safeguards or diminish the public's ability to protect itself have to be rejected.

We are not urging rejection at this point, of course. We are simply saying that, as the task force's recommendations head for the full commission, we want the panel to keep in mind that time and cost are not the prime consideration. Public safety is.

THE KANSAS CITY STAR

Kansas City, Mo., January 24, 1983

The Nuclear Regulatory Commission has gone along with some—not all—changes proposed by the administration to streamline the licensing of nuclear power plants. It remains to be seen whether Congress will enact these into law, or indeed whether the new one-step process will trim the eight- to 10-year lead time now required under two-step licensing to get a nuclear plant on line.

It seems only logical that if the NRC approves construction of a multi-billion-dollar project it will issue it a license to operate. It is wasteful of time and money to have to hold a second series of hearings to listen to the same witnesses enter the same objections before granting an operating license. In any case the commission provided an escape hatch by allowing a second hearing if the opponents could find a legitimate safety issue, such as improper construction of the plant. The head of the commission task force which worked up the proposals believes this will gut the one-step process.

There are some good ideas in the proposals: permitting the NRC to approve possible sites in advance and standardized designs for the reactor and other components. The wheel doesn't have to be reinvented in the design of each new plant if it matches one previously approved.

But the NRC goes too far when it seeks to bar opponents from challenging the need for a plant, or administrative judges who conduct the hearings from raising their own safety questions. If the licensing procedure is to be reduced to a single action then opponents should have the right to be heard, even if offering nothing more constructive than their fears of nuclear energy and reluctance to have a plant sited in their neighborhood.

Nuclear power is in trouble for several reasons of which the lengthy licensing process is only one. Recession, reduced electric demand, cost overruns, high interest rates until recently and public doubts on safety are others. With 83 reactors operating and 59 under construction, no new ones have been ordered since 1978. Shortening the federal approval time will help some, but this should not be done at the expense of arbitrarily denying opponents the right to be heard that one time.

The Idaho STATESMAN

Boise, Idaho, May 29, 1983

The Department of Energy took the wrong approach when it wrote regulations to address security problems that in truth might be much less severe than the agency claims.

The department's rules, published April 1, would allow the secretary of Energy to withhold from the public non-classified information about nuclear-plant design, security plans and weapons components.

The rules provide for withholding of information that would "increase the likelihood" of theft or sabotage by terrorists, but anti-nuclear groups say the Department of Energy's intent might be to squelch embarrassing demonstrations against reactor development and nuclear shipments.

One problem with the rules is that they might be used to keep harmless but important information out of the hands of the public, which these days is hungry for data about the construction of nuclear plants and the transportation and storage of nuclear materials.

Also, it's a little hard to believe that rules are needed to withhold from the public information that the Pentagon has seen fit not to classify.

Gov. John Evans, who chairs the nuclear subcommittee of the National Governor's Association and has been an outspoken critic of the DOE's handling of waste at the Idaho Nuclear Engineering Laboratory, plans to protest the rules. His attorney, Pat Costello, correctly says, "It's really impossible to tell what might be withheld under the rule because it allows such broad discretion."

DAYTON DAILY NEWS

Dayton, Ohio, January 21, 1983

The Nuclear Regulatory Agency has found a way to end all the hassling over nuclear power plant licenses: Give the public fewer opportunities for dissent.

Up till now, licensing has been a two-step procedure which has allowed dissenters to cross-examine witnessess and introduce new evidence. NRC is planning to ask Congress to adopt a one-step operation which will allow license construction and operations at the same time. There is a clause providing for a rehearing if dissenters can prove there was a safety problem which hadn't been raised before construction began.

NRC's package has a couple of other proposals to stop the anti-nuclear folks from getting too much information to work with. It wants to forbid administrative judges and NRC's Atomic Safety and Licensing Boards members from posing their own questions about potential safety problems during licensing hearings. And NRC thinks that approving lots of potential building sites before they're needed, plus standardizing designs for reactors and plant components parts, will cut off repeated public debates over design safety and feasibility. Unfortunately, it also would discourage new technology and improvements.

If the Energy Department could add its two cents worth, the NRC also would no longer force utilities to tack on new safety measures after a reactor design has been approved. That's a slam at post-Three-Mile Island edicts that forced all utility companies to put on new safety equipment even though they've had no problems.

Such thinking may save money and get power plants on line faster, but it also points out why the public is wary of nuclear power. Ratepayers want to know that everything possible is being done to protect their families. Utility officials, however, insist that they should be the final judges of what's enough.

Figuring out ways to shorten the licensing procedure is not going to build public confidence in nuclear power. If electric companies expect the public to pick up the tab for their building projects, they should offer the public a full say in what is being built.

The Hartford Courant

Hartford, Conn., August 17, 1983

Question: When is public information a secret?

Answer: When the federal Energy Department says so.

That's no joke. That is the gist of a proposal by the department to limit public access to a broad spectrum of nuclear information, even though the information is not classified.

The motive behind such restrictions — to make it more difficult for terrorists to make a nuclear weapon or to attack a nuclear plant — is commendable. But for most intents and purposes, the cat is out of the bag.

No restrictions imposed at this late date could possibly prevent anyone who wanted to discover how to build a nuclear weapon, or how to damage a nuclear plant, from digging up the information. It has been available from too many sources for far too long.

Among other things, the government would like to restrict access to reports and documents regarding plans for protecting nuclear material in transit, certain design information about bombs that is now public and unusual occurrences that might make nuclear materials more vulnerable.

The DOE proposal also includes a catch-all provision that gives the department the right to restrict virtually any other unclassified nuclear information.

If the government is intent on creating a nuclear priesthood through administrative fiat, then it is taking the right course. Not only is a large segment of the public already left out of nuclear issues because of their highly technical nature, but the rules would ensure that only a select group of the initiated would have access to certain sacred documents.

As many university administrators have pointed out, the restrictions would effectively close off material now used in standard courses in physics, electrical engineering, material sciences and even political science. It would, further, erect serious impediments to basic and applied research.

States could be shut out from information that affects their citizens, such as nuclear plant leaks and accidents, the health impacts of nuclear weapons tests and shipments of radioactive wastes within their borders.

Even if it were possible and constitutional to deprive the public of access to information previously available to all, nuclear safety would not necessarily be enhanced.

One has to believe that nuclear safety is better served with close monitoring and study by a variety of individuals and organizations, public and private, than by handing over all knowledge to a small group of nuclear shamans.

Or one has to believe that technology has outgrown the capacity of a democracy to control.

THE SACRAMENTO BEE

Sacramento, Calif., July 3, 1983

The Reagan administration has proposed to Congress a broad revision of the federal regulatory procedures for licensing new nuclear power plants. The proposal, according to Secretary of Energy Donald Hodel, is "an effort to permit a more responsible, more rapid approach to the construction of those plants without sacrificing ... safety." If that's what the administration intended, it has failed.

The licensing reforms that the Reagan administration is urging would gain speed — and cut nuclear industry costs — by allowing the Nuclear Regulatory Commission (NRC) only one review of a new nuclear power plant before it goes into operation, and that would occur before construction even began. In one step, when the design of the plant is first approved, the operating license would also be granted. Thus, no time would be "wasted" by NRC investigators checking to see if the approved blueprints had actually been followed, or if the utility involved had the management organization in place to safely operate a nuclear power plant. Under the Reagan proposal, the utility would simply certify that it had met all safety requirements.

There is no way to think of this "reform" except as a major sacrifice of safety. As NRC Commissioner Victor Gilinsky recently testified before the Senate, "Giving a utility an operating license on the basis of a paper review, before the start of construction, makes about as much sense as handing an incoming freshman his college diploma on the basis of his course outline."

There have, in fact, been several cases recently — including the case of the Diablo Canyon nuclear power plant in California — where utilities have made major construction errors, despite perfectly acceptable blueprints. In the Diablo Canyon case, the utility itself discovered these errors before operations began, but the need to pass a second NRC review undoubtedly increased the utility's vigilance. And the errors themselves raised doubts about the utility's quality-control procedures that can be resolved only in a full-scale licensing hearing. Moreover, as even such a strong nuclear proponent as NRC Chairman Nuncio Palladino has said, there are many nuclear utilities so ill-managed that they would never have noticed such errors themselves, or bothered to correct them in the absence of NRC oversight.

There is more sense in the administration's desire to cut down on the number of design reviews the NRC conducts before a construction permit is awarded. But it is not the NRC's regulations that in the past have prevented one-step design reviews. The utilities themselves, according to NRC testimony in Congress, have rarely submitted complete designs that can be reviewed all at once. Requiring the NRC to speed up its design reviews puts the cart before the horse, and could end up compromising public safety as much as one-step operational licensing.

"The conventional view," Gilinsky correctly warned the Senate, "is that utilities are so discouraged with the system of safety regulation that they cannot be expected to order nuclear plants; and that it's up to the regulators to cheer them up with a more accommodating licensing process ... (However) the key to the future of nuclear power is not easier licensing but improved performance of the reactors in operation today, and of the ones under construction." The Reagan administration has mistaken speed for progress.

ST. LOUIS POST-DISPATCH

St. Louis, Mo., August 13, 1983

Congress, ever obliging when the need for secrecy is linked to the magic phrase "national security," has recently been falling for the next best thing. Under an amendment to the Atomic Energy Act, the nation's lawmakers recently gave the Department of Energy new authority to keep "sensitive information" related to nuclear materials from the public, lest it fall into the hands of terrorists, extortionists and the like.

It all sounded reasonable enough — it always does — until the censors at DOE published the new secrecy regulations. Now librarians, environmentalists, state officials and academics from Stanford to Harvard are complaining that the new rules are vague, unnecessary, arbitrary and a secrecy grab, which indeed they are.

The proposed rules give Energy Secretary Donald P. Hodel broad powers to define what kinds of information related to nuclear materials are to be kept from the public. Nuclear materials, it seems, are anything he says they are. Therefore, under the guise of curbing terrorists, the rule could be used to conceal nuclear safety problems, mismanagement and a multitude of sins. Censorship is like that. A little goes a long way. No wonder the DOE rules have set off such an uproar or that the department has wisely, if begrudgingly, agreed to hold up on putting them in force. It has even conceded that it is "quite possible" the rules will be modified.

Of course it is possible. But whether the revised secrecy rules will be an improvement remains to be seen.

Albany, N.Y., August 26, 1983

Despite the best of intentions, the U. S. Energy Department is proposing new restrictions on nuclear information that are neither practical nor prudent.

Alarmed at the increasing likelihood that terrorists may use data about the transportation of nuclear material, the design of atomic weapons and the security of nuclear facilities, the department wants to classify thousands of volumes of information that are now readily available to government contractors and universities. The restrictions would also apply to data on the public's exposure to radiation, thereby affecting the outcome of civil suits against the government.

No one should want to make it easy for terrorists or deranged misfits to build a nuclear device and hold a large segment of the population hostage to their threats. Nor should anyone minimize the dangers of over-zealous protestors using information to sabotage nuclear plants. But if the Energy Department really wants to prevent such possibilities, it should cease trying to change the past and concentrate instead on drafting sounder precautions for future data.

Once information is out — as the nuclear data the Energy Department wants to keep secret is — there's almost no chance of reclaiming it. As Gerald J. Lieberman, dean of graduate research at Stanford University, notes, much of the material that would fall under the proposed restrictions is now included in basic and advanced course in physics, electrical engineering and material sciences.

It's safe to assume the same courses are offered abroad, where the Energy Department would have no jurisdiction. Under the circumstances, any attempt to recapture such widely published information would be a foolish gesture. At its worst, it might lead to an underground information network here to smuggle data to saboteurs.

Ironically, the very restrictions the department claims are necessary to protect the public good would expose the public to new dangers. That's because states would no longer have access to important information on radiation.

James I. Barnes, the head of Nevada's Energy Department, speaks for all states when he protests the federal restrictions because they might "deny access to the public to radiation exposure data for . . . test site workers" and " . . . could affect the ability of the state to know about shipments of defense high-level radiation waste moving through the state."

What kind of restrictions are they that might harm the public in the name of saving it? Very clearly, ones that ought to be dropped.

Detroit Free Press

Detroit, Mich., August 23, 1983

IF THE Department of Energy is truly serious about restricting access to nuclear data, it should drop its piecemeal approach and simply shut down the universities and libraries of America. That, in essence, is where the department is headed when it finds a threat to security in the unclassified information that is already part of the standard undergraduate curriculum in nuclear physics. The department's plan would have the effect, if not the intention, of classifying much ongoing research and putting off limits library selections of the better universities.

The plan is a reflection of the administration's general effort since last spring to control what was once considered public information. But at issue in this case is who would be hurt most by the curbs. We are not talking about restricting legally classified data, in this instance. It's not simply a case of blocking the Russians from access to nuclear data. It would be American citizens who, once they lost access to the relevant information, would also lose the ability to question the decisions of government officials regarding nuclear policy.

The stated intention of the DOE regulations is to reduce the possibility of espionage and sabotage of nuclear plants and weapons sites. So broadly written is the proposal, however, that it would require reviewing a tremendous amount of information already in the public domain — whether in textbooks or the archives of other governmental agencies — for reclassification. It would allow federal officials to declare material that cannot be classified under current law to be "unclassified controlled" — that is, unavailable.

It could also make it harder for workers in nuclear energy plants and weapons testing sites to determine on-the-job hazards. It could prevent states from monitoring shipments of high-level radioactive waste through their territory. It could sharply decrease the free exchange of scientific information necessary to pursue basic nuclear research.

The United States owes much of its political and technical success to its unparalleled ability to produce and disseminate knowledge. Damming the flow of unclassified information would only mimic the most stultifying aspects of the Soviet system without substantially adding to national security.

THE TENNESSEAN

Nashville, Tenn., August 22, 1983

AFTER a barrage of protests from states, universities, unions, environmental groups and government scientists, the Reagan administration is revising its proposal to make sweeping restrictions on nuclear information. It certainly should.

When the regulations were first proposed they were generally criticized as vague and overly broad. They would let the secretary of Energy seek fines of up to $100,000 and prison terms up to 20 years for anyone who allows access to information that might "provide important insights into muclear material production and processing."

The stated aim of the proposals was to prevent terrorists from getting documents that could help them make a bomb or sabotage a nuclear weapons plant. That may be a worthy idea, but the breadth of language raises a number of concerns.

Colleges and universities complained that the definition of "unclassified nuclear information" was broad enough to include material used in basic physics and electrical engineering courses. States and enviornmental groups said the language would weaken the ability of citizens and states to monitor the safety of nuclear programs.

Library associations said the rules would put an undue burden on employees in libraries possessing hundreds of thousands of declassified documents on nuclear weapons.

A spokesman for the Oil, Chemical and Atomic Workers Union called the regulations a "shotgun approach" that would make it impossible to identify and get information essential for protecting the health of nuclear workers.

Mr. Herman Roser, assistant energy secretary for defense programs, said much of the criticism was based on misunderstanding of the proposals. He said, "We do not intend to impose restrictions on materials which have received widespread dissemination." But if the regulations are carefully read, they are vague enough to allow the most sweeping interpretation.

It is curious why the government feels impelled to issue such regulations in the first place. It already has the power to classify information to keep sensitive documents out of the reach of terrorists. Besides, information on nuclear materials is generally available in most other Western countries.

In any case, a nuclear bomb is not something terrorists could put together in a warehouse somewhere, even with the most detailed outline of how to do it. It is a complicated and enormously expensive undertaking, requiring resources that normally only governments could provide.

It is encouraging that the government plans more restrictive language, but it would be better off to rely on the powers it already has and forget the idea of new regulations.

Part II: Nuclear Waste

The future of nuclear power depends to a large degree on the safe disposal of radioactive wastes at a reasonable cost. Concern over nuclear wastes ranks second only to fear of a catastrophic accident among popular worries about nuclear power. Radioactive wastes are usually classified either as low or high-level. In the first category are lightly contaminated materials such as briefly exposed clothing or equipment which are discarded by hospitals as well as nuclear plants. Spent nuclear fuel, sometimes referred to as medium-level waste or potential high-level waste, is radioactive enough to require special handling but not to generate high temperatures. High-level waste is created when the spent fuel rods are dissolved in acid during chemical reprocessing to recover unused uranium and plutonium; the remaining acid solution, containing heavy concentrations of radioactive isotopes, is referred to as high-level waste. These wastes generate a tremendous amount of heat and can remain toxic for hundreds or even thousands of years. They must be artificially cooled, possibly for decades, before they can be permanently stored. It is these wastes which pose the greatest technical problems in devising adequate storage facilities. Of more immediate concern in the U.S., however, is the storage of the growing amounts of spent reactor fuel awaiting reprocessing. If reprocessing is never undertaken, of course, similar permanent storage sites will have to be found for the spent fuel rods.

Scientists in many countries are currently conducting research with the goal of developing the best available technology for high-level waste storage. Geologic containment of the wastes appears to be the most promising of the options, although disposal in space or in oceans has also been considered. Deep underground salt domes, clay deposits, and granite or other stable rock formations are all possibilities for permanent geological storage sites. Before storage, most experts now predict, the wastes would be incorporated into solid glass blocks, through a relatively simple vitrification process already in use in France. There are uncertainties still to be resolved, however, about the potential effects of heat generated by the wastes, seismic disturbances, future climatic changes or other unknown factors on the geologic structures of the proposed sites. In the United States, where the primary focus of such research has been on natural salt beds in about half a dozen western, mid-western and southern states, there is some uncertainty about the possible future interraction of circulating ground water with the waste and surrounding salt formations. Researchers are optimistic that these potential problems can be overcome within the next decade or two. But the Energy Department announced in December 1983 that President Reagan would not be able to recommend a site for the nation's first high-level waste burial dump until 1990, three years after the deadline established by the Nuclear Waste Policy Act of 1982. A Department spokesman said the need for more technical studies would also cause a two-to-three-year delay in meeting the 1998 deadline for having a repository in operation. Some of the buried wastes could remain toxic for centuries; it is difficult for scientists to predict what geologic changes might occur in that time to expose the wastes, threatening future generations.

A more immediate stumbling block to the permanent disposal of high-level wastes in the U.S. is the continuing unwillingness of individual states to accept such a site within their borders. States' rights have played a large part in almost all aspects of the nuclear waste issue, including the transport of wastes across state lines. A recent Supreme Court decision—upholding a California ban on new nuclear plants until a federal solution to the waste problem has been found—confirmed the powerful influence of state governments on this and other important aspects of the nuclear energy dilemma.

Two of Three U.S. Low-Level Radioactive Waste Sites Closed

National attention was focused on the problem of radioactive waste storage in 1979 when two of the nation's three nuclear waste dumps were closed. The three sites were in Hanford, Wash., Beatty, Nev., and Barnwell, S.C. Washington Gov. Dixy Lee Ray closed the Hanford nuclear waste disposal site Oct. 4 because improperly packaged waste material was found in trucks bound for the plant. Ray said, "I am ordering the site closed until I can get a guarantee from the Nuclear Regulatory Commission that they will enforce proper packaging. I will not tolerate sloppy procedures." On Oct. 22, Nevada Gov. Robert List closed the low-level dump south of Beatty because technicians had found five barrels of waste buried outside the fence. Both dumps were operated by Nuclear Engineering Co. of Louisville, Ky.

The Nevada and Washington dumps were the only ones in the U.S. handling liquid radioactive chemicals, the type used in U.S. cancer research, and their shutdowns threatened to put a stop to much of the nation's cancer research.

South Carolina Gov. Richard W. Riley Oct. 31 ordered a reduction in the amount of low-level nuclear waste buried at the nation's third site. Gov. Riley said that within two years, the amount of waste stored annually at the Barnwell facility would have to be cut in half, to 1.2 million cubic feet from 2.4 million cubic feet. The 240-acre facility was now the only commercial nuclear waste disposal site in operation. Riley said the Barnwell plant would continue to accept all low-level nuclear waste generated in South Carolina, but would reduce shipments from other states. He said he hoped the decision would precipitate action on the national level to solve the nation's nuclear waste problem.

The Seattle Times

Seattle, Wash., July 25, 1979

GOVERNOR Ray is one of the nation's most enthusiastic supporters of nuclear power, and that's why her recent confrontation with the Nuclear Regulatory Commission raised a few eyebrows. Happily, the governor not only got the N.R.C.'s attention, but she got it to take some needed action.

In a letter to the commission this month, Dr. Ray and the governors of the other two states with dumping grounds for low-level nuclear wastes (Nevada and South Carolina) virtually threatened to shut down those facilities if the federal government failed to strengthen its inspection and enforcement efforts by August 1.

The three governors complained of "serious and repeated disregard for existing rules governing the shipment of commercially generated low-level nuclear wastes," and a "total lack of corrective measures" by the N.R.C.

Their action followed two accidents involving nuclear wastes in Nevada in two months — a truck leaking unsafe radiation at a desert dump site, and the exposure of 10 persons to low-level contamination when a truck carrying nuclear medical wastes burned.

Last week, we were pleased to note, the commission chairman, Joseph M. Hendrie, promised to impose stricter handling regulations on low-level nuclear wastes in the three states.

Hendrie said the N.R.C.'s regulatory improvements would include:

— Regular inspections of both sources and collection sites of nuclear wastes, including opening of containers.

— "Consistent and uncompromising" enforcement of sanctions for violations.

— Monthly inspection reports to the three states.

— A joint statement by the N.R.C. and the Department of Transportation to all generators of wastes, listing details of the new plan.

The transportation and storage of wastes is one of the most troubling and controversial aspects of nuclear-energy development, and it is encouraging that the governor is taking such a firm stand on this particular safety issue.

The Des Moines Register

Des Moines, Iowa, July 17, 1979

Two protests on nuclear waste — one from the governors of three states, the other from the leaders of 12 South Pacific nations — call attention again to the unresolved questions about disposing of the radioactive leftovers of nuclear power.

At the National Governors' Conference in Louisville, the chief executives of Nevada, South Carolina and Washington joined in a threat to close nuclear waste dumps in their states unless the federal government tightens the rules on packaging and shipping hazardous materials and then steps up enforcement efforts.

Nevada's Gov. Robert List temporarily closed the disposal site in his state after a truck with leaking containers arrived there. Two months ago, a truck carrying nuclear waste caught fire at the same site. In both cases, violations of proper loading procedures were found.

The three states are the only ones with designated sites for storing low-level nuclear waste. If those sites were closed, nuclear power plants would have no place outside their own sites to store wastes.

The outcry resulting from such a situation would likely produce quick remedial action in Washington.

The South Pacific protest focuses on a U.S. plan to dump nuclear waste on uninhabited Palmyra Island, lying just north of the Equator between the Hawaiian Islands and Samoa. As leaders of the 12 nations opened their annual conference, New Zealand Prime Minister Robert Muldoon summarized their feelings by saying, "We don't want nuclear waste in the South Pacific."

Given a choice, most people would not want nuclear waste anywhere, certainly not in proximity to where they live.

The fuel crisis has set the mood for full-speed-ahead approaches to the energy shortage. However, expansion of the nation's nuclear-power generating capability should be approached with considerable prudence in view of the dilemma posed by the potentially long-term risks from the disposal of radioactive wastes. It is not a question that should be left for another generation to solve.

CHARLESTON EVENING POST

Charleston, S.C., July 24, 1979

Gov. Dick Riley is pushing the federal government to develop a coherent policy on nuclear waste disposal, and his efforts should be applauded. The governor has given notice to the Nuclear Regulatory Commission, and to President Carter, that regulations for high-level and low-level nuclear waste should be forthcoming; otherwise, states like S.C. might balk at storing wastes in the future.

Out state is shaping up as one of the major recipients of nuclear waste. Allied General's plant in Barnwell seems a likely respository for wastes from nuclear power plants; Chem Nuclear Services near Barnwell already is a major receiver of low-level waste material; the Savannah River plant has stored nuclear wastes in its tanks for years.

Gov. Riley's tough tone is prompted by excessive delays by government in facing the nuclear waste problem, as well as by the reluctance of other states which have nuclear power generating facilities to accept responsibility for waste storage. His suggestions about regional waste disposal sites and reciprocal arrangements for low-level waste storage met with a cool reception at the National Governor's Conference held recently in Louisville, Ky.

Gov. Riley emphasized that he "absolutely" wants to see a reduction of waste storage in South Carolina by the end of his term in 1982. His uncooperative attitude should stand as fair warning to federal officials that our state will not become a nuclear dumping ground for the rest of the country. Eighty percent of the nation's low-level radioactive waste is now stored in South Carolina. Gov. Riley believes that's considerably more than a fair share. He's right.

Columbia, S.C., July 15, 1979

THE STATE encourages Gov. Dick Riley to persevere in seeking a national policy on the storage of low level radioactive wastes — and to get tough about it if necessary.

Low level waste is literally atomic garbage, things like clothing, animal carcasses, containers — generally anything exposed to radiation but which isn't from a nuclear reactor. These wastes come from industries, research laboratories, hospitals, etc. The plastic shoes President and Mrs. Carter wore at Three Mile Island are examples of such wastes.

Unlike high level radioactive materials out of the power reactors, the low level radioactive wastes are not subject to a federal policy on disposal. The federal government may turn the responsibility over to a state by agreement. South Carolina is an "agreement" state.

South Carolina has a ground storage facility for low level radioactive wastes in Barnwell County adjacent to the Savannah River Project, the federal nuclear defense materials installation.

The Barnwell site is operated by Chem Nuclear Services, a profit-making company which has been licensed by this state. At present, almost all of the radioactive garbage east of the Mississippi River is shipped to South Carolina for burial. The only other storage locations are in Nevada and Washington.

As the amount of the atomic garbage grows, will become a South Carolina problem unless other states provide for storage. Commendably, Governor Riley has tried to sell his fellow chief executives on the idea of a regional burial place, but he's not getting anywhere.

Governor Riley has rightly concluded that it will take something dramatic to force other states to accept some responsibility for their atomic wastes. Last week he informed the Nuclear Regulatory Commission, along with the governors of Nevada and Washington, that there must be some action on the matter.

Just as the NRC has been put on notice (State Energy Director Lamar Priester will be negotiating with NRC officials this week), our sister states should be given fair warning that South Carolina will not indefinitely accept the East Coast's atomic garbage.

Governor Riley has not yet set any deadline to place an embargo on the trucks which bring the materials to Barnwell County. He should make it clear, however, that a deadline is going to be set and South Carolina will stick to it.

From what we know of the problem, it is no big deal for other states to handle. They can dedicate some acreage to the storage of atomic wastes. It is a lucrative business, we hear, and an enterprising company can be found to take charge although the state will eventually inherit the burial site.

Chem Nuclear is required to pay 16-cents per cubic foot of stored materials into a state reserve fund. The money will be used to cover South Carolina's costs of maintaining the site when it is filled. There is about $1 million in the fund now, and the fee will be raised soon.

A national policy notwithstanding, we hope it will not be necessary to turn back shipments from other states at the South Carolina state line to get attention on the problem. But if thats what it takes, then South Carolina should do it when the time comes.

Columbia, S.C., October 28, 1979

GOVERNORS of South Carolina, Washington and Nevada have followed through on the nuclear waste warning they issued jointly last spring. Having forced the issue, the governors should not back down, despite strong pressures that are being exerted on them.

Govs. Dixy Lee Ray of Washington and Robert List of Nevada have closed down the low-level nuclear waste burial grounds in their states, leaving South Carolina with the only such despository in the nation still doing business.

And Gov. Dick Riley, true to his word, is not allowing any new sources of wastes to use the Chem-Nuclear site at Barnwell. He also is preparing to go even further and cut back on the volume of waste coming into the dump.

Pressing problems are at hand because of these developments, particularly in the field of nuclear medicine. Some centers and hospitals have no place to send their waste material and say they may have to curtail treatment of patients and other activities by Dec. 1 unless additional disposal sites are found. Utility companies and other sources of nuclear waste also are experiencing problems.

While some exceptions might be in order if treatment of the sick is threatened, the governors should stand firm on their present policies to make sure the attention of other states and the federal government is not lost.

The warning the three governors sounded last spring is coming into focus. It is time for each state to assume the burden of disposing of its own low-level nuclear garbage. This approach is better than having the federal government do it, because still more political unhappiness could occur.

We should dispel some of the alarm about this near crisis. The kinds of nuclear wastes involved — low-level — pose little or no health threats with proper handling. Years ago, some medical institutions literally buried this stuff in their backyards.

This point makes it even more imperative that individual states bear the responsibility of dealing with their waste. Furthermore, with the demand for disposal sites, there are potentially profitable commercial enterprises to be had.

The federal government should have taken a more active role in the low-level waste issue in the past, but if it is forced to do so now, the problem might well fall back in the laps of Nevada, Washington and South Carolina. The United States operates nuclear defense installations in these states, and they offer tempting alternatives for burying the low-level commercial waste.

The three states have absorbed more than their fair share of the nation's nuclear waste, and it is time for others to contribute. Governor Riley and his colleagues should stick by their course of action until they get some desired results.

The Providence Journal

Providence, R.I., October 29, 1979

The closing of two of the nation's three dumps for low-level radioactive wastes by state officials has dramatized a problem that has bothered them for some time. But it is not an insoluble problem. The loose practices in packing and transporting — and even in burying — these wastes can be corrected. Americans are finding that nuclear power plants require a higher degree of inspection, training and care. So too do the nuclear waste dumps.

When Gov. Dixy Lee Ray closed the low-level dump at Hanford, Wash., she was concerned because the firm that ran the dump was not observing a high enough level of care in its operation. Gov. Robert List of Nevada had a similar complaint, and it went back a lot further than the discovery this week of five barrels of waste buried outside the dump fence.

First of all, it must be understood that these dumps and one at Barnwell, S.C., are only for materials that are not "hot" in the usual sense. They include isotopes that have been used in medical treatment or research — small amounts that can easily be contained — and all kinds of materials that have been contaminated in any way at nuclear installations — rags, clothing, utensils, machinery, etc. that has absorbed some radiation. It is easier to bury the stuff than to try to decontaminate it; some can't be decontaminated.

But this material is of an entirely different order than the high-level radioactive materials that are a by-product of used fuel from a nuclear power plant, or the highly toxic wastes left over from the manufacture of plutonium for nuclear bombs — or even in the reprocessing of fuel used in nuclear submarines. As is well known by now, facilities haven't yet been developed for storing the high-level wastes. Sites haven't yet been selected, nor have any plants been built for converting millions of gallons of liquid waste into solids that can be transported to eventual burial sites.

The low-level wastes have never been considered a serious problem. The sites, comparable to landfills, should be in soil that has a minimum of leaching into waterways. And they should be determined not by political maneuvering but on the basis of hard scientific facts.

If the packaging and handling have been slipshod, then certainly the federal government, through the Nuclear Regulatory Commission or other agency, can see to it that responsible firms do the work. Nor should any one state, or handful of states, become the goat of such a program. The burden of accepting and policing such sites should be spread as widely as feasible, although the more sites there are, the harder the program is to police.

It's ridiculous for the medical use of isotopes to be held up by disputes over waste sites. President Carter, the NRC and the Department of Energy ought to be able to solve the problem, in cooperation with the states, quickly and efficiently. Storage at six federal laboratories, as suggested, would be only a temporary, stopgap, solution.

Officials ought to be pressing hard, moreover, for the creation of test storage sites for the high-level wastes. That's the real headache for this nuclear age. Nearly 100 million gallons of liquid or semi-liquid wastes are being stored at military installations around the country. Getting them into permanent storage where they won't leak out to contaminate the environment is a multi-billion-dollar undertaking. And despite the cost, it ought to be expedited.

The Charlotte Observer

Charlotte, N.C., July 19, 1979

Governors of the three states where the only operable sites for burial of low-level radioactive wastes are located — including South Carolina's Dick Riley — have sent an ultimatum to the Nuclear Regulatory Commission.

They say they've discovered "serious and repeated disregard" of rules governing shipments of commercially generated nuclear wastes to the burial sites in their states. They say the NRC hasn't tried to correct these problems. They're unhappy that their states (South Carolina, Nevada and Washington) have become the repositories for all the nation's low-level waste.

They make a good point. One problem with nuclear power now getting wide recognition is that its use has precededdevelopment of a well-defined, reliable plan for protecting people from the waste products.

The nation's first commercial burial site for low-level radioactive wastes opened in 1962. The idea was deceptively easy, safe and cheap:

We would "dispose" of material contaminated with small amounts of radioactivity by burying it. The stuff included nuclear workers' rubber gloves, medical preparations laced with tiny amounts of radioactive material and the like. (It did not include such highly radioactive materials as spent fuel — another problem.) If we put this slightly radioactive junk in boxes and barrels and covered it with dirt, we'd end the problem, we thought.

In the mid-1970s, problems emerged. Radioactivity was leaking from some of these nuclear landfills. Three of the six low-level burial sites were shut down.

Low-level wastes are the least scary of radioactive byproducts, but they've been so badly handled in many places as to endanger people living near such burial sites.

The angry governors don't want that to happen in their states. They're not-so-subtly reminding other states that Americans can't continue consuming nuclear energy and technology while tossing the refuse into their neighbor's backyard.

Governors Riley, Dixy Lee Ray of Washington and Robert List of Nevada, want a detailed plan for upgrading inspection and enforcement of proper packaging and shipment of wastes. They want the plan delivered by Aug. 1 and implemented by Sept. 1. They should get it.

Charlotte, N.C., October 22, 1979

South Carolina's Gov. Dick Riley is going to fool around with nuclear wastes until he starts a real discussion. Good. Carolinians will be wiser for it.

Events in South Carolina last week suggest that the nation's leaders — in government and the nuclear industry — are likely to be drawn into the debate. The result could be a clear national policy on nuclear wastes, an achievement Gov. Riley could be proud of helping to bring about.

Right now, not even scientists can agree how to manage wastes, how dangerous they are or even how long they must be isolated from people and the environment. But they agree that wastes present a problem requiring special care.

Essentially, that's what the discussion in South Carolina was about: how careful the nuclear industry must be in disposing of wastes, and how much the government must spend to assure that wastes are disposed of properly.

Early last week, Gov. Riley forbade an Illinois firm from sending more low-level nuclear wastes to South Carolina for disposal. He said the company has a history of sloppy waste shipments, and the ban would demonstrate the tough approach he is taking with a nuclear industry previous state officials had courted.

But later in the week, state officials acknowledged they had stopped full-time inspection of waste shipments to the state's burial site at Chem-Nuclear Systems Inc. near Barnwell. Regular state inspections were begun last April, shortly after the accident at Three Mile Island. They were stopped because there weren't enough inspectors to monitor the site six days a week, from 7 a.m. until 6 p.m.

David Reid, Gov. Riley's energy aide, acknowledged Friday that the less frequent schedule of inspections could be unsettling. But he said Gov. Riley's ban early in the week was aimed at a company which had not lived by the rules.

That isn't the case with Chem-Nuclear, which inspects every shipment it accepts, he said. The state's periodic inspections will ensure that Chem-Nuclear also continues to live by the rules. But right now, officials have no problems with the company. "If we did, we'd shut them down," he said.

But the problem brings home a debate rising nationwide over how much inspection is needed and who should provide it. To that end, Gov. Riley and the governors of Nevada and Washington will meet with experts at the Nuclear Regulatory Commission Nov. 6 to press for federal commitments to work with the three states now handling all the nation's low-level wastes. Let the debate continue.

Oregon Journal

Portland, Ore., October 26, 1979

Two down, one to go. The three dumps in the country equipped to handle low-level nuclear waste are now reduced to one — in Barnwell, S.C. The governor of Nevada has just closed Beatty.

The federal government has said it would take responsibility for handling this waste — but saying isn't doing. No federal site has been designated, no plan agreed upon.

Meanwhile, the stuff mounts up and must be dealt with. The governors of the states which accept the waste have closed the dumps in protest over improper handling or improper preparation of the waste.

The French government has begun a glassifying system of containing and stabilizing its nuclear waste. Some system had to be adopted in the face of the intense development of nuclear energy in France.

A number of disposal options exist, none of them perfect. But almost any of them would be better than vacillating back and forth among them while more and more waste accumulates.

Does the federal government simply hope the problem will go away while it marks time? Does it hope that states will take the matter into their own hands and deal with it?

Every time a substance has to be handled, it makes one more chance for somebody to do something wrong, for a container to be damaged or for the transport to break down.

If every state is finally driven to taking a large piece of land to store nuclear waste for its own industries, it's a poor use of a scarce commodity. If Hanford will accept nothing but waste from the state of Washington and Beatty nothing but that which comes from Nevada, all the other 45 contiguous states will have to take their waste to South Carolina, which is a mighty far piece from Oregon.

South Carolina has just announced that it will accept no waste originally intended for Hanford or Beatty.

In the minds of the public, deciding this matter should have had a high priority in Washington D.C. Periodically it is discussed with vigor, and then the difficulty of the decision saps away the energy to make firm arrangements and carry it.

If you're having cracked crab in May, you can freeze the shells until the day the garbage man is coming to collect. And if he's not coming at all, you just don't have cracked crab.

Nevada State Journal

Reno, Nev., July 25, 1979

Governor Robert List and Human Resources Division chief Ralph DiSibio are to be commended for their tough stand on nuclear waste in Nevada.

Following two episodes of careless handling of radioactive waste shipments to Nevada, the governor closed down the Beatty dump. He barred shipments from some industries. He has insisted on stringent controls and inspections. He has rallied the support of the governors of the only two other states which accept radioactive waste shipments - Washington and North Carolina.

And in a somewhat remarkable display of executive power, DiSibio ordered a leaking load of hot wastes to be sent right back to its home plant in Michigan. If Michigan was not going to bother to see the wastes were packed and shipped safely for Nevadans, then let's see how it likes getting them back, he reasoned.

It was a strong and gutsy stand for the state to take and one which shows our top officials are acting responsibly in safeguarding the health and safety of Nevada residents.

It is a stand we want to see continue as others seek to bring radioactive products to Nevada.

A case in point is the recent application of an Arizona-based corporation to do business in North Las Vegas.

Those reviewing the application of American Atomics Corp. should do so with a wary eye on the company's past performance in Arizona.

The firm, which manufactures products containing radioactive materials, has applied to move its operation from Tuscon. American Atomics Corp. manufactures illuminated traffic signs and digital watch components which contain tritium, a low-level radioactive material.

The corporation was recently charged with 17 radiation safety violations by the Arizona Atomic Energy Commission. The Environmental Protection Agency had found food and water to be contaminated with radioactive material from the plant and said monitoring of radiation at the plant was inadequate. The firm has also been charged with falsifying documents related to testing of tritium-filled tubes prior to shipping them out of state.

Furthermore, the corporation surrendered its license to Arizona during a hearing which included charges that the company violated 11 state laws governing radioactive matter.

During the Arizona hearing concerning the EPA findings, the corporation denied all charges. But then, in a surprise move, it announced its intention to terminate its license with the state and move elsewhere. A Nuclear Regulatory Commission investigation of the firm is also underway.

North Las Vegas city manager Ray Schweitzer said American Atomic officials had not told him about the Arizona problems, and had said the radioactive elements were "completely harmless."

That's a line of reasoning that Nevadans, who are becoming more sophisticated about matters of radiation, are not likely to accept. Neither should state officials.

Evidence is mounting that low levels of radiation can have a cumulative adverse affect within the bodies of human beings.

Although much is yet to be known about these effects, one theory produced by the federal Interagency Task Force on Ionizing Radiation is that: "there is no threshold dose level below which radiation exposure has no carcinogenic effect." In other words, any exposure to radiation - even low-level - may have a potential for causing cancer.

In Tuscon, EPA tests showed that a sample of cake taken from a central school kitchen which served 40,000 students in the city was contaminated with tritium, used at the American Atomics Corp. plant. Tests also showed tritium in some swimming pools and in the urine of some residents.

We don't know what the outcome of the Nuclear Regulatory Commission hearing will be, of course. But based on the news from Arizona, we hope Nevada officials will take the same strong, cautious attitude regarding this company as it has with other atomic industries.

North Las Vegas officials, upon hearing of the firm's Tuscon troubles, accused company representatives of a cover-up.

Councilwoman Brenda Price acknowledged she shared the attitude of many that North Las Vegas is "very hungry" for new industry.

But after hearing details of the Arizona hearings, she added: "But we're not hungry enough to eat radioactive cake."

We couldn't have said it better.

THE DAILY HERALD

Biloxi, Miss., October 24, 1979

One of the persistent problems nuclear energy advocates have not solved is the credibility gap of scientific experts. What they tell the public often isn't so.

The Nuclear Engineering Division of the American Institute of Chemical Engineers had the following assurings about storage of nuclear wastes of low or intermediate levels of radioactivity in one of its pamphlets:

"Radioactive wastes that are highly dilute or intrinsically of relatively low activity are disposed of by release and dilution or by shallow burial under careful controls, much as are other industrial wastes. There are no significant risks or effects from these disposals and the cost of more complete containment is generally not justifiable."

Sounds reassuring, doesn't it? Nothing to worry about, right?

Wrong.

Recent events reveal that not everyone shares the complacency of the chemical engineers.

Nevada Governor Robert List this week ordered a Nevada radioactive waste dump closed after a U. .S. Geological Survey team found what was called "irrefutable evidence...of gross mismanagement" at the site. It was the third time this year Gov. List ordered the dump closed.

Washington Governor Dixy Lee Ray temporarily shut down her state's commercial disposal site at the Hanford Nuclear Reservation following discovery of defective packaging of radioactive waste.

With the closings of those two nuclear waste dump sites, the nation is left with only one site open, the South Carolina Barnwell site, the largest of the three.

The governor of South Carolina, Richard W. Riley, has said he will not let waste from the closed dumps be diverted to South Carolina. Where then will the wastes go?

Most of the low level waste comes from hospitals and universities using radioactive materials for cancer treatment and research. The nation needs the benefits of such activities, but it also needs protection from the by-products.

Mississippi salt domes, some located only 60 miles from the Gulf Coast, are still being studied as possible sites for storage of high level radioactive wastes, which are more risky than the low level variety. We have urged state officials to be cautious in monitoring those studies and making any commitments we may all later regret.

It is obvious from the events in Nevada and in Washington that current management practices, even for low level wastes, are subject to carelessness, despite the optimistic view of the American Institute of Chemical Engineers.

The carelessness emphasizes again the need for this nation to begin providing safe containment of the radioactive wastes we continue to produce.

Long Island, N.Y., October 28, 1979

Because some states are accepting less radioactive waste for dumping, New York might have to find room for more. The government may ask Brookhaven and five other national laboratories to store wastes generated elsewhere. Or it may try to persuade the governors of the states that produce the most—New York is one of them—to set up temporary storage facilities.

What's needed, though, is not a temporary solution but a permanent one.

We don't blame the people of Nevada for being worried about radioactive wastes that have been improperly stored at the state's Beatty nuclear dump. Nor do we blame Gov. Bob List for closing it down; politically speaking, he had little choice.

But the implications for the country are grave. The situation is aggravated by the temporary closure of another commercial dump, at Hanford, Wash. That leaves only one open site, at Barnwell, S.C., where the governor won't accept waste diverted from Beatty or Hanford. Officials of hospitals and universities that generate radioactive materials are justifiably alarmed.

Meanwhile, high-level wastes and spent reactor fuels are being stored in temporary facilities at commercial power plant sites all over the country pending a federal decision on a permanent disposal site.

As the governors' actions demonstrate, the disposal problem is not only technological but political. People don't want a radioactive disposal site in their state, much less in their backyard. We doubt that Long Islanders are very pleased with the Brookhaven prospect, for instance. Unfortunately, that won't make the problem—or the wastes—go away.

Faced with political pressures of its own, Congress is thinking about granting states veto power over nuclear dumps. That would be a grievous mistake. When a method and a site are finally chosen—and we urgently hope that will be sooner rather than later—the choice should be based solely on the suitability of the site from a geological, geographical, security and transportation point of view. If politics is allowed to interfere, the country may never get a site operational.

We suspect that a good many governors realize this. But if Congress gives the states the power to veto sites, they will have no choice but to use it. The federal government must be able to make the choice and enforce it. There's simply no other way.

The Miami Herald

Miami, Fla., October 29, 1979

RADIOACTIVE wastes remain the Achilles heel of the nuclear industry, posing a growing risk to the public's health and safety. Chances of a spectacular explosion or a "China syndrome" meltdown are remote compared with the possibility that the environment will be contaminated by radioactive wastes.

The problem has been building for years as the applications of radioactive materials multiplied. Federal policies have encouraged, even fostered, the expanded use of such materials in fields ranging from medicine to the generation of electricity.

Yet the Federal Government has defaulted on its responsibility to provide for the safe shipment, storage, and disposal of nuclear wastes. Moreover, Federal policy reversals have left private industry with a serious and unanticipated problem.

Now that problem has been exacerbated by the closing of two of the nation's three commercial dumpsites for low-level nuclear wastes.

In Nevada and in Washington, state governors ordered dumps closed. This leaves a dump near Barnwell, S.C., as the only one for that purpose in the United States. Predictably, the governor of South Carolina has warned that he won't permit his state to become the nation's dumping ground for nuclear wastes.

The Federal Government, then, must face up to a short-term crisis and a long-term problem wrought by its own policies. In the short run, a place must be found for the wastes diverted from the dumpsites now closed. Telling users to keep the wastes on site is no solution at all; it multiplies the risk of a mishap.

In the long run, if the Federal Government is going to remain committed to nuclear power and to other applications of radioactive materials, then it must show as much interest in the technology of waste disposal as it does in the technology of weaponry.

The problem is not as dramatic as the accident at Three Mile Island. But unless a safe solution can be found for the disposal of nuclear wastes, the future use of nuclear materials will be hampered and mankind will be denied the many potential benefits."

San Jose, Calif., October 29, 1979

RATHER than bursting into the headlines in spectacular, Three-Mile-Island fashion, the latest crisis of the nuclear age sneaked up on us quietly and inconspicuously. But in its own way this quiet crisis is no less dangerous. Unless it's dealt with intelligently, it could jeopardize essential medical research all over America.

Radioactive isotopes are used in many types of research and therapy at virtually all major medical institutions. These activities produce radioactive residues that must be disposed of at approved sites.

There used to be three such sites in the country: one in Nevada, one in Washington state, the third in South Carolina. Three weeks ago the one in Washington shut down, and last week Nevada closed the second one. That left only the site in South Carolina, which won't accept radioactive materials that formerly were sent elsewhere.

The result: instant chaos at medical research facilities all over the country.

Stanford University Medical Center, which is better off than many because it recently sent a shipment off to Washington and has the capacity to store its wastes for at least several months, faces no real emergency yet, but has advised researchers to use radioactive materials sparingly until the situation is resolved. Stanford says there's no immediate danger of eliminating therapy that utilizes radioactive elements, but that could happen at some hospitals.

There are hopes that the Washington site will soon reopen. But Nevada Gov. Robert List has indicated he doesn't want the radioactive wastes stored in his state. That means the country will be left with a total of two approved disposal sites — hardly a comfortable margin.

Stanford wants Gov. Brown to aid in developing a disposal site in California. In general, the university takes the position that each state should be responsible for disposing of its own medical wastes.

That's a philosophically sound position, but perhaps a politically naive one.

Medical radioactive wastes are far less potent and long-lived than commercial wastes such as spent reactor fuel; there's no rational reason to fear a properly managed disposal site for medical wastes. But voters — and the people they elect — don't always react on the basis of pure reason. Radiation has become one of the great scare words of 1970s politics. Pressures against permitting disposal sites would be immense, and many governors might not have the fortitude to withstand them.

Some prodding is going to have to come from somewhere, and the logical entity to provide it is the federal government. It could contribute by setting standards for disposal of low-level radioactive wastes, defining the states' responsibilities for handling them, and possibly helping with the job of policing the sites.

The events of the past week demonstrate that we've let this problem take care of itself far too long. If we continue to delay developing and implementing a coherent national policy, we shouldn't be surprised to find another, and possibly worse, crisis sneaking up on us.

The Hartford Courant

Hartford, Conn., November 5, 1979

Of all the nuclear-related problems facing the industry and the nation, none should be easier to solve than the disposal of low-level radioactive waste. The fact that the nation faces a crisis at the moment on how and where to put such wastes is an indictment of the federal regulatory process for nuclear energy.

How can citizens have faith in government and industry promises to safely dispose of extremely toxic, long-lasting, high-level wastes, at a time when discarding dirty rags and old wrenches proves such a chore?

Only three federal repositories for civilian disposal of nuclear wastes have been established, and two of them have been closed down by state officials in Washington and Nevada. The remaining dump site in South Carolina is not accepting the overflow from the closed facilities.

The two closings are not expected to be permanent; the decisions are dramatic protests, based on sloppy procedures discovered at both locations.

When and if they reopen, however, the federal government must eventually encourage, even demand, that regional dump sites be established around the country. Until recently, Hartford Hospital was shipping its nuclear medicine wastes all the way to Washington for disposal. Northeast Utilities, the area's largest producer of nuclear wastes, sends about 300 shipments a year, by truck, to the South Carolina facility.

More — and better — dumping facilities must be established, but first the federal government must assure state officials that proper procedure exists, and will be enforced. No governor will welcome such a dump site, and federal watchdogs must offer convincing evidence that the storage can be done safely.

Typical is the reaction of Gov. Ella Grasso, who matter-of-factly rejects the notion of burying low-level wastes in Connecticut, because of a recently passed state law prohibiting it. Connecticut generates a large amount of low-level waste. Who is to be responsible for it?

The immediate concern is not so much for the commercial nuclear plants, although there is a serious long-term disposal problem at those facilities. For now, Northeast Utilities is storing the low-level waste on its considerable real estate surrounding its nuclear facilities. But hospitals and other medical facilities have neither the inclination nor the space to store their wastes for extended periods of time. If the situation does not ease within a short period of time, governors on a regional basis should resist the urge to do nothing, and designate some temporary sites for storage.

The low-level wastes range from tools and uniforms, which represent little appreciable danger, to certain resins and filters that require more careful packaging and storage. While they are not of the significance of the highly dangerous fuel rods at the nuclear plants, the "low-level" remnants are a high-level embarrassment to a federal bureaucracy that seems incapable of finding adequate solutions to any of the waste-related problems involving radioactive material.

The News and Courier

Charleston, S.C., November 2, 1979

Gov. Riley had two solid reasons for ordering a 50-percent reduction in the amount of low-level nuclear waste that can be buried in South Carolina. Both relate to promotion of public safety.

The commercial waste disposal site at Barnwell is the only one of its kind in the United States. Similar facilities in Washington and Nevada have been closed down in the last two months. Their closings meant that South Carolina was on its way to becoming the nation's nuclear waste dump. All the low-level waste that other states did not want to contend with would have wound up here. With it would have come all the attendant worries over radiation dangers.

There is nothing fair about that sort of arrangement. Why should South Carolinians worry more so people in all the other states can worry less? There is no reasonable answer. Gov. Riley recognized the unfairness — and hazard potentials — of the situation before the other sites shut down. When he took office back in January he put other governors on notice as to his intentions. Now he has done what he said he would do about cutting back nuclear waste shipments.

The second reason behind his order is the governor's hope that he can pressure federal and state government officials into formulating and adopting a reasonable policy for dealing with low-level nuclear waste nationwide. Mr. Riley believes each state should be responsible for disposing of waste generated within its borders. He believes the federal government should be responsible for seeing that the states meet their obligations. Finding effective means of disposal is a challenge that can't be shrugged off safely year after year. Gov. Riley was trying to make that point when he issued his order.

The Wichita Eagle-Beacon

Wichita, Kans., October 27, 1979

The reasoning that has led to the shutting down of two of the nation's three low-level nuclear waste dumps is understandable. Kansans, especially, can appreciate the situation that confronts the people of Nevada, South Carolina and Washington state — it was in the 1960s that the federal government singled out Kansas as a high-level radiation disposal site.

No state wants to become the radioactive trash heap for the rest of the nation. That sentiment, voiced loudly and insistently, finally paid off when the government abandoned its idea of converting an empty salt mine at Lyons into a high-yield atomic waste disposal pit.

Unfortunately for the people in the other three states, they must deal with the worries of radioactive waste dumps after the fact — disposal sites have been in business there for years. But Washington's governor, Dixy Lee Ray, and the governor of Nevada, Robert List, have closed down the disposal facilities in their states, citing unacceptable safety risks to their constituents.

That may increase the pressure on South Carolina to take more of the irradiated material turned out by hospitals and laboratories across the country, and may give new incentive to a Carter administration plan to establish regional low-level radiation dumps around the country. That is where the Lyons salt mines come back into the picture — a request to approve one of the mines there as a low-level radiation disposal site remains before the Kansas Department of Health and Environment.

Radiation technology is an important, valuable part of modern medical diagnosis and treatment, as well as research. Not having working disposal sites available could complicate health care and negatively affect other industries that produce low-grade radiation wastes. But even at that, there is good reason not to give blanket endorsement to the practice of disposing of such wastes simply by burning them in out-of-the-way places.

Just as we believe that a better answer than that must be found for high-level wastes of nuclear reactors, we think a less prosaic, but more effective, way of handling low-level wastes must be developed. Worries about the toxic effects of radioactive materials finding their way back into the human environment have been justified all too often in the form of spills and leaks from hazardous-waste disposal sites.

The time has come to stop covering up such dangerous wastes and to stop putting off serious, concentrated research that will provide an acceptable way of neutralizing them.

Rockford Register Star

Rockford, Ill., October 31, 1979

One of the nightmarish concerns involving the rapid growth of nuclear energy has suddenly come to life in Illinois — and nobody has much of any idea what to do about it.

Until last week, Illinois was living in what was just about the best of all possible worlds. Commonwealth Edison's early commitment to nuclear power generating plants has provided Illinois with electrical power while other states suffered major shortages. That has kept our factories running and protected our jobs. And, while this was going on, Illinois escaped that deadly by-product of nuclear plants — radioactive waste — by simply sending it off to other states to bury.

But then citizens in those other states started asking questions, and the governors of those other states started to worry about their own political futures. First the states of Washington and Nevada decided to take no more Illinois nuclear waste. Then, last week, South Carolina closed its boundaries to Illinois waste.

That did it. There was no place else to turn.

And, although Illinois is one of the largest generators of both nuclear power and nuclear waste, it has no sites to dump the waste.

There are temporary storage areas at the major nuclear plants in Zion, the Quad Cities and Dresden — but their capacity is very limited.

Dresden might be able to store its own waste for another six months, but Zion and the Quad Cities have only four to six weeks of storage capacity left.

That's it. After that, new sites must be found.

The federal government isn't any help. For that matter, although the federal government is far and away the largest producer of nuclear waste through its military programs, Uncle Sam is asking the 50 state governors for help. There is no federal plan for nuclear waste disposal.

Illinois Gov. James Thompson is seriously alarmed, as he should be. So should be every resident of Illinois.

In the raging dispute over nuclear energy in this country, the anti-nuclear forces' best argument long has been that nobody knew what to do with the deadly wastes being created in massive amounts. The weakest defense of the pro-nuclear forces has been that "somebody will think of something."

Now the nightmare has come alive. The nuclear waste is piling up and "nobody has thought of anything."

It obviously isn't a problem any governor or any state can depend on having some other state solve.

And it obviously isn't a problem which can be left up to the 50 governors individually.

A solution — if a solution is possible — must be found at the federal level. It must be a solution which will treat each state equally. Until such a solution is found, the entire nuclear energy industry, and the immediate economic future of the nation, is very seriously threatened. And, because of heavy reliance on nuclear power in Illinois, it's a very immediate threat in this state.

The Houston Post

Houston, Texas, November 7, 1979

The Nuclear Regulatory Commission warned more than a year ago that the nation faced "urgent" problems in disposing of low-level radioactive waste. But "urgent" became "acute" last month when the governors of Nevada and Washington closed two of the nation's three privately operated nuclear waste dumps. The governor of South Carolina, site of the third commercial dump, compounded the disposal problem by banning any wastes that would have normally gone to the Nevada or Washington dumps.

Govs. Dixie Lee Ray of Washington and Robert List of Nevada cited unsafe handling and disposal of waste materials as justification for their closure orders. Ray, a former chairman of the old Atomic Energy Commission, said trucks hauling radioactive waste, as well as some of the packaging, were defective. List said several drums of waste material were found buried outside the Beatty, Nev., dump site.

Though the dumps were shut down to protect public health and the environment, their closure not only threatens nuclear power plant operations, but has paradoxically created a potentially critical problem for medical diagnosis and treatment that use nuclear materials. The producers and users of radiopharmaceuticals warn that if the dumps are not reopened or some other disposal sites made available, there may soon be no place to dispose of the wastes generated in the manufacture of these materials. The wastes are fast piling up in temporary storage.

Though the radioactivity in these wastes is not of the high level associated with reactor wastes from nuclear power plants and atomic weapons programs, many questions have been raised about the degree of health risk low-level radiation poses. Unfortunately, we may not have the answers to these questions for years, if ever. In the meantime, however, if something isn't done to relieve the problem of storing these low-level radioactive wastes, a wide range of medical research, diagnosis and treatment could be severely hampered. At M. D. Anderson Hospital and Tumor Institute here, for example, an average of 75 diagnostic tests are administered daily using radioactive materials.

In citing the problems of the commercial nuclear storage dumps in July, 1978, the NRC added that "there is no prospect of opening new commercial burial grounds in the near future." It recommended that the Department of Energy develop a contingency plan under which the government would handle disposal of low-level radioactive wastes. The closing of two of the nation's three waste disposal sites lends fresh impetus to the search for a solution to the storage and disposal problem — if only on an interim basis. Whether the government does the job or individual states make sites available, especially for their own wastes, the experience of Washington and Nevada underscores the need for more stringent policing of disposal procedures at the dumps. A remedy must be found for this particular facet of the nuclear waste problem that threatens to hamstring both electrical power generation and medical science.

The Louisville Times

Louisville, Ky., November 5, 1979

" . . . we don't have to start doing anything with the waste from nuclear power plants until nearly the middle of the 1980s . . . . we have got at least five years still to conduct research and maybe improve the techniques that are available today."

Dixy Lee Ray
Governor of Washington

When Gov. Ray, a former chairman of the Atomic Energy Commission and a proponent of nuclear power, made that statement in 1976, there were six sites around the nation that accepted what is termed "low-level" nuclear waste.

Last month, Gov. Ray announced that one of them, Hanford Reserve at Richfield, Wash., would be "temporarily" closed until difficulties with packaging of the wastes could be worked out.

That leaves only the site at Barnwell, S. C., still taking low-level waste. And last Wednesday, South Carolina Gov. Richard Riley gave notice that Barnwell henceforth will accept only half of what it was taking.

Low-level waste such as that going to Barnwell comes from nuclear power plants, hospitals and research institutions. It includes materials ranging from workers' contaminated coveralls to barrels of water. It is called "low-level" because it is only supposed to be dangerous for a few decades, rather than the hundreds of thousands of years that "high-level" garbage from spent fuel rods and bomb production will be deadly.

That is misleading, however, as is shown by what happened to the "low-level" dumps which closed. Despite all that vaunted technology that we were assured would protect us, "safe" radioactive waste at Maxey Flats, Ky., and West Valley, N.Y., leaked out into the soil. At Sheffield, Ill., erosion exposed some of the waste containers. Nevada's governor closed the site at Beatty last month after explaining that he and his constituents had had enough of the perils of transporting nuclear waste.

But "closed" only means they aren't taking anything new. They have plenty to handle with what is already there. At Hanford, for instance, so much plutonium leaked into the soil from one "low-level" waste trench that it had to be hurriedly cleaned out, at a cost of $1.9 million, because of the danger of a spontaneous chain reaction. If that sounds implausible, there is a report from an exiled Soviet scientist that hundreds were killed and thousands suffered radiation sickness in 1958 when atomic waste which was buried in the Ural Mountains exploded.

Yet, despite such horrible reminders, we still try to apply an "out-of-sight, out-of-mind" disposal philosophy to nuclear waste. Some of this stuff is going to be around for eons, and we haven't even been able to contain it for the first 25 years of the atomic era. We sank it in the ocean, and it leaked; we buried it in the ground, and it leaked. Now scientists are talking about shooting it into space (remember Skylab?), dropping it into volcanoes, parking it on remote islands or old salt mines.

It won't go away. We must learn how to dispose of it, not hide it. But until we do, we must learn to live with it as safely as possible. And, since transporting it thousands of miles to bury it only compounds the hazard, it might be fairest to require each state to furnish a storage site within its own boundaries or join in a consortium with its neighbors to develop and operate disposal sites.

And it would be the best incentive to make us face the problem instead of shutting our eyes to it.

Chicago Tribune

Chicago, Ill., November 3, 1979

Our newspaper recently printed a series of articles by Casey Bukro and James Coates entitled "Nuclear waste: a growing crisis." Their documentation of this unsolved problem was especially timely because of the release, during the series, of the report of the President's Commission on the Accident at Three Mile Island. This report was sharply critical of the laxity of industrial and governmental controls over nuclear power plants. The message is clear: our society's use of nuclear material is no subject for complacency.

Nuclear power occupies a central place in the military arsenal of the United States, and an important and, for Illnois at least, an essential place in the generation of electrical power. We are 37 years into the nuclear age. But there still is no satisfactory answer to the unavoidable question: What do we do with the radioactive wastes that are the end products of both military and civilian uses of nuclear energy?

The series by Reporters Bukro and Coates consisted for the most part of the recital of troublesome residues from past nuclear processing in several states. A couple of maps summed it all up: One, of storage sites for military and civilian wastes, showed all but one of the civilian sites now closed; another showed the location of many commercial nuclear reactors where burned-out fuel is stored at the power plants because it has nowhere to go.

The Presidential commission's report said, "After many years of operation of nuclear power plants, with no evidence that any member of the general public has been hurt, the belief that nuclear power plants are sufficiently safe grew into a conviction. . . . The commission is convinced that this attitude must be changed." The commission of course was talking about the siting, construction, and operation of nuclear power plants. But it issued a stern warning particularly applicable to the storage of wastes.

Radioactive materials combine acute danger, when artificially concentrated, with tremendous longevity. They require more respect than has been given them. At the beginning of the nuclear age people had the excuse of ignorance, and no doubt that will long remain the best excuse for much that will—or will not—be done. But by now enough is known so that the planning and control of present and future nuclear projects must be dominated by caution rather than by complacency.

We've already noted with satisfaction, in these columns, that the President's Commission on the Accident at Three Mile Island did not call for a moratorium on further construction of nuclear power plants. Our need for energy and the expense and looming scarcity of supplies of oil make nuclear power necessary for the present and near future both in the United States and abroad.

What we can do and must do is to proceed with caution, realizing that radioactive materials are never safe and that we should not recklessly multiply them before we learn how to dispose of their end products.

'By the way' how do we limit the nuclear threat to us from ourselves?'

Sentinel Star

Orlando, Fla., October 30, 1979

THE closing of nuclear waste dumps in Nevada and Washington is threatening to seriously hamper nuclear medical research and even some critical medical procedures because hospitals and laboratories are suddenly without any place to dispose of their liquid nuclear wastes. While this situation is in itself serious, it is a precursor of even worse problems as the government and nuclear industry attempt to find permanent storage sites for nuclear power plant wastes.

Most of the national debate on the development of nuclear power plants currently focuses on the safety of those facilities. The near disaster at Three Mile Island earlier this year helped focus that debate. But the radioactive waste from those plants is critically dangerous for at least 500 years and potentially harmful for tens of thousands of years. So far, science has not produced a fail-safe method for disposing of it.

With an element of uncertainty about the safety of disposal methods and questions about securing dumping sites, the political complications in finding acceptable storage locations will continue to grow. The immediate problem is that no state wants to become a nuclear dumping ground. Washington Gov. Dixy Lee Ray suspended disposal at the Hanford, Wash., site because there was no way to assure the safety of even these low-level wastes.

It seems certain states will continue to demand a greater voice in deciding where the federal government locates dumping sites. More than 30 states have already passed laws attempting to establish a foothold. There is sentiment to let each state assume responsibility for disposing of its own waste products. That is not a viable alternative in disposing of power plant wastes simply because of technical problems in handling and securing high-level radioactive materials.

Energy officials now say it probably will be another 10 years before they come up with what they believe would be, technically, an acceptable site for dumping radioactive wastes from power plants. Even then, the political hurdles would still remain. In the meantime, wastes are being stored at the various power plants and reports indicate that within four or five years, the storage facilities at some of those plants will be exhausted.

At Florida Power Corp.'s Crystal River plant, about 35 tons of spent fuel is stored in water in stainless steel-lined, reinforced concrete pools. Florida Power and Light Co. has almost 10 times that much at its Turkey Point facility south of Miami.

The scope of the waste disposal problem is immense and threatens the development of nuclear energy as a critically needed power source. But as the current disruption in getting rid of even low-level wastes again illustrates, there is still no solution, either scientific or political.

States Given Responsibility for Storing Low-Level Waste

Congress failed to formulate a comprehensive nuclear waste policy bill in 1980, grappling unsuccessfully with the issues of the disposal of high-level nuclear waste and the storage of spent fuel from reactors. Legislation was passed, however, which gave the states the responsibility for burying low-level radioactive waste. The only three sites for such waste—dumps in Washington, Nevada and South Carolina—would soon be insufficient to handle the vast amounts of low-level waste being generated, and all three states had threatened to close or reduce the intake of their sites. As signed into law by President Carter in December, the legislation made the disposal of commercial low-level waste a state responsibility. States could build their own dump sites or join with other states to establish a mutual burial site. In the case of a regional compact, which was subject to approval by Congress, the states which had joined together could refuse wastes from states that did not join the compact. The deadline set by Congress for formation of the compacts was Jan. 1, 1986.

Since passage of the bill in 1980, two regional compacts have been agreed to by all the state legislatures involved. These two are the Northwest compact (Alaska, Hawaii, Idaho, Montana, Oregon, Utah and Washington) and the Central compact (Arkansas, Kansas, Louisiana, Nebraska and Oklahoma). Five other compacts are still under discussion, with some but not all of the states resolved to join: the Western compact (Arizona and California), the Rocky Mountain compact (Colorado, Nevada, New Mexico, Wyoming, and possibly Arizona and Utah), the Midwest compact (Illinois, Indiana, Iowa, Kentucky, Michigan, Wisconsin, West Virginia, South Dakota, Ohio, Minnesota, Missouri and North Dakota), the Southeast compact (Alabama, Florida, Georgia, Mississippi, North Carolina, South Carolina, Tennessee and Virginia) and the Northeast compact (Connecticut, Delaware, Maine, Maryland, Massachusetts, New Hampshire, New Jersey, New York, Pennsylvania, Rhode Island and Vermont.) Of the remaining states, Texas has decided to build its own site, and Washington D.C., Puerto Rico and the Virgin Islands are as yet unaffiliated with any regional compact. Some of the compacts may change slightly in membership before they are presented to Congress for approval; for instance, the Western and Rocky Mountain compacts may become one compact. Congress has not yet approved any of the compacts, four of which are far enough along in state legislatures to be submitted for their consideration. One potential problem is the provision in the original bill that would allow states belonging to compacts to refuse waste from non-member states beginning in January 1986. This would create a storage problem for those compacts where there is not already a waste storage site; the estimated time required for the development of a site is four to five years.

The Dispatch

Columbus, Ohio, January 3, 1981

REGARDLESS OF the uncertainty that besets its action, Congress was wise to give the states responsibility for terminally storing low-level radioactive wastes. That job, simply, is beyond the practical capability of the federal government.

The main advantage in assigning the operational role to the states lies in the checks and balances it will impose on the two levels of government.

Definitions of what low-level nuclear wastes are and standards by which they must be managed for safety's sake necessarily rest with the central government in order to provide uniformity. But inasmuch as each state generates much low-level waste, the volume is so large and the selection of storage sites so controversial that only the states, or interstate compacts, can handle this unique trash properly.

By this arrangement, state and federal authorities can look over each other's snoulders to see that the safest methods of processing, transport and storage are actually carried out.

Initially, some experts will not be pleased. They rightly observe that the term, "low-level," implies these wastes are not very dangerous compared to high-level and transuranic wastes and spent nuclear fuel. The term derives from the sources of these wastes, not from the degree of their radioactive hazards in some instances.

However, state authorities will have to answer to their own residents for safe handling. Thus they should be able to successfully demand appropriate definitions and standards from the Nuclear Regulatory Commission. Federal handling in the past has been unsatisfactory.

The sooner this issue is settled, the better. There can be no acceptable disposition of radioactive wastes — whether low or high-level — in the end without mutual federal-state cooperation. This is a good place to begin.

Detroit Free Press

Detroit, Mich., August 10, 1981

MICHIGAN enacted a ban on the dumping of nuclear wastes in 1978, when the state was trying to fend off proposals to turn the salt caverns near Alpena into a nuclear dump. Even though the federal government would probably have been able to override the ban if it chose, the prohibition had some symbolic value in dramatizing the unsolved problems of radioactive disposal and of relying too greatly on nuclear power.

But as a practical policy, Michigan's ban is increasingly untenable, at least as it applies to low-level wastes. Some 100,000 cubic feet of low-level waste are produced here each year, about two-thirds of it from the state's three operating nuclear reactors, another third from hospitals, industry, universities and research facilities.

Low-level waste is produced by cancer research facilities, by the pharmaceutical industry, by hospitals that use radioactive isotopes for diagnosis and treatment. The waste materials include contaminated clothing, hospital dressings and instruments, machinery parts and solidified cooling fluid from the reactors. Michigan is 10th in the nation in the generation of low-level wastes and, after Illinois, the second largest producer of such wastes in the Midwest.

In the early 1970s, there were six commercial nuclear disposal sites in the U.S. Today there are three, but the need for safe, tightly regulated disposal facilities is more acute than ever. Last year, Congress enacted the Low-Level Radioactive Waste Policy Act, which encourages the formation of regional agreements on the handling of low-level waste.

Beginning July 1, 1986, a disposal site can refuse to accept wastes from a state that has not joined such an interstate compact. Michigan has five years either to establish its own licensed low-level waste facility or to join other states in developing one. Otherwise, we face the possibility of being abruptly barred from existing dumps and forced to devise a hasty, temporary — and most likely inferior — solution.

The present low-level waste sites are in Washington, Nevada and South Carolina, but all three are getting grumbly about continuing to accept wastes from distant producers. The transport of the wastes is expensive. And a state that ostentatiously bars nuclear shipments across its bridges, as Michigan did not long ago, is hardly being consistent if it goes on shipping its own radioactive garbage across country.

Since the construction and licensing of a disposal facility takes six to seven years, we are already behind schedule. Representatives of several Midwestern states, including Michigan, are trying to devise a regional agreement to be presented to state legislatures in January. It could be that Michigan will eventually be found to be the safest, most logical choice for a low-level disposal site. The Legislature should be prepared to accept that possibility.

Accepting the need for a low-level disposal facility does not imply an endorsement of the expansion of nuclear power or an abandonment of safety concerns. It does recognize that low-level radioactive materials already exist here in abundance — and that a state that enjoys the benefits of nuclear medicine and nuclear research, and accepts the power generated from nuclear plants, bears some responsibility for the handling of the wastes they produce.

The Wichita Eagle-Beacon

Wichita, Kans., September 18, 1981

A much-improved set of hazardous waste disposal laws stands as the major accomplishment of the last session of the Legislature. As a direct result of that needed updating, it was decided no new low-level radioactive waste sites would be licensed by the state until a functioning interstate compact dealing with that type of waste disposal has been created.

Both were wise moves, and Kansas now is at work, with seven nearby states, to form such a compact. They include Arkansas, Missouri, Oklahoma, Nebraska, Iowa, New Mexico and Louisiana.

In the past, few states have sought the dubious honor of becoming someone else's dumping ground for hazardous materials. Formation of a regional compact, which would allow the participating states to refuse non-participating state's wastes at their approved disposal sites, should help alleviate that problem.

By having a much larger geographical area from which to select a site best suited for a particular type of disposal, and by formally agreeing to share each other's appropriate facilities, the odds for providing an effective, safe solution to the problem of hazardous waste disposal should be much improved by the interstate compact approach.

The Providence Journal

Providence, R.I., November 20, 1981

State governors are threatening to close all three of the dumps in the United States now receiving low-level nuclear waste. If they did so, what is now only a moderate nuisance could be turned into a major environmental issue. Because such radioactive waste material is generated in hundreds of industrial plants and scientific institutions, a clear understanding of the risks is needed.

If the three existing dumps were shut down, Washington would have to order all such waste shipped to the federally run Nevada Test Site. An order directing all such waste to Nevada would create a bottleneck on the roads leading to the site, as low-level waste shipments converged from all parts of the country. The order also would represent an end run around action of the Nevada governor to close the dump now operated, with state approval, at Beatty, Nev. It would do nothing, however, to keep waste shipments flowing to the other two current sites at Hanford, Wash., and Barnwell, S.C.

This low-level waste should not be confused with the high-level waste that is contained in spent fuel from nuclear power plants. The spent fuel is still being stored in big swimming-pool-size tanks at the power plants, pending study and decision on where and how to store it for the hundreds and thousands of years of its dangerous half-life.

Low-level waste consists mostly of tools and equipment of all kinds contaminated by use at reactor sites or in laboratories, hospitals and industrial plants. Nearly 130 million cubic feet of such material is expected to be produced over the next 20 years.

Virtually all of it is in solid form, easily transported and not highly dangerous if accidents should break open the containers it is shipped in. Yet enough accidents have occurred to raise the ire of the governors in the states now receiving the waste.

Moreover, as the volume increases over the years, containment of the waste and possible radiation from it may become a problem. The Nuclear Regulatory Commission has just issued a report recommending improved techniques for burying and storing such waste and guarding it for at least 100 years, with safeguards for protecting inadvertent intruders on the sites in the subsequent several hundred years.

These recommendations should insure against harmful impacts on the public, reduce long-term care costs and increase confidence in the performance of the facilities, the NRC said. Present procedures, it added, leave doubts on all three points.

The most important challenge is to meet state objections to continued shipments of low-level waste. If the governors follow through on their threats to close the dumps, then the federal government will have to decide what dumps are most necessary to national welfare.

But the federal government also faces problems with the 14 low-level nuclear waste dumps of its own. The total amount of its waste, from nuclear manufacturing and bomb fabrication plants, is far greater than that from civilian sources. While the NRC has no jurisdiction over these sites, the same standards should apply to all, state or federal, military or civilian. The source of the nuclear-tinged garbage doesn't matter. The NRC rules should be applied to all these dumps.

The Seattle Times

Seattle, Wash., December 21, 1981

NORMALLY, a 300 per cent increase in fees by any private industry or government agency would seem exorbitant and unjustified.

In this case, however, it's completely in order and long overdue.

The fees for shippers dumping low-level nuclear waste at Hanford, on a site leased by the state from the federal government and operated by US Ecology, Inc., of Bellevue, have been raised from 57.5 cents to $2.30 a cubic foot.

Now the Hanford site's fees will be comparable to those charged at the only other low-level waste dumps in the nation — at Beatty, Nev., and Barnwell, S.C. And those who must dispose of nuclear wastes may not be so inclined to ship them all to this state in the future.

That's fine with us. Hanford is a good dump site, geologically and geographically, but it should not be expected to take almost half of the nation's total low-level nuclear waste, as it's now doing.

Other regions must develop their own sites to deal with these wastes, whose volume can only be expected to grow as more nuclear-power plants come on line and nuclear medicine and industries expand in decades ahead.

That was the clear message of Initiative 383, which aimed to ban most out-of-state nuclear waste in this state. Passed last year by an overwhelming margin, it was almost certainly unconstitutional, but put other states on notice that they could not keep shipping their wastes here indefinitely.

A Northwest regional compact on low-level nuclear waste, limiting the use of Hanford to eight Western states after 1983, has been submitted to Congress.

Other regions are still lagging behind in their efforts to establish new dump sites and get compacts in force, however.

The steep fee increase should help force action in other states on what is a difficult, but clearly solvable, problem.

Arkansas Gazette.

Little Rock, Ark., January 26, 1982

A little more light has been cast upon the issue of low-level radioactive waste disposal in Arkansas by presentations last week at a meeting of the legislative Joint Energy Committee. In legislation passed last year, Congress gave the states a choice of either providing their own low-level radiation waste disposal sites or joining together in regional compacts for common disposal. In either case, the states' deadline is 1986 and the decisions cannot be put off for long.

Arkansas has been invited to join eight other states — Louisiana, Oklahoma, Missouri, Minnesota, Nebraska, North Dakota, Iowa and Kansas — in the proposed Central Interstate Low-Level Radioactive Waste Commission. The legislatures of each of these states will have to accept or reject, without amendment, the proposed arrangement before each can participate. The commission will determine which state or states should become the site of waste disposal for all nine. Any state refusing to accept the commission's decision would be rejected from membership and would have to prepare its own disposal site.

Members of Arkansas's legislative Joint Energy committee learned last week that Arkansas would be an unlikely site because the state is near the southern edge of the region and public opinion isn't considered favorable. Arkansas, for the moment, is the largest producer of such wastes but it expected to rank only fourth by 1985. As more information flows into the legislative mill, Arkansas can make a better decision whether, in effect, to gamble and join the arrangement with others or to strike out alone.

WORCESTER TELEGRAM.

Worcester, Mass., November 10, 1981

The prospect of locating a chemical treatment plant in Warren or some other Massachusetts community is touching off a guerrilla war of claims, accusations and countercharges.

But that controversy will look like a church supper compared to the nuclear waste issue just around the corner. The Warren plant would *not* accept nuclear waste.

Last week, representatives from nine northeastern states agreed to decide how to pick a site for a regional nuclear waste disposal site by December, 1982. That will leave little time to spare. The nine Northeastern states, which generate 40 percent of all civilian nuclear waste in the nation, have been sending the stuff to the three nuclear disposal sites in Washington state, Nevada and South Carolina. But those three states have been authorized by Congress to close their gates in 1986 to industrial waste generated outside their regions.

The proposed disposal site will not be used for the toxic nuclear waste generated by nuclear power plants. It will accept only low-level wastes — gloves and clothing used in laboratories, used containers, the byproducts of the hospitals and medical industries — things like that.

These waste products are mildly radioactive, and nobody wants them in his back yard. Already, more than 60 communities in the western part of Massachusetts have tried to outlaw or restrict radioactive facilities within their borders. Legislation before the Massachusetts Legislature would make it possible for the state to override local objections in locating a nuclear waste facility in an area. A lobbying group in Greenfield has denounced the bill as "radiation without representation." There will be more scare slogans and tactics.

But the Northeast has little choice. Either it sets up a facility to deal with the problem or it poleaxes some of its most advanced industrial and medical research. Industries and laboratories that employ thousands would be crippled. High technology companies would go elsewhere. Medical care would suffer.

The coming battle for a nuclear waste disposal site is going to be rough, noisy and bitter. But it has to be joined. A safe way to get rid of the wastes in this region can and must be found.

Roanoke Times & World-News

Roanoke, Va., April 27, 1982

WITHIN less than four years, Virginia must find a new place — probably within its own borders — to dispose of the large amounts of low-level radioactive waste generated in the state. That is not much time at all to deal with a potentially dangerous situation.

We are not talking here about spent fuel. That presents a disposal problem all its own to places like Louisa County, one of two nuclear-plant sites in Virginia. That is considered high-level waste.

Low-level waste includes materials like leftovers from medical treatment (e.g., cobalt therapy), filters and control rods from power plants, and tools, equipment, gloves, rags, etc., that have become contaminated at those plants. It is called low-level, but it can be quite hazardous — the most radioactive of such matter must be stored in leak-proof containers and buried in deep clay trenches for hundreds of years.

There is also quite a lot of it. By 1980, 23.6 million cubic feet of low-level waste had been buried at six commercial sites in the United States; at minimum, another 3.3 million cubic feet is expected to accumulate every year. Among the states, Virginia is one of the biggest generators of low-level radioactive waste. It comes from nuclear power sites in Surry and Louisa counties; the Babcock & Wilcox plant at Lynchburg; and more than 300 hospitals and research institutions.

Wastes from Virginia still are being shipped to a dump in South Carolina. But time is running out. In 1979 that state's governor, Richard Riley, called attention to the looming shortage of disposal capacity and served notice that South Carolina would begin limiting shipments. In 1980 Congress assigned to states the responsibility for commercial radioactive wastes and authorized them to form regional compacts for disposal.

Fine so far. But after 10 months of negotiating with Southeastern states, Virginia got voted out of any compact they form. Effective at the end of 1985, that shuts the commonwealth off from South Carolina's dump.

To meet its needs, then, Virginia may have to provide its own low-level disposal site. And because of federal legislation, it may have to build its dump big enough to admit radioactive waste from North Carolina, Maryland and other mid-Atlantic states. In a recent report to the General Assembly, Virginia's Solid Waste Commission warned that the clock is racing: "From a practical viewpoint a site cannot be located and permitted under current federal law within this time frame."

Timing considerations aside, such a v[illegible]ture would be very expensive: from $7 million to $12.5 million to build, and perhaps several millions more in annual operating costs. The latter could be recouped by disposal charges on waste generators; but the dump wouldn't pay for itself unless it collected a much larger volume of wastes than is produced within the Old Dominion.

So Virginia is caught between a rock and a hard place. If other states play a waiting game, Virginia's urgent need may force it to accept the role of site for its own and others' wastes. That would mean large quantities of potentially dangerous radioactive material would be shipped on the state's roads and rails.

To where? The Piedmont, with its flat and dry terrain, seems a prime candidate. But no political jurisdiction in Virginia will want to open its borders to such a facility, even a small one. Yet a little one won't do. In the Old Dominion's case, the dump had better be built big enough to take care of low-level wastes a generation or so from now, when those nuclear power plants in Surry and Louisa reach the end of their useful lives and are shut down.

That's the dilemma. It's far from unique. States all over the country are scrambling to make new arrangements for low-level waste disposal — in some other state, of course. Virginia at least has company in its misery. But not, we may be sure, much sympathy.

THE KANSAS CITY STAR

Kansas City, Mo., February 3, 1982

None too soon, the eight-state agreement on setting up a regional dump for low-level radioactive wastes has set in motion the long process of review prescribed for such a project. Missouri and Kansas are particularly involved members of the group because both will have their first nuclear power plants operating by the mid-1980s. Disposal of contaminated clothing, tools and other materials from nuclear sites, hospitals, medical laboratories and some industries is becoming an increasingly difficult task. The only three states presently operating such sites—Nevada, Washington and South Carolina—have shown growing restlessness at having to be their brothers' keeper in this regard.

The 1980 federal law providing for regional compacts to set up additional dumps permits member states, after January 1986, to refuse low-level radioactive wastes from non-members, so no state wants to be left outside of an agreement by that date. Missouri, Kansas and their six colleagues—Minnesota, Iowa, Arkansas, Louisiana, Oklahoma and Nebraska—have spent months writing the 22-page draft compact which now faces a lengthy series of approvals: by the various state task forces named for this purpose, by the legislatures, the governors and Congress.

Then comes the hard part. If no state volunteers to host a dump site, the compact commission must advertise for an operator, and applicants will be stringently examined for financial security and where they plan to locate. The Nuclear Regulatory Commission itself must give the final go-ahead. All this procedure obviously is going to take quite a while, so 1986 is not too far off in these terms.

Especially when, at some point, one or more of those eight states is going to have to agree to provide a dump site for radioactive waste, an honor which, so far, no one seems to be willing to enjoy. The ultimate test of the regional compact's worth may well be its ability to enforce a choice.

The Atlanta Journal AND THE ATLANTA CONSTITUTION

Atlanta, Ga., February 14, 1982

NOW AND AGAIN some good news comes out of the General Assembly. We can all be pleased to learn that Georgia's legislators seem to be accepting the idea of joining a seven-state compact for the disposal of low-level nuclear wastes in South Carolina.

The cooperative effort would be of great benefit to Georgia for two reasons. It would give us someplace to put the radioactive waste materials created by hospitals and other industries that use nuclear substances, and it should reduce the amount of such wastes traveling through Georgia.

In the past the Barnwell, S.C., disposal site has accepted low-level wastes from all over the country, and much of that has passed through Georgia on its way to be dumped. When the new compact is created, Barnwell will be used only by the states which are members. That means wastes from the Midwest and Central states won't be passing through Georgia anymore.

That doesn't mean there won't be any radioactive materials on our streets and highways at all, though. If the federal government decides to create one long-term disposal site for high-level radioactive wastes from nuclear reactors, and if it picks Barnwell or some of the salt domes along the Gulf Coast, we may see a lot of that kind of traffic.

We'd like to see the government consider the possibility of regional compacts for these high-level wastes, too. It may be that geological formations suitable for storing spent nuclear fuel for thousands of years don't exist in some parts of the country, but we'd hope that some can be found someplace other than the South. We're willing to be responsible for cleaning up our own mess, but we don't relish the idea of having everybody else's trucked through our state every few days.

That's a decision yet to be made, however. For now, we're pleased that the Georgia General Assembly is moving to cooperate with our neighbors on the low-level wastes. It's a wise move that can make all our lives safer.

The Detroit News

Detroit, Mich., May 19, 1982

Disposal of low-level nuclear waste is not a subject to grip the spirit. Nevertheless, the state must decide what to do with potentially harmful refuse from hospitals, universities, industries, and nuclear-power plants.

If the Legislature fails to act, Michigan may be forced to stop the use of radioactive materials for medical diagnosis and therapy, and shut down nuclear electric plants and some university and industrial research labs.

Each year, state institutions generate 100,000 tons of low-level nuclear wastes, such as clothing worn by workers, sludge from filters, dead laboratory rats, industrial residues, and so forth. Of the total, 61 percent is from electric utilities, 26 percent from medical diagnosis and academic research, and 13 percent from industry.

Fortunately, the wastes are not as intensely radioactive as spent power-plant fuels, and thus disposal is less complicated.

Congress made the states responsible for such disposal in 1980, authorizing them to form regional compacts to cut costs by building a single facility to serve member states.

Presently in the United States, there are only three operating pits that receive low-level materials. They are in South Carolina, Nevada, and Washington state. All are being melded into regional compacts that exclude Michigan.

The Michigan Energy and Resource Research Association (MERRA) recognized the crisis early on, and formed a state committee to propose a solution.

The state has five choices: temporary on-site storage (no hospital, power plant, industry, or university has room for that); a Michigan disposal site (far too costly); disposal in another state (Michigan is about to lose that option); a ban on all nuclear materials (unthinkable in this age); or membership in a regional compact.

Representatives from 16 Midwest states, including Michigan, have drawn rules and conditions for a compact that will exist as soon as three or more legislatures mandate membership. Michigan's bill is now before the House.

Two deadlines hang over the issue. On Jan. 1, 1986, the present disposal sites will not be available to Michigan because "compact" states can exclude wastes from non-member states. Next, states have until July 1, 1984, to join a compact. After that date, membership could be gained only by the unanimous approval of member states.

There are other important considerations.

As Michigan moves toward development of high-technology industries to diversify its economy, more such wastes will be generated, notably in bio-molecular research. If the disposal problem isn't solved, vast areas of research won't be possible in Michigan.

The Legislature doesn't have to budget for a disposal site, which is developed by a private contractor under the control of the compact's commission, with all costs recovered from user fees.

Finally, no site has been identified by the 16-state committee, although Illinois may offer its Sheridan disposal pit, thus using the fees to clean up and improve that facility.

Michigan's proper course is clear: It should join the 16-state compact quickly.

The State

Columbia, S.C., March 20, 1982

THE S.C. Legislature shouldn't hesitate to enact the proposal joining South Carolina and six neighboring states in a new compact to regulate disposal of low-level radioactive nuclear wastes.

Without being funny, we can say that there couldn't be a compact without South Carolina since the only low-level storage facility east of the Mississippi is at Barnwell.

Furthermore, this compact is a direct result of the labors of Gov. Dick Riley and a federal commission which he headed under the Carter and Reagan administrations. The commission developed proposals for a series of regional compacts for disposal of these not-so-dangerous wastes and the Congress has authorized the agreements. The wastes include clothing, carcasses, containers, etc., from hospitals and research facilities.

This is the first such proposal in the nation, and it is a good example for the rest of the states. The purpose behind such pacts is to cause other states to take responsibility for their own low-level wastes, so they will no longer send them to South Carolina.

That is most desirable since South Carolina was getting the reputation of being "the nation's nuclear dumping ground." The pending bill is by Rep. Harriet Keyserling, D-Beaufort. It deserves expeditious handling by the Legislature this year.

ARGUS-LEADER

Sioux Falls, S.D., December 21, 1982

AVAILABLE: Site for low-level nuclear wastes. In return South Dakota community wants reputable national firm to build $5 million facility, offer 100 plus jobs. Operation should generate millions of dollars for the local economy.

That's the opportunity the Edgemont Chamber of Commerce sees from the quest of a Bellevue, Wash., firm — Chem Nuclear Systems Inc. — for a national or regional disposal site. The chamber asked the firm to consider the Edgemont area.

As Don Hanson, an Edgemont chamber member said: "We did a study of what we had to offer industry and found the only thing we've got...is isolation and poor quality land." The chamber recommended an abandoned Army munitions depot at Igloo.

K.C. Aly, a Chem-Nuclear Systems spokesman, said of the site selection process, "This kind of thing is so sensitive through the news and politics...politics may preclude opening of another site somewhere. It's not a popular subject." He said his firm wouldn't go ahead with expensive survey work and environmental tests unless it picks up support from state government.

We remind Aly that the process has become sensitive because the public has discovered (1) it cannot leave such decisions solely to the bureaucracy and (2) South Dakota and its people will have to live with whatever is decided. The process deserves the closest scrutiny and the widest possible publicity.

Gov. William J. Janklow says he remains generally opposed to creating a low-level nuclear waste dump in South Dakota, because the state is such a small contributor of low level wastes.

This newspaper has long taken the position that South Dakota should not become a dumping ground for someone else's nuculear waste, whether low-level or high-level radiation.

There are still reminders in Edgemont and area of the first uranium mining boom which extended from 1956 to 1972. Some homes and buildings in Edgemont failed federal radiation standards in 1981.

Radioactive mill tailings from an inactive uranium processing mill are still piled in a large area near Edgemont's downtown.

Moving those tailings — 2.5 million tons plus up to another 3 million tons of underlying earth which may be contaminated — to a prepared site about 2½ miles southeast of Edgemont will take several years. Removal possibly will start in the 1984 construction season. Work is expected to start next year on a private haul road and site preparation.

Cleaning up Edgemont's problems will be undertaken by Silver King Mine, a contractor for Tennessee Valley Authority which owns the old mill. It is being decommissioned.

The project is under the supervision of the Nuclear Regulatory Commission (NRC). The Environmental Protection Agency has responsibility for setting standards applicable to proper disposal of the mill tailings. As we note in another editorial today, EPA is having problems with Congress about sensitivity regarding superfund dump sites. The Edgemont cleanup, however, is not in that category and is under the NRC's oversight.

Edgemont and Fall River County already have one nuclear disposal site in their future. They shouldn't ask for a second one. Edgemont should reconsider its offer to Chem Nuclear Systems. State government should help Edgemont seek an alternative industry or development.

Portland Press Herald

Portland, Maine, November 22, 1982

Everyone wants a solution to the problem of disposing of low-level radioactive wastes. But nobody's willing to provide that solution by offering to host a dump in his own backyard.

The stuff's got to be put somewhere, however, and a regional plan for handling low-level radioactive wastes makes far more sense than individual attempts to resolve the problem in each state.

The Coalition of Northeastern Governors is sponsoring discussions among 11 Northeastern states aimed at developing a regional plan. But talks hit a snag after Massachusetts voters approved a measure prohibiting the state from entering into a regional compact without a ratification referendum.

Representatives of the other states are so upset with the vote that they're talking about expelling the Bay State from the project. Massachusetts, after all, produces more low-level wastes than any other Northeast state but appears to be saying in advance that it won't be likely to host any regional disposal site.

The lesson to be learned by the rest of the states—including Maine—is that nuclear waste disposal must be dealt with as a serious scientific and public health issue, not as a political one.

The solution will prove to be highly technical and can come about only through the concerted efforts of experts and well-informed representatives of each state. To throw this issue to the people in referendum—as, indeed, at least one Maine lawmaker has suggested in this state—is both impractical and unwise.

Voters here and in the other Northeastern states—like those in Massachusetts—naturally will vote for a waste site "anywhere but here." And that, of course, is "nowhere."

Herald News

Fall River, Mass., December 27, 1982

As predicted, the consortium of 11 northeast states considering where to place a low level, radioactive waste dump was confronted with a demand to expel Massachusetts from its meetings.

The motion to expel this state was tabled, but only in the face of demands that Massachusetts do whatever it can to amend the siting law approved by a referendum vote in November.

Ever since the law was approved, it has been obvious that it would create very serious difficulties for Massachusetts as part of the consortium of states trying to find a suitable site for a waste dump.

Rep. Thomas C. Norton of this city, the House chairman of the joint Legislative Committee on Energy, as well as co-chairman of the state's Special legislative Commission on Low Level Radioactive Waste Disposal, has pointed out that the new law would be an obstacle to the state's participation in a joint project such as the consortium hopes to propose.

The new law would require voter ratification of any site selected for a waste dump in this state.

Certainly it severely restricts the power of the state to enter any agreement with the other northeastern states in terms of establishing an official dumping place for radioactive waste materials.

How quickly or effectively the new law can be amended is not yet known. Meanwhile, Massachusetts drifts along in a highly unsatisfactory position.

On the one hand, it is the third largest producer of these radioactive waste materials in the nation.

On the other hand, it has been frustrated in its various efforts to set up an official dump for these materials by the refusal of every locality approached to accept its presence within its borders.

Now it has joined the consortium of 11 northeastern states in the hope that a solution to the problem of disposing of these materials would emerge.

The new law approved in November acts as a real stumbling block to this hope and could in the end mean that this would leave Massachusetts all alone to solve its own problem.

This could very well mean that it would eventually have to establish its own dumping ground in the face of considerable opposition and at much greater expense than it would have if it shared the expense with 10 other states.

There is an element of unreality about the new law, as if somehow refusal to give the state government the power to establish a dumping place would eliminate the radioactive waste materials themselves.

The new law is a genuine embarrassment to the state government in its attempt to find a satisfactory solution to a difficult problem.

Since the motion to expel Massachusetts from the consortium was tabled, the state has gained some time to find a way to amend the new law in the name of common sense and the best interests of the Commonwealth.

But there are times when postponement is a real detriment to solving a knotty problem.

The state government must go to work at once to try to find a way of preventing the new law from damaging a possible way of establishing a low level, radiocative waste dump on a joint basis.

To be fair, the resistance to approving such a dump is understandable. It is, however, shortsighted, since the absence of an approved, official site leaves every locality open to illegal dumping, which is far more dangerous.

The threat to expel Massachusetts from the consortium should make the people of this state aware that the new law must be amended as soon as possible.

RAPID CITY JOURNAL—

Rapid City, S.D., March 2, 1983

Given the unknowns, the uncertainties and the conflicting testimony offered as to the effects on South Dakota of membership in a regional low-level waste dump compact, it's best that the state Legislature rejected a proposal that the state join such a compact immediately.

With all the other business confronting them, legislators were hard pressed to determine the responsibilities membership in a compact would place on the state and the protections such membership would afford.

The nuclear waste issue is rife with emotion. There is no safe way to deal with the waste products of the nuclear cycle according to the most rabid critics. That's not so according to those who recognize the value of the peaceful uses of nuclear energy and are working on ways to handle the waste from such usage.

Between those two views is the middleground which has little upon which to base an informed opinion. The vast majority of South Dakotans, including legislators, fall into that category. All they can do is to try to balance the arguments and opinions of the spokespersons on both sides.

That's the kind of dialogue that should take place in an interim legislative committee study conducted between now and the next legislative session. Such a study would allow sufficient time for reasoned debate and an opportunity to determine what membership in a waste dump compact would actually mean.

For instance, there seems to be a question as to whether South Dakota's participation in a compact would require the state to accept waste from other states whether it wanted to or not. It's also uncertain whether South Dakota would be better off to establish a dump for the small amount of wastes it produces or, by virtue of belonging to a compact, to ship it to a depository in another state. Also, whether it would have be be part of a compact to avail itself of the latter option.

These are questions for which, hopefully, answers could be provided before definite action is taken that would obligate the state for responsibilities which could continue for hundreds of years.

With the deadline for participation in a waste disposal compact three years away, there's time to make a careful appraisal and reasoned decision based on facts.

The Boston Globe

Boston, Mass., June 3, 1983

The fears which sparked last November's Question 3 campaign – winning approval of a petition requiring statewide referendums on proposals to site a nuclear waste facility – are not far in the background as Massachusetts moves slowly toward membership in a regional nuclear waste compact.

A special legislative commission on low-level radioactive waste is holding hearings on the issue, while the Dukakis administration is preparing a specific compact proposal for submission to the Legislature which must approve it before it goes to the referendum required by Question 3.

The compact proposal is a sensible one with a key section leaving decisions on exactly how and where to site a waste facility up to whichever states are chosen as their locations.

Despite that, the fears which produced the 67 percent vote for Question 3 have been very evident at the commission's hearings in Pittsfield and Holyoke which attracted many of the western Massachusetts activists who were involved in that campaign. A prime concern is that there might be an attempt to sidestep the referendum process in order to join the compact.

The logic of the arguments on behalf of Massachusetts joining a regional compact is overwhelming.

Because of its nuclear power plants – as well as the medical and scientific enterprises located here – Massachusetts is a major producer of low-level nuclear waste. Most of that waste is now transported to a facility in Barnwell, S.C., but Congress has set Jan. 1, 1986, as the deadline for states to join regional compacts or prepare their own waste sites – and after that date Barnwell will become the waste site for a southeastern regional compact.

If Massachusetts does not join in a northeastern regional compact – and possibly, but not certainly, find itself chosen as a site for a regional waste facility – it will find itself with absolutely no place to dump the nuclear wastes from its hospitals and research facilities, and be forced to go it alone.

There is no logic on the other side of this issue. The problem, however, is that referendums on complex issues lend themselves to emotional manipulation – with the Question 3 campaign a prime example. It is not difficult to foresee the question of joining the northeastern compact put on the 1984 state ballot and be defeated by an emotionally appealing Shays Rebellion campaign.

Fortunately, many leaders of the Question 3 campaign now understand the dangers which may result from their victory and are willing to help structure the compact and work on the subsequent question of where and how to site the facility.

The risk of a referendum defeat for the compact is still very much present and the Dukakis administration and the legislative commission should not ignore that risk until after the compact is adopted by the Legislature. The administration and the commission have few options: given the virtual impossibility of repealing the referendum section of Question 3, they must convince people that joining the compact makes sense.

The Dispatch

Columbus, Ohio, June 12, 1983

VALUABLE BARGAINING power may be lost to Ohio if it fails to get in on the ground floor of a Midwest interstate compact charged with determining the future of low-level radioactive waste disposal in this section of the country.

Although the $50,000 fee for membership in the Midwest Interstate Compact on Low-Level Radioactive Waste may seem high on the surface, failure to join before the compact legally comes into being July 1 could carry a much higher price tag.

Four states — Michigan, Indiana, Iowa and Minnesota — already are members and six others, Ohio among them, are considering membership.

Because Ohio is a major producer of low-level radioactive waste, a voice in the decisions about where and at what cost this disposal process will take place is most important.

Low-level waste materials include protective clothing, contaminated paper and rags, and other laboratory and medical discards which have been exposed to some form of radioactive energy. Although these materials are only mildly radioactive and often only for a short time — unlike high-level contaminants such as those directly a part of the nuclear power generating process — disposal of them is strictly regulated and costly.

Power plants, hospitals, industry and research centers in Ohio ship their radioactive waste materials to commercial disposal sites in Hanford, Wash., and Barnwell, S.C.

Last year, Toledo Edison Co., which operates the Davis-Besse nuclear power plant in Port Clinton, Ohio, spent $600,000 to dispose of 9,300 cubic feet of waste at the Barnwell facility while OSU spent $100,000 to dispose of 4,000 cubic feet at Hanford. The volume of low-level radioactive wastes from OSU hospitals and laboratories has gone up 99 percent in five years, according to Walter Carey, the university's interim radiation safety director. Disposal costs have increased 500 percent over the same period.

Regional compacts, under the 1981 federal law that established them, may refuse waste shipments from non-member states beginning Jan. 1, 1986. A more pressing consideration, however, may be what the Midwest compact will do in the meantime.

The Ohio House Finance Committee began hearings on an affiliation bill recently, but a similar bill died in the House last year. Gov. Richard F. Celeste's administration has not pushed membership in the compact. Many politicians have preferred to steer clear of nuclear waste issues.

If Ohio refuses to join the team dealing with the low-level radioactive waste disposal responsibility, it cannot expect much say in the game plan. Carey warned that the price of coming in late might be acceptance of the regional disposal site in Ohio.

We believe the legislature should enact the affiliation measure promptly.

ST. LOUIS POST-DISPATCH

St. Louis, Mo., June 6, 1983

With only a few days left in the 1983 session of the Missouri General Assembly, a final decision is yet to be made on a bill that would have this state join with others in an interstate compact to dispose of low-level nuclear waste. It is a bill that should be rejected because there are more disadvantages than advantages for Missouri in the compact proposal.

For one thing, the statutory wording of the compact — which must be the same for all participating states — would invalidate all state laws or parts of laws in conflict with its provisions. It would thus deprive Missouri of the power to protect the safety of its own citizens as it saw fit. The compact would also forbid the state or any of its subdivisions from discriminating against waste generators from another member state. Although Missouri, if it is chosen as the site of a waste dump, could withdraw from the compact within 90 days, that process might be much more difficult than refusing to participate in the first place.

As a state that generates far less nuclear waste than other states that would be in the compact, Missouri in effect would be joining an effort to relieve these states of their waste problem, one created mainly by nuclear power plants — a kind of hazardous waste generator that this state does not yet have, although one is unwisely being built.

If Missouri stays out of the compact, it would still have to dispose of its own nuclear waste. But that task would be minimized, especially if the Union Electric Co. nuclear plant in Callaway County does not go on line. Even if the plant is activated, Missouri's nuclear waste would be far less than that of other states like Illinois. We cannot believe that if Missouri chose to take care of its own waste, it might also be required by federal law — as some have warned — to accommodate the waste of other states.

Chicago Tribune

Chicago, Ill., June 15, 1983

Illinois, a big producer of low-level radioactive wastes, is facing a choice of ways to dispose of them safely: either by itself, or in cooperation with other Midwestern states. In spite of a pressing deadline the decision has been postponed until fall—which might not be all bad, if the state makes use of the time to do some needed homework.

As presented to the General Assembly, the choice is either to join a multi-state regional compact or to go it alone, choosing an "Illinois-only" disposal site and trying to keep other states from using it. Either way has dangers; the Senate Agriculture committee voted Tuesday to hold in committee a bill to make Illinois a party to the compact.

The Midwest Compact would include Indiana, Iowa, Michigan, Minnesota and probably Missouri, whose legislature is soon to vote on it. The House last month voted to join it, and the arguments for joining are impressive.

Whatever Illinois does about the compact, we will have the problem of disposal and only a limited time in which to solve it. Under federal law the states will have to take over responsibility, as of Jan. 1, 1986, for disposing of their own low-level wastes. On that date the two disposal sites now available to the whole nation—in Hanford, Wash., and Barnwell, S.C.—will be able to close their borders to all radioactive wastes from outside their own regions.

That leaves Illinois with a particularly urgent problem. Unlike high-level radioactive wastes (mostly spent fuel from nuclear reactors), low-level waste consists of objects that have been contaminated by radiation—rags, gloves, clothing, tools, animal carcasses and so on. Illinois, because of its high concentration of nuclear power plants, research laboratories and medical facilities, is the third largest producer of it in the nation.

One of Congress' purposes in approving the 1980 law was to encourage states to form regional compacts. Since it takes at least five years to establish a burial site for low-level radioactive wastes, Illinois and its neighbor states have already missed the deadline. That is not quite an emergency; if they can show that good-faith efforts are being made and are producing some results, the Washington and South Carolina sites will go on accepting low-level wastes while the new sites are being established.

Still, time presses. The Midwest Compact is to be formed this July by at least four states. If Illinois is not a member, it will be formed without us. We might join later, but the framework would already be in place—one not designed with Illinois' needs in mind.

The chief advantage in joining is that member states share the costs and benefits of managing radioactive wastes. Each member state will take a 20-year turn as the "host" for the whole group; one of the first tasks of the compact will be to find the best and safest storage sites in the region. Each state will designate and control its own storage site, and will charge the others a disposal fee to meet the costs of perpetual care after the site has closed. Any state has the right to withdraw from the compact on five years' notice.

Illinois might well be the first host, since it produces 40 to 60 percent of all the radioactive wastes in the region. Even so, the choice apparently is between providing a regional storage facility for 20 years or providing an intrastate one forever. One way or the other the state will have to find a permanent burial site for these contaminating wastes, which are now building up at temporary, limited-capacity sites.

So stated, the problem seems to have a simple answer: Hurry up, pass the bill and let's join. Unfortunately that answer is exactly backwards. It means signing the agreement first, then looking around for ways to live up to it.

The real problem is not to agree on a set of organizational charts and timetables; it is to find the right site. That means finding the best possible combination of geological conditions and location: an area far from population centers, with no nearby rivers or aquifers and a minimum of seepage. That job should not be subjected to the pressures and panic of deadlines.

But it can be started. Over the summer the state should begin serious survey work to pinpoint the likeliest sites for radioactive waste disposal, so that the most essential part of the job will not be put off till last.

The Midwest Compact is an attractive idea. In dealing with radioactive wastes, however, the name of the game is not speed; it's safety.

The Kansas City Times

Kansas City, Mo., September 28, 1983

On a dismayingly long list of proposed legislation to be considered at the Oct. 19 special session of the Missouri General Assembly, the low-level radioactive waste compact is a "must do" item. Missouri's choices are either to join a disposal compact of as many as 15 Midwestern states — taking a long-odds chance that it would become the regional repository site — or abstain and face the absolute certainty it would have to operate its own site.

A 1980 federal law required that all states join a regional disposal compact by Jan. 1, 1986 or be left to establish their own individual dumps. A Missouri governor's task force recommended joining the Midwestern pact (Kansas opted for another group). Iowa, Indiana, Michigan and Minnesota already have approved membership in the Midwestern combine and plan to begin writing the rules next July 1. If Missouri wants to have a voice in those early decisions, time is running out.

The House approved such legislation earlier this year, but it died at adjournment time in the Senate when an opponents' filibuster threat endangered other key bills. The opponents object to a compact which would supersede state laws and are concerned about the host state's liability in damage suits arising from transportation or disposal of the wastes. The task force met that problem by endorsing companion legislation to require the operator of a possible Missouri disposal site to carry liability insurance.

Low-level radioactive wastes consist of contaminated clothing, tools and other materials from hospitals, X-ray clinics and laboratories. They are not nearly as toxic as high-level wastes from nuclear power plants or weapons production, but they are hardly more welcome in a given neighborhood. Missouri, on the western edge of a Midwestern compact area, is a relatively minor producer of such wastes compared to Illinois with its Chicago area. It would be an unlikely choice for the regional repository in terms of transporting the materials. But if the General Assembly doesn't get the state into the compact soon, it will have the sure burden of operating and regulating its own dump, with insufficient volume to produce the sustaining fee income.

Pittsburgh Post-Gazette

Pittsburgh, Pa., September 20, 1983

As the home of Three Mile Island, Pennsylvania has had more than its share of bad publicity in the nuclear age. So there ought to be an extra incentive here for dealing responsibly with the disposal of low-level nuclear wastes.

Pennsylvania ranks first among the northeastern states in the volume of low-level waste produced, and the uncertainty surrounding the safe and effective disposal of it means that state officials, especially Gov. Dick Thornburgh, have an important opportunity to provide leadership that will keep the problem from growing to crisis proportions.

Disposing of low-level radiation wastes involves technical complications that probably help to explain why the governor's office has been relatively quiet on the issue. It does require study and reflection. But finding an acceptable disposal site is also a political potato that sometimes seems hotter than radioactive wastes themselves. In devising an approach, the governor has to think a great deal about educating the public to the need not only for disposal sites but also to the idea that the wastes can be handled safely.

Thus it's crucial to keep in mind that the horror stories often embodied in public perceptions of atomic debris don't have to cloud the future of low-level radioactive waste disposal. As the label implies, this particular category of nuclear wastes embodies a very slight threshold of danger to the public.

For decades, no one bothered to provide the special handling that radioactive wastes require. Consequently, the debate over atomic disposal is fueled — and properly so — by the still unresolved problems of towns like Canonsburg, Pa., where atomic wastes were dumped without any thought of long-term safety.

New low-level radiation waste-disposal sites can assure that problems like those found too late in Canonsburg never occur again. And because Pennsylvania not only produces more waste than other states in this part of the country but also has more suitable space for a disposal site, the governor shouldn't feel too reticent about proposing that low-level radiation wastes be disposed of in this state under a regional agreement.

Those who want to spurn other states in this effort have to face the fact that Pennsylvania would have still to construct its own waste site. It may as well enjoy the financial benefits and political controls of a shared effort. Then the problem becomes one of designing a site that will permit the maximum control over atomic wastes.

Establishing a site with those features will stem a crisis that is now growing with present inventories of undisposed radioactive wastes. The governor and Legislature can make a lasting contribution by finding a place to bury them.

The Burlington Free Press

Burlington, Vt., July 9, 1983

For Vermont and most other states, it's a Catch-22.

Because South Carolina and Washington no longer want to be dumping grounds for storage of the nation's low-level nuclear waste, Congress in 1980 passed a law requiring all states to have in place by 1986 a plan for safe disposal of the low-level nuclear waste their businesses and utilities produce. States were given the option of devising their own plans or of banding with neighboring states to set up regional waste dumps.

As a result of the new law, representatives of Vermont and 10 other northeastern states have been meeting for more than a year to shape criteria for choosing a host state for a regional site. And this week, the Vermont State Nuclear Advisory Panel endorsed Vermont's participation in the regional compact. It will be up to the Legislature next year to determine whether membership is a good idea.

Assuming that all 11 states agree to the regional approach, the fun will begin once the states get down to selecting an actual site. Until now, the talk has been devoted to developing criteria for selecting a site.

Because Vermont generates only 1 percent of the low-level waste in the 11 state region, Vermont Human Services Secretary Lloyd Novick, one of two people who represented Vermont during the negotiations, believes there is little chance of Vermont being designated as the host state, but adds: "I don't think any state has an absolute guarantee of not hosting the site."

Although reasonable people may feel that New York, the second largest generator of low-level waste in the area and the only state with an existing nuclear waste dump, is the logical choice to host the dump, it's certain New York isn't going to volunteer to accept nuclear trash from other states.

The New York site is located about 35 miles south of Buffalo but has been closed since 1975 after radioactive chemicals were discovered leaking from the site. A cleanup is now underway.

Earlier this year, New York Gov. Mario Cuomo indicated he has "not yet made a decision" on whether to send the compact to state lawmakers for consideration, which may mean New York could decide to not participate and accept responsibility only for its own waste.

Of the 11 states in the region, Pennsylvania is the largest producer of nuclear junk. After New York, the next largest generator of waste is Massachusetts, with New Jersey and Connecticut close behind.

Although conceding that they produce the most low-level waste, the larger states favor putting the dump in Vermont, New Hampshire or Maine where there are fewer people to be harmed if something goes wrong.

Obviously, most Vermonters don't want the dump here. Neither does Maine nor New Hampshire, because they too generate little of the region's waste. Assuming that fairness prevails once the 11 states agree to the compact, it would seem unlikely that any of the three states would be chosen for the facility.

Unfortunately, however, as Novick points out, there are no guarantees. Nevertheless, Novick is correct when he says it would be a mistake for Vermont to reject the regional approach.

Failure to approve the compact might force Vermont to build its own plant, since there would be little chance of shipping waste from here to a site developed by states participating in the regional approach. The cost alone of developing and maintaining our own facility for the small amount of low-level waste in Vermont would be prohibitive. And, of course, the state would be stuck with exactly what it hopes to avoid: a waste dump within its boundaries.

Assuming the proposal endorsed by the Vermont State Nuclear Advisory Panel doesn't place Vermont at a disadvantage once the 11 states begin the bargaining process for site selection, the Legislature ought to approve the regional plan even though there is some risk that the state could wind up with a low-level nuclear waste facility.

The Evening Gazette

Worchester, Mass., June 9, 1983

Ask any thousand voters where nuclear waste should be disposed of, and 900 will say "Anywhere but near here." Make that 975.

That is the dilemma facing those who are trying to figure out where and how the nation's industries can get rid of their mildly radioactive residues such as used gloves, clothing and containers that once held isotopes. The stuff used to be dumped into the ocean. More recently it has been shipped off to South Carolina, Washington or Nevada, which have the only commercial disposal sites for radioactive waste products. But a federal law says that those three sites won't have to accept any more of the stuff after Jan. 1, 1986. The same law encourages the states to form themselves into regional compacts for the purpose of setting up disposal sites and procedures. A Northeast compact is being set up. Gov. Dukakis has filed a bill to include Massachusetts in it, even though this state may not qualify until a new law is changed. That will take some doing.

Last November, Massachusetts voters overwhelmingly endorsed a ballot question to require a statewide referendum on any nuclear waste site in the state. The other states in the Northeast compact believe — correctly — that the Massachusetts law effectively shields Massachusetts from ever locating a site within its borders. The other states want the Bay State to give up that unfair advantage and take its chances with the others.

Parodoxically, if Massachusetts does not take its chances with the others, it may eventually be forced to set up a disposal site itself. Either that, or see an exodus of scientific and medical industries that provide thousands of jobs. Massachusetts produces almost half of the radioactive waste in New England. Enormous amounts of research and thousands of jobs are based on the use of nuclear isotopes. Unless the wastes can be safely disposed of, the state's economy will be badly hit.

No one expects the deadline to be met by most states and regions. It seems doubtful that any new sites will be set up by 1986. Congress will have to approve an extension of the current system. However, a decision will have to be made sooner or later.

NELRAD, a consortium of medical, research, university and industrial members located in New England, is trying hard to inform the public about the issues.

The hearing at the University of Massachusetts Medical Center last week showed some of the divisions of opinion. Some want the Northeast compact to be reduced in size. Some want Massachusetts to go it alone. Some think the new state law, Chapter 503, will have to be changed. Others don't think so.

The road to a state or regional disposal site for radioactive wastes is going to be long and rough. Dukakis' proposed law is only the first step.

Maine Sunday Telegram

Portland, Maine, November 27, 1983

☐ There's no easy way to determine how Maine should dispose of its low-level nuclear waste. The risks inherent in each of the likely choices—whether to make nuclear waste storage an in-state responsibility or join with others in a three-state or 11-state compact—are like the wastes themselves: they neither diminish readily nor conveniently disappear.

Yet just as there's no easy way, there's no single way which so far has emerged as clearly preferable to the others. And until one does, Maine should listen, learn and deliberately preserve its options.

Low-level nuclear waste consists of radioactive debris, exclusive of nuclear fuel, from nuclear power plants like Maine Yankee as well as clothing and equipment from nuclear medical facilities, industries and research facilities. The object is to store it or, when possible, dispose of it altogether.

OK. That describes the problem. Dealing with it requires that priorities be established. Cost is one, so is reliability of service and transport. Priorities might even include the philosophical considerations of private vs. public development.

Fine. But once all such priorities are established, one consideration alone—safety—must outweigh all of the others. Environmentally safe storage must be the dominant factor that determines Maine's choice just as it should determine the choices of our neighboring states.

So let the emphasis be placed on safety and the counterpoint be the issue of costs. Vermont, unfortunately, seems to lean toward an 11-state compact largely because of financial concerns. A study in that state says a three-state compact, which began as a Maine proposal, wouldn't be economically feasible and would require "a large up-front capital investment."

Certainly cost is a relevant factor, both to the states and to the users who will pay waste dump storage fees. But it's not so relevant as the issue of safety. And the Vermont study suggests a three-state radioactive waste dump could run into security problems if it's open only part of the year "to save costs."

That order of business puts the storage cart before the nuclear horse.

It's not an order that should appeal to members of Maine's Low-Level Radioactive Waste Siting Commission who must keep their options open and their priorities straight.

"...AND IF YOU DON'T QUIT LOITERING, I'M GONNA HAVE TO SHUT YOU DOWN FOR CREATING A PUBLIC NUISANCE...."

THE ANN ARBOR NEWS

Ann Arbor, Mich., November 4, 1983

Finding disposal capacity for low-level nuclear waste became a major national issue in 1979, spurred by the near-accident at Three Mile Island.

Low-level nuclear waste refers to materials generated by medical diagnostic and scientific activity as well as nuclear power plants. High-level waste is defined under Nuclear Regulatory Commission regulations as spent uranium fuel rods and wastes from the reprocessing of these rods.

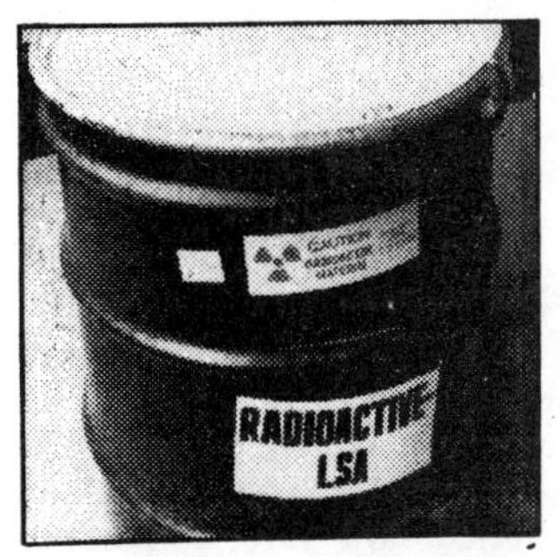

In response to the need for additional disposal capacity, Congress in 1980 passed the Low-Level Radioactive Waste Policy Act.

In effect, Congress said to the states you're on your own. The act made each state responsible within its borders. It also encouraged regional waste management by inviting states to enter into compacts for regional waste disposal.

The states in the Midwest regional grouping beside Michigan, according to the newsletter published by the National Governors Association, are Iowa, Minnesota, Illinois, Kentucky, Indiana, Missouri, Ohio, North Dakota, South Dakota and Wisconsin. These states generate about 15 percent of the nation's waste.

But in ratifying this compact, Michigan hasn't had much by way of companionship. Only Indiana, Iowa and Minnesota are the other co-signers which leaves the Midwest's largest generator of low-level nuclear waste, Illinois, militantly unratified and still worried about its likely candidacy as a host state for a dump site.

One way of saying, we suppose, that he who generates is host.

Some of that fighting Illini spirit is popular in Michigan, too. Gov. James Blanchard last week opposed siting high-level radioactive nuclear wastes in Michigan because of "our proximity to the Great Lakes and (our) high water table."

Michigan has been eyed by the feds as a potential waste storage site because of its favorable geology, i.e., Upper Peninsula granite overlays and Alpena-area salt formations. What Gov. Blanchard is worried about is an accident of some kind which might leach lethal wastes into groundwater supplies and into the recreational resource of the Great Lakes.

There isn't time to debate the issue much longer. A mountain of the stuff is growing out there, day by day. And of course, handling, transshipment and storage of the high-level wastes are extremely dangerous assignments.

There are some favorable signs. Obviously we're doing a better job of monitoring nuclear wastes of both varieties. The states are making progress via the regional approach. The technology exists to resolve nuclear waste problems, for example, burning at extremely high temperatures.

The barriers of politics and expense are still formidable, however.

Clearly, the administration was within its rights in telling the states that if you generate the stuff in quantity, you must also assume responsibility for its safe disposal. And that doesn't mean rich states paying to have the stuff removed, as if this were some routine garbage pickup.

Just as clear though, is the national administration's role. A generally pro-nuclear administration should take the lead in helping to devise solutions, establishing reasonable timetables for the regional compacts to have their affairs in order and making available more research dollars for safer, better and less costly methods of disposal technology.

High-Level Waste, Spent Fuel Storage Provided for in 1982 Bill

President Carter sent a message to Congress in February 1980, outlining a program to have at least one permanent storage site for high-level nuclear waste operational by the mid-1990's. Carter characterized the plan as the nation's "first comprehensive radioactive waste management program." In addition, Carter asked Congress to buy or build one or more facilities where spent nuclear fuel could be stored on an interim basis, dubbed AFR sites because they were to be located away from the reactors. The spent fuel rods contained plutonium and other valuable radioactive isotopes which could be removed through reprocessing, and thus were not technically "waste." But since Carter had placed a moratorium on fuel reprocessing in 1977, the nuclear plants had been storing spent fuel rods in shielded "swimming pools" near their facilities, and the rapid exhaustion of their storage capacities threatened to lead to a wholesale shutdown of nuclear reactors by the mid-1980's. (See pp. 144-153.)

The Senate passed a bill in July 1980 authorizing the Department of Energy to create short-term storage sites for high-level radioactive waste, as well as away-from-reactor storage sites for spent fuel. The bill foundered, however, over the question of whether states should be given veto power to reject nuclear waste, particularly the military waste from production of nuclear weapons and from naval nuclear reactors.

Finally, in December 1982, Congress cleared a comprehensive bill on the disposal of nuclear waste, authorizing the Energy Department to choose potential sites for both permanent and long-term storage. The permanent sites, for high-level waste, would have to be thousands of feet underground in salt domes or granite or basalt caverns; the first site was to be chosen by 1987. From three to five sites for "monitored, retrievable storage" facilities, to hold spent fuel available for reprocessing, were also to be recommended by the Energy Department by June 1985. A key amendment added to the bill in conference committee allowed a state to veto the choice of a storage site within its borders unless both houses of Congress voted to overturn the veto.

The Providence Journal

Providence, R.I., July 12, 1979

At the dawn of the nuclear age, and for years afterward, no one paid much attention to one inevitable and ugly byproduct of man's tampering with the atom. The byproduct: radioactive metals, sludges and other waste material discarded by nuclear-weapons plants and atomic power stations. The stuff remains "hot" and thus dangerous for a very long time (in some cases as long as thousands of years). It cannot be discarded just anywhere, but until recently Washington paid scant attention to the question of disposing of it safely.

Although this poisonous trash has accumulated in substantial volume, the United States still has no permanent site for its disposal. The weapons program alone has generated nine million cubic feet of waste. At the nation's nuclear reactors, some 15,000 spent fuel assemblies are being stored in nearby pools, pending a federal decision on a permanent solution; and this stockpile of radioactive fuel-rod bundles is growing at the rate of 5,000 a year.

Initially, the used cores were to be reprocessed into reusable nuclear fuel; but in 1977 President Carter cancelled all such reprocessing work in order to discourage the spread of technology that could be used to produce nuclear weapons. Meanwhile, though, the continued delay on an ultimate disposal decision could mean that some operating reactors, their space for nuclear wastes used up, will have to shut down.

Despite accelerated federal research into safe methods of disposal (current spending on the subject: $450 million a year), a decision is proving elusive. History affords no answer, for no one can say how this waste may affect the health of future generations. The problem assuredly will not go away: Regardless of what course nuclear power follows in the future, existing wastes must be isolated in a way to be secure for perhaps hundreds of years.

Shooting them into space is expensive. Burying them in the Greenland ice cap would buy time, but they might emerge, ages hence, embedded in icebergs. Burial at sea has its advocates — some, indeed, urge depositing them on remote Pacific atolls that already are radioactive from earlier atomic tests — and this deserves further study, although the long-term stability of such storage seems questionable.

The Carter administration, like the Ford administration, continues to concentrate on schemes to bury the stuff (properly solidified and placed in sealed containers) deep in layers of salt, basalt or granite. The most-favored sites are in the West — New Mexico, Nevada and Washington. Carter aides are split: some want to proceed promptly with development of a site in an underground salt bed near Carlsbad, N.M.; others argue for further study to choose the best among five or six sites.

The preferred course would be a Carter compromise: developing the New Mexico site now to avoid further delay, while continuing the research on how the waste containers behave in other types of underground formations. Whatever site is chosen, court challenges are likely; and it would be well to join the issue promptly and get it resolved. Washington already has dragged its feet far too long on a problem that will get more serious until decisive action is taken.

The Kansas City Times

Kansas City, Mo., April 24, 1979

One of the more telling arguments of opponents of nuclear power is that, after years of studying the problem, the federal government has not decided on a method of disposing of the radioactive wastes from spent fuel rods. Most of them remain in temporary storage in cooling ponds at the sites where they were used.

It is not, however, that the energy engineers haven't been busy examining the various disposal possibilities; they just haven't made a final choice or undertaken the considerable task of selling the public on the safety of whatever method is selected. Their evident preference seems to be for deep burial in rock or salt formations. The salt option was tested some years back in the Lyons, Kan., salt beds and more recently in New Mexico, evoking in both cases the predictable local resistance.

Now, in an environmental impact study reviewing no fewer than 10 possibilities, the Department of Energy again finds that burial in rock or salt formations "has an edge over other options" and could be carried out with little harm to the environment or risk to health. Grading the various measures on a 1-to-5 scale, this one received the highest rating on "status of technology."

The lowest rating on state-of-the-art went to the idea of launching the junk into outer space, aimed at the sun. Other possibilities are burial after pretreatment to make the wastes resemble rock or dilute them with cement, burial in melted rock, beneath ocean islands or under the ocean floor, burial under the arctic and antarctic ice sheets and transmutation into less-radioactive substances. Obviously the researchers have gone far afield in quest of a way to get rid of this hot garbage as far away as possible.

The DOE study claimed that nuclear wastes in general pose less risk than some of the toxic, non-radioactive chemicals "now being handled routinely by society." Certainly train derailments involving such hazardous substances have taken many lives in recent years. But radioactivity, as Three Mile Island demonstrated so graphically, inspires a special kind of public fear, an irrational terror that the body is being poisoned by unseen waves despite official assurances that monitoring shows negligible levels of radioactivity.

So the doubters postulate earthquakes or heat melting whereby buried nuclear wastes, even in deep, inert salt and rock formations, could leak into underground water supplies. And the energy planners, in this study, say more tests are needed to resolve some uncertainties. The final answer may well be that there can be no 100 percent, fail-safe disposal method, just as now seems to be the case with nuclear electric generation itself. As with flying in airplanes and riding in cars, you can only do everything conceivable to reduce the odds of mischance as low as possible.

Arkansas Gazette.

Little Rock, Ark., October 17, 1979

There are those who insist that nuclear wastes do not pose a significant danger to human health because the level of radioactivity in their view does not warrant special concern.

Most of us, even so, consider the wastes to be, at least potentially, a massive environmental problem that will grow worse until some permanent disposal site is found and developed. It is the conclusion of the federal Energy Department, in any event, that the best method of disposal would be burial of wastes deep underground. This is the conclusion of its draft environmental impact statement, and it is expected to form the basis for a final decision on disposal methods and sites by next summer.

The Energy Department report gives some idea of the dimension of the disposal problem, for now as well as the future. The 72 licensed commercial nuclear reactors in the United States now generate about 50,000 megawatts of power, such producing between 8,000 and 14,000 cubic feet of wastes annually. By the year 2000, nuclear generating capacity — assuming no disruptions — will be at least 225,000 and 400,000 megawatts, with proportionate amounts of wastes. By 2040, the assumption goes, the reactors will have been phased out. By that time, however, the nation will have been left with as many as 1.3 million cylinders of high-level waste, 1.7 larger containers of intermediate-level waste, and two million 55-gallon drums of low-level waste. Permanent burial of all this, says the report, would require from 6,000 to 20,000 acres of salt, granite, shale or basalt.

The sheer size of this challenge is impressive, for the radioactivity will be around for thousands of years after those of us who have enjoyed the benefits of the power it has produced are gone. This is not a problem to place on the shoulders of our great grandchildren. Let us be uncommonly careful about how and where these wastes are buried.

Minneapolis Tribune

Minneapolis, Minn., November 30, 1979

American power companies with nuclear plants have a problem: They are running out of space for radioactive used fuel because the federal government hasn't developed a new, permanent storage method. Unwisely, some opponents of nuclear power are fighting requests to expand existing storage capacity, hoping to force plant closings. A controversy along just these lines is shaping up in Minnesota.

Northern States Power Co. must expand the spent-fuel pool at its Prairie Island nuclear-generating plant at Red Wing or stop operating the big plant by 1983 or 1984. So the company has applied to the Minnesota Energy Agency for a certificate of need that would allow the expansion. And environmental and anti-nuclear groups almost surely will urge that the agency deny the certificate.

But Prairie Island now supplies nearly 30 percent of the electricity NSP's customers use every year. If the plant were shut down, the company would have to burn 3 million tons more coal and nearly 1 million barrels more oil each year in its non-nuclear plants — and the oil could well be in short supply. Additional purchases of electricity from other utilities would bring the total extra bill for Minnesota consumers, the company estimates, to $160 million in 1985, not counting the cost of prematurely closing an expensive investment. Higher rates would hit particularly hard at the poor — the very people the Legislature is about to help by subsidizing their heating costs. And NSP soon would need to supply replacement power; that would mean new coal-burning plants with significantly higher generating costs (and electric bills) than Prairie Island's.

In addition, the Energy Agency is prohibited, under its rules, from making "a decision which could reasonably be expected to result in a forced shutdown" of a nuclear facility. The reason is the federal government's pre-emption of nuclear regulation. So the most the agency could do, it says, is tell NSP to ship the spent fuel somewhere else.

It's one thing to oppose, as nuclear-power foes do, *new* nuclear plants until the difficult problem of radioactive waste disposal is solved. But closing plants that already are supplying needed electricity cheaply and efficiently is another matter. Blocking the expansion at Prairie Island makes no sense — legally or practically.

THE SUN

Baltimore, Md., August 5, 1979

It once was believed that the problem of disposing of radioactive wastes ("radwastes") from nuclear power plants was as good as solved. Storing them in dry, geologically stable salt mines seemed an excellent idea, and the now defunct Atomic Energy Commission suggested a few years ago that solving the problem was only a matter of finding the right salt mines.

Well, this has turned out to be quite an obstacle. Some objections to use of salt mines have been purely political, based on public fear of radiation. But there have been technical problems, too. As it turns out, not all salt mines are as dry as had been thought. And even when they seem to be, they may not really be; ordinary salt contains "water of hydration" which causes it to be crystalline. Geologists have discovered that heat from the radioactive wastes may cause this water to become free-flowing and capable of leaching away the radioactive wastes, carrying them elsewhere than intended. The answer to that problem may be to go deeper into the mines where the salt is "anhydrous" or water-free. But that costs more money.

Another issue, whether glass should be used to encapsulate high-level wastes, has generated incredible rancor among involved scientists, bureaucrats and corporate contractors recently. Some scientists believe glass is too susceptible to leaching, and favor encapsulation in some other substance, such as ceramics. But the scientists who favor glass say rejecting it could set the waste management program back 10 years—because far too little research has been done on ceramics and other materials. This controversy has generated so much heat that a National Academy of Sciences report on the subject has been delayed several times. Scientists and officials involved in producing and reviewing the report sometimes are so at loggerheads on this issue that they hurl insults at each other in public.

That's a brief sampling of the problems associated with radioactive waste disposal. The sampling is not meant to suggest that the problems are impossible to solve; the probability is that they will be solved. But the solution will not come soon, it will not be easy, and —almost certainly—it will cost a great deal of money. Our point is that there probably are no easy panaceas for the energy crisis on the production side—and that conservation of energy remains one of the nation's most attractive options for meeting the crisis.

Roanoke Times & World-News

Roanoke, Va., June 25, 1979

Whither nuclear power in Virginia? Its future may not be determined by anti-nuke demonstrations, by legal actions about site suitability, or even by the industry's questionable economic situation. Whether nuclear energy can survive in the Commonwealth, and elsewhere, may hang on the answer to one question: What can be done with the plants' growing stockpile of radioactive wastes?

Since nukes began operating in Surry and Louisa counties, the short-term solution has been to store the wastes in temporary holding tanks. In the beginning, there seemed no hurry about permanent disposal; this was a problem that would find a solution. So far, though, it hasn't. And all over the United States, these temporary tanks are running out of room. Nor is there, yet, any other place to put most of the spent fuel rods.

Doug Cochran, spokesman for Virginia Electric & Power Co., says Vepco still has "all the fuel we've ever consumed at our nuclear plants" in temporary storage. But even with recently expanded storage, Surry is reaching capacity; if there's no alternative by 1985, Cochran says, Surry units 1 and 2 may have to shut down. The same fate awaits North Anna 1 and 2 by 1988, if Vepco's request for extra storage capacity there isn't granted.

Some observers contend that several feasible choices are available for permanent disposal of these highly dangerous wastes, and that people on high just can't reach a decision. There's no doubt that the issue is politically hot; feedback comes not only from anti-nuclear ideologues but also from states and localities that want the power the atom gives but don't want to become the receptacle for its wastes — some of which will remain hazardous for hundreds of thousands of years.

There's ample evidence, though, that paralysis of official will isn't the whole problem. One study, by the thorough and reputable General Accounting Office, concluded nearly two years ago that several decades of work had not yet "demonstrated acceptable solutions for long-term storage and/or disposal of . . . high-level waste."

The search must go on. It is lent urgency by President Carter's decision, early in his term, to forego reprocessing of spent fuel because of the dangers associated with the masses of plutonium that would thus be extracted. But as the deadline nears — and a decade or so isn't a long time in which to choose and then implement a vast new technology — progress, if any, seems agonizingly slow.

The Evening Bulletin

Philadelphia, Pa., February 15, 1980

After two decades of government study of the problem, President Carter has come up with a plan to deal with both private and government nuclear waste.

It is perhaps typical of debates on nuclear energy that the Carter plan is already being criticized for having a time frame that is too long and too short.

President Carter's general timetable is to identify four or five sites for permanent storage of high-level radioactive wastes — spent fuel from reactors and the like — by 1985 and have at least one ready to receive the stuff by the early or mid 1990s. Meanwhile, a temporary waste disposal place would be picked to handle what cannot be accommodated at the on-site waste pools at power plants and other nuclear facilities.

Does that seem like stretching things out a bit? It might until you consider the number of environmental, geological and seismological (earthquake) safety factors that have to go into these decisions. We are talking about radioactive waste materials, some of which don't "cool" for thousands of years.

The timetable reflects at least three problems facing Mr. Carter on the issue.

The first is physical. Residues are piling up. Salem Unit No. 1 in New Jersey, for instance, has applied to expand its on-site waste capacity but as of now will exhaust that capacity in 1982. Peach Bottom No. 2 on the Susquehannah River in York County will run out in 1990; the undamaged unit at Three Mile Island in 1989. You find the same situation across the country.

And that's only private industry. The Federal Government, which from its weapons work, nuclear engines and research has produced 65 times as much waste as the private sector, represents the big need for expanded and permanent storage.

The second problem is one of public confidence. Fiascos like Three Mile Island followed cheery assurances of nuclear power's safety. The public is now thoroughly skeptical about *any* assurances on the subject of nuclear radiation. People will have to be convinced that the evil genie that is stuffed down a mine shaft won't come creeping out a generation later.

Finally, there is the political problem of picking sites. President Carter was wise, in our view, to name a commission, including a number of governors, for "consultation and concurrence" purposes. Not that they can be counted on always to concur, but the panel can smooth the path toward final decision.

Nonetheless, it is good to hear that Mr. Carter has established as a principle that the ultimate disposal of nuclear waste is a national problem and the Federal Government will do what is necessary — override a state if need be — to get the job done.

We urge all deliberate speed toward solving the waste problem. Neither Mr. Carter nor the country can have a nuclear power policy that means anything without that solution.

The Christian Science Monitor

Boston, Mass., February 14, 1980

In 1957 the National Academy of Sciences warned that radioactive waste disposal "is a major problem in the future growth of the atomic industry." In 1980, more than two decades later, a United States administration has at last set forth a coordinated program for disposal in keeping with the magnitude of the problem.

The question is whether the program will go forward quickly and effectively enough to allow "future growth" of nuclear power in the United States.

The certainty is that the program must go forward to handle the hazardous military and industrial nuclear waste that already exists and will continue to accumulate even if no more nuclear power plants go into operation.

Thus, for the purpose of preparing advice for President Carter's just-announced program, government officials have assumed neutrality on whether to expand nuclear energy – though, coming from various energy, environmental, and other agencies, they represent a spectrum of opinion on that subject. And the President has laudably decided that wholehearted tackling of the present urgent problem should be the responsibility of this generation and not left, perhaps tragically, to its descendants.

Is there a political element in Mr. Carter's coming out with the plan on the brink of the vote in New Hampshire, where nuclear energy is an issue, and when the Jerry Brown campaign is nagging him on it? We doubt that Mr. Carter, with his demonstrated concern for nuclear matters, both foreign and domestic, would need that kind of nudge on such an overriding safety issue. Yet it had seemed hardly excusable for him to take so long to outline action after the massive 1978 report to him by the Interagency Review Group on Nuclear Waste Management.

Beyond any incidental partisan political dividends for finally launching the waste management effort – or political criticism of its elements – he has taken a step toward meeting public doubts and fears about the nuclear enterprise. (Just between 1978 and spring of last year, according to a national survey, those believing the nuclear waste problem could be solved had dropped from 53 percent to 38 percent.) The program now must be carried out in a way to build public confidence.

This does not mean rushing along uncertain avenues. Indeed, Mr. Carter wins points for deciding to cancel a military-waste pilot project in New Mexico that had apparently been gone into too hastily. What is needed is a consistent policy of bringing the public into the informational and decisionmaking process with the kind of candor that is now beginning to be displayed.

The Interagency Review Group's operation was encouraging. It brought together people inside and outside the government, with hearings for the public.

Yet flaws appeared in providing advance notice about hearings, in eliciting various points of view fairly when some were those of groups with plenty of money to prepare them and others were not, and in assuring participants that their views had actually been taken into consideration and not just suffered for political window dressing. A recent Harvard study for the Energy Research and Development Administration suggests that funds for public participation might be made available on some sort of proportional representation basis. The polarizing of nuclear vs. antinuclear voices might be reduced through efforts to involve middle-of-the-road groups and individual voices.

One promising part of the Carter plan is formation of a state planning council for advice from governors and other elected officials on waste management issues. The siting of repositories for the waste, whose radioactivity takes long periods of time to fade, will require informed cooperation by the states involved. The proposed broadening of licensing of sites by the Nuclear Regulatory Commission also ought to bolster confidence.

The program valuably dispels the myth that the technology is already available and only political, economic, and social considerations remain. These latter considerations are important, but technical questions persist. The farthest the administration goes is to state a technical consensus that "no insurmountable barriers are known" to prevent permanent disposal of the waste.

Meanwhile, the Nuclear Regulatory Commission has the task of determining whether it has confidence that indeed the wastes *can* be disposed of safely. It is a note of caution that helps explain why, even after all these years of nuclear weapons and energy, the difficulty of making and implementing right choices pushes the estimated date of the first full-scale operational repository to the middle of the 1990s.

Detroit Free Press

Detroit, Mich., February 15, 1980

IN UNVEILING his long overdue plan for nuclear waste disposal, President Carter has finally pointed the way toward a reasonable disposal policy. Though short on specifics, the proposal is the first substantive indication that the Carter administration is prepared to do something other than wait for the problem to solve itself.

The plan, however, will not make the nuclear power industry's selling job any easier. In backing away from President Ford's plans for quick construction of a pilot waste depository, Mr. Carter has acknowledged that the nuclear waste problem is far from being resolved. Such an admission can only heat up the debate over nuclear safety, adding to the woes of an already troubled industry.

That admission was necessary, however, as a prelude to dealing with a problem that —whatever happens to commercial nuclear power—the United States and the world will have to face.

The president's plan is a blueprint for caution. It calls for four or five more years of research before committing the nation to any particular approach. And it establishes a planning council largely comprised of state officials, in recognition of the local political problems inherent in selecting any disposal site.

While an intra-agency review of nuclear waste was being carried out, President Carter put all planning for waste disposal on hold. Even after receiving the report, a report with which he was apparently unhappy, the White House remained silent for several months, leaving the fate of nuclear waste—and hence of the nuclear industry —very much up in the air.

Now, by imposing a timetable on the nuclear waste disposal problem, the president has removed at least some of the bothersome uncertainty surrounding it. His plan, because of its caution, is unlikely to satisfy either proponents or critics of nuclear power. But it is much better to finally have some kind of plan than to continue to rely on speculation about what the White House may one day decide.

The Philadelphia Inquirer

Philadelphia, Pa., August 2, 1980

Just as the Nuclear Regulatory Commission is about to resume licensing nuclear power plants, the U.S. Congress has set out to dismantle a cautioned and reasonable program for the disposal of the waste produced by nuclear reactors and defense programs.

In both the Senate and the House, a variety of measures are currently under consideration that will commit the United States to a waste disposal program that is ill-conceived and does nothing to address the crucial question of long-term, safe storage of deadly radioactive wastes. The Congress has chosen to ignore a responsible waste disposal program, proposed last February by President Carter, one which, ironically, required little in the way of legislative action. Instead, the Congress is responding to pressures from the nuclear industry to help the industry out of a jam.

Working from a promise to provide the United States with a waste disposal program "current and future generations" could live with, the President commissioned a two-year review of nuclear waste disposal technology by a panel of engineers, scientists, environmentalists and energy experts. The panel determined that four or five years of intensive scientific study were required before enough was known to safely select a permanent method and site to store nuclear wastes. The U.S. should not be committed to any technology until that study is completed, the panel recommended.

By its intervention, the Congress is ignoring those recommendations and legislatively establishing, and funding, a waste disposal program that is regressive at best and potentially dangerous at worst. The nuclear industry has done little or nothing to resolve its waste disposal problems. Instead, it has waited for the federal government to step in. Now it has told Congress it cannot wait four or five years; it needs to get rid of its nuclear waste, which is in the form of spent fuel rods currently stored at reactors.

In a vote earlier this week earmarking $300 million to build away-from-reactor storage sites, the Senate came close to giving the nuclear industry the ultimate gift it has sought from the federal government: free storage of the reactor fuel.

That plan was defeated, but the controversial away-from-reactor sites will mean that above-ground storage areas will be constructed and spent reactor fuel will be hauled to them, with utilities paying a one-time storage fee. "What the Senate has done is approve a concept for temporary storage that was abandoned by the old Atomic Energy Commission in the early 1970s as unsuitable," explained one environmentalist.

On the other side of Capitol Hill, a similar spirit of irresponsibility prevails. The House Armed Services Committee reversed Mr. Carter's decision to cancel the controversial Waste Isolation Pilot Plant in New Mexico, voting to authorize it and at the same time banning the NRC from exercising any licensing authority over it. The committee bill also prohibits New Mexico from having a voice in the project's development.

The Congressional activity in the area of nuclear waste programs takes on special significance as the NRC prepares to end its self-imposed 16-month licensing moratorium until a number of safety-related issues could be addressed. Although the NRC has utilized the time since the accident at the Three Mile Island nuclear plant to scrutinize its rules and regulations on the operation of reactors, and to upgrade some of those requirements in an effort to prevent another accident like the one at TMI, it has not devoted much attention to the nuclear waste storage problem.

Thus, new reactors will begin generating power, with each producing about 30 tons of spent reactor fuel annually, and the storage problem left unresolved.

Unquestionably, the answer to the tons of radioactive garbage piling up around this nation must be found — and it must be found quickly. Haste is not the wisest pace, however. The price of irresponsibility will be borne by current and future generations. If the Congress fails to realize the enormity of that price, and proceeds to enact legislation commiting the nation to an unsafe waste program, Mr. Carter must stand firm with an appropriate veto.

THE ATLANTA CONSTITUTION

Atlanta, Ga., August 1, 1980

The Senate has taken a wise step in voting to store nuclear wastes at interim sites while technology is developed to store them away for centuries until they become harmless.

We can debate on whether nuclear power is helpful or harmful until the cows come home, but the fact is nuclear power is here now and working. And as long as that is the case there will be wastes. Dangerous wastes. Those wastes will remain dangerous for centuries.

In recent years, controversies and disputes have erupted constantly about how — and especially where — these wastes would be stored. Some states and cities moved to prevent even the transportation of nuclear wastes across or within their legal boundaries. Some states with nuclear-disposal facilities said they didn't want wastes from other areas.

Whatever, it's become clear that the disposal of nuclear wastes is a federal problem, and the Senate-passed legislation now goes to the House where a similar bill is in committee.

Long Island, N.Y., July 31, 1980

The world's first nuclear explosion took place at Alamogordo, N.M., on July 16, 1945. Since then, radioactive wastes have been piling up all over the world, a source of potential harm for present and future generations.

The need to provide safe, long-term disposal has been recognized for decades in this country; the arguments over the best methods and sites have persisted for nearly as long. The result was described this week by Sen. Charles Percy (R-Ill.):

"Today, 7,700 tons of spent reactor fuel, 70 million gallons of high-level radioactive waste and 69 million cubic feet of low-level nuclear waste are scattered in storage pools and shallow dirt pits all across America. And we still have no assured medium for long-term disposal."

Percy made this observation during the Senate floor debate on the first nuclear waste bill ever to get that far in Congress.

That bill, which the Senate passed yesterday, is by no means perfect. By providing funds for temporary storage of commercial wastes, for instance, it might obscure the need to develop permanent disposal facilities. But the Senate managed to resolve one point that has obstructed federal nuclear waste management for years.

Before passing the bill, it agreed unanimously to a compromise that would give states the right to object to a nuclear waste dump within their borders—but not to veto one.

The secretary of energy would be required to notify states that are being considered for nuclear waste sites and to seek state "consultation and concurrence" throughout the process. Any disputes would ultimately be resolved by Congress, where either house could veto a dump for civilian nuclear wastes. But vetoes by both houses would be required to kill a military dump if the president declares it vital to national security.

Nuclear waste dumps, like other unpleasant realities of modern life, are not in demand as neighbors. States like New Mexico and Louisiana, where nuclear dumps have already been considered, have strenuously objected.

But the decision ought to be based on such criteria as geology, geography, climate, access and relative isolation, not on political clout or local objections. If states were granted veto power, there might never be a permanent site. In that sense, the Senate compromise is an important step forward.

The Tennessean

Nashville, Tenn., July 5, 1981

SCIENTISTS at the University of Tennessee and the Oak Ridge National Laboratory believe they may be on the way to solving the problem of disposing of radioactive wastes from nuclear power plants.

Under their proposed system, the radioactive waste would be bound chemically with a plentiful earth mineral called monazite and enclosed in a heat-resistant glass. Then the material would be placed in a stainless steel vault and buried.

The scientists say preliminary tests indicate the waste, which would be in a crystalized state, can be stored for possibly as long as a billion years — long after it had ceased to be radioactive — without leaking or posing any threat to living matter.

Whether or not the method will be satisfactory may be pretty much a matter of guesswork. Any process that is supposed to operate over such a long time can't be tested by actual experience. But it can be exposed to the expert opinions of atomic scientists — both those for and against the development of nuclear energy — and this should be done.

Those who say the danger of nuclear energy is greater than its benefits are automatically suspicious of risk-free techniques advanced by those who have a vested interest in the promotion of nuclear energy. Thus, the UT-Oak Ridge plan should be subjected to the critical review of all shades of scientific thought.

About the best thing that can be said for the plan at this point is that it appears to be far better than the present method of disposing of nuclear wastes by storing them in deep water pools or at government reservations in tanks and burial pits.

The present method is an invitation to disaster, since it is estimated that by the year 2000 there will be millions of tons of waste scattered around at various locations to pose health threats to people and other living things.

The nuclear industry seems bent on moving full speed ahead in the construction of facilities before the problem of disposing of dangerous wastes is solved. This seems foolish to those who think the disposal problem should be taken care of first. But the industry, with the support of the present national administration, seems to put quick profits above safety guarantees and above an orderly, first-thing-first plan of nuclear development.

As long as this attitude prevails, any discovery that may promise greater safety in the use of nuclear energy and the disposal of its wastes is welcome and should be given thorough consideration.

The Arizona Republic

Phoenix, Ariz., May 30, 1981

ARIZONA is among the states being considered by the federal government as the site for a major nuclear waste storage facility.

No sooner had the U.S. Department of Energy and the U.S. Geological Survey indicated that Arizona was targeted as a potential dumping ground — along with seven other Western states — than Sen. Barry Goldwater and Gov. Bruce Babbitt announced they would oppose the move.

This may be good politics, but their timing left much to be desired.

Babbitt attempted to explain his opposition by suggesting the federal choice would be a political decision — against the less populated Western states.

Goldwater simply fired off his salvo of opposition and stood behind it waiting for the smoke to clear.

Where would the country find itself if the leaders of all 50 states suddenly announced — without benefit of discussions — that their territory was off-limits to the storage of nuclear waste?

Where could radioactive material be stored? In the Atlantic? Pacific?

Babbitt probably is correct when he says that the risk of ground-water contamination and earthquakes may rule out Arizona for a site.

If that is indeed true, then the record will show it in time.

But to simply draw conclusions before the issue is debated and discussed is a form of seceding from the national interest.

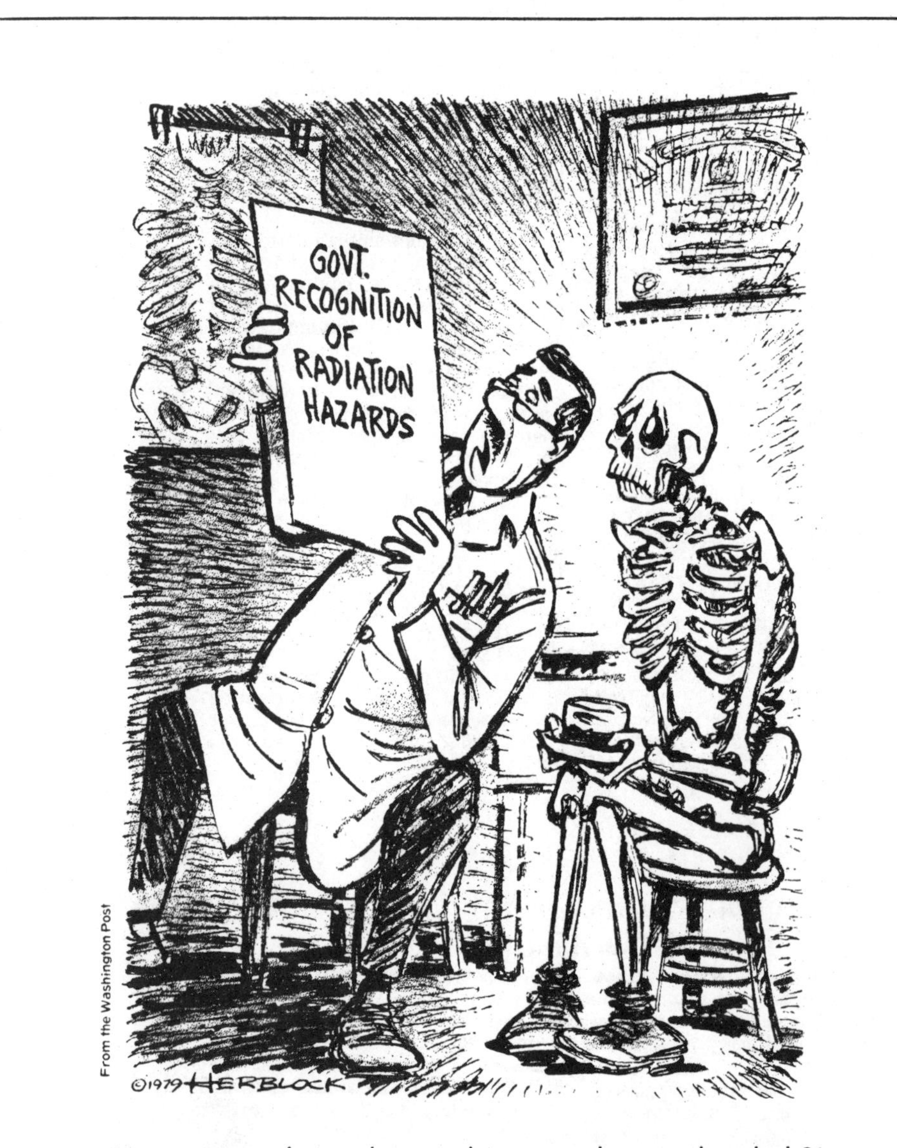

'You say you have this tendency to glow in the dark?'

THE LOUISVILLE TIMES

Louisville, Ky., August 3, 1981

President Reagan came into office arguing that more nuclear power is one answer to the country's energy needs.

But he didn't say what he planned to do about radioactive waste, the accumulation of which has become a major obstacle to the expansion of the nuclear industry. The President, preoccupied with economic issues, has yet to propose a comprehensive solution.

However, his administration is considering a scheme to use some of the material that is now temporarily dumped in power plant "swimming pools" after its value as fuel has ended.

The details would make Darth Vader's flesh crawl.

The idea is to use laser technology to separate the bomb-quality plutonium created in nuclear reactors from other, less-valuable wastes. The plutonium would then go into all those new warheads that are planned as part of the administration's military build-up.

Our friendly local utilities, including Public Service Indiana, could, according to one estimate, supply enough material for 10,000 nuclear weapons. Turning on the air-conditioner would thus become a patriotic act, since it would increase the country's capacity to incinerate potential aggressors.

But even if this process is developed — and many would-be nuclear powers are watching it with great interest — there will still be plenty of waste material that must be put away for hundreds, even thousands of years.

Yet, 3½ decades into the atomic age, there's still no disposal plan, much less a permanent disposal site. Without one, the construction of new nuclear plants is inconceivable. Indeed, some of those already built may have trouble operating at full capacity when their waste storage pools are filled.

There is widespread, if not universal, agreement that safe, long-term storage of radioactive waste is technically feasible. The federal government's responsibility for establishing disposal sites has also been generally accepted, although the taxpayers are likely to resist a plan that doesn't require utilities to pay a large share of the costs.

Congress, however, has yet to act. After failing to pass a waste bill last year, House and Senate are now considering several proposals. What's needed to get things moving is a strong push from Mr. Reagan.

The schemes knocking around various committees include a temporary storage program designed to prove government can do the job. Sen. Bennett Johnston of Louisiana wants to put the waste in shelters in Nevada. His purpose, of course, is to keep the stuff out of geological formations in his state.

The best bet, however, is Rep. Morris Udall's bill, which calls for permanent disposal and a detailed timetable for achieving it by the year 2000. The process of selecting sites, gaining local approval, meeting environmental rules and actual construction will easily take that long.

But that's all the more reason to get started. If Mr. Reagan can be persuaded that even the plutonium ought to be put in vaults rather than bombs, so much the better. If not, he must still lend his voice to those who point out that the nuclear waste dilemma must be settled promptly.

THE SAGINAW NEWS

Saginaw, Mich., July 19, 1981

Thinking Americans could do worse these days than to take a more active interest in what is rapidly shaping as a major concern for all of us: the problem of disposing safely of highly radioactive nuclear wastes.

Time is running out more swiftly than we like to consider. Wastes from both the nation's production of plutonium for nuclear weapons and from its growing numbers of commercial nuclear plants are multiplying annually.

Like the more than 240 tons of spent-fuel assemblies now in underwater storage at Michigan's three operating nuclear power plants, these are still looking for a permanent burial site. And a permanent containment form.

We are writing here about nuclear wastes that most scientists argue will remain a threat to human life for the next 200,000 to one million years — should man still survive on this earth in those far-off millenia. We are more concerned, of course, with the short-term future of our children, our grandchildren, and their children's children.

Some utility spokesmen argue, for instance, that the rate of radioactive decay actually accelerates under some conditions, making nuclear wastes a lethal threat perhaps for only the next 1,000 years. That is little comfort, we think, to the next 33 generations of humans.

The case of Michigan is an example of the nuclear waste problem close at hand.

We now have three nuclear plants operating near Charlevoix, South Haven and Benton Harbor. During the 1980s, two more are scheduled to come on line, one near Detroit, the other in our Saginaw Valley backyard at Midland, 20 short miles away.

Michigan is only one example. As of last month, this nation had 76 reactors in operation, 81 under construction, and 17 more on order. Last year, there were another 118 operational reactors in the Free World alone, not counting those in Russia, China and their satellites. The problem is global. But Michigan must, selfishly, be our major concern. And a draft from a special joint committee on nuclear energy of the Legislature last year warned that commercial nuclear wastes are increasing in the state, annually, at the rate of up to 90 tons.

When the proposed Fermi and Midland plants come on line, its estimate was that the annual production of nuclear wastes would rise to 170 tons a year. Even after the expedient of jamming the radioactive fuel rods closer together in reactor site pools, the draft warned Michigan could have filled its total temporary storage facilities before the year 2000.

That, in turn, would mean reactors would have to shut down shortly thereafter.

So your concern should be aimed now at federal efforts to: 1) find the best method of embedding radioactive wastes inside a shielding element that will prevent leakage for centuries to come; and 2) find an area inside the U. S. geologically stable enough to become the respository for such wastes.

There is hope on both scores. The Department of Energy is about to approve production facilities for perhaps two methods: immobilizing the wastes in borosilicate glass, or inside a synthetic rock armored with titanium and other oxides. Both offer a chance for safety over what DOE calls 'geologic time,' measured in galactic eons.

DOE's National Waste Terminal Storage Program is looking for permanent repositories. Projects under study include basalt formations near Richland, Wash., granite and shale formations in Nevada, domed and bedded salt formations along the Gulf Coast and in Utah and Texas.

Of more immediate concern, one study involves vast underground salt beds in the Salina Basin area — which includes almost all of Michigan, and parts of Ohio, Pennsylvania, and New York. Ohio and New York salt beds have apparently "won" the nod for further evaluation over Michigan and Pennsylvania.

Will they be safe against volcano, earthquake, the shifting of the earth's floating tectonic plates?

And your concern should also not overlook the current hassle over the transport of high-level nuclear wastes from Canada, through Michigan, to a recycling center in S. Carolina.

What Michigan authorities do now toward establishing reasonable regulations for strictly monitored and safe transport of wastes across Michigan, whether by truck, train, or boat, and maybe someday even by plane, can well serve to safeguard Michigan citizens when the time comes — and it is not far off — when our own nuclear wastes will have to find their way, some from the Saginaw Valley.

What can we as individuals do? We can become more informed on the pros and cons of nuclear power, by listening both to the utilities and the environmentalists — and making up our own minds on nuclear power need versus nuclear power danger. We can study the measures being taken to protect us when nuclear plants are, again not too far off, phased out and decommissioned.

We can also turn to our congressman and our legislators, in whose hands lies the nuclear future of this nation and state. Let them hear your voice of concern. Urge them to get on with it. Delay in the halls of government must not be tolerated — for the stakes are too great.

THE PLAIN DEALER

Cleveland, Ohio, December 5, 1981

Nuclear technology produces dangerous garbage that must be stored somewhere, but not — ever — in the salt caverns under Lake Erie, within whistling distance of the people of Greater Cleveland. There are so many good arguments against it that the idea is ridiculous.

Department of Energy officials say they have decided against using Cleveland as a storage site for nuclear wastes. They would, however, like to do a few tests, subleasing a portion of International Salt Co.'s under-lake acreage. They say these tests are innocuous and will give them information applicable to other salt beds in the West, where wastes will eventually be stored. Since Ohio is a nuclear waste generating state that expects to dispose of its effluent elsewhere, who are we to refuse permission for tests that would help make proper storage in the West easier, cheaper, safer?

Yet, the panic could not have been more immediate. No tests, none, should be done here, the reaction was, because tests lead to results, and the result in this case would be nuclear storage directly under Cleveland's doorstep. Opponents of the testing are being led by Rep. Mary Rose Oakar, who is vehement, clear and convinced in her belief that if the "tests" are successful Cleveland will become the leading contender in the storage sweepstakes.

In the absence of more than a vague, generalized suspicion that the federal government and, most particularly, that portion of it associated with the nuclear industry, is nefarious, tests should be permitted here. There are sound technical reasons why the International Salt beds are a good testing site.

But the people of Cleveland must keep a taut rein on the whole business. Every progressive inch of the testing, from the terms of the government's contract with International Salt to the equipment being used, from the early results to final recommendations, must be open to scrutiny and comment.

The government has laid itself open to all of the criticism it is receiving. It has a history of lying about nuclear tests, albeit decades-old tests, and decades-old lies, followed by decades-long cover-ups. The Cleveland testing revelation was done at a hastily called session the day before Thanksgiving, hardly an ideal time and one that invites suspicion. Spokesmen who presented the proposal were short on details, didn't know about the agreement with International Salt and had no opinion on whether some of the tests might not cause salt cavern collapses and subsequent cave-ins on the land surface, problems that subsequently have been explained away. Energy officials are just reaping the witchgrass they sowed by approaching the touchy issue so casually, and we hope they have learned something from the experience.

A cautious "yes" to testing would give Cleveland a lever with which to pry open every door concerning siting plans for the nuclear waste disposal, and might give the city early warning of undesirable developments. No technician, no politician or bureaucrat, no Department of Energy official must ever be allowed to think, even in his secret heart, that Cleveland and Lake Erie might be a nice place to bury all of that troublesome nuclear waste no one else will have.

The Cleveland Press

Cleveland, Ohio, December 4, 1981

Should the International Salt Mine here be used as a federal testing ground for radioactive nuclear waste storage?

In a word, the answer is "no."

The proposal for such tests here was made by Battelle Memorial Institute, which manages a nuclear waste program for the federal Department of Energy.

A whole clutch of public officials here had a strongly negative reaction to the proposal to determine what effect the intense heat from nuclear waste would have on the salt mine testing site.

Despite the protestation of the DOE that there are no plans to actually store nuclear waste here, we are concerned along with the critics of the project that testing would be a foot in the door.

Why would the DOE want to spend millions of dollars on a testing program in this area if there weren't some thought being given to dumping nuclear waste here some time in the future?

Under the proposal, huge electric heaters would be used to simulate the heat — around 200 degrees Fahrenheit — generated by radioactive nuclear wastes. If all went well with such tests, it seems to us that the DOE would then push for the actual storage of nuclear wastes here, despite the denials they are making now.

County Commissioner Virgil Brown and Rep. Ronald Mottl — two of the officials who hotly oppose the tests — make the point that there are many less populated areas better suited as possible sites than Cleveland.

Mottl has written to Energy Secretary James Edwards, asking that the nuclear waste proposal be scrapped immediately. Rep. Mary Rose Oakar also has her dander up, charging that DOE officials are "lying in their teeth" when they deny any intent to actually store nuclear waste here.

Maybe the DOE was launching a trial balloon to see which way the wind was blowing on this issue. If that is so, the fervor with which the balloon was shot down should tell them to take their proposed tests elsewhere.

The Boston Globe

Boston, Mass., August 29, 1981

Apart from war, waste disposal is the ultimate nuclear safety issue because the radioactive byproducts steadily accumulating must eventually be isolated from the biosphere for something like 100,000 years.

In the short term, waste storage has been manageable, and as a result the search for a long-term solution got little attention until the mid-70s. It was the kind of problem, said one nuclear expert, that you don't get around to thinking about until it's five in the afternoon and time to go home.

Now, for the first time, nuclear waste disposal will get sustained public attention as the result of a proposed rule which has been published for public comment by the Nuclear Regulatory Commission. In drafting the rule the NRC had two broad options.

The one favored by the nuclear power industry was to avoid specifics and simply state, in general, that any repository must meet Environmental Protection Agency standards for the emission of radiation. Then it would be up to industry to decide how to do that. The more cautious alternative was to set forth specific design criteria which a repository would have to meet before being licensed, and that is the route the NRC has chosen.

Nuclear waste will generate great heat during the first 200 to 300 years, and no one knows how the heat and radiation may affect surrounding rock. Thus the NRC proposes that the waste be stored in canisters engineered to contain radionuclides for at least 1000 years. The canisters would be be placed in vast underground chambers mined out of solid rock formations that have been stable since start of the Quaternary Period, 2 million years ago.

The chamber's design and the eligible rock formations – for example the virtual absence of groundwater – would mean that if the canisters failed after 1000 years, radiation would escape very slowly. Finally the NRC would require that the repository be designed so that, for 50 years after it is filled, it would still be possible to reenter the chamber, retrieve the waste and dispose of it in some other way.

The nuclear power industry, the interested public and the Department of Energy, which would operate such facilities under NRC rules, have until early November to respond to the draft.

The nuclear power industry will probably appeal for a less rigorous solution, and given the new Reagan-controlled majority in the NRC, it may get a sympathetic hearing. On the other hand some will argue that even this relatively conservative approach to handling waste is inadequate and is likely to fail in some unforeseen way.

The broad ethical question underlying the rule on nuclear waste is how conservative this generation should be in planning for the future life of this earth. The specific issue requires practical thinkers – engineers and policy makers – to try to conceive of cannisters and underground chambers built to last 1000 or 100,000 years. That requires a bizarre leap of imagination, but it is a real decision, not science fiction.

The Evening Gazette

Worcester, Mass., March 6, 1982

Unless Maine Yankee's nuclear power plant at Wiscasset can find a way to store more of its spent fuel rods, it will have to shut down in 1987.

Vermont Yankee at Vernon has only enough storage space to last until 1988. Massachusetts' Pilgrim 1 plant in Plymouth can run until 1989, along with Millstone 1 in Connecticut.

Those dates are only five, six and seven years away. The four plants have a generating capacity of more than 2,400 megawatts. If the utilities have to substitute power generated at coal or oil plants, customers will pay hundreds of millions of dollars extra every year.

Maine Yankee wants to store more of the spent fuel rods in its pool by methods known as "reracking" and "pin compaction." Reracking means moving the rods closer together in the pool and separating them with special boron metal plates to prevent any chain reaction. Pin compaction involves stripping the individual fuel rods from the assemblies and putting them back together, thus saving space. The Nuclear Regulatory Commission has permitted other nuclear plants to rerack their fuel rods, but Maine Yankee is the first utility to apply for permission to do pin compaction.

Either proposal is going to bring out the opponents. In fact, the state of Maine is formally intervening. The plant — the only nuclear plant in Maine — survived one public referendum to shut it down, but may face another one next year. Any plan that may increase the risk in the storage pools is going to heighten public fears.

The federal government has been of little help. President Jimmy Carter banned reprocessing of spent fuel elements because he feared the risks involved in producing plutonium. Carter wanted the government to operate a temporary repository for spent fuel rods, but Congress didn't vote any money.

The Reagan administration has reversed the Carter ban on reprocessing, but it wants the nuclear industry, not the government, to do the job. So far, the industry hasn't shown much interest.

In the meantime, the radioactive waste ticks away in 78 pools at nuclear plants around the country. Virginia Electric and Power Company's plant at Jamestown may have to shut down by 1985 if it fails to get permission to truck the stuff to another plant 75 miles away. The town in which the other plant is located wants no part of the Jamestown rubbish.

For all the controversy that swirls around nuclear power, it has been a godsend to consumers, especially here in New England. If four of our six plants have to shut down within the next seven years, electricity rates will skyrocket again.

Nevada State Journal

Reno, Nev., March 19, 1982

Over the hills and through the valleys, across a continent, you can see the trucks coming in determined procession.

What are those trucks carrying? High-level nuclear waste.

And where are they headed? Nevada, as sure as nuclear fission releases little gamma rays.

Anyway, that is the feeling one gets as Uncle Sam seeks a single dumping ground for all the high-level civilian nuclear waste from the shining Pacific shore to the quaint hamlets of Maine.

Of course, the bill which came out of the House Interior Committee Wednesday does mandate a thoughtful approach to the selection process, and it does give the chosen state the opportunity to appeal and veto. Also, several states besides Nevada are in the running for this honor: Washington, Texas, Minnesota, Louisiana and Utah. But one cannot help feeling that the beady eyes of the nation are zeroed in on the Southern Nevada desert.

True, Nevada could appeal to Congress and use federal funds to do its own research into safety. And its veto could be overriden only by a vote of both houses of Congress.

But let's be realistic. Nevada is not very powerful in Washington. It is a state of small population with a lot of vacant land. Under these circumstances, a veto override would be hard to prevent. It is doubtful that even Sen. Paul Laxalt, for all his growing influence, could stop an override.

The simple fact is that nobody wants the waste. Yet it must go somewhere, and the desert seems a logical place. Basically, that leaves Nevada and Utah; and the atomic industry's longtime love of Nevada burns unquenched.

Nevertheless, the House bill is decent enough, as such bills go. It requires a specific time frame, which is essential if this unpleasant matter is to be decided before the next Ice Age. The Energy secretary must recommend five sites by 1985, and then reduce this to one; the president must make his decision by 1987, and the license must be issued by Jan. 1, 1990. That seems a reasonable time period.

Also, the bill is fair, or as fair as such things can be. Public hearings are required at all the proposed sites, and the appeal process at least permits a cry from the wilderness. The firms which create the waste, mostly nuclear power plants, would pay the dump site operating costs. Population centers of 2,500 or more, or with a density of 1,000 or more per square mile, would be ruled out. All of this seems rational.

Still, there are questions. Persons in the nuclear field continue to assure us that transportation is safe and sure, but is it? Is it wise to create only one dump and require trucks to haul nuclear waste hundreds and thousands of miles? Wouldn't several sites be better — and shouldn't the government re-examine this issue?

What of military nuclear waste? Shouldn't a dump site selection for this waste be placed under the same careful process as civilian waste, as Nevada Rep. Jim Santini suggests? The military accounts for 80 percent of high-level nuclear waste, and states should have at least an opportunity to formally appeal and discuss its disposal.

Beyond this, it remains debatable whether any site, even in the desert, is fit for *permanent* disposal of an ever-growing pile of atomic refuse. Can we — and other nations — continue to dump this stuff indefinitely without creating an unexpected and perhaps fearsome hazard sometime in the future? Shouldn't we re-examine the use of nuclear fuel for energy, admitting that we leaped in before we had all the answers and that we still don't have the answers? And shouldn't these uncertainties be an additional spur toward reduction and perhaps eventual elimination of atomic weapons worldwide? And if we can't control nuclear use, what then?

These are the questions that disturb our thoughts, as we look over the horizon toward that never-ending line of trucks heading toward — Nevada?

The Des Moines Register

Des Moines, Iowa., March 17, 1982

"Unsubstantiated fearmongering," the head of the Atomic Energy Commission yelled 12 years ago at people who wanted construction of nuclear power plants delayed until somebody figured out how to dispose safely of their wastes. The problem, said the official, was "manageable."

Federal energy officials still are saying that today, although they are hedging a bit more: "The [Energy] department has stated that it does not attempt to prove that safe disposal of radioactive wastes, with the required approval of appropriate regulatory authorities, can be achieved today. Rather, the department has shown that such disposal can be achieved within specified reasonable times upon completion of its current research and development and site exploration programs."

Meanwhile, the wastes pile up. Already, some utilities are threatening to close plants that are running out of storage room for their spent fuel.

No one can agree on what should be done with nuclear waste, which remains toxic for centuries. While it piles up in pools of water, in metal-alloy cladding that is gradually corroding, policymakers send it back and forth across the nation, cities and states pass laws to keep it out of their territories and scientists argue about what kind of rock to store it in.

Congress failed to solve the nuclear-waste problem in 1980. Now, as it tackles the issue again, the most spectacular action is that of senators and representatives scrambling to keep the stuff out of their areas.

By contrast, the "fearmongering" of a dozen years ago was calm, and it certainly has been substantiated. Perhaps more of it would be helpful now, to let the Energy Department know that the "reasonable times" for achieving a disposal method have been "specified." It is no longer enough to call the wastes manageable. It is time to start managing them.

The Orlando Sentinel

Orlando, Fla., May 17, 1982

Come December, it will be 40 years since Enrico Fermi and his colleagues launched the nuclear age. Ever since, radioactive garbage has been accumulating because of our belief that technology would eventually find a way tc dispose of it safely.

The truth is that no one yet knows what to do with the stuff. Thousands of tons of nuclear wastes sit in "temporary" storage until Congress decides how and where to dispose of it. Meanwhile, the public becomes more disenchanted with everything nuclear and doubts the scientists who are supposed to solve these problems.

The Senate passed a nuclear waste bill earlier this month but the House is unlikely to complete action on it this year. The big problem is that everyone says, "Bury it, but not in my backyard." No matter how much we might enjoy low-priced nuclear energy or the comfort of nuclear arms deterrence, nobody wants to live next door to a radioactive dump.

But we must find a solution. "Temporary" dumps holding radioactive trash are getting full. They are a dubious legacy for future generations.

Spent fuel assemblies, for instance, pile up in "temporary" water cooling pools alongside each nuclear power plant. In six states — including Florida — some pools will be full next year. At dozens of uranium mills, toxic tailings quietly give off suspected carcinogenic gas.

The Senate bill calls for the president, after thorough surveys, to select two sites for storing all nuclear waste. It's a reasonable bill under which utilities would pay their share of the storage and handling costs. But even if the House acts positively this year, the sites won't be ready until the next decade.

Admittedly, not all the scientists are wringing their hands. Nuclear advocates cite studies showing that wastes from fossil fuels are much more dangerous to human health because of pollution given off when they are burned. And other industries use substances like mercury and arsenic, which stay poisonous forever.

But for most of us, the existence of such potent, silent killers, even in permanent storage, is a worry. Because much of the material will remain toxic for centuries, it shouldn't be stored near geological faults and it shouldn't be buried near minerals that might be needed in the future.

For 40 years our hopes of solving the nuclear waste problem have been pinned on naive faith in technology. Finally, we realize that technology has no magic wand. There's no nuclear garbage disposer. Like everything else, the problem has its costs and, in this case, they are huge.

The Houston Post

Houston, Texas, May 8, 1982

Recent Senate passage of a bill to establish a national policy for disposal of nuclear waste is a positive step in solving what has become more of a political than a technical problem. Clearly the long-term storage of nuclear wastes requires the attention of the nation's best scientific minds. But delays, midstream policy changes, resulting lack of credibility and concerns of individual states have become more of a factor than the technical ability to develop a safe program for disposal of nuclear wastes.

Nuclear power has not become the clean source of cheap, abundant electricity early proponents thought it would be. Safety concerns, some of them legitimate and some questionable, have added so much to the cost of nuclear plants that more than 40 reactors have been canceled or delayed indefinitely. No new ones have been ordered since 1978. The South Texas Project near Bay City has been delayed while the design engineer and contractor were replaced. A proposed plant near Wallis seems to have been shelved.

About 11 percent of the electricity generated in the United States comes from nuclear plants, a percentage that is likely to grow as more reactors come on line. While nuclear fission is no energy cornucopia, it is and will remain an important part of the energy mix increasingly necessary in an uncertain world. Nuclear proponents point out that there are, worldwide, about 200 major nuclear plants that have been operating an average of 10 years each without hurting anybody.

The Senate bill would provide for additional temporary storage space. Otherwise some plants might have to close by 1986 because many existing waste storage facilities will be full by then. Long-term storage facilities would be established and a national plan for permanent waste disposal would be developed. A state could veto proposed sites within its boundaries, but the president could override that veto if it were in the national interest.

A firm commitment to long-term action is needed. That commitment must come from the political leadership, which now finds the ball in its court. The Senate action is a step — however small, preliminary and delayed — in the right direction.

THE COMMERCIAL APPEAL

Memphis, Tenn., May 10, 1982

AFTER 40 YEARS of buck-passing and downright neglect, the federal government now appears ready to develop a policy for the permanent storage of high-level nuclear wastes. At the urging of the Reagan administration, the Senate passed a bill last week that — if approved by the House — would give the federal government responsibility for the disposal of nuclear wastes.

At this point any bill is preferable to no bill at all. Accountability must begin somewhere. But the bill, which has the backing of the nuclear industry, unfairly releases the industry from the responsibility of taking care of its own wastes on a temporary basis, falls short of providing the sort of safeguards needed to give states a strong role in the process and moves too quickly in the selection of storage sites.

Under the terms of the Senate bill, the federal government would select three sites in 1984 for the storage of dangerous radioactive materials, most of which will remain lethal for up to 240,000 years. By 1986 one of the sites would be selected by the president for the first repository. And by the mid-1990s the first facility would be in operation.

Also in the bill are provisions that would: (1) have the federal government assume the costs of providing temporary storage sites for the spent fuel generated by nuclear utilities; and (2) allow states to veto the construction of storage sites selected by the federal government only if either the House or the Senate passed a resolution backing the state.

Something must be done — and soon. With the wastes from commercial nuclear power plants now occupying 104,000 cubic feet of space and the liquid wastes from weapons production estimated to be about 77 million gallons, time is running out for the government to get a firm grasp of the situation. Forty years is long enough to "temporarily" store materials capable of the mass destruction of human life.

Those critics of the Senate bill who oppose it on the grounds that the federal government has no business providing facilities to private enterprise are missing the point. Quite clearly it is in the best interests of the public to have the storage of nuclear wastes tightly regulated and controlled by the federal government. Public health standards require no less.

But it is not in the public's interest to subsidize sites for the storage of nuclear fuel from commercial plants. On-site storage is an economic fact-of-life for the utilities and it is unreasonable to expect taxpayers to pick up the tab. Neither is it fair for the federal government to run roughshod over those states that object to having the sites.

THE SENATE BILL also is to be faulted for rushing plans for the construction of the sites. While it is imperative that action be taken, 1984 is simply too soon to select the locations for the sites, especially considering that a federal survey of potential sites will not be completed until 1986.

Sen. John Stennis (D-Miss.), who opposed the rushed schedule, asked that the selection be delayed for three years: "What is a delay of three years compared to a program of 10,000 years?" Surely that is a reasonable request, one that could be addressed by the House when it considers a similar bill.

While the Senate was right to address the issue by seeking to make the federal government accountable for what quite obviously is a national problem, it is to be hoped the House will adopt a more responsible version of the bill, one that will eliminate the weaknesses of the Senate bill and somehow find acceptability in the House-Senate conference.

THE SACRAMENTO BEE

Saramento, Calif., May 14, 1982

Although by law the federal government is responsible for safely disposing of the radioactive wastes from nuclear power plants, three successive administrations have failed to create any waste disposal facilities. Worse yet, the goals of the federal research and development effort have shifted with each change of administration — leaving the nuclear industry uncertain about the future it must plan for, and alienating the states and local governments with a string of broken promises about the protections that will be provided at waste disposal sites.

It's been obvious for some time that Congress would have to step into the fray, but until now Congress itself was divided on the subject. And while Congress stalled, the nuclear utilities were running out of temporary on-site waste storage space, and the current administration was trying to hold up development of permanent disposal sites for budgetary and other reasons. The states, meanwhile, were refusing to cooperate in federal site-selection studies for either permanent or temporary disposal facilities, because they feared the administration would shortchange the program — that once any permanent or temporary facility was built, further funding for the full program would disappear.

As the Congressional Office of Technology Assessment (OTA) reported, the remaining technological problems with nuclear waste disposal paled before the mistrust and inconsistency that the politicians had created.

In its research, however, the OTA discovered that there were measures that would satisfy all sides in the nuclear waste debate and get the program moving again. These have now been incorporated in two bills — one that just passed the Senate and another that is now being studied in the House.

The key to breaking the logjam was not technical. The technical community, the OTA found, has been in nearly unanimous agreement for some time about what must be done: The radioactive wastes must be permanently buried thousands of feet below ground in stable geologic formations protected by redundant natural and man-made barriers. What Congress had to do, the OTA concluded, was create an agency with an independent source of funding and a plan of action written into law so that all participants would be assured that these facilities would actually be built

The OTA proposed to tax nuclear power users for the costs of the program — and to write into law a commitment to build several facilities so that no one state is stuck with all the nation's radioactive wastes; a prohibition against relying on temporary storage facilities except in an emergency; a plan for simultaneous study of several alternative geologic sites, and a guarantee to the utilities that the federal government will assume liability for and physical possession of new wastes at a specific date. The date, the OTA suggested, should be set conservatively to reassure everyone that the government will not end up rushing into an inadequately planned project.

This is essentially what the Senate has now done. It is what the House should do as quickly as possible.

Perhaps it should not have been necessary to legislate so many specifics to solve what was supposed to be a mere technical problem. But the political situation has gotten so out of hand that, unless Congress succeeds in clearing the air, there will be no nuclear waste disposal program at all.

Oregon Journal

Portland, Ore., June 14, 1982

Congress finally is moving rapidly to establish a national policy for handling nuclear wastes, and it is about time. Nuclear wastes are a fact of life and a method of storing them for long periods of time must be found.

Some environmental groups oppose such legislation because they don't want future generations to have to deal with today's nuclear garbage. The point is well taken, but repositories must be found to store what already exists. With nuclear plant construction in disarray because of high interest rates, this lull — or perhaps it is a permanent halt — in the construction process should be seized to develop facilities for better handling of these dangerous wastes.

A Senate-passed bill lays out a schedule for building a permanent deep underground repository for highly radioactive wastes by the mid-1990s. Natural geologic formations would be considered, but — at the demand of congressmen from states with potential deep storage sites — so would man-made waste storage facilities. The bill now is before the House.

One possibly troubling aspect of the Senate bill is an amendment submitted by Sen. James McClure, R-Idaho, designed to remove state control over future nuclear reactor construction.

It may be that the federal government has pre-empted this area of regulation so the issue could be moot. However, the voters in Oregon approved an initiative in 1980 tying future nuclear plant construction to the existence of a federal disposal facility as well as voter approval of each site certificate.

McClure's amendment says that mere enactment of the bill by Congress shall satisfy local statutory requirements for the existence of a disposal site before future nuclear construction can continue. This seems designed to blunt the attempts by seven states, including Oregon, to restrict nuclear plant construction.

Because of the wording of the Oregon law, it isn't clear if McClure's amendment would apply. However, it would be well to remove the amendment anyway and allow state regulation to continue.

But the main focus should remain on finding waste storage sites. Millions of gallons of liquid wastes from military programs are stored in steel tanks in Washington, Nevada and South Carolina. Portland General Electric's Trojan plant stores wastes under water at the site of the nuclear generator.

Some resolution — one hesitates to say solution — of this problem is direly needed, and the ball is in Congress' court.

The San Diego Union

San Diego, Calif., July 11, 1982

A consensus appears to be forming in Congress on a national policy for disposal of radioactive waste from nuclear power plants. It's about time. As early as 1957 the National Academy of Sciences recommended that provisions be made for handling the by-products of the nuclear power industry then beginning to materialize.

Spent reactor fuel is accumulating at the 74 nuclear plants now in operation, and another 79 plants are under construction or on order. With a push from the Reagan administration, Congress is finally willing to tackle a problem that has been on the national agenda for 25 years.

The problem is more political than technical, since scientific and engineering work on safe storage of radioactive waste is largely behind us. What has led previous administrations and Congresses to procrastinate on bringing the issue to a head is the inevitable prospect of resistance by any state in which the federal government wants to locate a disposal site.

The 69-9 vote for a disposal bill in the Senate last April shows the high degree of bi-partisan support for a formula which promises to unravel the political knot. Like a similar measure that received a favorable vote in a House subcommittee last month, the Senate bill provides for "consultation and concurrence" by the states in the process of selecting sites.

If federal officials find an ideal disposal site in a state that doesn't want it, the state could carry its objection to Congress, where a vote by either house could sustain the objection and send the government looking for an alternative site. The burden would be on Congress, then, to decide whether the national interest demanded that a state's cold reception to a disposal facility should be overruled.

It may never come to that, of course. If emotion can be sorted out from the facts about the long-term safety of waste storage, some states may find good reason to accommodate such federal installations. The pending legislation also provides for financial assistance to alleviate any "economic and social impacts" from a disposal project.

The nation cannot afford to delay any longer in laying down a firm plan for nuclear waste disposal. The technology for both permanent and retrievable storage of radioactive material is in hand. The bills in Congress provide for a tax of one mill per kilowatt hour on nuclear-generated electricity to pay for the storage program, which must be calculated into future utility rates as one of the costs of nuclear power which has been avoided up to now.

The Senate bill calls for selection of a commercial-scale repository site by 1986, which means the facility could begin receiving spent fuel elements by 1990. Having such a timetable in place will remove one nagging uncertainty from the future of the nuclear power industry in America, which remains beset by a variety of other problems.

Despite an appearance of stagnation, nuclear power is inching ahead in relieving the country of its dependence on fossil fuels for generating electricity. Nuclear overtook oil as a source of electric power in 1980, and is expected to outpace both natural gas and hydro-electric power in 1982, placing it second only to coal as the source of the nation's electricity.

Those statistics are a compelling reason for Congress not to let its current session end without putting a nuclear waste disposal program into place.

ALBUQUERQUE JOURNAL

Albuquerque, N.M., May 1, 1982

For 37 years the United States has been producing radioactive waste without a plan for disposing of it. Now the Senate has taken a step that could lead at last to a broad national policy for temporary and permanent disposal of the waste.

The Senate passed a bill that reflects impressive groundwork done by the congressional Office of Technology Assessment, which recently issued a report, "Managing Commercial High-Level Radioactive Waste." The matter now is in the House, where a companion bill has been introduced.

President Reagan supports the effort "so that we may clear the way for continued development of peaceful uses of nuclear energy."

The "continued development" is clearly threatened. By the year 2000, some 72,000 metric tons of spent nuclear fuel is expected to be generated in commercial plants. Other wastes generated by the military are also accumulating.

The decay process for radioactive waste elements, OTA points out, "takes from minutes to millions of years, depending on the type of atom." OTA agrees with many scientists that safe disposal technology is available. But "false starts, policy shifts, and fluctuating support" — i.e., political problems — will continue unless a broad national policy is accepted.

After the vote, a spokesman for Sen. Pete Domenici, R-N.M., said the bill does not apply to the Waste Isolation Pilot Project planned near Carlsbad. That site, the spokesman said, is intended for low-level, transuranic (TRU) wastes only.

But the OTA report identifies TRU elements as "primarily fissionable plutonium," which is anything but "low-level." Could another political problem be rearing its head?

Nevertheless, the opportunity to work out a comprehensive policy is at hand. The policy is needed because thousands of tons of waste already have accumulated, and thousands more are on the way. That waste must be disposed of safely, or nuclear power garbage could jeopardize mankind.

THE WALL STREET JOURNAL.

New York, N.Y., May 4, 1982

It looks as if the administration and Congress have at long last agreed on a program to handle the disposal of nuclear waste. We, of course, have had the technical capability for years to dispose safely of this waste. But until now, the political problems have been far more intractable. Anti-nuke demagogy, with the eventual aim of shutting down America's nuclear power plants, had Washington in a political straitjacket. Now those clamps appear to be broken.

With administration support, the Senate last week passed by a 69-to-9 vote a three-stage plan to cope with radioactive waste. The first step would provide additional temporary storage space in order to avert a possible shutdown of some nuclear plants by 1986, when many of the existing bunkers will become full. The other steps involve building a longer-term handling facility and setting a timetable for finding deep underground sites that will safely contain the radioactive residue forever.

The plan is probably overly complicated. Responsible engineers and scientists have insisted for years that we have the technology to store nuclear waste permanently and safely. Even the Congressional Office of Technology Assessment has swung around to this conclusion; an OTA report published last week says there are no "insurmountable technical obstacles" to safe disposal. Nuclear experts have even selected several possible sites around the U.S. that have stable underground structures.

While the scientists agree on the safety of radioactive waste disposal, there are still some zealots who would like to make nuclear power as expensive as possible in hopes of pricing it out of the market. Given the history of their successes against nuclear power, it's not surprising the administration and Congress want to tread softly. Only last week, Washington Public Power Supply System voted to suspend construction of a 61%-completed nuclear power plant at Hanford, Wash. In the last year, the utility has decided to stop work on three out of five nuclear plants under construction.

Despite the new disposal plan, which still requires House approval, all the problems haven't been solved. For instance, several states cited as potential storage sites have raised objections. These might be overcome by a reverse auction, i.e., putting the site in the state that asked for the least amount of money. Another problem regards sharing costs between utilities and the federal government's nuclear military program, which has produced the bulk of the nation's radioactive waste.

None of these managerial—or, for that matter, technical—obstacles is insurmountable. All that is necessary is the political will to address the disposal problem squarely and to defuse the environmentalist demagogy. The administration and the Senate last week took a long-awaited step in that direction.

RAPID CITY JOURNAL—

Rapid City, S.D., July 15, 1982

The technological problems are minor compared to the political problems involved in disposing of nuclear wastes that have been building up for 35 years.

By law, the federal government is responsible for disposing of the radioactive wastes. Three successive administrations, however, have failed to create any waste disposal facilities. The goals of the federal research and development effort have shifted with each change of administration. The nuclear industry has been uncertain about the future it must plan for. States and local governments have been alienated by a string of broken promises about the protections that will be provided at waste disposal sites. There was a fear that once a permanent of temporary facility was built, further funding for the full program would disappear.

Research by the Congressional Office of Technology Assessment (OTA) found there were measures that would satisfy all sides in the nuclear waste debate and get the program moving again. These have been incorporated into two bills — one that passed the Senate in late April and another that's being studied in the House.

The key to breaking the logjam wasn't technical. The technical community has been in nearly unanimous agreement for some time about what must be done: The radioactive wastes must be permanently buried thousands of feet below ground in stable geologic formations protected by redundant natural and man-made barriers. What needed to be done was to have a plan of action for short-term, long-term and permanent storage sites written into law so that all participants would be assured that the facilities actually would be built.

That plan is incorporated into the bill the Senate passed and which the House is considering.

The big problem that remains is that everyone says, "Bury it, but not in my backyard."

That attitude was reflected this week by Gov. Bill Janklow who said he won't allow a disposal site even for low-level radioactive wastes to be established in South Dakota. The governor took that position even though he says he doesn't know how dangerous such a disposal site would be because he's never studied the issue.

Janklow's reaction is a political one and is based on emotion rather than on technology. It's hoped the governor hasn't put himself into a position he can't retreat from if a disposal site in South Dakota is shown to be in the national interest and can be accommodated without threatening the state's population or environment.

The State

Columbia, S.C., May 6, 1982

THE U.S. SENATE passed a nuclear waste disposal bill April 29 which could be bad news for South Carolina.

There is no question that the nation sorely needs a national policy on the disposal of highly radioactive wastes from civilian reactors as well as the military's. At present, millions of gallons of the military's nuclear wastes are being "managed" in tanks at federal facilities like the Savannah River Plant.

Spent fuel elements from the utility-owned reactors are being held in storage pools at the reactor sites. The utilities say, however, that they will run out of space soon. They want the federal government to provide another temporary storage place.

What is most needed is a *permanent*, federally owned storage (disposal) facility. (You can get into an argument as to whether both the government's and the commercial nuclear wastes should be stored together. The bill lets the President decide that.)

The good news in the bill the Senate passed is that it sets a timetable for the federal government's provision of a permanent repository. By 1984 three sites will be picked for serious study. The President will select one in 1986 for the nation's first nuclear waste repository. It is to be ready by the early 1990s.

The bad news is that the Senate decided to allow the federal government to build or buy facilities to store temporarily the spent fuels from the utilities' reactors. This is called away-from-reactor storage, or AFR. The Barnwell Nuclear Fuel Plant, which was designed to recycle spent nuclear fuel but never has operated, is one of three prime sites for AFR storage. The others are in New York and Illinois.

Sens. Strom Thurmond and Fritz Hollings led the fight on the Senate floor against the AFR provisions, but they lost by narrow margins, 47-43 and 46-43.

The Senate also declined to give the states a strong voice on the location of the permanent repositories. If a state objects, it can appeal to Congress where the objection would stand if either the House or Senate passed a resolution supporting the state.

South Carolinians must now look to their members of the House of Representatives to try to eliminate the AFR provisions of the Senate-passed bill. This state is already host to one-third of the nation's high-level radioactive wastes at SRP. Let the utilities expand their storage pools and keep their wastes until the permanent repository is ready.

Rockford Register Star

Rockford, Ill., May 3, 1982

Nuclear waste. It's the by-product of nuclear energy plants. And it's growing by leaps and bounds, a radioactive generator that can remain dangerously active for more than 1,000 years.

The United States has managed to locate nuclear energy plants across the country. It can't muster an agreement on where the waste should be stored.

Ironically, in a Reagan administration that devalues the role of federal government, the Republican-controlled Senate found a place where the feds could be useful: in operating the nuclear storage sites, if and when they can be agreed on.

In fact, the Senate resurrected the Department of Energy (DOE) for this supervisory role.

Although the Senate vote was 69 to 9 to establish a national policy on nuclear waste, most of the hard decisions lie ahead.

Quite logically to Illinois, Illinois does not want to be the dump site. Nor does Wisconsin. Or Michigan. And on.

The decision for two such burial sites, according to the Senate action, would rest with the DOE and be due in five years. Then it would be up to the Nuclear Regulatory Commission to select the first such burial site by 1990.

Right now, there are 72 nuclear energy plants in operation across the nation, each with temporary waste storage facilities. The new act would place a 12-year limit on such temporary locations.

Also, the nuclear energy industry would be authorized to pass along the cost of storage to its customers.

Should a state object to the ultimate location of a permanent site, either the House or the Senate could overrule the objections and force the dump site on the chosen state.

We think the Senate acted about as bravely as circumstances warrant. A national policy must be established. The House ultimately must concur. A lot of politics lie ahead. But it is that thicket that has delayed the issue until now, when an element of desperation underscores the priority that nuclear waste sites must be given to protect public safety.

St. Louis Review

St. Louis, Mo., August 13, 1982

The dumpers of nuclear waste are knocking on our door. The U.S. Department of Energy wishes to move additional radioactive waste material to the area of four lakes already contaminated at Weldon Springs in St. Charles County.

Some of this waste is already in the St. Louis area where it has lain for 30 years. Over and above this local material, the Department of Energy proposes to add additional nuclear waste from four other states over 10 years. Their plans call for covering this material with a somewhat impermeable layer of clay to be finished off with topsoil and seeding. Alternative plans similar to this are hinted at.

Touring the area of the lakes and listening to the experts' explanations was quite an experience. A vast amount of scientific data was spelled out in "curies," "mr's" and "half-lives." For the average layman making a sound judgment based on this kind of technical information is next to impossible.

It is only when we realize that we are acting for the future of all humans and in their name that we begin to wonder. If a substance will still be half as radioactive 10,000 or 200,000 years from now we realize that we might be acting beyond our rights.

A point needs to be made early in what will be a media argument of months, maybe years. This will be scientific matter to only a small degree. In the main it will involve faith and trust in science on the one hand and distrust of science and emotion on the other. Most citizenry will remain ignorant of the scientific evidence, but many will trust scientists and believe that sometime in the future they will find ways of neutralizing the radioactive dangers of this material.

On the other side are those distrusting science, big business and federal pronouncements. They present evidence of leakage from these ponds into nearby wells and dangers from accidental exposure to this material. These citizens doubt that adequate methods of decontamination can be found in time to prevent serious permanent harm to lives and the environment.

In all of this we look for more information on similar sites around the world. The early public outcry should warrant development of possible alternatives. We realize that nuclear waste must be kept somewhere, but is the plan for Weldon Springs really that adequate? In much of this we realize that the future must belong to God since it is obviously beyond our control.

Post-Tribune

Guarding Your Interests Daily

Gary, Ind., August 13, 1982

Congress is having a difficult time trying to establish a national policy for burial of nuclear waste. The main problem seems to be that nobody wants the stuff ending up in his own back yard.

The House Energy and Commerce Committee finally approved a bill, but only after four days of deliberations in which most members of the committee tried to make sure the burials wouldn't take place in their states. Now several members of the Rules Committee, which decides what bills reach the House floor, say they will oppose the bill unless certain states (theirs) are protected.

You'd think, the way they're acting, that these brave leaders fear nuclear waste. Of course, most didn't show any such concern when promoting the nuclear industry. The waste has to go somewhere. It's up to them to find the safest method and the most appropriate sites. Nothing involving such hazardous waste is going to be the ultimate answer, but it's a little late to think of that.

Reno Evening Gazette

Reno, Nev., July 30, 1982

Rep. Jim Santini no doubt is attracting good numbers of potential voters through his efforts to block any federal effort to put the nation's first permanent nuclear dump in Nevada. Gov. Robert List, and to a lesser extent, Sen. Howard Cannon, are right behind him.

But a member of Congress has an obligation to serve the nation as well as the home district. Santini would serve the nation better, and Nevada more honestly, if he turned his attention to the conditions under which any nuclear dump would be managed, rather than simply shouting "no."

If every state had a veto power over the location of a nuclear dump, as Santini has proposed, the nuclear waste problem might never be solved. At the very least, the time and political horse trading that would be required to get Congress to override a veto by any state could set the nuclear waste program back several years.

And nuclear waste is a problem that cannot be set aside much longer — even if no additional nuclear power plants were ever built; even if a freeze on military weapons went into effect today.

Here are just a few indications of the problem:

□ More than 2,500 metric tons of spent fuel have accumulated as a result of commercial nuclear power plant operations. This amount is steadily increasing. Most spent fuel is presently in temporary storage near the reactors. And some of those storage pools are beginning to leak.

□ Approximately 80 million gallons of high-level radioactive waste have been generated by government-operated nuclear reactors, principally for weapons production.

□ Military activities also have generated 65 million cubic feet of low-level radioactive wastes.

□ Approximately 600,000 gallons of liquid high-level radioactive wastes stored at the Nuclear Fuel Services reprocessing center at West Valley, N.Y., are beginning to leak into Lake Ontario. They must be solidified and permanently stored somewhere.

"Somewhere" does not necessarily have to be Nevada, of course. But the fact is that there are few places in the country where these wastes can safely be put. The Nevada Test Site, whether we like it or not, may turn out to be one of those places.

It's not as if the government were planning another MX for Nevada. That project was unnecessary or even dangerous to the security of the country. And at any rate it would have meant massive economic, social, and environmental damage to the state.

The atomic waste program clearly is needed. And it is a project of a much smaller order. It would have nowhere near the effect that MX would have had.

So if Santini and our other representatives in Congress insist simply on hollering "no," the rest of the country might just get together and jam it down our throats. Just this week, a House committee voted down Santini's proposal to give states the right to veto a dump site, though he says he will bring it up again before the full House.

A constructive approach would be better. If it is finally decided that the waste is going to come here, let's make sure Nevadans are in a position to have some say over transportation routes and over rules and regulations that will govern operation of the center once it is opened. We ought to make sure the federal government has adequate preparations to handle any accidents that may happen, and we ought to to be pushing plans for nuclear recycling that could eventually make dumping less of a necessity than it is at present.

It's easy enough for our representatives to say, "Put the waste in somebody else's back yard."

But if they don't come up with something better than that, the rest of the country is just not likely to listen.

The Salt Lake Tribune

Salt Lake City, Utah, August 5, 1982

Finding a safe and sane way to dispose of nuclear wastes is beginning to resemble the frantic search for a secure basing mode for the intercontinental ballistic missile known as MX.

In the case of the missile, some 30 possibilities have been seriously studied and discarded because of one weakness or another. The missile is soon to go into production (unless Congress comes to its senses and stops funding) even though nobody knows what will be done with it as it comes off the factory floor.

It is much the same with nuclear waste. The stuff is piling up at temporary sites and at nuclear facilities around the country. It is reaching dangerous proportions but agreement on safe disposal is hard to come by.

The U.S. Department of Energy, over the protests of Gov. Scott Matheson, is pushing ahead on a study of a possible Utah waste repository (or dump) just outside the boundary of Canyonlands National Park. In addition to the unfortunate selection of a possible site so close to a national park, other questions about the Utah facility seem to be surfacing with regularity.

First, there was the suspicion that samples of the salt formation in which the repository would be situated, showed the presence of moisture in sufficient quantity at some levels to make the site unsuitable for its dangerous mission. These fears, publicized by Friends of the Earth, a conservation organization, were dismissed by the DOE as without foundation.

Monday, a respected Utah geologist, Dr. William Lee Stokes, in a memorandum to Genevieve Atwood, director of the Utah Geological and Mineral Survey, raised the possibility that oil and gas seepage in the repository area could cause explosions and related dangers. If Dr. Stokes' questions cannot be resolved, the Utah site seems marginal at best and would probably have to be abandoned. DOE is checking its data in light of Dr. Stokes' observations.

The DOE is looking at several other possible repository locations and we have heard nothing on what, if any, similar shortcomings are being found at them. Considering that ground water and gases are not uncommon occurrences, it will be surprising if any of the potential sites are found to be without some imperfections. As with the MX, it may be next to impossible to find a perfect, foolproof storage site.

The country then will be faced with selecting the site that presents the fewest negative considerations, realizing all the while that even an imperfect repository is better than leaving the radioactive waste sitting in makeshift holding areas throughout the country. To put it in a more down home way, it is better, and safer, too, to have a city's garbage buried in an unsightly landfill than to let it stack up on the streets.

THE ANN ARBOR NEWS

Ann Arbor, Mich., July 21, 1982

Nuclear waste is literally a hot issue which governments have been slow to address. The chief agreement is that the stuff is dangerous.

But when it comes to transport, proper means of detoxification (if any) and safe disposal, agreement frequently breaks down and bickering over rules and costs sets in.

Defense-generated nuclear waste — allegedly because of its implications for national security — is being placed in a special category. A House Armed Services subcommittee last week approved a measure that would allow the military to dispose of its nuclear waste without public scrutiny.

"There are possibilities," said Herman E. Roser of the Energy Department, "for the loss of what we consider to be very important secrets in the handling of nuclear wastes."

The measure approved last week would remove military wastes from the jurisdiction of the Nuclear Regulatory Commission and specify that environmental impact statements are not required for military waste dumps.

ONE OBVIOUS QUESTION here is why, if military waste is disposed of carefully and properly in the first place, is there danger of defense secrets being compromised?

The appearance is one of a sensitive public health issue being handled without benefit of public overview. Given the nuclear consciousness raised by the freeze issue, disarmament rallies and a best-selling book by Jonathan Schell, the no-scrutiny legislation is ill-advised.

The diminished authority of the NRC in this case is in keeping with the administration's penchant for deregulation, but the question is whether the public really wants fewer safeguards where something as dangerous as nuclear waste is concerned.

Military waste is just as dangerous as civilian waste. An excellent case is made for handling both under the same regulatory standards, including environmental reviews, public hearings and close cooperation with state officials.

There should be no double standard where nuclear waste is concerned. What applies to military waste applies to civilian waste and vice versa. The public should know whether waste is disposed of at sea, in outer space or in its locality. Hasn't Love Canal taught us anything?

The News and Courier

Charleston, S.C., December 3, 1982

Lame Duck congresses have their uses. They are particularly useful for passing legislation that congressmen would not like to have to explain to their constituents before elections. A case in point is the federal away-from-reactor (AFR) nuclear storage program which was finally passed by the House on Tuesday. It is a bill that hardly anybody in Congress wants, that is not popular with voters but the utilities have been pushing it and the administration has been backing it.

The utilities claim that unless they are provided with temporary federal facilities to store nuclear waste, they will have to close down their nuclear power stations. Nobody believes them and even their own p.r. handouts are not convincing. Their argument, that unless there is AFR storage available by 1990, 389 nuclear power reactors might have to be closed down is merely a threat, aimed at getting the government to do what the utilities themselves should be doing. There is no present urgency for away-from-reactor storage for the waste generated at nuclear power plants. It can be handled perfectly adequately at the reactor sites. Of course, it will cost money; but investment in the technology needed to handle nuclear waste on site should be part of the operating costs of a power plant and was so considered until the Reagan administration revealed that it has a soft spot for the utilities.

The bill, like most ill winds, may blow some good. It should increase pressure on the government to come up with long overdue plans for a permanent waste repository. In the meantime, all those states which have federal nuclear waste storage facilities will be living with the fear that they will be dumped upon. In this regard, Rep. Butler Derrick, D-3rd District, has carried out an heroic struggle to stop the rest of the country dumping more commercial nuclear waste on South Carolina. He says he has secured a commitment that Barnwell will not be used and has managed to get enough safety clauses written into the bill to protect the state, although the federal Savannah River nuclear weapons plant could become the site of an AFR.

But, as Rep. Derrick and the entire South Carolina delegations in both House and Senate contend, the involvement of the government in providing away-from-reactor storage for commercial nuclear waste from the reactors of private utilities is bad news. Every effort must be made to see that the permanent waste repository is on line as soon as possible. Then the hopeful forecast of House Interior Committee Chairman Morris K. Udall, D-Ariz., may come true and AFRs will not be needed. But the sorry record of the government suggests that this is wishful thinking and that we all may have cause to regret the fact that this lame duck House gave in to administration pressure and the lobbying of the utilities and laid this particular egg.

AKRON BEACON JOURNAL

Akron, Ohio, September 6, 1982

A ROUNDUP by Walter Sullivan of the New York Times of the progress of Western European nations toward solving the problem of radioactive waste disposal indicates that they still face difficult technical, geological and political problems. But those nations have come a great deal further than the United States.

Few need reminding that the U. S. Department of Energy and the Congress have so far been unable to reach agreement on any plan for permanent safe storage of such wastes, some of which will remain dangerously radioactive for thousands of years.

Less familiar is the looming effect of this failure: For lack of any disposal strategy, some of this country's 73 functioning nuclear plants, in a nuclear program already in deep trouble for other reasons as well, may have to be taken out of operation as early as 1986. And the DOE and others in the field predict a full-blown storage crisis by 1990 if adequate disposal sites have not been provided.

Political opposition

Even research work aimed at identifying the kinds of geological deposits that would be best and safest for long-term entombment of nuclear wastes has run into fierce political opposition virtually wherever it has been undertaken or proposed.

Western Europe, with 85 nuclear plants operating in 10 countries and more planned, also has yet to put its first pound of medium- or high-level radioactive waste into a long-term repository. But it has at least arrived at some agreed plans and is well into the research work necessary for implementation of those plans.

Five of those 10 nations have decided on ways to store their own more troublesome nuclear waste and that of some of their nuclear fellows on the continent and are well along in work toward implementation of those plans. At least two others appear close to decision.

• France, with 29 reactors now and plans to have a total of 55 furnishing 55 percent of its electricity within the next three years, is working out the details of a plan to store waste deep in the rocks under the mountains of its southern Massif Central.

• Britain, with 32 reactors in operation, is working on similar deposition of waste deep in granite.

• Sweden's people decided in a 1980 referendum to close down all nuclear power generation there by 2010, even though their nation depends on its nine nuclear plants for almost half its electricity. Working with scientists from seven other nations, including some from the United States, Canada and Japan, it has made considerable progress to dispose of its own high-level wastes in the granite vaults of an old mine. Less dangerous medium-level wastes are to be stored in granite a mile out under the Gulf of Bothnia.

• West Germany, with 12 nuclear plants in operation, has done much of the preliminary work toward ultimate deposition of its more troublesome waste deep in salt domes.

• Belgium, with three, is nearing readiness to start putting its high-level waste deep in the clay of an old seabed, a layer starting 525 feet down beneath the nuclear research center at Mol and reaching down another 360 feet.

The Dutch have been watching the West German work in salt and are considering traveling the same path And Italy is considering following the Belgian example, using one or more of its own large clay deposits.

Each method has its drawbacks and uncertainties so far unresolved.

Granite studied

Granite is still being checked for its vulnerability to water leakage and its ability to carry away the considerable amounts of heat generated by high-level nuclear wastes. Clay domes are believed to have reached their present locations by extrusion, rising because they were lighter than surrounding rocks; the possibility that they are still rising and might eventually surface is being studied. Clay is a poorer conductor of heat than granite, and in deep deposits keeps trying to close up any chambers cut out of it. And the construction difficulty and costs of an undersea chamber like the one Sweden contemplates are huge.

But these are at least fairly firm plans and much has been done toward their execution.

No such luck here — and the job still has to be done. The administration and the Congress ought to get to it, and promptly. And Americans must understand the pressing necessity for an early solution, whether they favor or oppose nuclear power generation. It isn't just the future wastes; already thousands of tons of it are waiting in temporary storage, ticking, ticking, ticking. It must be dealt with.

The Dispatch

Columbus, Ohio, December 6, 1982

THE U.S. House of Representatives is making progress on a necessary bill aimed at establishing a permanent disposal site for this nation's nuclear waste.

Last week, the House approved an amendment that would limit a state's ability to block the creation of a disposal facility within its borders. It gives Congress the ultimate authority to approve or disapprove the selection of a site — something that is needed if the country is ever to dispose of its nuclear waste in a safe and secure way.

It is understandable that many states would oppose the placement of such a facility within their borders. It is conceivable that if the states were given the final say, no site would ever be selected and no solution would ever be achieved for disposing of the radioactive waste. Rep. William Dannemeyer, R-Calif., put it well when he said: "This is an issue of national significance that transcends states' rights. I agree that states should have input, but somebody has to decide."

The nuclear waste is now being stored at temporary facilities at nuclear power plants, but the storage space is running out. For the first time since the start of the nuclear age, Congress is close to passing a law that would solve the disposal problem in a realistic and workable way. Even if the bill is passed this year, it will be several more years before a facility is actually in operation — a time lapse that makes prompt action even more important.

The House should be encouraged to continue to work on this vital legislation and passage should be secured this year.

The Dallas Morning News

Dallas, Texas, December 1, 1982

CLEARLY Texas wishes to be something more than a dumping ground for nuclear waste. Just as clearly, nuclear waste must be disposed of somewhere. Both priorities seem in fairly good shape as Congress polishes up legislation setting national policy in the matter.

The federal government is considering Texas' Palo Duro Basin as a prospective storage site; the facility would be started by 1989. Gov. Bill Clements has accordingly worked hard to make sure that whatever state may be finally chosen is not unduly imposed on.

Chiefly the governor has wanted Congress to say that a state's veto of a disposal site would stand unless overturned by both houses of Congress. The House Monday refused to say that; it said instead that a veto could be overturned by either house, acting alone.

The standard (the same one adopted earlier by the Senate) is obviously looser than Clements'. Not that this should greatly perturb Texans.

The urgent thing is that congressional review of some sort should cap the site-selection process. Making a state veto too easy could, on the other hand, ruin prospects for putting the disposal site anywhere.

A little perspective about nuclear waste is plainly in order. To begin with, nuclear dumps are not little atom bomb factories waiting to spread contamination throughout the vicinity. Nuclear waste can be disposed of safely.

The waste would not be moved until most radioactive material had decomposed. The material would be compressed into a very small volume. And it would be immobilized — rendered incapable of moving around by itself. In short, nuclear waste and public safety are compatible concepts.

Part of the political problem here is semantical. The very word "nuclear" is the most emotive of the modern age. The word "waste" suggests heaps of rubbish being carried off by trash trucks; the word "dump" calls to mind — what else? — the city dump. One is led to think of spent nuclear products thrown somewhere out in the country: which, as we say, isn't the case at all.

This is why Fred Singer, an energy expert at the Heritage Foundation in Washington, D.C., suggests dumping the word "dump." Just say "storage facility," Singer suggests; or "repository." Likewise he finds the word "waste" misleading. Today's waste can be tomorrow's energy resource as scientific investigation proceeds apace. Not so long ago, no one knew what to do with uranium. It would have qualified as pure waste.

A "nuclear dump" by any other name might smell as rank to those who fear and dislike nuclear power, period. But accuracy never comes amiss. If "resource repository" doesn't exactly sing, the phrase tells it as it is.

THE DAILY HERALD

Biloxi, Miss., December 1, 1982

Americans, except those in Mississippi and five other states, will have little trouble understanding, and probably agreeing with, the House decision to lessen state authority to veto the nation's nuclear waste site.

The House voted Monday to require that any state veto of a nuclear waste site must be ratified by one house of Congress within 90 day to be effective. Otherwise, the veto will be ignored.

Mississippians, and the residents of Utah, Texas, Louisiana and Nevada who may have to live with those nuclear wastes, might entertain another view. They might believe a state ought to have a stronger veto in the process that will decide where the nation will deposit its nuclear waste.

The sixth state under consideration as a possible site for the nuclear dump — Washington — takes a different position; it welcomes anything nuclear.

Rep. Sid Morrison, who represents that part of Washington, has already told his colleagues that his district would welcome the nuclear dump.

If selecting a nuclear waste disposal site were simply a matter of pleasing people, Congress could skip over a lot of details and select the Hanford Federal Nuclear Reservation in Washington and that would be that. The folks in Richland, Kennewick and Pasco, the cities nearest the Hanford Reservation, have been part of the nuclear scene since the Manhattan Project of 1943. According to one estimate, quoted in the Oct. 4 issue of *Newsweek*, two thirds of the local economy depends on Hanford.

Unfortunately for South Mississippians, pleasing people isn't the only criteria involved in selecting the nation's nuclear dumpsite. Pleasing people isn't even very high on the criteria list. In fact, people seem to rate lower than geological formations on that list.

Rep. Trent Lott, R-Miss., our Fifth District congressman, tried to add an amendment making population a consideration. He would have disqualified any location containing 1,000 or more people per square mile. That amendment, had it passed, would have eliminated the Richton Salt Dome, a site the Department of Energy seems to covet highly, as a possible dump site.

Other Congressmen fussed that Lott was simply trying to keep the dump out of Mississippi. They weren't wrong, but at the same time, we believe that population considerations ought to be at least as important as geological considerations.

Another Mississippian, Rep. G. V. Montgomery, articulated another truth during Monday's debate: "When a state is picked as the site, the other states are going to be ganged up against it." The reason his prophecy rings with truth is that, save for Washington, no other state (or its representatives in Congress) wants the rest of the nation's hazardous nuclear discards, which are certain to ruin the neighborhood.

The nation has been in the nuclear age for 40 years now and the only nuclear disposal operations are temporary ones. America needs a permanent nuclear repository. But it shouldn't sacrifice the safety of some of its citizens for convenience of others.

The Seattle Times

Seattle, Wash., December 26. 1982

AFTER ignoring or avoiding the problem for nearly 40 years, Congress at last has established a national policy for dealing with high-level nuclear waste from the country's nuclear-power plants.

The measure that cleared both House and Senate last week — and went to the White House for President Reagan's signature — looks like a commendable compromise on a controversial issue that politicians have kept at arm's length for decades.

The need to do something finally became overwhelmingly clear. The nation's inventory of commercial nuclear waste, now about 8,000 tons of mostly spent fuel from power plants, will grow to more than nine times that amount by the year 2000.

The bill, which wisely does not put all the nation's nuclear-waste eggs in one basket, requires the federal Department of Energy to nominate five sites around the nation and recommend three to the president by 1985. He will then pick one. The DOE must go through the same exercise by 1989, leading to selection of a second site.

The Hanford Nuclear Reservation in eastern Washington, home of a test project to store nuclear waste in deep tunnels in underground basalt-rock deposits, is a likely candidate. So are federal properties in Nevada and Utah, and salt deposits in Texas, Louisiana and Mississippi.

If geologic and other tests prove that nuclear waste could be isolated safely for the thousands of years it takes for its radioactivity to decay, we see no reason why Hanford should not be one of the sites selected. But it should not be the ***only*** site.

The bulk of the nation's commercial nuclear waste is generated in the East and Midwest. Other storage sites closer to the source should be found and designated. We're pleased that this legislation seems to recognize the basic-fairness issue, and we'll be watching to make sure the letter and the spirit of the new law are followed.

The Times-Picayune
The States-Item

New Orleans, La., December 27, 1982

Louisiana apparently will escape inclusion in the list of states under consideration as sites for nuclear waste storage. If the decision stands, it must be considered one of the more positive developments in the state during 1982.

A North Louisiana salt dome ranked high on the Energy Department's list of possible storage areas for the nation's growing stockpile of nuclear waste. The department persisted even though serious environmental questions were raised about the use of such domes for nuclear waste.

Legislation passed at the tail end of the special post-election session of Congress provides what Sen. J. Bennett Johnston, D-La., calls "maximum protection" for the state. The legislation apparently will cause the Louisiana site to be dropped during preliminary screening in the dump site selection process.

Nuclear waste disposal is a nagging, growing problem for the United States. The question has been debated in Congress for five years. Action taken by Congress at the session just ended represents forward motion. The legislation sets up a program leading to construction of long-term or permanent nuclear waste depositories by the 1990s. The Energy Department will conduct a 2½-year study leading to a decision on a location for a "monitored retrievable storage" site where waste could be stored on a non-permanent basis while permanent deep geologic sites are chosen and prepared. Some analysts believe that the retrievable storage sites will eventually become the permanent sites.

Louisiana has substantial environmental problems. The state does not need a nuclear waste disposal site that has not been proven safe.

THE KANSAS CITY STAR

Kansas City, Mo., December 7, 1982

On the day the House passed a nuclear wastes disposal bill, observers noted that it was the 40th anniversary of the first controlled atomic chain reaction, at the University of Chicago. The House action, however, may not turn out to be a landmark: Congress has only two weeks left in the lame-duck session to reconcile sharp differences between this bill and a Senate version passed last April. Similar legislation died in a lame-duck conference committee two years ago.

The bill sets up a schedule for building a permanent repository for radioactive wastes—spent fuel rods—from the nation's nuclear power plants. The Department of Energy would recommend five possible sites by July 1984; the president would recommend one of these to Congress by April 1987 and the Nuclear Regulatory Commission would accept or reject it by 1989. The state chosen for the site could veto its selection if it could persuade either house of Congress to support the veto. Altogether it would be the mid-1990s before actual storage began.

The 8,000 tons of wastes—expected to quadruple by 1990—would be removed from present underwater storage at reactor sites, solidified into a marble-hard compound and buried in canisters designed to hold up for 1,000 years. They would be placed 2,000 to 3,000 feet deep in a stable geologic formation.

The disposal legislation makes no provision for wastes from the military weapons and nuclear Navy programs, which make up 90 percent of the volume but only 3 percent of the radioactivity. Some critics fear commingling the two types of wastes could expose U.S. nuclear weapons secrets. Both bills authorize the president to stipulate a separate military storage site if deemed necessary for national security.

The two versions differ sharply on the possible need for temporary interim storage away from the reactors and the desirability of monitored, retrievable storage. Congress has sporadically looked at the nuclear waste disposal problem for 35 years without a solution. The likelihood of reconciling two such different bills in two weeks hardly is encouraging.

Lincoln Journal

Lincoln, Neb., December 6, 1982

Almost 40 years to the day after physicists brought off the world's first controlled nuclear chain reaction at the University of Chicago, the House of Representatives finally approved a bill designed to establish permanent nuclear waste garbage sites.

That act should be reckoned as a much-welcomed advance. It's not as if the problem of what to do with the radioactive waste from military and civilian nuclear programs hasn't been recognized for at least the last 35 years — and nothing substantive has been done.

Nevertheless, differences between the House bill and a comparable measure previously approved by the Senate are of such magnitude that the whole enterprise may fall between the cracks of the lame-duck congressional session. Compromises may not be able to be struck in conference, Washington insiders warn.

Beyond the language differences, there is a piece of extra freight added by the Senate but knocked off by the House on a 241-to-148 vote — an extraneous provision limiting uranium imports. New Mexico Sen. Pete V. Domenici, backer of the import restriction feature, is said prepared to stall the entire measure unless the House sees things his protectionist way.

Environmental groups also are unhappy over basic provisions. They contend the House measure specificially writes out too many environmental health protections which should be there and prevents subsequent judicial review.

The basic thrust of the competitive measures is about the same; that is, set in motion a process for selecting one or more permanent nuclear garbage burial sites in stable geologic formations, granite or salt. While this search is being conducted and storage points picked by the end of the 1980s — whether the people on the surface like it or not — there is the increasing problem of interim nuclear waste storage above ground.

Electric utilities which operate light water reactors [the Nebraska and Omaha Public Power Districts, for example] very much want to move spent fuel rods away from those power plants to temporary storage elsewhere. Say to Morris, Ill.

The desire is propelled by the crowding of on-site spent fuel storage, and by a wish that if at all possible, somebody else be stuck with paying long-term, downstream storage costs.

You should not be more than barely optimistic compromises can be effected and a bill put before President Reagan for signature yet this month. But optimism is justified by noting both chambers have come further this year than ever before. And the reality this problem will not go away.

If the 97th Congress in its fading hours fails to agree to a plan for nuclear waste disposal, then the 98th Congress should pick up the competitive bills, wring them through the process again and press anew toward the compromises which will have to be made eventually.

THE DENVER POST

Denver, Colo., January 1, 1983

CONGRESS finally — after years of trying — has passed a nuclear waste storage and disposal act. We have had the technology for processing and permanently disposing of high-level wastes. All we needed was congressional authorization.

The bill sent to President Reagan for signature is far from perfect. It gives the states too much power to hold up effective storage procedures. But it is a sound start and — if properly funded — will at least get the program moving. Rep. Morris Udall, D-Ariz., summed up: "This is a bill the industry can live with and that environmentalists can live with."

The new law addresses three levels of need. One is away-from-reactor storage needed as a secure, temporary cooling-off place for highly radioactive fuel rods taken from power plants. Over a few years, they will lose enough radioactivity to be reprocessed or sent away for disposal.

Another need addressed by the bill is retrievable storage, an underground location for wastes of all kinds which might someday be reprocessed. The third area addressed is deep, permanent storage where high-level wastes will be locked in a matrix of boron glass, or even in rock itself.

To finance the program's role as garbageman to the nuclear power industry, the law levies a small charge on kilowatts generated. It's estimated to raise $300 million a year. Nuclear power furnishes about 12 percent of U.S. electricity.

The law is still a long way from fruition. The Department of Energy will nominate five sites for a permanent repository and then, in 1985, recommend three of them to the president. Several Western states are likely prospects. Colorado's uranium mining town of Naturita has, in fact, asked to be considered as a disposal site for wastes of fuel mined there originally.

The law allows a state to override the president's final site choice unless both houses of Congress override the state. This is an odd exercise of national political power, but if funding is adequate the disposal plan should work. A great deal of thought has gone into it, and it's long overdue.

The Miami Herald

Miami, Fla., January 13, 1983

BEHOLD! The United States finally has a policy on how to manage permanent disposal of the nation's nuclear waste. Considering that these radioactive wastes have been building up for more than three decades, one might say that it's about time. Yet one month ago there was no such policy.

Then, so fast that it almost escaped notice, the lame-duck Congress whipped one through just before adjournment. President Reagan signed it. Though it differs in important ways from versions approved initially by either house of Congress, the final draft improves on both.

Most important, the policy outlines procedures and target dates for choosing where and how to dispose of the wastes. By Jan. 1, 1985, the Secretary of Energy must recommend to the President three potential sites geologically suitable to house permanent nuclear-waste repositories. Another five sites would be studied, and three more recommended, by July 1, 1989. The President would have to recommend at least one site to Congress by March 31, 1987, and a second by March 31, 1990.

In the meantime, the Department of Energy would be authorized to provide up to 1,900 metric tons of temporary storage capacity for waste from civilian nuclear-power reactors. This solves a most pressing problem for many electric companies whose own storage facilities are nearly full and can't be expanded. Utilities would pay to use such temporary Federal facilities.

The proper procedural safeguards are included in the statute. Each potential repository site must pass extensive environmental review. Public hearings will be held in the vicinity of each site under consideration. No repository could be placed in a highly populated area, or dangerously near one.

Most important, states would have to be consulted closely throughout the process. A state veto of a site within its borders would stand unless both houses of Congress overruled the state.

True, this is a drawn-out process with many built-in stumbling blocks. Yet it necessarily must be so. No one wants his backyard to become the nation's nuclear dumping ground, yet the waste exists and must go somewhere. Arranging how to decide what is the safest and best place is less a matter of science than of politics. A victory of politics produced this policy.

It'll take two victories — of science and politics — to produce a storage site that's both safe and acceptable.

Los Angeles Times

Los Angeles, Calif., December 24, 1982

Many years after the fact, the United States has a plan for disposing of the radioactive waste generated by its power plants and weapons plants.

The approach will not satisfy nuclear power's severest critics because not much will change in the next few years. But it is the first such plan on which Congress has been able to agree, and it does follow the outlines of the most thorough analysis of the problem to date. It is, at worst, a case of better late than never, and the plan may well work.

Under the plan, the stockpile of hot garbage that has accumulated since the dawn of the Nuclear Age will be kept above ground in existing or expanded storage tanks at power plants while geologists look for places to bury it permanently.

In many ways, the nuclear-waste program, one of the few solid achievements of the lame-duck session of the 97th Congress, is more a triumph of accommodation than of technology.

Geologists have until the year 2000 to prepare vaults that will seal in radioactivity until the year 12,000. But Congress says it is willing to settle for 100 years of temporary storage above ground in case the technicians miss their deadline.

States whose geology makes them prime sites for permanent disposal vaults have broad veto powers under the plan. But the alternative is something akin to a nuclear dictatorship that would give Washington the right to force a nuclear dump on any community. Congress' approach is preferable.

The nuclear-waste program is bound to expose the weakness of various interest groups for finding broad philosophical meaning in essentially minor events.

The nuclear industry, for example, will interpret the program to mean that the last barrier to expanding nuclear power has been overcome. Environmentalists will point to the temporary-storage feature as evidence that there is no way to resolve the nuclear-waste problem. Neither view is valid.

The most important feature of the plan probably is that it acknowledges that the nation has waited so long to settle its nuclear-waste issue that nothing can be done overnight.

Following generally the findings of its own Office of Technology Assessment, Congress would have power plants store their own spent nuclear fuel rods until permanent vaults are ready. That would avoid the expense of building temporary central storage facilities as well as the risks of moving the spent fuel around too much.

Power plants have accumulated 8,000 tons of nuclear waste so far, and will generate 16,000 more tons by 1990. The plan assumes that there is room at most power plants to expand storage facilities until permanent burial sites are chosen.

The focus of the plan is on nuclear waste from power plants, but the disposal vaults would also be used for waste from weapons manufacture unless a President declared the vaults inadequate.

The weak point of the bill is a loophole that opens the way for putting off permanent disposal for as long as 100 years with interim depots for above-ground storage. Over the next several years, Congress should emphasize to the energy bureaucracy that will find and build the permanent vaults that it does not plan to use the loophole.

The strong point of the bill is that it finally gives federal officials both the authority and the money to get serious about a project that they have wanted to run with for years. That does not solve the nuclear-waste problem. But it makes it possible to try, and that is a major step forward.

THE ANN ARBOR NEWS

Ann Arbor, Mich., January 10, 1983

Too hot to handle. Until last month, that seemed to be Congress' longstanding policy on highly radioactive nuclear wastes.

Tons and tons of toxic nuclear waste have piled up since the first nuclear power plant went on line in the 50's and before that, the beginning of the atomic age. Mostly, the wastes are stored in pools at nuclear plants around the country.

Years in the making, the Nuclear Waste Policy Act of 1982 finally addresses the urgency of the disposal problem as summed up by the familiar "don't do it in my backyard but please do it."

The Act establishes criteria for building the first permanent underground storage dump by the year 2000 and constructing interim dumps to store waste until then.

UNDER GUIDELINES DRAFTED by the Environmental Protection Agency, the wastes would be buried in geologically stable rock formations 2,000 to 3,000 feet below ground to prevent any release of radioactivity for 10,000 years.

The permanent facility would be financed by a tax on nuclear power plants and other generators of the wastes.

The legislation requires the Energy Department to nominate five sites for the first permanent repository and recommend three of them to the president by 1985. Another five sites for a second repository would be recommended by 1989.

This legislation, said Rep. Morris Udall, D-Ariz., "is something the industry can live with and that environmentalists can live with."

But in seeking and achieving compromise legislation, backers approved a potentially lethal fishhook in the form of gubernatorial veto.

At the behest of Wisconsin Sen. William Proxmire, the Nuclear Waste Policy Act included an amendment permitting any governor to veto siting of a nuclear waste dump in his state unless majority votes in both the Senate and the House overrule him.

THE HAMMER OF STATE veto over what is essentially federal policy is one Congress may live to regret. It is known that the prime sites under consideration are all in states of the west and southwest; that makes sense as far as population density is concerned, but it tends to excuse the states which generate much of the toxic stuff themselves from any responsibility for its safe disposal.

That doesn't seem fair. The "out of sight, out of mind and out of state" approach of the past only encourages states to pay others to store their wastes. This mercenary-mindedness is not good public policy and it may not be environmentally sound either.

There is at least one other major drawback to the Nuclear Waste Policy Act. The transportation mode of waste disposal is left unaddressed. Trains are one means, although the state of American railroads and the bumpy roadbeds on which they ride quickly give rise to second thoughts.

The interstates, then? Well, what about the crumbling sections of our highways and all those unsafe bridges?

CLEARLY, THE NUCLEAR WASTE disposal problem is at the critical stage. Some agency or unit of government needed to take charge and make its recommendations stick.

A long-delayed nuclear waste policy has been marred by the potential mischief in state veto power. A court test of this provision may be what's needed to make sure a dirty job gets done well.

The Honolulu Advertiser

Honolulu, Ha., December 31, 1982

Among the few accomplishments of the 97th Congress' lame-duck session was passage of a nuclear waste disposal bill that had been stalled for four years. President Reagan is expected to approve the measure.

The bill calls on the Energy Department to recommend within two years five potential sites for permanent, underground storage of high-level nuclear waste and spent fuel. It also allows for increased capacity for temporary storage of spent fuel from civilian reactors.

What to do with nuclear byproducts is a serious and growing problem. Years of inaction have resulted in a near-crisis situation, with many temporary sites located next to reactors filling fast. Now, at least, planning for permanent facilities can get underway.

(In a similar vein, passage of a two-year ban on the ocean dumping of low-level radioactive wastes, which was a rider to the gas-tax bill, is also welcome.)

Unfortunately, Congress also agreed to amendments which may make it difficult to build permanent structures once scientists and government officials agree on the best storage method. States (and Indian tribes) were given veto power over any federal decision to build repositories within their borders. That could make it extremely difficult to actually construct them.

Another weakeness in the bill was the exemption of wastes from defense-related sources. Military reactors produce a majority of the country's nuclear wastes.

Still, passage of any bill has to be seen as progress. The hope is that work over the next two years will not be for naught.

BUFFALO EVENING NEWS

Buffalo, N.Y., January 17, 1983

It would be an inflation of language to contend that the law just signed by President Reagan will solve the dreaded problem of how this nation safely and permanently disposes of the radioactive nuclear waste that has been accumulating since World War II.

But Mr. Reagan's signature marks a historic step because now, for the first time, agreed-upon timetables and procedures exist for selecting, licensing, financing and regulating deep underground sites for the burial of these lethal residues by the mid-1990s.

Western New York, home of the West Valley nuclear waste repository laden with 600,000 gallons of radioactive fluid at a reprocessing facility closed 10 years ago, had a large stake in this legislation.

Thanks to efforts by Sen. Daniel P. Moynihan, D-New York, and Reps. Jack Kemp, R-Hamburg, and Stanley N. Lundine, D-Jamestown, among others, the outcome was favorable. The final legislation includes language that will effectively bar West Valley from being considered as a possible interim site.

Whether or not any interim sites are eventually created, the exclusion of West Valley from that possibility is logical enough. Including it could have delayed and disrupted the important demonstration project by the federal Department of Energy at West Valley, involving processes by which radioactive fluids can be solidified before removal to permanent storage areas.

Geologically, too, West Valley is unsuitable as a permanent burial vault. All sites under consideration for such a facility are located in six southern and western states.

With the law now passed and signed, the Department of Energy must review five possible permanent sites and recommend three by 1985 to the president, who then must submit his single choice to Congress by early 1987. The sites must not be adjacent to populated areas, and states can veto a choice within their borders that can be overturned only by votes in both the House and the Senate. The latter compromise provision resolved a sticking point that had foiled adoption of similar legislation in past years.

Between now and mid-1990s, much can still go wrong, of course. Delays could occur. But at least a consensus now exists where none did before. Responsible procedures are set in motion to cope with a problem that simply must be faced.

It lurks along the shadowed edge of the nuclear frontier, but the existing 6,700 metric tons of spent nuclear fuel, the 77 million gallons of radioactive wastes like those at West Valley and the 175 million tons of uranium mine residue are unmistakably there.

The issue is not whether we are to live in a nuclear environment. We already live in one. Rather, it is how to make that environment as safe as possible, and this includes devising and executing responsible policies for the safe disposal of the dangerous byproducts of nuclear power and weapons production. Now, at least, a substantial first step has finally been taken to accomplish that goal.

Wisconsin State Journal

Madison, Wisc., January 17, 1983

Like many bills forged through elaborate compromise, the nuclear-waste disposal measure passed by the lame-duck Congress and signed into law by President Reagan is satisfying, but only partly.

It broke a marathon failure by the federal government to deal with the problem of permanent storage for high-level nuclear wastes. Since 1957, when nuclear power came into use, 8,800 tons of toxic nuclear waste have accumulated in temporary storage facilities. By the turn of the century, an estimated 86,900 tons of spent fuel will need disposal.

Under the new law, the Department of Energy must recommend five potential disposal sites to the president by the beginning of 1985, and five more potential sites by mid-1989. In turn, the president must select and recommend two sites, one by 1987, the other by 1990.

(Informed speculation is that Wisconsin is likely to be among the states recommended for nuclear-waste disposal, probably on the list submitted to the president in 1989.)

Because of pressure from Sen. William Proxmire, D-Wis., the law contains a strong state-veto provision. A governor or legislature can nix a site within their state's borders. The veto can be overriden only by a majority vote of both houses of Congress.

This leverage is expected to force the Energy department to reverse its poor record of communication and cooperation with Wisconsin and other states being studied as potential hosts of nuclear wastes.

On the minus side, the bill exempts military nuclear wastes from most of its provisions, although military wastes make up 90 percent of the growing pile of nuclear garbage.

Nor can the bill assure that safety rather than fear and power politics will govern the selection of disposal sites. Safety and politics are related, certainly; sites which studies determine will offer the safest disposal are likely to be the most politically palatable.

Yet, the danger remains that site selection will be a battle decided by which states have the most powerful influence in Congress. Because few people want nuclear wastes buried in their back yards, selecting disposal sites may continue to be a process of negative pork-barrel, with congressmen fighting hard to keep a facility *out* of their districts, scientific findings about safety notwithstanding.

The bill's stipulations about public hearings, environmental reviews and state veto powers will help encourage public involvement and education about storing nuclear wastes, and that might allay some fears.

But what really is needed to make safety the primary factor in finding disposal sites is a new level of maturity among citizens. We continue to generate electricity in nuclear power plants; we must take responsiblity for disposing of the deadly wastes — even if that means tolerating the location of a site close to home.

That message is particularly pertinent in Wisconsin. Although state-government analysts predict that nuclear power will provide a decreasing share of electricity in the years immediately ahead, Wisconsin remains one of the nation's chief nuclear states, with a quarter of its electricity and almost a tenth of its overall energy coming from nuclear power plants.

THE SUNDAY OKLAHOMAN

Oklahoma City, Okla., January 16, 1983

ALMOST unnoticed in the recent lame-duck session of Congress, admittedly more notable for what it failed to do than what it accomplished, was one productive enactment of potentially major significance.

Little publicity attended President Reagan's signing last week of landmark legislation that resolves the long-standing controversy surrounding the question of managing nuclear waste disposal.

Finally, after more than a quarter-century of nuclear power plant operation, the nation has in place a coherent system and schedule for managing nuclear waste that accommodates most of the views of both supporters and critics of nuclear power.

The political hurdle that had stymied past efforts — the degree of state participation in the siting process — has been cleared by including views of the National Governors Association. State consultation is required at every step.

More important, a vote in both House and Senate is required to override any state's objection to a proposed site within its boundaries.

The new act establishes schedules and timetables aimed at providing deep underground repositories for nuclear waste in the mid-1990s, to be funded by a user fee of one mill per kilowatt hour on nuclear power generation. In the interim, the measure authorizes storage at existing federal facilities for spent fuel of utilities that are running out of on-site storage space.

By the end of this year, DOE must come up with three or more recommended sites for a test-and-evaluation facility, then five sites for a permanent geologic repository by July 1984. The president is to select the first site by March 1987, by which time testing will be completed and techniques approved. A second site will be picked in 1990.

There also is provision in the act for eventual storage of military nuclear waste, which represents a larger volume than commercial wastes, at the same sites.

Hailing the new law as a "major step forward," Energy Secretary Donald Hodel said it "provides long overdue assurance that a safe and effective solution to the nuclear waste problem is in hand."

This is welcome news to the troubled nuclear power industry, beset with continuing environmental opposition, plant cancellations due to rising costs attributable largely to the undue length of the licensing process and a slowdown in the growth of electricity demand.

Ample, reliable supplies of electric power are essential to any sustained recovery and resumption of sound economic growth. And whether the anti-nukes like it or not, kilowatts from uranium fission will play an increasing role in providing that energy.

The more than 80 licensed commercial reactors last year generated nearly 13 percent of the nation's electricity, and the Atomic Industrial Forum points out that 59 others hold construction permits, with five more on order. The AIF forecasts issuance of 39 more operating licenses within the next four years.

Rabid no-growth and environmentalist opposition notwithstanding, preliminary estimates indicate that 1983 will be the year when nuclear power moves into second place behind coal as the nation's producer of electric energy.

Houston Chronicle

Houston, Texas, January 5, 1984

A year ago, the passage of a federal law dealing with nuclear power plant waste was noted with pleasant surprise. After long debate, the measure setting forth a schedule for selecting a suitable repository slipped through Congress almost routinely.

One year later, that program is — amazingly — more than three years behind schedule. The Energy Department says it will be December 1990 before any recommendation is sent to the Oval Office on a permanent dump site. The bill passed by Congress calls for such a recommendation to be made by March 1987.

The delay is caused, the Energy Department people say, by trouble preparing guidelines and a need for more at-depth testing of potential underground sites. Environmental groups have also thrown up barriers. In the meantime, studies are continuing on possible above-ground storage facilities. These are supposed to be temporary, but are looking more permanent all the time.

The technical problems of handling these wastes can be managed, and the wastes will have to be dealt with sooner or later. But if the schedule has slipped three years in only one year, the question is whether the decision-making process will function at all this side of year 2000.

Plans to Resume Ocean Dumping of Radioactive Waste Thwarted

Ocean dumping as a method of disposal for low-level wastes has not been practiced by the U.S. since 1970. From time to time, however, it has been suggested as the most economical method of disposal. Interest was shown by the Environmental Protection Agency in again permitting ocean dumping of low-level wastes in 1981 and 1982. (Under the Marine Protection, Research and Sanctuaries Act of 1972, the EPA had the authority to issue permits for the ocean dumping of waste materials.) In the last days of the 97th Congress, however, as part of a bill funding highway construction and increasing the gasoline tax, legislators passed a measure imposing a moratorium on any such dumping for two years. The two-year moratorium, passed in December 1982, was intended to give Congress time to investigate Navy proposals to drop decommissioned nuclear submarines into the ocean.

The Kansas City Times

Kansas City, Mo., April 23, 1980

You've seen those pictures of a submarine-launched missile popping out of the ocean to begin its deadly flight. How about a missile plopping *into* the ocean, diving swiftly to strike the sea bottom at 120 mph and bury itself hundreds of feet deep in the mud? Whatever for? To dispose of radioactive nuclear wastes, that's what for. Seabed burial of the controversial nuisance residue of nuclear weaponry and power generation is being weighed as a distant option to the presently contemplated underground storage in geologically stable formations such as salt beds.

The Department of Energy will spend between $5.9 and $15 million a year the next five years studying this disposal method, but figures that any test firings are at least 20 years away at the end of the century. And environmental critics of the idea already are suggesting that's too soon. The search for a permanent storage scheme for the mounting store of "hot" trash which is such a burden and embarrassment to the nuclear age has proceeded with just such studied deliberation.

The violently negative official and public reaction to past tentative proposals to bury the wastes in salt beds in Kansas or New Mexico will become even more so under the Three Mile Island mentality. No government official wants the final assignment of having to say, "OK, here's where we're going to bury the stuff and this is how we're going to do it." President Carter in February at last promulgated a 10-year study schedule leading to such a decision, to be made well beyond his own White House tenure.

The ocean missile disposal concept has more problems than just technological. A 1972 ocean dumping treaty which this country ratified prohibits spilling high-level radioactive wastes into the sea, but legal loophole proponents of the plan point out it doesn't say anything about burying them in the ocean bottom.

Kansas City, Mo., January 18, 1982

It is disquieting to hear that the Environmental Protection Agency is preparing rules that would allow the dumping of low-level radioactive wastes in the ocean. The Marine Protection Act of 1972 does not prohibit dumping of such materials, but until now it has been federal policy not to permit it.

These are not the highly radioactive wastes of weapons production or nuclear power generation, but rather such things as contaminated clothing, rags and so on produced by hospitals, laboratories and certain businesses. The general public, however, is just as wary of having such material discharged in local landfills or off our seacoasts.

The oceans, because of their vast size and depths, were once considered as limitless pits for all sorts of trash. But the proven damage to shellfish beds and other marine life in areas where such pollution has occurred have forced a rethinking of the wisdom of unrestricted ocean dumping. The United Nations Law of the Sea treaty, if it is ever successfully completed, will contain some international controls on polluting the seas.

The explanation given for the new line of thinking at EPA is the growing reluctance of state and local governments to permit dumping of radioactive wastes within their borders. But this issue of permanent storage of both high- and low-level wastes — inherent in the weapons and power programs and modern medicine — is one with which the federal government is slowly coming to grips after lengthy delay. Last year's law on low-level wastes would require the states by 1986 to form regional compacts to designate repositories. A lengthier schedule is in place for making the hard decisions on types of storage containers and sites for the high-level hot garbage.

In the meantime concern continues over the growing volumes of such materials temporarily stored at government sites, underwater at nuclear power stations and in low-level state repositories in Nevada, South Carolina and Washington. Residents of those states would like to stop being their neighbors' trash bin as soon as possible. But going back to old-fashioned ocean dumping to ease the pressure is environmentally regressive and could only provide a further excuse for dodging the tough choices that must be made on permanent storage. The ocean is a source of life on this planet. Don't poison it.

The Honolulu Advertiser

Honolulu, Ha., January 16, 1982

Eleven years ago, the Environmental Protection Agency banned the dumping of nuclear waste in the ocean. The practice had been carried on for about 30 years by military and civilian agencies. But leaks were discovered in some of the barrels which stored the waste, and sloppy records left many others unaccounted for. (Hawaii is the site of a number of off-shore nuclear dump sites.)

Now the EPA is considering dropping that prohibition. A spokesman said "draft regulations" to that effect are "under consideration." And a network news report added EPA officials would have "the final word on whether to make exceptions for any waste which failed to meet the established criteria and. . .allow nuclear reactors from obsolete submarines or more likely the entire submarine to be scuttled."

IF ANYTHING should be scuttled it is the draft regulations.

Too little is known about the effects of extended exposure of marine plants and animals to low-level radiation.

Until ongoing experiments examining methods to bury waste into the sea bed are completed — and that may take some time — there really isn't a proven way to safely dispose of wastes at sea. It still basically comes down to placing nuclear material in reinforced drums and dropping them into the ocean. That hasn't worked well before; there's no reason to assume it will now.

The EPA's problem, of course, is that no state wants to turn itself into a nuclear dump. Since West Coast sites were closed to "outside" waste, there has literally been a growing problem. It even affects Hawaii, which produces low-level waste in research labs and hospitals.

BUT WHILE IT may appear so on paper, sea disposal is not an easy out. It is at best a short-term solution to where to put the waste, but it doesn't answer the really serious question of what effect it might have on the environment and the food chain.

Experience has shown that sea dump schemes have produced serious safety questions. In this matter safety rather than convenience should be the EPA's main concern.

Long Island, N.Y., January 27, 1982

No radioactive waste from the United States has been dumped in the ocean since 1970. That's as it should be: Entrusting an unknown danger to the sea is a shortsighted policy with long-term peril.

But encasement technology is now better than it was when ocean dumping began. Land sites, never popular, are falling under regulations that have made them more difficult to use. And the users of radioactive material are pressing for ways to dispose of it.

So again the question is being asked: Why not the sea? Unfortunately, the Environmental Protection Agency appears ready to allow low-level radioactive waste to reach the seabed.

Between 1946 and 1970, about 87,000 containers were dumped off the U.S. coast. The long-term effects haven't been satisfactorily determined—in part because the dumping was haphazard and the sites were poorly monitored—but researchers have found leaking containers and higher than normal levels of radiation.

That material was nothing compared to what will end up on the bottom if the gates are reopened. But the potential for damage remains real, and the best of today's technology still can't offer any guarantees.

Disposal of radioactive waste is a global problem. And even though European nations and Japan are already dumping, the United States shouldn't revise its rules until the evidence is more reassuring.

The fact that the nation is running out of tolerance for dumping on land is no reason to rush headlong to the sea.

THE SACRAMENTO BEE

Sacramento, Calif., January 24, 1982

The Rules Committee of the state Senate has unanimously passed a resolution calling on the president to ban ocean dumping of radioactive wastes and committing California to keep a close eye on any ocean dumping plans. The resolution was introduced by Sen. Barry Keene after he learned that the Navy was studying the possibility of offshore burials for 100 nuclear submarines that eventually will have to be decommissioned. While there is no chance the president will enact such a ban — and a ban, indeed, may be proved by future research to be inappropriate — the committee's swift and strong action was the right response to the Navy's news.

The problem is not that the Navy's plan poses any immediate threat to California. As yet, the Navy is only considering ocean burial as one alternative, and it is planning environmental impact studies of both ocean and land disposal schemes before choosing between them. Even after that decision is made, the Navy will have to complete additional studies of the specific disposal sites chosen, obtain approval from the Environmental Protection Agency, and (in the case of ocean dumping) consult with the other nations that have signed the London Ocean Dumping Convention. If offshore burial is more dangerous than land disposal, these procedural steps should prevent its being adopted.

What does justify concern, however, is that the Navy's record on nuclear waste disposal does not inspire trust in its procedures, as is clear from the recently declassified Navy documents reported in The Bee the other day. Those postwar documents show the Navy to have been more concerned with keeping the public ignorant of its nuclear waste disposal programs than with discovering the effects of radioactive contamination. And the Navy's recent response to inquiries from Sen. Keene suggests, as state Health Services Director Beverlee Myers testified, that "a consistent undervaluing of long-term consequences (still) pervades most analyses of ocean disposal of radioactive wastes."

The quantities of low-level waste currently under consideration for ocean dumping are unprecedented. The Navy is talking about eventually scuttling 100 ships, and every one of them — even with their nuclear fuel removed for separate disposal, as the Navy plans — could still contain an amount of radioactive material equal to half of all the radioactive waste dumped off the Atlantic, Pacific and Gulf coasts since ocean dumping began. With the ocean such a vital resource and the effects of low-level radiation still so poorly understood, California has every right to be wary of Navy assessments that call this radioactivity "insignificant."

It is thus appropriate for the California Legislature to send a clear message at the beginning of the long study process, announcing its intention to carefully scrutinize the Navy's research. As Myers testified, the best protection for California is to "maintain constant (independent) surveillance over any environmental impact analysis prepared by the Navy." There is no reason as yet to assume the Navy will do something dangerous to the coast of California, but there is every reason to pay close attention as its plans are developed.

The Charlotte Observer

Charlotte, N.C., January 19, 1982

The federal Environmental Protection Agency admitted last week that it is considering reversing a 10-year-old ban on dumping nuclear wastes in the ocean.

On the face of it, the proposal is frightening. And it should be of particular concern to Carolinians, because part of the impetus for the proposal is the Navy's desire to sink five decommissioned but still radioactive nuclear submarines off the North Carolina coast.

The EPA hasn't said yet how radioactive the material is, or whether the radioactivity of the objects to be dumped is the kind that will disperse and become harmless within a short time — that is to say, weeks or months. Responsible people will wait until they have answers to those questions before assuming the EPA wants to carelessly — and permanently — irradiate the North Carolina coast.

But the proposal raises an interesting question about the mindset of those officials, elected and appointed, whom Americans trust to make public policy on the issue of nuclear waste. At times it seems their primary concern is not so much to find the safest depositories for radioactive materials, but to find the least conspicuous — as though the operative theory is "out of sight, out of mind."

For example, EPA officials say there's a growing need for ocean disposal of low-level nuclear wastes not only because land disposal sites are filling up, but also because plans for new land sites meet strong public opposition.

The EPA's response to that pressure seems to focus only on the problem of where such unpopular trash might be stuck with a minimum of protest. In that respect, the new ocean-dumping proposal seems akin to another plan revealed a few years ago to shoot nuclear wastes into space to get them off the planet.

Experts nixed that idea, pointing out that rockets sometimes go off course; other experts reminded us that objects in space — including meteorites and, recently, America's space lab — sometimes fall back to Earth. We suspect similar objections might be raised about ocean dumping.

An even more basic question, of course, is this: Should our society support a technology that generates waste that makes us so uneasy we have to keep trying to hide it?

The Chattanooga Times

Chattanooga, Tenn., April 12, 1982

Nearly 30 years after the first U.S. nuclear submarine, the U.S.S. Nautilus, was built, the Navy is faced with the serious problem of what to do with decommissioned nuclear warships. Powered by highly radioactive reactors, the nuclear ships can't simply be stripped and cut up like conventional ones.

Currently, the Navy has more than 130 nuclear-powered aircraft, cruisers and submarines in service. Since those ships can remain operational for approximately 25 years, the Navy will be decommissioning five or six a year over the next 20 years or so. How should it dispose of the vessels?

There are only two workable methods available now — burial at sea or on land. The Navy says it's cheaper to sink the vessels at sea, and it has two sites for such disposal, one about 200 miles southeast of North Carolina, the other roughly 150 miles southwest of California. If the ships are buried on land, they much be cut up and shipped to dumps in Georgia or Washington.

The latter method seems to us more preferable, despite the greater cost. After all, land burial means that the decommissioned ships could be dug up and reburied elsewhere if necessary. Granted, the sea burial sites are approximately 15,000 miles deep. Nevertheless, the environmental impact of dumping radioactive vessels at sea is not fully known. There is no telling what effects such ships could have on sea life and, ultimately, on humans.

The Houston Post

Houston, Texas, April 20, 1982

This country suspended ocean dumping of low-level radioactive wastes 12 years ago, after the federal Council on Environmental Quality urged that the wastes be buried at sea only as a last resort. The action was essentially precautionary since no consensus exists in the scientific community on what hazards, if any, the dumping might pose. Nor did our restraint stop other countries from dumping their radioactive wastes in the world's oceans. But if a proposal by the Navy to sink its old nuclear submarines offshore wins approval, environmentalists fear it would open the door for resumption of the practice by this country.

The Navy plans to remove the subs' high-level radioactive fuel before disposing of them. Some scientists contend that the low levels of radiation emitted by the subs would pose no environmental or health threats. But Charles Hollister of Woods Hole Oceanographic Institute in Massachusetts believes that, while there is only a remote chance that the sinking of the vessels would be harmful, "it needs to be proven that it is safe."

This country dumped low-level radioactive wastes in the ocean from the 1940s to 1970. Two years ago the House subcommittee on the environment, energy and natural resources, which studied the effects of the dumping, reported that no accurate records were kept of where and how much waste was dumped. The available records did show, however, that there were more than 50 dump sites containing around 100,000 drums. In addition, an unknown quantity of military nuclear wastes was also disposed of at sea. The military did not need civilian regulatory agency approval to dump the waste.

A report by the old Atomic Energy Commission further noted that ships under contract to haul wastes to designated disposal sites sometimes dumped the material elsewhere when the weather was bad or equipment malfunctioned. Inspection of 250 drums in one site a few years ago showed that a fourth of them were leaking. Though no environmental damage was detected, an Environmental Protection Agency official credited this to luck, not planning.

The Navy is preparing an environmental impact statement for the EPA on its submarine disposal plan. The report is due this summer. But before a final decision is made, those responsible for making it should consider the precedent it might set for the future dumping of low-level nuclear wastes at sea, as well as the long-range harm the accumulated dumping by many nations might have on the marine environment.

In view of the seas' vital importance to life on Earth, we should not risk contaminating them, however small some scientists think the risk would be. Oceanographer Hollister has offered sound advice on this score. We should have *proof* that the radioactive waste will not harm the oceans before we resume a practice that could adversely affect generations yet unborn.

Lincoln Journal

Lincoln, Neb., April 13, 1982

When the Manhattan Project developed the first atomic bomb back in the 1940s, and opened the nuclear age, it also, naturally, produced some radioactive contamination.

Today — would you believe this? — the Department of Energy has 100,000 cubic yards of radioactive soil left over from the Manhattan Project stored in a parking lot under a plastic cover in Middlesex, N.J.

Nothing illustrates so well the fact that the problem of nuclear waste won't just go away. Not for millenia, at any rate.

That pile of dirt, the DOE hastens to assure everyone, is only slightly hotter than natural background radiation. Still, it would like to get rid of it. The department is considering asking the Environmental Protection Agency for permission to dump it in the ocean.

As America's nuclear waste accumulates in temporary storage areas on land, the EPA is receiving inquiries about sea-disposal from a growing number of sources — state governments, drug firms, hospitals. And not least from the U.S. Navy, which wants to sink old nuclear submarines.

Not since 1970 has the United States used the seas as a dumping ground for radioactive waste. Some European nations still do. Now the EPA is under pressure to end the ocean-dumping moratorium.

Even here in the landlocked Midwest, that seems like a highly risky business.

Obviously there's a hazard in keeping nuclear waste on land, too, until somebody can figure out a foolproof way of disposing of it. But at least on land, or under land, the waste can be got at more easily if problems develop. It has the potential of being moved to safer locations, or shielded, or dealt with in some manner.

Retrieving waste from the ocean floor or stemming its contamination is another kettle of (irradiated) fish.

Scientific opinion is divided about the safety of putting nuclear waste in the sea. When experts disagree, common sense argues: err on the side of caution.

In its recent high-handed attitude toward the Law of the Sea Conference, the United States has seemed to have scant regard for humanity's stake in the earth's oceans. If it now returns to dumping nuclear waste in them, one can only conclude that to our government the seas are simply civilization's cesspool.

DAYTON DAILY NEWS

Dayton, Ohio, April 14, 1982

The U.S. Navy has come up with a plan to bury its discarded radioactive submarines at sea. It's a plan that should be scuttled before it is ever launched.

There's no doubt unwanted nuclear submarines are a growing problem for the United States. There are 100 in the fleet now, and they are to be decommissioned at a rate of four a year for the next 30 years.

Do we really need thousands of tons of our radioactive garbage in the deepest parts of the Atlantic and Pacific oceans? The Navy's says it will take every precaution to protect the environment. But if we haven't the technology to deal with our existing nuclear waste pile, how can we know what precautions are enough?

A decade ago, thousands of drums of toxic wastes were buried throughout this country with what were presumed to be "adequate" precautions. Today, we are just beginning to learn the extent of what has been done to once usable water and habitable land. And those dumps at least are reachable — on land, not three miles under water.

The bottom line to the Navy's proposal is that sea burial would be cheaper than land burial. Cheaper today. Tomorrow?

Dayton, Ohio, December 30, 1982

Tacked on the bottom of the 5-cents-a-gallon gasoline tax was a fine amendment which the administration isn't going to like. It puts a two-year hold on U.S. Navy plans for ocean disposal of the radioactive parts of decommissioned nuclear submarines.

The amendment came from Rep. Glenn Anderson, D-Cal. He's concerned not only because one of the dump sites is off California but also because he'd like all costal waters to be safe for living things.

The Navy points to the accidental sinkings of two nuclear subs as proof it can be done safely. Two is one thing. The Navy, however, wants to dump 100 over the next 30 years. At-sea disposals of radioactive wastes were halted in 1970 when the Council on Environmental Quality reported it posed a threat to marine life. And no one really has satisfied that concern.

Land burials are a possibility. But they're expensive — $7.2 million versus $5.2 — which means a lot to a cost-conscious administration.

The administration should accept the two-year moritorium gracefully and use the time to get some hard facts about what it wants to do. That includes long-term effects of low-level radiation on aquatic life. If at-sea burials are going to be made on a large scale, the safety of that should be assured.

Oregon Journal

Portland, Ore., April 17, 1982

The U.S. Navy plans to sink tired old nuclear submarines, after the nuclear fuel has been removed, in the ocean. The Navy and some scientists say that deep holes in the ocean — 14,000 feet deep off Northern California, for example — are the best resting place for the old subs, but the military had better be 100 percent sure.

Until 1970, the United States dumped nuclear wastes in the ocean, but since then this country has stored them in nuclear landfills, like the nuclear storage facility at Hanford, Wash., or left them in water at the site, like at PGE's Trojan nuclear plant.

Water disposal always has been a potential for storage of low-level wastes, which is what the Navy says would be left after the nuclear reactors were removed. This is an important point. If the nuclear wastes were removed, and all that was left was the sub that had plied the seas with its crew of sailors, then disposal in the briny deep is an alternative for consideration.

G. Ross Heath, dean of the School of Oceanography at Oregon State University, says the radiation risk in storage of old submarines is "relatively minuscule" either in water or on land. The option comes down to a political one, not a technical one, he adds.

Heath says that a disposal site of 14,000 feet, which OSU is helping locate, is far deeper than the existing fishery. The question becomes what is the "absolutely worst case" — if fish are caught and eaten? He says European nations have dumped low-level material into the North Atlantic, and biology sampling never has detected any damage. A site under study is 185 miles southwest of Cape Mendocino off Northern California.

Another view comes from Clifton Curtis, an attorney representing a dozen environmental organizations. He says the sinking of submarines could be "just the thin edge of quite a large wedge" of deciding to use oceans as dumping grounds for radioactive material.

The main question is the level of the radioactive material. If it is the hull of the submarine that otherwise would sit in U.S. coastal ports such as Seattle, Portland, San Francisco or San Diego, that shouldn't cause any problem. But high-level nuclear wastes could be something else. The federal government has failed to decide how to dispose of high-level wastes, and that question needs more study before the oceans become the garbage disposal for more than a few submarines.

ST. LOUIS POST-DISPATCH

St. Louis, Mo., January 25, 1982

In planning to permit a resumption of ocean dumping of so-called low-level nuclear waste, the Environmental Protection Agency is exhibiting scant awareness of the serious drawbacks of such a disposal method. The main problem with the EPA proposal is that no one knows the long-range effects on marine life and on the human food chain (via edible fish) of nuclear dumping that has already occurred. But enough is known to justify great caution in resuming the practice.

High-level and low-level nuclear wastes (about 89,000 barrels) were dumped at 50 undersea locations off the Atlantic, Pacific and Gulf coasts of the United States from 1946 to 1970, when public concern prompted a moratorium. In the 1972 Marine Protection Act, Congress banned ocean dumping of high-level nuclear waste but said dumping of low-level waste could be resumed under EPA guidelines. Guidelines, however, were never issued and the moratorium has remained in effect. Now the EPA — because land sites in the United States are getting full and because citizens across the country oppose land sites near them — is proposing to issue rules in April that would allow the ocean to be used as a dump for the garbage of the nuclear industry, which the Reagan administration has been trying to protect from its multiple problems.

The dumping proposal appears also to be related to the Navy's desire to scuttle retired nuclear submarines in the ocean after removing the nuclear reactors and fuel. One critic pointed out that the steel in the reactor shells would remain highly radioactive. Between 1948 and 1951 the Navy scuttled 12 vessels that had been contaminated by radiation in atomic tests at Bikini Atoll in the Pacific in 1946. Records show that these ships, scuttled off California, Washington and Mexico, were sunk near migratory fish routes. To resume such scuttling, the Navy would now have to get EPA clearance.

Although the EPA would permit public comment before putting the proposed rules in effect, its schedule scarcely allows time for essential studies before any change is made. By general agreement, the government's inventory of past dump sites has been sloppily kept. There has been no systematic attempt to monitor known dump sites. What is known is that many of the sunken drums containing radioactive garbage have ruptured and that radioactive levels at some of the sites are surprisingly high. Plutonium levels in ocean sediment cores from the Farallon Islands dump site (near San Francisco) were found to be eight to 2,208 times higher than expected. Edible fish in waters off California have been found to be contaminated with radiation.

The assurance that only low-level wastes will be dumped provides little consolation. Some waste is low-level only in the sense that it is diluted. If enough of it with a toxicity life of thousands of years is concentrated in one spot, it would be a matter for serious concern. Moreover, the oceans contain globally shared resources. The United States should not hastily adopt them as nuclear garbage dumps without thought for other peoples who use those resources or for future generations of Americans who will use them.

The Wichita Eagle-Beacon

Wichita, Kans., January 12, 1983

Americans may have understandable safety concerns about the underground disposal of nuclear wastes, but expected authorization by the Environmental Protection Agency for ocean dumping of radioactive materials certainly isn't the solution. Legislation to prohibit this questionable action currently awaits President Reagan's approval and should be signed into law promptly.

The EPA has given a clean bill of health to waters off the coast of Massachusetts that were used as depositories for large amounts of low-level nuclear wastes dumped between 1947 and 1958. There may be as many as 50 such offshore sites around the country, and barrels in at least one site have been found to be leaking. The Navy — which currently has four nuclear submarines in need of disposal — is expected to be the first applicant for renewed nuclear waste dumping in U.S. coastal waters.

In 1972, after 91 nations, including the United States, banned the dumping of high-level radioactive wastes and regulated low-level waste deposits, Congress invoked a two-year moratorium on nuclear waste dumping in ocean waters. During the last session of Congress, a bill was passed ordering a new two-year moratorium and placing restrictions on any future dumping after the moratorium. That measure awaits the president's signature.

It's ironic that only three months after the Netherlands finally bowed to international pressure and ended a 15-year policy of dumping nuclear wastes into the Atlantic Ocean, the United States is giving serious consideration to enacting such a thoughtless and shortsighted policy. The oceans of the world are no place for the disposal of nuclear wastes that will remain toxic for thousands of years. No container is guaranteed leak-proof for such extended periods of time, and none provides adequate protection for the three-quarters of the earth's surface that is water-covered and all the creatures that depend on the oceans for survival.

What to do with increasing amounts of nuclear waste is a problem of sufficient magnitude to rival any facing this nation. Short-term solutions — such as underground burial — may be necessary while long-range procedures are developed. The oceans, however, are far too important to the total environment to be included in these experiments. Other options must be found, and President Reagan's signature is needed to ensure that a grave mistake is not made in the name of expediency.

Community Bans on Nuclear Waste Shipments Overruled

Some of the public opposition to nuclear energy is associated with the fear of an accident involving the vehicles used to ship spent fuel and other radioactive waste from nuclear plants to waste storage sites. By 1983, more than 200 communities had passed laws restricting the transport of nuclear waste.

Two recent court rulings have called these bans into question. A federal judge in Spokane, Washington ruled in June 1981 that the state's ban on the shipment of out-of-state nuclear waste was unconstitutional. The ban had been approved by voters through a state ballot the previous November; nearly all the atomic waste stored at a low-level nuclear waste site in Hanford came from out of state. (See pp. 70-77) The court ruled that the ban interfered with interstate commerce and the power of the federal government to regulate such commerce. A federal appeals court upheld this decision in 1982.

A similar federal appeals court decision in August 1983 overturned a 1976 New York City ban on shipments of nuclear waste. The decision cleared the way for hundreds of truckloads of spent nuclear fuel from a non-functioning reprocessing plant in West Valley, N.Y. to be shipped back to nuclear plants in Wisconsin, Illinois, New Jersey and New York. The ruling upheld a federal statute permitting the Nuclear Regulatory Commission and the Transportation Department to override local objections in selecting transportation routes.

The Star-Ledger

Newark, N.J., June 28, 1980

Come November, the federal Department of Transportation intends to have rules in effect that would restore a semblance of sanity to the emotionally charged problem of shipping radioactive materials.

Trucks carrying the high-risk loads would have to avoid heavily populated areas and reduce travel time to a minimum by using limited-access Interstate roads or other preferred highways, to be designated by the states.

Shippers would be required to prepare advance route maps listing all stops and emergency telephone numbers for drivers. And the drivers themselves would be obliged to undergo special training every two years.

Congress' General Accounting Office reports there were about 910,000 shipments of radioactive materials for medical use and about 215,000 shipments for industrial purposes in 1975. By 1985, the number of these shipments is expected to double — to 1.7 million and 560,000 respectively.

Reported accidents have been few and far between and none of them, says the DOT, "resulted in radiological health consequences as severe as the consequences reported sometimes to result from the behavior of flammable liquids in transportation accidents."

But the shipment of radioactive cargo continues to stir controversy and command the attention of local and state officials. In the absence of federal rules, cities and states have seized the initiative by banning or severely restricting shipments. New York City enacted the first local ban in 1976; 79 other municipalities around the country followed suit.

Shipment of low-radiation waste — such as contaminated clothing and tools from nuclear plants — are still being trucked to disposal sites, but the major share of radioactive material in transit is in no way connected with the production of nuclear power.

The DOT was given authority to act in this sensitive area five years ago. It is lamentable it has taken so long to act. But it would be more lamentable to allow the irrational pattern of local rule to continue. The shipment of radioactive materials is properly a national concern. It is expected that the federal rules, when finally issued, will give full consideration to improved safeguards and standards of safety.

NEW YORK POST

New York, N.Y., June 14, 1980

Four years ago New York City banned the shipment of dangerous nuclear wastes through our streets. The move represented an elementary act of civil defense against the appalling threat of nuclear spill.

Now the U. S. Transportation Dept., in an indefensible display of remote bureaucratic omniscience, is proposing to wipe out the city's protective measures and supersede them with its own flabby "regulations" on radioactive traffic.

Mayor Koch has served notice that he will fight this unwarranted intrusion. He has stated the case clearly and bluntly:

"Enactment of these federal regulations and consequent nullification of the City Health Code regulation would generate the potential for a public health disaster of the first magnitude which could occur within our city on any day of the week. Statistics show, unfortunately, that the possibility of such an accident occurring is very real indeed."

Koch pointed out that the Transportation Dept. bureaucrats would not only "allow the relatively unrestricted transport of radioactive materials through the city of New York" but also deprive the city of the ability to safeguard itself against the perils created by such transport.

The Mayor has never been identified as a ritualistic environmentalist prone to exaggerate the hazards of existence. This is no phantom or implausible danger he is talking about. It is a deadly menace.

Let the U. S. Dept. of Transportation cease and desist; it is playing games with too many lives, and simultaneously flouting elementary principles of home rule.

The Charlotte Observer

Charlotte, N.C., November 7, 1980

The Atomic Safety and Licensing Board made the right decision last week in barring Duke Power Co. from moving used nuclear fuel around the Carolinas. Each movement of used fuel involves risk for people who live on or near its route, and such risks should be kept to a minimum.

Beyond that, the decision may force government policymakers, after almost four years of fence-straddling, to make some tough decisions about how nuclear wastes ought to be handled.

In April 1977, President Carter deferred government plans to license reprocessing plants which accept used fuel, saying he feared reprocessing might give terrorists easier access to material that can be used to make bombs.

Instead of reprocessing plants, the president said, there would be a permanent disposal site for used fuel from commercial reactors. Until such a site could be found and prepared, he said, utilities should "maximize" storage space at reactors, and the government would provide "temporary" storage space for emergencies.

But Congress has not allocated money for temporary storage areas. All of that leaves utilities to wonder: Should they spend millions, perhaps pushing up utility rates needlessly, to expand on-site storage capacity? Or will the government build temporary storage before utilities run out of storage space and have to shut down reactors?

Utilities had built relatively small storage pools to hold used fuel temporarily before sending it to reprocessing plants. Duke Power is running out of that kind of storage space at its Oconee nuclear plant in South Carolina, so it requested permission to move 300 bundles of used nuclear fuel from Oconee to storage at the new McGuire plant north of Charlotte.

By denying that request, the federal licensing panel in effect, is saying the government must either provide the promised temporary storage or Duke and its customers must pay to expand storage space at reactors.

We think the government should forget the temporary storage concept and get on with providing a permanent disposal site. Meanwhile, utilities should store wastes on site, despite the cost. That way the used fuel will have to be moved only once.

Setting up "temporary" government storage sites and meanwhile allowing Duke to shuttle fuel from one reactor to another in its system adds two unnecessary opportunities for transportation accidents.

The chances of such an accident are small, but the consequences of one could be large if it resulted in a release of radiation, which could make an area uninhabitable for days or even weeks. People shouldn't be subjected to such risks more often than absolutely necessary.

DESERET NEWS

Salt Lake City, Utah, February 7, 1981

The movement of radioactive material in interstate traffic is becoming increasingly difficult as more and more states and local governments enact restrictions or absolute bans on such shipments.

The restrictions pose problems not only for nuclear power plants, but also for scientific and medical researchers, hospitals, and other institutions using radioactive materials. The volume of shipments is large and is growing. Last year an estimated two million packages of radioactive material were shipped in the United States. In 1982, that figure is expected to reach approximately 3.5 million packages.

The Federal Department of Transportation has concluded that the public risk in transporting radioactive materials is too low to justify the state and local restrictions being enacted. Therefore, the department is proposing a rule to become effective in February 1982, that would encourage states to designate routes for radioactive shipments within their borders.

In setting such routes, the states would be required to consider traffic accident rates, population densities, and local emergency response capabilities.

If the states fail to designate routes for the radioactive shipments, the interstate highway system automatically would become the designated route system in each state. Such routes would override any local restrictions, according to the rule.

Some opposition has arisen to the plan and legislation has been introduced in Congress to overturn the proposal. The passage of such legislation would be a serious step backward. The need for radioactive materials is too widespread and important for this industry to be left completely at the mercy of 23,000 local jurisdictions, each no doubt preferring that its nuclear wastes go elsewhere.

As technology advances, better ways likely will be found to make the use, transportation, and disposal of radioactive materials safer. Certainly, a strong effort should be directed toward that goal.

In the meantime, the Transportation Department's plan seems to be the fairest, most sensible answer to the problems of moving radioactive materials to their destinations.

THE TENNESSEAN

Nashville, Tenn., June 13, 1981

CONTROVERSY over the shipment of nuclear wastes from Canada to South Carolina through Tennessee is a reminder that the U.S. still doesn't have a coherent plan for disposing of its own nuclear wastes, to say nothing of Canada's.

Public officials in New York and Michigan temporarily blocked a highly radioactive shipment of wastes through those states from Canada's national atomic research center this week. A dozen tractor-trailer loads of spent nuclear fuel were scheduled to travel to the only East Coast nuclear waste dump, located at Dunbarton, S.C., passing through East Tennessee. But they were halted by the Ogdensburg, N.Y., Bridge Authority, which voted recently to ban all "highly-explosive and radioactive" cargoes from the bridge on the U.S.-Canadian border.

Michigan officials questioned the safety of the Canadian shipments and said the National Regulatory Commission had not announced an approved route.

Although this incident probably won't produce any lasting hard feelings, it is an indication of the magnitude of the problem of disposing of nuclear wastes and the difficulties it can create with bordering nations. The Nuclear Regulatory Commission and other concerned federal agencies need to find a more satisfactory solution to the problem.

There is surely a better way than transporting such dangerous material by truck over long distances, through cities and towns and alongside automobile traffic on the highways.

State officials apparently do their best to regulate the transport of these cargoes within the limits of their jurisdiction. But they are hampered by the lack of authority.

Tennessee Public Service Commission Chairman Frank Cochran said the last shipment of nuclear waste from Canada that passed through Tennessee was stopped at the state line because the truck had faulty brakes. The shipment was allowed to proceed only after the brakes were repaired on the spot.

The PSC chairman said state inspectors meet all trucks carrying radioactive materials at the border and check every aspect of the trucks, including checking for radiation leaks with a Geiger counter. But, he said, the state is limited to inspecting the trucks and not permitted any say in other aspects of the shipments such as deciding the safest routes.

Despite the states' efforts, there seems to be little assurance that over-the-highway shipments of radioactive nuclear wastes are safe.

There is a great deal of public uneasiness about the storage of wastes and about the transportation of wastes over long distances. It may be hard for many people to believe that places cannot be found in Canada to store Canadian wastess and that shipping such dangerous material from Canada to South Carolina is necessary.

The nuclear industry has a long way to go before it satisfies the public that it can handle the wastes from nuclear plants without endangering the people's health.

Long Island, N.Y., January 19, 1981

Five years ago, New York City became the first municipality in the country to ban shipments of high-level radioactive wastes on its streets. By the time the Department of Transportation overrode that prohibition last week, 50 other communities had enacted similar ones of their own.

The fact that one of those communities was New London, Conn., effectively put a cork in Brookhaven National Laboratory's nuclear bottle. High-level wastes from Brookhaven's research reactor couldn't leave Long Island either by road or by ferry, so they've been piling up in storage pools at the lab. By the time the new federal rule goes into effect, in about a year, present storage capacity will be almost exhausted.

While Long Island's unique geography made the transportation problem particularly acute here, trucks carrying nuclear wastes were forced into circuitous routes elsewhere. And chances are that more and more communities would have barred radioactive shipments if the federal government hadn't stepped in.

Under the new rule, states and localities are encouraged to pick the routes they prefer for nuclear waste shipments, taking into account such factors as travel time, accident rates, population density and availability of emergency vehicles. If they refuse, the trucks would be required to use interstate highways.

City health officials are expected to continue the battle in the next administration. But these rules were two years in the making, and little can be gained by starting the process over again.

Instead of fighting an eminently sensible regulation, New York City and state officials should start selecting the best routes. Hauling highly radioactive materials from Brookhaven along the Long Island Expressway and through the streets of Manhattan is obviously not a good idea; taking them across Staten Island to the Outerbridge Crossing might be.

THE PLAIN DEALER

Cleveland, Ohio, December 10, 1981

A simple, sensible state law requiring proper state officials be notified before hazardous nuclear wastes are trucked across Ohio will be nullified in February — not by direct congressional action or presidential decision, but by bureaucratic fiat. Such a casual dismissal of state law in unconscionable.

Waste is an inevitable result of nuclear technology; it must be transported as quickly and safely as possible to proper disposal sites. It must move across state lines, and the federal government has a legitimate interest in seeing that a tangled skein of local laws doesn't unnecessarily complicate the process. Some towns, frightened by anything with the label "nuclear," have passed profusions of ordinances with the underlying intent of discouraging all nuclear waste trucking in their vicinity. Intervention by the federal Department of Transportation was certainly proper to prevent truckers from being forced into circuitous, time-consuming routes that might, potentially, endanger even more of those who lived or worked by the roadside. Ohio's localities were no worse than any others, but in the federal-local potshotting, the state law got caught in the crossfire.

The state has at least as compelling an interest as the federal government in the transportation of hazardous nuclear wastes. Federal regulations that supersede state laws should be at least as protective, if not more so. That is not the way it will be after February if the Department of Transportation doesn't change these new rules; no notification will be required at all, so state officials won't be able to provide special escorts or reroute the nuclear waste trucks around hazards. The notification system worked smoothly in the past, and no one is saying it did not. Why change it?

Atty. Gen. William J. Brown has filed suit in federal court to keep Ohio's notification law alive; it shouldn't take a full-blown court case to make officials at the Department of Transportation understand that this particular cavalier usurpation of state authority is wrong.

The nuclear star is rising; more wastes are being created — and shipped from place to place — every year. That is the reality. But the vital interests of the people of Ohio lie in assuring that nuclear wastes travel from border to border under layers of protective legislation, both state and federal.

THE ANN ARBOR NEWS

Ann Arbor, Mich., June 30, 1981

WASHTENAW COUNTY has every right to be as concerned over the nuclear waste issue as Gov. Milliken and other state officials.

Those potentially lethal wastes come within a few miles of Ann Arbor and Ypsilanti City Hall on their journey down US-23 to the south.

Now we (the state) have told Canada, from whence these wastes come, that the Mackinac Bridge is closed to shipments of spent nuclear fuel.

That doesn't end the matter, of course.

DURING SENATE debate on the subject, Sen. Richard Allen, Republican of Ithaca, said that banning these shipments through Michigan only increases the risk because then an even longer route south must be found.

Perhaps more to the point, Senate GOP leader Robert Vanderlaan of Kentwood said Michigan ships its own nuclear waste out of state and has no right to ban the transports along its highways.

Canada lacks disposal sites for nuclear wastes. The U.S., which generates far more nuclear material than Canada, has shown a paucity of leadership when it comes to safely disposing of nuclear waste.

We appear to want the benefits of nuclear energy without the responsibility of following through on the safety aspect.

CLEARLY, Michigan cannot have it both ways on the nuclear issue.

We can't generate the stuff ourselves, ship the waste away from our borders out of sight, out of mind and at the same time declare our territory off limits to shipments passing through.

Within a few months, Michigan's tough new rules will be in conflict with the federal government, which claims preemptive authority over the disposal of radioactive materials.

That preemptive role is a clear call for action on Washington's part to get cracking, pronto, on the twin issues of nuclear waste transport and the development of safe disposal sites.

In the meantime, we in Michigan can't be too squeamish about waste shipments. It's a problem shared by all and we are not an island unto ourselves.

A peninsula or two, yes, but not an island.

The Pittsburgh Press

Pittsburgh, Pa., December 11, 1981

The Thornburgh administration has dismissed as "simplistic" and "grandstanding" former Gov. Milton J. Shapp's demand that it oppose federal plans to ship nuclear wastes across Pennsylvania.

Certainly, it would be premature for the present governor to take any such action.

For the state's own 15-member Hazardous Substance Transportation Board, an adjunct of PennDot, is still studying this problem.

★ ★ ★

At present, radioactive wastes from atomic power plants are stored on site — because there are no approved dumping grounds for them elsewhere.

But beginning in the mid-1980s such wastes are to be consolidated at a number of temporary storage sites now under consideration by Congress.

According to a study by the National Academy of Sciences, four of Pennsylvania's interstate highways (78, 80, 81 and 84) would be prime candidates for such nuclear-waste transport.

However, Mr. Shapp contends that if "this hideous plan" is adopted "Pennsylvania might well go out of business as a state safe for families to grow and for safe industries to flourish."

He feels Pennsylvania should join with Ohio and New York in a suit to prevent the federal government from implementing this plan — on grounds it would pre-empt state laws governing the movement of high-level radioactive wastes.

★ ★ ★

Of course, neither the Thornburgh administration nor anyone else should acquiesce in any disposal plan that would compromise the safety of Pennsylvania citizens.

But Mr. Shapp is being unduly alarmist.

Nuclear wastes can be transported safely, as attested to by both the on-going Three Mile Island cleanup and by the safe shipment of such material across Pennsylvania even when Mr. Shapp himself was governor.

The fact is that if Pennsylvania and other states are to house atomic power plants, some provision must be made to ship their ashes to designated storage sites. And that means moving nuclear wastes over some highways, and across some state lines.

Indeed, the ill-fated Three Mile Island plant could never be decontaminated for safe operation again if other states were to block off their roads to the radioactive gunk that still must be removed from there.

All things considered, the Thornburgh administration is quite right to view the fears expressed by Mr. Shapp as exaggerations typical of nervous nuclear Nellies.

The Dispatch

Columbus, Ohio, January 10, 1982

OHIO HAS DEVELOPED an excellent system of alerting proper agencies when hazardous nuclear materials and chemicals are shipped across the state. Now a federal agency wants to invalidate this and similar laws in other states.

The Ohio law requires notification to its Disaster Service Agency prior to trucking a load of nuclear wastes or chemicals.

The agency then reports the route of shipment to county sheriff offices along the way, the State Highway Patrol, the Public Utilities Commission (PUCO) and the state's Environemental Protection Agency (EPA).

Unless the Ohio Attorney General William J. Brown is successful in a lawsuit filed in federal district court in Cleveland, a rule adopted by the U.S. Department of Transportation (DOT) will go into effect Feb. 1 and will pre-empt the Ohio law.

The federal regulation says nuclear shipments must be made on interstate highways. We don't disagree with this and we are aware that numerous communities, frightened by trucks carrying dangerous materials, have enacted ordinances to control routes of such shipments through their municipalities.

But invalidating Ohio's safety notification law does not make sense except to give DOT total control.

Who knows more quickly than police authorities about a danger ahead on a highway? An accident might be blocking an interstate roadway. Notifying the driver and then rerouting the truck away from the accident scene could save lives and time.

Ohio's alert and notification system has been successful. We think the DOT rule runs counter to it and are grateful that Brown is attacking this blatant infringement on states' rights.

Federal and state regulations are necessary for the safety of the public. It would be far better if the DOT would augment and not usurp Ohio's fine safety regulation.

The Burlington Free Press

Burlington, Vt., September 1, 1982

Mere mention of the word nuclear conjures up fears of radioactivity, mushroom clouds and mutations in the minds of many people.

While authorities often go to great lengths to dispel such apprehension, there is evidence to indicate that a large segment of the public cannot be dissuaded from thoughts of catastrophe when the issue of disposal of the nuclear byproducts is raised.

That shipments of nuclear waste from a Canadian research reactor have been made through Vermont this summer has aroused the ire of anti-nuclear groups and people in towns which have passed referendums calling for a ban on such traffic in their communities. The controversy has taken on a political tone with charges being hurled back and forth by Lt. Gov. Madeleine Kunin and Gov. Ruchard A. Snelling who are likely to square off in the battle for the governorship after the September primary.

Kunin has attacked Snelling for being too willing to allow highly radioactive wastes to be shipped through Vermont. Snelling has responded by saying that the wastes are being routed through the state in accordance with hazardous waste regulations that were adopted by Vermont three years ago. He said Vermonters were not in any danger from the shipments.

Kunin said she favored advance notification to the towns through which nuclear wastes are to be shipped. But Snelling's Transportation Secretary Thomas Evslin replied such a step would violate "federal law and common sense." Indeed such notification would serve a doubtful purpose because there are no steps local officials could take to protect residents even if authorities were aware of the shipments. There simply are no precautions that can be taken to prevent an accident that might happen and, chances are, would not occur. Since little can be done in advance of a tragedy, there seems to be little point in notification.

At the same time, such a procedure would open up a wide range of possibilities for demands that communities be notified in advance of shipments of any hazardous materials on highways that pass through their territory. Notification could become a bureaucratic nightmare at the state level.

While Evslin's defense of the shipments made some valid points, his effort to link Kunin with Abbie Hoffman, a leader of New York State resistance to nuclear waste shipments, was little more than a hysterical reaction to her remarks.

Evslin quoted Kunin as "supporting a call by former underground leader Hoffman for advance notification of such shipments." But Kunin made her remarks without knowledge that Hoffman would be interviewed by a reporter. Evslin's charge was ill-conceived and unwarranted. Had he avoided the trap, he might have strengthened his case against the notification proposal.

Even though an estimated 38 Vermont communities have passed referendums banning the shipments, a state law which was passed in January on designation of routes of travel in the state for high-level nuclear waste supersedes local ordinances, according to an unofficial opinion by the attorney general's office. Until the state constitution is changed, the state will have the last word on the subject.

What may be bothering some Vermonters is that the shipments originate in Canada. They may think that Vermont should have no responsibility to provide access routes for Canadian waste which is on its way to South Carolina for reprocessing. And it would certainly seem that citation of state and federal law to justify another country's shipment of nuclear wastes through Vermont is a misapplication of the statutes. Vermonters might not have objections if the waste was being shipped from American nuclear reactors.

Because of the questions that have been raised as a result of the shipments, the 1983 Legislature should appoint a study committee to explore the issues of safety and insurance liability and consider possible measures for dealing with the problem.

If public sentiment is strongly expressed against such shipments, lawmakers might consider the possibility of a statewide ban on the waste traffic.

As it stands now, however, the shipments conform to the requirements of state law.

THE LINCOLN STAR

Lincoln, Neb., March 5, 1983

A month ago we plainly stated our objections to a proposed city ordinance which would have made transport of nuclear waste shipments through Lincoln very difficult, if not impossible.

At the same time we urged the Nebraska Public Power District to provide adequate notice to the city when it sends the 30 shipments of nuclear waste through the city over the next five years.

NPPD will notify the governor's representative, Col. Elmer Kohmetscher of the State Patrol. Kohmetscher has said he will notify Lincoln Police Chief Dean Leitner of the shipments. Leitner has said he will release the information to Civil Defense and Fire Department officials, but not to the general public.

That's progress, but not enough.

The Nuclear Regulatory Commission says release of information on such shipments should be judged on three factors: a request in advance, the person's need to know and the person's capability to respond in case there is an accident.

Both Kohmetscher and Leitner cite NRC rules and the possibility of sabotage as a reason for limiting the release of information to the general public.

There's more than enough paranoia about nuclear waste to go around. The industry fears sabotage and emotional criticism. Opponents to nuclear power fear incompetence and accidents.

The first step to breaking down the mistrust and suspicion is openness and candor.

We think city officials should notify citizens who express a concern, particularly those who have homes and families who live along the railroad corridor, about the shipments.

We urge the action not because we think the shipments are unduly dangerous or that citizens should take special precautions, but simply because if there is an interest or concern, the city should respond to it. The first tonic for fear is simply information.

And sabotage? If the NRC and law-enforcement officials really want to increase cooperation and reduce public anxiety, we suggest they stick that scare word back in their dictionaries.

The Courier-Journal

Louisville, Ky., March 17, 1983

THE IMPENDING SHIPMENT of radioactive wastes via the federal highway system, as *The Courier-Journal's* Mike King reported from Washington Monday, is a touchy enough issue without federal insensitivity to local concerns adding to the problem. Kentucky should be particularly worried, since a fair portion of whatever wastes are shipped will go to Oak Ridge, Tennessee — much of it over this state's federal highways.

As things stand, a Department of Transportation regulation supersedes all state and local laws governing these shipments, even laws as simple as those requiring that police and fire departments be notified when a dangerous shipment is going through. The regulation has been challenged in court and one federal judge has agreed that it goes too far, but that ruling is on appeal and wouldn't be binding in other judicial districts, in any event.

A patchwork of state and local laws would delay shipments, and localities obviously couldn't be permitted to ban trucks carrying nuclear wastes altogether, as some have tried to do. But shutting localities out of the entire process is overreaction. DOT or, if necessary, the courts or Congress, should set reasonable regulations requiring notification of the cities and towns through which waste is being shipped, and allow local governments to suggest appropriate times and, if necessary, alternate routes. No one wants trucks loaded with casks of dangerous nuclear wastes rolling along I-75 through Lexington at rush hour, for instance.

On the other hand, this shuttling of dangerous wastes from place to place won't solve the essential and most pressing problem, how and where to safely store this radioactive garbage forever, so it can't hurt anyone or anything. Oak Ridge and the other two temporary storage sites, in Idaho and South Carolina, will accept wastes only on a short-term basis, and final disposal is up to the utilities generating it in their nuclear power plants.

A Department of Energy spokesman says the government expects very little waste to be shipped for temporary storage, simply because of the added expense. Not everyone agrees; a just-released study by the Council of Economic Priorities contends that many utilities are running out of on-site storage space and will be forced to ship wastes elsewhere.

The government is working on regulations to govern the disposal of radioactive wastes. But experts generally agree that it will be the 1990s before any permanent disposal sites are designated and prepared for use — and then only if the necessary technology has been perfected and if some state, somewhere, will accept everyone else's nuclear cast-offs.

Meanwhile, the U. S. is has come up with a solution to the temporary storage problem that may be both workable and safe. Letting state and local governments in on the action — not to veto but to coordinate — should help make it both.

EVENING EXPRESS

Portland, Maine, January 5, 1983

More than any other single industry, nuclear power plants have a public obligation to meet special standards of safety—inside the plants themselves and when shipping radioactive waste.

That's why it's disturbing to learn South Carolina has imposed a $2,500 fine on the Maine Yankee Atomic Power Co. for shipping into that state a container of low-level radioactive waste which was discovered to be giving off excessive radiation.

Maine Yankee has agreed to pay the fine, imposed with a loss of its permit to dump radioactive waste in Barnwell, S.C., for 30 days. Those were strictures South Carolina imposed.

But Maine people, in return for voting their confidence in the plant in two nuclear power shutdown referendums, have a right to impose a stricture of their own: and that is the clear expectation that Maine Yankee will meet, or exceed, every safety standard relevant to a nuclear plant.

Granted, Maine Yankee has operated with an enviable safety record. There has been no nuclear accident at the Wiscasset plant. But there have been minor incidents—four in the past two years, including the shipping incident that drew legal penalty in South Carolina.

For this industry, meeting safety standards—almost—can never be good enough. Familiarity with nuclear power must not breed familiar shortcomings—overconfidence or a sense that safety standards may be safely hedged.

Judy Barrows, a spokesperson for the Maine Nuclear Referendum Committee, may not be a person Maine Yankee officials like to listen to. But Ms. Barrows spoke rightly when she said, "A mishap that might be merely inconvenient in some other type of plant can be disastrous in a nuclear plant." The same might be said for the waste that leaves a nuclear plant.

The Des Moines Register

Des Moines, Iowa, September 5, 1983

On Feb. 3, 1982, it was announced that radioactive nuclear wastes would be shipped across Iowa on the Burlington Northern railroad. That afternoon, a train crossing Iowa on the BN derailed at Creston, spilling chemicals. Hardly a reassuring omen.

Not to worry, said the railroad, the Nuclear Regulatory Commission and the Nebraska Public Power District, whose hot wastes will traverse Iowa for the next several years. The wastes — more than 1,000 bundles of spent atomic-fuel rods from the Cooper Nuclear Power plant at Brownville, Neb. — are being shipped to a General Electric storage facility at Morris, Ill., in casks so tough, the shippers say, that they have withstood being dropped 30 feet and being crashed into walls by trucks going 60 mph.

But as a veteran railroader asked, can those tests simulate the crushing effect of the crash of a 5,000-ton freight train?

Rolly Kuhlmann, manager of hazardous materials for Burlington Northern, said last year that accidents pose no danger; the real danger, he said, is sabotage. If the precise shipment dates were announced, "We'd be wide open," he said. Could the secret be kept from a deranged but determined terrorist?

The Illinois attorney general, who looks forward to the arrival of the hot wastes in his state about as enthusiastically as the uneasy residents of Iowa cities along the route, has tried to block the shipments until the NRC's nuclear shipping regulations are in effect. But NPPD says the wastes will begin moving as soon as workers can be trained to handle them — probably in October.

Illinois, upset with the GE plant's becoming a dumping ground for hot waste from other states, adopted a law banning the import of the wastes. But the U.S. Supreme Court ruled against Illinois; authority to regulate safety in handling nuclear material rests with the NRC. No quarrel there; a federal agency should be better equipped than a state to make technical judgments.

But the question remains: Is this trip necessary?

•

The fuel rods that power nuclear plants are far more radioactive when they are spent than when they enter the plant. They remain deadly — and must be kept isolated from the human environment — for thousands of years. There are now more than 30,000 spent fuel assemblies stored in water, most of them at nuclear plants, awaiting the day that they can be finally and safely disposed of.

The problem is, 40 years into the nuclear age, science still hasn't found a safe method of handling hot garbage, despite the soothing assurances of nuclear-power promoters. There isn't an ocean deep enough or an underground rock formation or salt mine impenetrable enough to assure that the deadly poisons won't surface or pollute an aquifer. And some of the present storage barrels are leaking.

There is no reason to believe that when and if a proper nuclear graveyard is found, it will be in Morris, Ill. Wherever it is, the wastes presumably will have to be shipped to it — across Iowa, again?

The Cooper plant in Nebraska has the storage capacity to handle its spent fuel rods until 1990. That at least would give atomic science seven more years to work on the disposal riddle so the waste, when moved, could be destined for its final resting place.

AKRON BEACON JOURNAL

Akron, Ohio, September 22, 1983

SPENT NUCLEAR fuel rods will soon be trucked from upstate New York through Ohio on their way to the two Midwest nuclear plants that produced the waste in the first place.

Movement of the 144 truckloads may start as soon as next week, despite all the shouts of protest from Cuyahoga and Medina county officials and legal efforts by Attorney General Anthony J. Celebrezze Jr. to keep the trucks from our borders.

Shipments of nuclear material occur throughout this country. Ohio has no reason to expect to be immune from them when other states are not.

But Ohio's protest was not wasted energy. It most likely helped convince the Nuclear Regulatory Commission that the radioactive material should move through Ohio as far from the Cleveland suburbs as is possible.

The NRC says the trucks should enter Ohio on Interstate 80 in Mahoning County and then hook up with the Ohio Turnpike at Gate 15 — a relatively unpopulated route.

They had wanted to enter the state on I-90, travel on I-271 through Cleveland's eastern suburbs, and then down Route 8 in Summit County to enter the Turnpike at Gate 12.

The NRC also approved this second route but said it is to be used only if unexpected repairs or events make the first route unavailable.

We think the NRC's language should have been tougher: Take I-80. If you can't use it, wait until you can.

Nonetheless, the agency's message to shippers is basically: We don't want you traveling on roads that take you dangerously close to thousands of Ohio homes and businesses.

The turnpike route, which includes northern portions of Summit County, is the safest course. The State Highway Patrol can easily protect the trucks and their dangerous cargo and usher them quickly through the state.

It has happened before. Ten years ago, spent fuel from the two nuclear plants in Wisconsin and Illinois was shipped through Ohio to a nuclear fuel reprocessing plant in West Valley, N. Y. And last year, there were 128 large shipments of nuclear waste moved through the state.

Now a federal judge has ordered the spent fuel be removed from storage at the defunct West Valley plant and returned to their plants of origin.

Despite the NRC action, Gov. Celeste still wants to keep the shipments out of Ohio. That position may score points with the public but it is an irresponsible one for a state with its own nuclear plants and a future waste disposal problem.

The issue of what to do with nuclear waste won't be resolved until there is a permanent U. S. disposal site. After years of ducking the controversy, Congress last year passed a law that set the machinery into motion for creating a site by 1990.

Of course, there is still the unsettled question of which state will get stuck with the radioactive burial ground. And nuclear waste must still be transported once there is a permanent disposal site.

This country, for better or worse, has adopted nuclear power as a source of its energy. As long as nuclear waste is shipped as safely as we know how, no state — including Ohio — can fold its arms and say, "Don't move it through here."

CHARLESTON EVENING POST

Charleston, S.C., April 5, 1983

It is too early to say for sure, but Gov. Richard W. Riley's idea for taxing low-level nuclear waste shipped into South Carolina for burial might yield a dual dividend. It possibly could, as aides to the governor have explained, produce revenue to provide low-cost loans to local governments for environmental studies or to build water or sewer systems that would help attract industry. It possibly could serve, too, to discourage the shipment of nuclear waste.

Imposing a tax of $10 a cubic foot on waste shipped to the Chem-Nuclear plant at Barnwell could raise as much as $12 million annually, according to a member of the governor's policy staff. Such a yield is not to be sneezed at, even in these big-money days.

When the proposal is introduced in the General Assembly, legislators should remember, however, an important point made by a policy staff spokesman. A tax on nuclear waste would be a finite source of money for the state. It would run out if South Carolina bands together with other Southeastern states — as it hopes to do — in a regional compact and refuses to accept any more nuclear waste shipments. No one, including lawmakers or the beneficiaries of a nuclear waste tax, should balk at trading off $12 million in tax revenues for a ban on further nuclear garbage deliveries. South Carolina has now more than its share of nuclear waste to worry about.

THE BLADE

Toledo, Ohio, October 2, 1983

SHADES of the zany days of the nut-and-berry gang! Proponents of a ban against transporting spent nuclear fuel rods on Ohio's highways turned up at a press conference the other day and indulged in their customary fear-mongering.

They conjured up the specter of radiological disasters, disintegrating fuel casks, and other potential accidents as the spent fuel moves through this state. They paint the side effects of shipments of such material in the grimmest terms, while urging public officials to oppose transporting nuclear fuel across this state.

Unfortunately, Ohio attorney general Anthony Celebrezze, Jr., was one who succumbed to the temptation to make political hay while playing on the emotions of the public. The attorney general was unsuccessful, it is good to note, in his noisy attempt to block in the courts the scheduled shipments of spent fuel rods from New York across Ohio to nuclear power plants in Illinois and Wisconsin.

More recently, the Lucas County commissioners — Democrats all — weighed in with a statement concurring with Mr. Celebrezze's ill-considered stance — a collective opinion worth its weight in hot air.

What the vocal opponents of these nuclear shipments ignore is the fact that moving radioactive materials through this state is nothing new. In the past two years alone, there have been literally hundreds of shipments of such material — including spent fuel — in Ohio without accident.

More significantly, the spent fuel that is to be moved westward soon was shipped across Ohio to a facility near Buffalo 10 years ago without incident and without any outcry from huffing politicians or the cause-of-the-month club. That same fuel today contains *less than 1 per cent* of the radioactivity it had when it was trucked originally across Ohio.

There is no reason for any qualms about transporting nuclear wastes in Ohio, especially under the rigid regulations imposed by the Nuclear Regulatory Commission. The casks that will contain the spent nuclear rods have been proved in destruction tests to be virtually indestructible, even in the most serious transportation mishap imaginable.

When the shipments reach the Ohio-Pennsylvania border, they will be inspected there by the Ohio Disaster Services Agency before they will be permitted to proceed. The trucks then will be escorted by the Ohio Highway Patrol on I-80 to the Ohio Turnpike and then westward, with patrol cars equipped with radiation-detection equipment. That is it.

What this boils down to is that since the dawn of the nuclear age 30 years ago there have been thousands of shipments of radioactive materials among nuclear facilities without any instance of container disintegration, explosion, or leaking. There is no reason to believe that unblemished record will not continue considering all the precautions that will be taken when the spent rods hit Ohio.

It is a disservice by this state's attorney general or anyone else to try to stir up public fears needlessly. What does Mr. Celebrezze think Ohio will have to do in 1990 when spent nuclear rods at the Davis-Besse nuclear plant near Port Clinton will have to be transported to established repositories? Float them over Ohio in balloons or across Lake Erie in barrels?

California Moratorium on Nuclear Power Plants Upheld

States won a victory in 1983 in their battle to acquire more authority over the construction of new nuclear power plants. The Supreme Court ruled unanimously in April, 1983 that states could prohibit the development of commerical nuclear power for economic reasons. The case, Pacific Gas & Electric v. California, involved a 1976 California law that had been challenged by two utility companies with interests in nuclear energy. The law placed a moratorium on the building of new nuclear power plants in California until the state's energy commission was satisfied that a cohesive federal policy on the storage and disposal of high-level nuclear waste had been developed. Eight other states had similar laws, and 30 states supported California in its court battle.

The utilities argued that the Atomic Energy Act of 1954 and subsequent federal laws gave the federal government the exclusive authority to regulate atomic power. The Supreme Court, however, upheld a previous appeals court finding that federal laws on nuclear power superseded state laws only in matters related to "protection against radiation hazards." The court acknowledged that a new federal law set national policy on the disposal of nuclear waste, but said the law was not enacted to preempt state laws. (See pp. 88-107.) Justice Byron R. White, writing for the court, stated that Congress "had preserved the dual regulation of nuclear-powered electricity generation: the federal government maintains complete control of the safety and nuclear aspects of energy generation; the states exercise their traditional authority over the need for additional generating capacity, the type of generating facilities to be licensed, land use, ratemaking and the like." The Nuclear Regulatory Commission, White noted, did not have any authority over "economic considerations," leaving a "regulatory vacuum" that could only be filled by individual states. The Supreme Court accepted California's claim that the law was based on concern about the economic viability of new nuclear power plants built in the absence of a permanent solution to the waste disposal problem, rather than on concern about radiation hazards associated with the plants.

The Wichita Eagle-Beacon

Wichita, Kans., December 1, 1982

The trend toward returning more power from the federal to state governments suffered a setback at the hands of Congress this week. The bill that did this was a revision of legislation that would have given a state the ability to keep a nuclear waste facility from being located within its borders — unless both houses of Congress voted to override that decision within 90 days. The new wording that passed in the Senate earlier this year, and was approved by the House on Monday, makes any state veto ineffective unless one house of Congress also approves.

The new version of the bill severely limits the ability of any state to stop the creation of nuclear waste sites, regardless of the sentiment of a majority of its residents. In Kansas, for example, the case of the off-again, on-again controversy over the use of abandoned salt mines at Lyons for low-level nuclear wastes could be decided without the feelings of Kansans being taken into consideration — unless the Senate or House of Representatives could be persuaded to sanction any "no" vote by the Legislature. Even more importantly, if there were a determination to upgrade a facility in Lyons to handle high-level nuclear wastes, the fate of the project would rest in the hands of legislators from 49 other states.

The bill passed by a margin of 190-184. Only Rep. Dan Glickman, D-Kan., voted against the measure from the Kansas delegation. Republicans Jim Jeffries, Pat Roberts, Bob Whittaker and Larry Winn all voted in favor of the bill. Had the rest of the Kansas delegation seen fit to vote in the negative, the legislation would have been defeated. Kansans should let their representatives know how they feel about the matter — though, if the bill becomes law, how they feel will be a moot point.

THE KANSAS CITY STAR

Kansas City, Mo., December 1, 1982

Considering the lengthy schedule Congress has proposed for dealing with the problem of nuclear waste disposal, it hardly seems necessary to resolve the issue at the present lame-duck session. But some members are eager to settle one basic question now: the right of a state to veto its selection as an ultimate repository for radioactive wastes from the nuclear weapons and power programs.

All the proposed ritual of preparing three burial test sites in different types of geologic formations, making the necessary environmental impact studies and having the president make a final choice by March 1987 will be meaningless if the governor of the state or states chosen can veto the selection. Rep. Morris K. Udall, Arizona Democrat, introduced a bill whereby such a veto would stand unless overridden by both houses of Congress.

But now the House has weakened that veto power, on a narrow 190-184 vote, with an amendment that the veto must be ratified by one house of Congress to be effective. Similar language is contained in the Senate version of a waste disposal bill passed on a 69-9 vote last April.

Nuclear wastes with a radioactive life of thousands of years have been accumulating in temporary storage since World War II. Some states have banned further construction of nuclear power plants until the waste question is worked out. Government and other scientists have exhaustively researched how to process the wastes into less toxic forms, such as glass blocks; what type of containers to put them in and where to bury them safely.

But with the unreasoning public fear that prevails over having radioactive wastes stored anywhere in the vicinity, this has become more of a political than a scientific issue. The Udall bill faces a host of amendments, mostly proposed by states that have been or are likely to be chosen for the three test sites. But unless the final version of nuclear waste disposal legislation gives the federal government substantial right to make the repository site choice, this nagging nuclear-age problem will not have been solved.

Oregon Journal

Portland, Ore., August 11, 1982

Congress has regained its senses, at least temporarily, on the troublesome problem of nuclear wastes. Now, if the House can keep the meddlesome fingers of Sen. James McClure, R-Idaho, out of the waste can, it might make some progress on this difficult issue.

Oregon and six other states have said that no more nuclear plants can be constructed until a safe method exists for disposing of nuclear wastes.

McClure, chairman of the Senate Energy Committee, became a bit tricky in May. Then he asked unanimous consent of his senatorial colleagues to add several "technical" or clarifying amendments to a bill on the Senate floor. After he gave his word that the amendments weren't substantive, no objections were made and the amendments were approved by a voice vote.

These clarifying amendments actually changed nuclear waste policy in Oregon and six other states. McClure's amendments pre-empted the states' laws. The effect of the McClure amendment was to pre-empt the U.S. Nuclear Regulatory Commission's longstanding authority to determine whether a safe disposal method exists.

Some of his colleagues, including Sen. Mark Hatfield, R-Ore., didn't like McClure's tactics, which certainly were a breach of senatorial etiquette. Now the House Energy and Commerce Committee, of which Rep. Ron Wyden, D-Ore., is a member, has voted for an amendment to remove McClure's handiwork. Wyden said the McClure amendment "insinuates that if we pretend a safe disposal method exists, we can make it so. Oregonians are not that naive, and Congress shouldn't be either." Wyden proposed, and the committee adopted, the amendment.

The battle over whether states can regulate nuclear wastes and nuclear plants now moves to the House Rules Committee, which must pass on the Wyden amendment. There may be some enemies of the amendment lurking in the bulrushes around the Rules Committee. Fortunately, the committee's chairman is Rep. Richard Bolling, D-Mo., a venerable veteran who usually can be counted upon to do the right thing. Let's hope Bolling does the right thing this time

The Providence Journal

Providence, R.I., December 2, 1982

House adoption of an amendment making it harder for any state to veto location of a nuclear waste depository within its borders was accomplished by a narrow vote, 190 to 184. The Senate bill passed earlier this year is even tougher on the states — as it should be. Even if the House adopts the waste depository bill today, the measure will still have to go to a conference committee. But a compromise between the two must be found; the country can no longer postpone action on this issue.

Until a firm stand is taken by Congress to have depositories for high-level nuclear wastes located where they will be most secure, nothing will be done. Every state designated as a prime location on the basis of scientific evidence will find some pretext to veto the decision; everyone wants to duck the responsibility. Only strong action by the federal government can get the disposal program moving.

It isn't as if the decision hadn't been delayed, postponed and put off again for more than 20 years. The failure to make provisions for this most deadly of all hazardous waste is a national scandal. It is going to become a cause of crisis some day not too far off: the nuclear power program will come to a screeching halt because there is nowhere to put the used fuel that has been piling up in pools of water at the sites of the power plants.

The program for disposal of low-level wastes is less urgent only because such wastes can be contained in less rigidly designed areas — and are not in themselves so objectionable. But even there, the states that now have burial facilities for low-level wastes are threatening to bar shipments from the rest of the country.

The bill before Congress admittedly doesn't go far enough. It makes no attempt to deal with the nearly 100 million gallons of high-level nuclear waste produced by the military program for building nuclear weapons. The volume of that waste is far greater than the spent fuel from nuclear power plants. It is more difficult to handle, too, because it is in liquid or sludge form, while the spent fuel is solid and therefore more easily transported.

Until Congress takes a firm stand on disposing of nuclear waste, nothing will be done

Under the House bill as amended, a state could block a waste depository only if it could get one house of Congress to support its action. Previously, the bill provided that the state action could be overridden only by a vote of both houses of Congress.

The issue here is whether a state can be forced to accept a depository against its wishes. State champions shout about states' rights. But the federal government must be able to overrule the states or the country will be faced with an impasse in which huge amounts of highly radioactive waste will have no final burial place. And the states where it is then located will be in far more jeopardy than those where carefully planned depositories are built in rock or granite impervious to water, volcano or earthquake.

Congress has got to face up to this decision or run the risk of blame for irresponsible handling of one of the most deadly materials ever produced by man.

The Kansas City Times

Kansas City, Mo., July 7, 1982

A ruling probably is a year away, but the Supreme Court has agreed to consider a case which could determine the future of commercial nuclear power. At issue is a challenge by two California utilities, Pacific Gas & Electric and Southern California Edison, to that state's 1976 law imposing a moratorium on building new nuclear plants until the question of disposal of radioactive wastes is settled. A federal district court overturned the moratorium, but an appeals court reversed that ruling.

Since five other states — Connecticut, Maine, Maryland, Massachusetts and Oregon — have enacted similar laws, judicial sanction of the California statute could lead to a rash of state legislation restricting or even banning nuclear power. The California moratorium was adopted just days before a statewide referendum to ban all nuclear plants, including existing ones. The law was considered a more moderate alternative to this Proposition 15, which was rejected by the voters.

The two California power companies contend that the 1954 Atomic Energy Act gave the regulation of nuclear power to the federal government and barred individual states from controlling their construction or operation.

This lawsuit imposed a difficult choice on the Reagan administration, given its support for nuclear energy but also for states' rights. The Justice Department ended up intervening for the plaintiffs, saying the moratorium "poses a serious obstacle to the development of nuclear power as a source of electricity."

Certainly the embattled nuclear power industry, already beset by problems of high interest and construction costs, plant breakdowns and harassment by anti-nuclear groups, can ill afford a tangle of conflicting and overlapping state restrictions. The weight of federal regulation by the Nuclear Regulatory Commission, properly concerned as it is with safety factors, is heavy already. State shackles on nuclear energy could drive away the last wary investor. But the waste problem remains. If the court upholds the act of 1954, Congress may consider an act of 1983 that would be considerably different.

The Atlanta Journal AND THE ATLANTA CONSTITUTION

Atlanta, Ga., December 4, 1982

Nuclear wastes from military and commercial nuclear-plant operations are a national problem. And the ultimate decision about where to store these potentially deadly wastes has to be a national — not a state — decision.

The House acted wisely Monday in approving a proposal which would require either the House or Senate to sustain a state's objection before that state's presidential selection as a nuclear-waste site could be canceled.

The vote is part of a congressional process that is working toward establishment of the first national policy for permanent disposal of the nuclear wastes which have been accumulating in temporary-storage sites since World War II.

Opponents of the legislation contend that a state should have the absolute right to reject federal designation of a nuclear-waste site within its borders. But that makes neither pragmatic nor constitutional sense.

Given the nature of nuclear wastes, political pressure to reject a waste site would be strong in any state, but permanent sites must be established.

The pending legislation provides a state chosen by the president as a location for a disposal site a fair and sufficient opportunity to challenge the selection. If neither body of Congress upholds the state's objection, the state still may appeal to the courts.

That leaves plenty of room for any state to argue that the methods chosen for disposing of the nuclear wastes are improper, or the characteristics of the site selected are not safe for the disposal methods. Those are the only proper considerations.

No one wants nuclear wastes as a neighbor. But they are here, in the millions of tons and gallons, and increasing daily. They have to be safely and permanently stored somewhere.

Newsday

Long Island, N.Y., April 25, 1983

The unanimous Supreme Court decision upholding California's ban on construction of new nuclear power plants until their radioactive wastes can be permanently stored may well have beneficial effects beyond the scope of the ruling itself.

In the absence of a national energy plan, states have been left to fashion their own. So the court concluded that they had a right to determine for themselves whether nuclear power plants were economically appropriate. The ruling might also offer opportunities for opponents to contest a proposed plant by contending that a state didn't need the energy the plant would produce.

Chances are the court's decision won't affect Long Island's Shoreham plant, which is part of New York State's master energy plan. The state sanctioned construction at Shoreham years ago by issuing a site permit, and it was already well under way when former Gov. Hugh Carey imposed an administrative moratorium on new nuclear plants until the waste issue was resolved.

Three Mile Island notwithstanding, it's economics, not safety, that's choking the nuclear industry. No utility has sought permission to build a new nuclear power plant since 1978, and many plants have been canceled since then because of skyrocketing costs and much lower demand for electricity than expected. Safety has been an economic factor, though; continual improvements in design, engineering and construction of new plants and retrofitting of older ones have added greatly to the utilities' costs.

California pinned its law to the lack of permanent disposal facilities for nuclear wastes, stressing the economic uncertainty this creates. But nuclear waste is just as clearly a safety problem — one that the high court specifically singled out as a federal responsibility — and it will remain urgent even if no additional nuclear installation is ever built.

Spent radioactive fuel is piling up at every plant site in the country. The military has been accumulating its own nuclear wastes for decades. After 25 years of indecision, Congress finally passed a nuclear-waste disposal law last year, but its timetable is so stretched out that the problem is unlikely to be resolved before the end of the century.

The Reagan administration has tried to speed up the process a bit, recognizing the implications of further delay. The Supreme Court decision makes it all the more urgent for the industry to assist in that effort. What's more, the need for a burial site for radioactive wastes is one issue on which those who favor and those who oppose nuclear energy should find common ground. Every living American and millions not yet born would benefit if, in this instance, the efforts of both groups were devoted to the same goal.

DAYTON DAILY NEWS

Dayton, Ohio, April 28, 1983

The Supreme Court struck a blow for states' rights that has knocked America's nuclear energy industry on its atoms.

It ruled that California (and the five states that followed suit) were within their rights to ban new nuclear generating plants until the feds come up with a way to dispose of used atomic fuel.

California based its case on economic issues, a criterion not covered by the Nuclear Regulatory Commission's authority to deal with safety matters. The court ruled that a state had the right to refuse establishing sites for any more nuclear power plants as long as there was a danger the plant would be closed prematurely because it couldn't dispose of spent fuel. Such closings would cause an economic hardship on those who had forked over money for the plant and who would depend on the power.

Congress did pass a nuclear waste storage bill in December, but that bill is disliked by environmentalists who see it as industry's way to get rid of nuclear waste in any way possible so construction of power plants can proceed apace.

The recent ruling could be used to the advantage of both sides of the nuclear issue. Instead of debating the subject in general, they now can focus on one issue: Safe disposal of radioactive waste.

Most people accept the fact that nuclear energy has its proper place in this world, but they want it used with care.

The Louisville Times

Louisville, Ky., April 22, 1983

Now that the U. S. Supreme Court has handed down a decision that could seriously hinder a revival of commercial atomic power, it is important to keep in mind why the nuclear industry is in such a serious bind.

The tendency among nuclear promoters is to blame "zealots" who have raised serious economic and safety issues. The fault actually lies with the atomic wizards who were all too willing to believe their own claims that the public had nothing to worry about.

The issue was a California law, passed in 1976, that forbids any new nuclear construction until a state energy resources board is satisfied that the technology exists for disposing permanently of dangerous radioactive waste. Seven other states have since enacted similar moratoriums.

Two California utilities challenged the law on the ground that the federal government has the exclusive right to regulate nuclear energy. Along with the Reagan administration, which sees nuclear energy as the wave of the future, they argued that restrictive state laws subvert the federal goal of promoting atomic power.

The justices, long reluctant to curb nuclear development, surprised everyone by deciding unanimously to reject these untenable claims. Justice White, speaking for the court, drew a logical distinction between safety and economic regulation.

Congress, he said, reserved control over radiation hazards for the federal government. But states have the power to decide whether various means of making electricity are economically desirable or necessary. The federal government, he said, cannot force states to accept nuclear power.

Since the lack of waste sites and the uncertainty about future storage of spent fuel have a bearing on the cost of new plants, the California law is economic in nature, even if legislators backed it for safety reasons.

Nuclear enthusiasts may consider this a fatal blow to their expansion plans. Other states will surely pass their own moratoriums. This trend, the U. S. Justice Department warned the court, "poses a serious obstacle to the development of nuclear power."

But the blame should not be heaped on countless citizens who have become disillusioned with a power source that was supposed to provide unlimited energy at insignificant cost.

The "nuclear priesthood" has had 40 years to come up with plans for storing waste. Even though federal law calls for construction of a disposal site to start by 1989, doubts remain as to how and whether radioactive waste can be successfully contained. Simple prudence demands that new building be put off until a solution is in sight.

Earlier, the industry won a victory when the court decided the government could allow restarting of the undamaged Three Mile Island reactor without considering the "psychological impact" on nearby residents. That makes sense, since worthy projects might be tied up by claims of vague effects on psychological health.

But a strong case can be made that nuclear plants cause unnecessary anxiety because Congress has set a $560 million limit on their liability in the event of an accident. Their neighbors therefore have reason to fear they will not be fully compensated for their losses.

These issues may be moot since new power plant orders are down to zero. Still, it is essential that the industry be made to confront these problems *before* trying to bless us with more reactors.

Detroit Free Press

Detroit, Mich., April 25, 1983

THE REAL moratorium on nuclear power development in this country has been imposed neither by state governments nor the vocal critics of nuclear power development, but by the recent history of the industry. What the U.S. Supreme Court did, with its ruling last week declaring that California has the power to declare a moratorium on nuclear power plant development, is to permit the states to give force to that de facto delay in the country's movement toward nuclear power development. It is not a province reserved to the federal government.

For a long time the industry tried to portray itself as merely misunderstood and abused, by the Jane Fondas of this world and by an excitable press. And there has been some shrill criticism of nuclear power and of the danger of a nuclear accident. Thus far, even after the incident at Three Mile Island, much of the safety concern is essentially theoretical and potential.

What is not theoretical, though, is the industry's inability to control its costs, its failure to confront the issue of what is to be done with nuclear waste, its lapses on design and site selection and its failures on operator training and fail-safe procedures. The horror stories of mismanagement, cost overruns and inability to assure the sustained and safe operation of many of the plants have multiplied across the country. The state regulatory commissions have not been out front in challenging the utilities on these problems; they have probably, on the whole, been entirely too passive.

Most of these problems, perhaps all of them, are not inherent features of nuclear power development. But those who swept aside the cautions and the concerns about the pace and direction of nuclear power have now done so much to put the industry in disrepute that it will be decades recovering. That could be a serious problem as the country seeks to build a varied base from which to meet its energy needs. It will be a problem created more by the friends and advocates of nuclear development than by its avowed enemies, or even by the state agencies who now have Supreme Court sanction to delay nuclear power development in certain circumstances.

Albuquerque Journal

Albuquerque, N.M., May 6, 1983

At first blush it may appear the U.S. Supreme Court recently made contradictory rulings in cases involving the nuclear power industry. But there was good reasoning behind both rulings.

First the justices upheld a state's right to ban new nuclear power plant construction until the federal government devises a way to store waste safely. Then the court rejected a state's right to prevent shipments of waste from passing through their borders, or in some cases within their borders.

The federal government has exclusive power to regulate safety of nuclear power plants. But states can protect their rights when the subject is broadened to include economics. The court buttressed that right in the new-construction case: states could slow or even stop new construction "for economic reasons," said Justice Byron White.

After all, if a safe waste repository is not available, states may be left with tons of nuclear garbage and no place to put it. And electrical power generation may be interrupted, causing economic problems for customers.

However, the government has expended a lot of money on research and testing in search of a safe repository design. By Jan. 1, 1985, the federal government is supposed to recommend a high-level waste repository site and is studying nine sites in six states. One community, Andrews, Texas, already has volunteered to house a waste repository, despite Democratic Gov. Mark White's opposition. And Energy Secretary Donald Hodel has scheduled hearings this month in three Texas cities to discuss placement of a site in the Texas panhandle.

The court rejected state laws that ban transport of waste, saying such laws are unconstitutional. There are equally strong reasons to back the court's decision not directly related to constitutionality.

If a power plant's storage area for spent fuel fills up, the power company has three options: build another storage bay nearby to use until a permanent repository is built, ship the waste to another storage site, or close the facility. Any of the three options would raise costs.

As tests at Sandia Laboratories have shown, a great deal of damage can be done to vehicles transporting nuclear waste without causing significant damage to containment casks. Meanwhile, accidents in transporting oil, gas, chemicals, coal and other energy-related materials have killed thousands of people over the years.

The public acceptance of injuries, evacuations and deaths caused by accidents during transport of other energy-related materials stands in stark contrast to the state laws that attempt to ban shipment of nuclear waste in relatively safe casks. As energy consultants Mark Mills and Donnamarie Mills have pointed out, "A transport ban does not significantly enhance public safety but could inflict financial penalties on consumers."

States that elevated the panic of some citizens about nuclear waste into legal form could have spent useful time peeking at the Constitution. At the same time they could have considered the consequences of such panic for other citizens when measured against actual risk. Those opposed to nuclear power plants may view the court decision as a setback, but it is not anything of the kind: it merely restores a sense of proportion to a debate that is far from finished.

Los Angeles Times

Los Angeles, Calif., April 22, 1983

For nearly three decades most Americans accepted the notion that once Washington declared a nuclear power plant safe there was no way to stop its construction. That was how the nuclear industry interpreted the 1954 Atomic Energy Act and that was the way it almost always turned out.

On Wednesday the Supreme Court set the record straight in a decision on a California case that probably means that the sick nuclear industry cannot get well—at least not until it makes basic changes in its way of life. It is a decision as sound as it is far-reaching.

The court upheld a 1976 California law that banned further nuclear construction until the federal government produced "demonstrated technology" for the permanent disposal of waste materials—the nuclear equivalent of ashes from a fireplace. Some nuclear wastes would harm human beings who came in contact with them for tens of thousands of years.

California argued that its moratorium had nothing to do with safety; the state said it was concerned that billions of dollars in nuclear investments would be wasted if power plants were forced to shut down because they could not get rid of their nuclear garbage.

The justices agreed unanimously that California and other states have a right to make such decisions on economic grounds. Justices Harry A. Blackmun and John Paul Stevens went further, writing that states can block plants for no more reason than fear of catastrophe.

The decision strikes directly at the structure of the nuclear industry, which has always used its considerable heft to get what it wanted in Washington on the premise that the states would then fall in line. From now on, the industry must divide its energies among all of the states in which it wants to build.

The effects will be uneven. Some states are comfortable with nuclear power and deeply committed to it, as in Illinois and New England. But the decision describes state powers in such broad terms that almost any state that has doubts of any kind about the wisdom of nuclear power should have no trouble blocking projects.

There is irony in the outcome. The nuclear industry lobbied hard for the California law as the lesser of two evils from its perspective. The greater evil was a ballot proposition that called for a complete ban on nuclear construction. The ban failed to pass.

The real effects of the court ruling lie far in the future. No utility company has bought a nuclear power plant for years, and about 70 orders for plants have been cancelled since the mid-1970s. But as the price of fossil fuels rises and the supply falls, all industrial nations will be forced to some alternatives. If techniques for generating massive amounts of electricity with solar energy have not been perfected by then or if thermonuclear power is still out of reach, nuclear power could be the only fallback for energy-oriented societies.

The nuclear accident at Three Mile Island gave the nuclear industry a chance to tighten quality controls over what the investigating commission called an inherently dangerous technology. The industry's claims that it followed the advice are not persuasive. In a back-handed way, the Supreme Court now has given the industry a second chance to put its house in order by elevating state governments to the status of bodies that the industry must persuade that it knows what it is doing. The industry may not get a third chance.

THE DAILY HERALD

Biloxi, Miss., April 22, 1983

Mississippians concerned that this state will become the nation's nuclear dumpsite can find encouragement in the Supreme Court's latest ruling on a nuclear topic.

The court held, without dissent, that states have the authority to ban new nuclear plants until the federal government devises a safe method of radioactive waste disposal.

If states have that much authority concerning nuclear generating plants, isn't it reasonable that states have similar authority to ban nuclear disposal sites? It is a question the Supreme Court will likely be asked by the state eventually selected to host the country's nuclear trash.

The salt dome formations in Perry County of Mississippi are still being considered by the Department of Energy for possible selection as the location for the country's first — and maybe only — nuclear dump. President Ronald Reagan signed a law in January promising the federal government will have a system to begin disposing of radioactive waste by 1998. DOE officials have said they would like to quicken the timetable and the department has now scheduled hearings in states where sites are under consideration.

Mississippi has objected that these hearings should not be held until after the Energy Department completes its technical guidelines for choosing the first site. The state's position is reasonable; unless the guidelines are available before the hearings, the hearings will be perfunctory and impractical.

California enacted a moratorium on new nuclear plants in 1976; the moratorium was challenged and Wednesday's ruling settled the challenge in favor of California. The decision is a setback for the nuclear power industry, probably more so than the Three Mile Island accident four years ago. The outlook for expansion of nuclear power dimmed when a reactor at TMI failed; with the Supreme Court ruling, it has become absolutely bleak.

Since the TMI incident, there have been no new orders for nuclear power plants; some old orders have been canceled. The TMI cleanup, costing about $1 billion, is not yet completed and may take another five years. A companion reactor, not damaged in the incident, is still shut down.

The Supreme Court placed a heavy emphasis on economic reasons as the basis for states to slow or stop nuclear power developments. However, two of the justices, Harry A. Blackmun and John Paul Stevens, said they would go even further and permit states to block nuclear plants for safety reasons. Those two gentlemen have their priorities in a more reasonable alignment than do their brethern.

Safety ought to be the prime consideration in all things nuclear, not just the generation of power and the disposal of waste. Economic considerations ought to hold a secondary place as a determinant in siting nuclear facilities.

And if safety is the prime consideration in the nuclear dumpsite search, the nearly three-quarters of a million people residing within a 60-mile radius of the Richton salt dome will supply more than sufficient reason for the federal government to seek a more isolated location.

CHARLESTON EVENING POST

Charleston, S.C., April 22, 1983

The tendency in some quarters is to talk down the impact of the U.S. Supreme Court ruling that states can ban construction of nuclear power plants until the federal government devises a safe way to dispose of radioactive waste. Because states have the legal right to block construction does not necessarily mean they will exercise that right, nuclear industry spokesmen note. Besides, say some, the effect of the ruling is likely to be limited in light of recent congressional action mandating establishment of federal waste storage sites.

However accurate such assessments are, they do not alter other, related consequences of the ruling. There is little doubt, for example, that the ruling has put states in a much stronger legal position insofar as protection of the health of their citizens is concerned. Moreover, it could generate pressure from another direction for a quicker resolution of the nuclear waste disposal problem.

For years now people in the nuclear business, science and government have been saying that the technical aspects of permanent storage of radioactive waste have been mastered; that the real problem is political. Waste could be put in lead containers and buried in salt formations, for instance. But which state will be the repository?

Pressure for a political solution has been coming from the few states — such as South Carolina — which already have been shouldering more than their share of the waste burden and from those states which make no bones about not wanting to become nuclear waste burial grounds. The Supreme Court ruling just might have the effect of bringing industry pressure to bear on those seeking the elusive political solution, and such added pressure could only be welcomed by all concerned about public safety and the future of commercial nuclear power.

SYRACUSE HERALD-JOURNAL

Syracuse, N.Y., April 21, 1983

Even for those who do not oppose the development of nuclear power, there still is the lingering, gnawing question of how and where to dispose of the "hot" nuclear wastes that result from the generation of power in nuclear plants.

The United States Supreme Court addressed that question this week, and came to the conclusion there are, in fact, no answers ... at least, not yet.

This does not come as news to the nuclear power industry, nor to the federal government or the more than casual observers of the growth of nuclear power.

Justice Byron R. White, writing the decision for the high court, which was unanimous in its vote, supported the California contention that "the promotion of nuclear power is not to be accomplished 'at all costs'." The cost, in this case, is the possible danger to this and future generations who are not protected properly from the radioactive effects of the spent fuels.

The Reagan administration responded that the ruling could jeopardize the growth of nuclear power as a source of electricity in this country. That could well be true.

The president also has signed into law, this year, a promise that a system for disposing of radioactive waste will be in place by 1989.

Environmentalists, and now the Supreme Court, recognize that this promise is not, indeed, a guarantee that such a disposal system will be developed — or even if it is possible.

The court's decision puts the responsibility clearly and heavily on the federal government and the nuclear developers to realign their priorities.

The emphasis must be on the safe and permanent storage of nuclear waste if, in fact, that has not been the case.

While it's been five years since a utility sought permission to build a nuclear plant in this country, Laurence Tribe, the Harvard professor who argued the case in Supreme Court for the state of California, is convinced the decision is "independent of the question whether the plant has begun construction or not."

In other words,Tribe and other anti-nuclear power people — whatever their reasons — will most certainly insist that the ruling also applies to such plants as Nine Mile II, which is well along the way to completion. In New York, as we understand it, that would require a new law, comparable to the California statute that was tested yesterday.

Presumably, the battle lines are being drawn already in Albany, the nuclear power producers on one side and the anti-nukes on the other, seeking to convince state legislators they should or should not adopt legislation that would, in effect, put the nuclear power plants out of business in New York.

Be assured, though, that such a law also would be subject to court tests, perhaps, again, going all the way to the Supreme Court.

But in the confusion and controversy surrounding this week's decision, their is a real danger that its meaning may be lost: The key to the future of nuclear power in the United States is the disposal of its waste. The Supreme Court left no doubt about that fact. That responsibility is in the hands of the federal government and the power companies, not the states.

The options are limited.

ST. LOUIS POST-DISPATCH

St. Louis, Mo., April 25, 1983

The nuclear power industry has won one and lost one in the Supreme Court recently. But on balance, the industry comes out a loser, because the case it won is far less significant than the one it lost.

In a suit related to the accident at Three Mile Island, the court ruled unanimously that the Nuclear Regulatory Commission does not have to take into account "psychological stress" of residents living nearby when making a decision on the environmental effects of operating nuclear plants. The issue arose in the NRC's deliberations on whether to allow the restarting of TMI 1, the sister plant of the one that remains a billion-dollar cripple as a result of the 1979 accident.

The court acknowledged that pyschological perception of risk may have real effects and "may be an important public policy issue." But it is not something Congress expected the NRC to consider as part of an environmental review, the court said. And we would agree. Mental stress is a concept much too vague to serve as a guide for environmental regulation — or for a serious challenge to nuclear power.

Many other arguments are more powerful and to the point. Take, for example, those raised in the other case the court decided, which dealt with a 1976 California ban on nuclear power.

Utilities and the nuclear industry wanted the court to rule that California and the five other states with similar bans have no such authority, because the Atomic Energy Act gives the federal government sole power to regulate safety issues related to radiation. But the court upheld the ban, because California's concerns were based on the economic viability of nuclear power, not its safety. California argued that nuclear power is an "unpredictable and uneconomical source of energy," one that is vulnerable to unpredictable costs and potential disruptions, in part because the federal government has failed to provide for safe and permanent disposal of the dangerous radioactive wastes nuclear plants create.

The court's decision may put new pressure on the government to deal with the nuclear waste issue — a political and technological controversy it has consistently put off resolving. But the immediate practical effect is likely to be minimal. There has not been a U.S. order for a new nuclear plant since 1978 and more than 30 orders have been canceled since then.

But what about nuclear plants that are operating or under construction, such as the one Union Electric Co. is building in Callaway County? Justice White, writing for the court, said state efforts to regulate construction or operation would be "clearly impermissible," so perhaps UE is correct in asserting that the opinion has no application here. But it can be read other ways.

By handing the economic issue to states, the court has delivered a blow to nuclear power — and maybe even Callaway — where it is most vulnerable. The ruling sends state legislatures and regulatory agencies a clear signal that they have the power to protect their citizens from the economic burdens and uncertainties inherent in nuclear technology.

Missouri regulators, alas, have been reluctant to do that. Yet the economic questions surrounding nuclear power in general are associated with Callaway in particular. There is no place to store its waste, either. And the limited insurance available is arbitrarily low and clearly inadequate should there be a major accident. So it seems there is an economic basis for the state to act, if it can only marshal enough evidence and find the will to do so.

Portland Press Herald

Portland, Maine, April 22, 1983

A Supreme Court decision granting states power to ban construction of new nuclear power plants carries a stern message to Washington: The federal government must live up to its obligation to establish permanent radioactive waste disposal sites.

That obligation exists regardless of whether or not another nuclear power plant is ever constructed in the United States. The wastes being churned out by the 70-odd existing plants must be permanently—and safely—stored somewhere.

The court's unanimous decision gives states the power to ban construction of any new nuclear plants until the federal government establishes permanent waste disposal sites.

While the decision only affects new plants, the actual impact may be more illusory than real. No new plants have been proposed in more than four years, and it is problematical whether any will be built.

But new plants are one thing, existing plants another. The federal government has been derelict in failing to develop permanent storage sites for long-lived radioactive commercial and military nuclear wastes.

Maine Yankee, for example, is an older plant unaffected by the Supreme Court decision. Yet the plant is rapidly running out of on-site storage space for its spent nuclear fuel rods. Unless given permission to expand on-site storage capacity, it will be forced to close within five years.

From that standpoint, debate as to whether the Supreme Court decision means the death of the commercial nuclear power industry is immaterial. The fact is, permanent storage areas for the tons of wastes being generated by *existing* plants must be developed in any case.

DESERET NEWS

Salt Lake City, Utah,
April 21-22, 1983

The U.S. Supreme Court made it considerably harder Wednesday for Americans to meet right away the growing demand for more electricity by building more nuclear power plants.

But the high court ruled in such a way that the public should be able to feel at least somewhat more comfortable about nuclear power in the long run when more nuclear plants are eventually built.

In a unanimous ruling, the Supreme Court upheld a California law banning the construction of new nuclear power plants until a safe method is found for storing their dangerous radioactive wastes.

The ruling can be expected to encourage other states to adopt laws similar to the one in California. Connecticut, Maine, Massachusetts, and Oregon already have such laws on the books. A number of other states are considering them.

But the same ruling should also add impetus to efforts to find safe ways of storing or otherwise disposing of nuclear wastes, whose radioactivity retains cancer-causing and other dangerous properties for thousands of years. Although the federal government is looking at underground salt mines and other possible repositories for such wastes, including a site in Utah, no storage method has yet been approved. With the new court ruling, by all means let's step up the search for such repositories as long as it does not mean taking dangerous short-cuts with the studies involved.

Sixty-nine nuclear plants are now in operation throughout the U.S., and another 68 are in the early stages of design. Fortunately, the country is in a better position than it used to be to go slow on the new plants.

Until fairly recently, the demand for electricity was increasing faster than the population was growing. But then demand slackened as the recession set in. Besides, as the Supreme Court put it this week, promotion of nuclear power is not to be achieved "at all costs."

Moreover, if the California law had not been upheld, the high court would have cast a shadow over the validity of state regulatory systems throughout the country. About half of the states require that nuclear power plants, like all other electric generating plants, be approved by state regulatory agencies. There's no reason for the federal government to have a monopoly in this field.

As the recession ebbs, the demand for electricity can be expected to increase — along with the need for more nuclear power plants. When those plants are built, they should generate more power than problems. That wasn't always the case until the Supreme Court issued this week's ruling.

THE SAGINAW NEWS

Saginaw, Mich., April 27, 1983

Opponents of nuclear power are trying to stretch a Supreme Court decision to say something the court, by its own notation, did not say. The reasons are obvious.

The court last week upheld a California ban on new nuclear plants. It thus gave states the right to judge proposed new plants on economic grounds; safety was not an issue in the case.

While that clarifies the rights of states, it is meaningless as a practical matter. No new nuclear licenses have been issued since 1979. For good reason, most utilities, including Consumers Power Co., have decided on their own that atomic power isn't worth it. Given the $4 billion example of the Midland twin-reactor project, most stockholders and ratepayers would heartily agree.

Since new plants are no issue, some anti-nuclear activists are trying to make old ones, like Midland, into one.

Hugh Anderson, assistant attorney general, sees the court ruling as ending the federal monopoly on atomic plant regulation. State Rep. H. Lynn Jondahl, D-East Lansing, takes it one step farther — and about two steps too far.

Jondahl is sponsoring a bill to clamp tough new limits on nuclear power in Michigan. He would demand that utilities show there's no cheaper or safer way to provide power before starting a plant — or before continuing or operating one already started.

As we've noted, Consumers itself today regrets its original decision on Midland. But that's based on what's happened since. So is Jondahl's proposal. Both amount to wishful thinking.

The legislation, however, also amounts to second-guessing of a kind that is at least unfair and perhaps illegal.

Any technology can be overtaken by a newer, better one. A plant that can be proved to be crucially needed in 20 years, for instance, may be ready to come on line just as cheaper, cleaner fusion power is made feasible.

Applying new laws to past decisions also smacks of "ex post facto" legislation — saying it was all right at the time, but now we're going to punish you for it. The Constitution speaks unkindly of such laws. That's why the Supreme Court specified that state regulation affecting current construction "would pose a different case."

As Consumers has pointed out, the state does have economic authority over utility decisions. When and if the Midland plant goes on line, there's probably going to be a real donnybrook, in the Public Service Commission and maybe the courts, about how much ratepayers should be charged for it.

We may yearn wistfully for no such charges at all. But what's done is done, however mistakenly. The Supreme Court has given states the right to prevent future mistakes, but not to pretend that old ones never happened.

The Boston Globe

Boston, Mass., April 22, 1983

In ruling that states have the right to ban nuclear power plant construction for economic reasons in the absence of a definitive plan for disposal of wastes from those plants, the US Supreme Court has made the same mistake which Congress has made on the issue. The problem, which can almost certainly be solved, should not have to conform to a fixed timetable.

That nuclear wastes are dangerous and must be handled safely no one denies. That the handling of those wastes has proved more complex than was originally envisioned is equally true. That the answers have been slower in coming than people generally thought is also true, but yields a different lesson than the one gained either by Congress or the court.

It is precisely the complexity of the problem that argues in favor of approaching it slowly. Three main issues dominate the question. First, a technique for containing the longest-lived of the wastes has not yet been selected although there is considerable promise in "vitrification" by suspending it in a glassy, non-soluble material.

Second, study is still being made of the most promising kind of geological structure for burial, if that technique is ultimately adopted. Early assumption that subterranean salt beds were best suited seem to have yielded to granite structures. Europeans and Americans are conducting these studies and may have much to learn from each other.

Third, no decision has been made between "permanent" burial and temporary. Strong arguments have been made for being sure that wastes can be recovered relatively easily for future reprocessing should that become possible or desirable.

By forcing the issue, as Congress did last year in setting a schedule for the waste issue, the court in effect leans against the fullest possible investigation of the range of disposal alternatives.

There is no need for haste. Spent fuel is being kept in "temporary" storage facilities at existing nuclear plants throughout the country pending a decision on its disposal.

With adequate caution, these facilities will serve the purpose until thorough study and the development of waste disposal have been completed. They could even be expanded if the decision takes longer than any of the current estimates.

To some extent the court had a moot question on its hand. It allowed (not required) states to ban new plants pending solution of the disposal question, provided the grounds for the ban were economic uncertainty rather than a determination of the safety issue, a federal prerogative. In practice, no new plants are being ordered and no such new orders are in sight.

While the court said its rule did not apply to existing plants or those under construction, that view has already been disputed by lawyers who assisted in the suit. That argument confirms once again the fact that, whatever else it does, the Supreme Court guarantees that there will be more work for lawyers.

The Miami Herald

Miami, Fla., April 26, 1983

CALIFORNIA scored a ringing victory for federalism and states' rights the other day in the arena of the Supreme Court. The issue was the legality of a 1976 California moratorium prohibiting new nuclear power plants.

The statute blocked new nuclear plants until the state certified that a proven technology exists to dispose of nuclear waste permanently. No such technology exists in the United States. Last year Congress set a timetable directing the Federal Government to select and implement a permanent waste-disposal method.

The Federal Government joined two California utilities in challenging the California statute. They argued that the state's moratorium intruded illegally on established Federal prerogatives to promote and regulate nuclear power.

The Supreme Court ruled unanimously, and meritoriously, that California was squarely within its rights. The legal argument turned on interpretation of the Atomic Energy Act of 1954.

In that act, Congress asserted promotion of nuclear-powered electricity generation as a national priority. It reserved to the Federal Government responsibility for nuclear-plant safety regulations.

The Court agreed unanimously that the 1954 act does not require the states to build or authorize nuclear plants. Nor does the act prohibit states from blocking such plants or attaching conditions to their authorization.

The sole point on which the Justices disagreed is intriguing. California asserted that its 1976 statute was based upon economic concerns. With no assured program of nuclear-waste disposal, nuclear plants could threaten unpredictable future costs, the state argued. Because economic regulation of electric utilities is a traditional state responsibility, the unanimous Court found such reasoning to be consistent with the 1954 law.

Yet the Court's opinion drew a distinction that two Justices rejected, though they concurred with the overall judgment. The Court majority said that Federal law reserves to the Federal Government sole authority to regulate nuclear plants for safety purposes. Therefore, the majority said, if California's moratorium on nuclear plants had been based upon fear that nuclear power is unsafe, it would have been illegal.

Justices Blackmun and Stevens dissented. They contend that Congress made promotion of nuclear power a priority, but did not make it a requirement. Hence, in the dissenters' view, states may reject the nuclear option for any reason they might choose, including safety.

The practical effect is the same. States now clearly hold the responsibility to weigh the long-term costs and benefits before allowing new nuclear power plants. Such decisions are complex, for no method of electricity generation is without substantial costs, both economic and environmental. Yet the Court was right to leave the burden of such decisions on those who must live with the consequences — the states.

The Oregonian

Portland, Ore., April 23, 1983

The U.S. Supreme Court, in upholding a California law that imposes a moratorium on building new nuclear power plants until solutions to the radioactive waste disposal problem satisfy state officials, has turned up the heat on the federal government, forcing it to move faster with its long-delayed disposal plans. That is a definite plus. But the court also has greatly damaged nuclear power development.

In unanimously rejecting claims that only the federal government may regulate nuclear power, the court has thrown a massive roadblock against nuclear plant expansions. The decision will greatly lengthen the construction lead time, and thus the costs of reactors, by allowing expensive, multistop licensing permits. The decision may permanently cripple the nuclear industry in the United States, forcing it to look for more foreign markets or totally lose out to foreign competition.

The court's decision did not pre-empt federal authority to regulate reactor safety, but the opinion, written by Associate Justice Byron R. White, said Congress in 1959 had allowed state and local agencies to regulate nuclear plants "for purposes other than protection against radiation hazards." White wrote that "Congress has allowed the states to determine — as a matter of economics — whether a nuclear plant vis-a-vis a fossil fuel plant should be built."

While disposal of radioactive materials would appear largely to be a safety issue, except for disposal costs paid by utilities, the court chose to make the California moratorium an economic issue.

The court affirmed a state's right to take actions that would protect the rates paid by consumers, a power long exercised through public utility commissions.

While voters in both California and Oregon were rejecting outright bans of nuclear power plants, perceiving that in normal times they offered ratepayers the cheapest thermal way to produce power, the voters were rightly concerned about the government's delays in finding a safe way to dispose of reactor waste, by far the greatest amount of which comes from the bomb program.

The administration hopes to have disposal sites selected by the end of the year, two having been nominated in Washington (near Hanford) and Nevada on federal reservations and a third yet to be selected. It is not clear what point in the construction of these sites must be reached to cause a moratorium to be lifted.

At the present time, the high cost of building a nuclear plant, coupled with a surplus of electrical energy, has stopped construction in the United States.

Nevertheless, nuclear power work is booming in many parts of the world, including Canada, where plans are afoot to build nuclear plants solely dedicated to the export of power to the United States to take advantage of soaring American costs due to regulations.

Nuclear power has to be a vital part of the nation's future as its fossil supplies wind down. Local governments that exact moratoriums, ban shipments of radioactive wastes and otherwise impede nuclear development are doing little to promote nuclear safety and a great deal to increase the future costs of electricity.

Post-Tribune

Guarding Your Interests Daily

Gary, Ind., April 27, 1983

Our opinions

Having to pay for much of NIPSCO's losses in the Bailly nuclear plant fiasco is not a pleasant legacy for its customers. But at least the question of to build or not to build is settled. If that question were still open, it's hard to imagine the confusion a recent Supreme Court decision would have caused here.

The court's 9-0 decision upheld a California moratorium on nuclear plants until the federal government proves it has a way to dispose of nuclear waste. Now, we know that bizarre legal and political developments bloom continuously in California, but this case has a magnitude that reaches beyond the state's borders.

For one thing, it's a door-opener to suits and to delays. Northwest Indiana won't be touched directly, but southern Indiana will. Already, a group plans to protest licensing of the Marble Hill plant at Madison, on the Ohio River, when that time finally comes. The Public Service Company of Indiana, the builder, says the ruling won't affect Marble Hill, but who knows?

The Indiana plant is one of 57 being built across the country. It takes about 10 years, if things go smoothly, to get a plant on line and operating — although in the Bailly case, the end wasn't even in sight after a decade.

This new ruling touches a basic issue in nuclear plant operation, and it should force the government to get serious about that deadly problem. It has the public frightened, logically or not, and it has to be dealt with. The government has not done that with urgency or clarity. The system approved by Congress might be in place by 1990, not soon enough to deal with the uneasiness.

It's not fair to blame all these delays on environmentalists or "obstructionists." The California law was passed in 1976. Seven years ago. A ruling comes in 1983.

The country just does not have its nuclear act together. The prescience of California's lawmakers is commendable.

AKRON BEACON JOURNAL

Akron, Ohio, April 23, 1983

THE SUPREME Court's surprising 9-0 concurrence that states have the power to forbid construction of new nuclear power plants adds one more woe to those already besetting that industry.

But at base this one is no different from the rest: However convoluted the contributing factors, the decisive question is economic. It is this that has caused orders for new nuclear plants to drop from highs of 30-odd a year in the early 1970s to zero from 1978, and cancellations of orders to climb sharply from 1973 on.

What seemed like an almost unbelievable bargain in the halcyon early days of nuclear power development — power almost as cheap as air — has become so expensive that the old ardor for it has cooled. It is boardroom considerations of corporate profit and loss, not sign-toting protesters, that have slowed the nuclear race to a crawl.

Decisive factor

And here again, although two of the justices called the basis too narrow, the deciding factor in the minds of seven justices was economic.

Current law reserves to the federal level all safety-related decisions on such plants, they held, but the states retain their normal power to decide whether proposed plants are economically sound and thus in the interest of power consumers.

In this case two power companies had challenged the propriety of a 1976 California law forbidding the construction of new nuclear plants until the state's Energy Resources Commission is convinced that a demonstrated technology exists for the permanent disposal of nuclear waste. (Eight other states have similar moratorium laws.)

The stated reason for the California law was not safety concerns but economic ones. Absent an assured way to get rid of their wastes, the lawmakers argued, plants face the risk of having to shut down when their temporary waste-holding facilities filled up; this could turn them into costly and useless mistakes for which consumers would have to go on paying.

A district court ruled for the companies; an appeals court reversed the finding. The Supreme Court upheld the appeals ruling.

Justices Harry Blackmun and John Paul Stevens agreed that the California law is proper, but disagreed that a state's reason for enacting such a law must be economic. The Congress, they said, "simply made the nuclear option available, and a state may decline that option for any reason" — including safety concerns.

Whatever the effect of the decision on future nuclear development, it underlines the urgency of getting a satisfactory technology for safe disposal of the wastes into place. Unless this is done, the plants already operating and those now nearing completion will face precisely the problem cited in passage of the California law.

The Congress, after a quarter-century of inconclusive debate over the problem, has finally set up the legal mechanism for such a technology.

But ahead still lies a long and tortuous path potholed with technical and political problems before we can have burial places to protect humanity from the radiation threat those tons of waste will present for thousands of years.

The obstacles must be overcome, with all deliberate speed — not so much to clear the way for further development that may or may not materialize, but to accommodate the waiting waste from what we have already done and the tons more accumulating as existing and soon-to-be-operating plants continue to operate.

Could block completion

As to the possible effect of the court ruling on plants already under construction, legal opinions differ. The language of the majority opinion appears to leave open the possibility that states could block their completion if the reasoning were economic.

This seems unlikely to affect the fate of the three reactors being built in Ohio. One of the two at Perry is 85 percent complete, the other about 50 percent complete. The one near Cincinnati, 97 percent complete, is at a presumably temporary standstill pending solution of some problems.

In none of the three cases does it seem likely that consumers would ultimately lose more by letting the power companies complete the plants and bring them into service than by ordering construction halted.

The State

Columbia, S.C., April 24, 1983

IT HAS been almost 40 years since the atomic bomb ushered in the era of nuclear energy, and the nation is still sorting out the ramifications of the resulting technology.

Among the societal adjustments to this strange and powerful source of energy are laws and regulations to control its development by the private sector, and to protect the public health from its potential dangers.

At the outset, the federal government became responsible in the main for the public's safety, and that is proper. Licensing authority must be centralized so that comprehensive safety standards will be consistently required in construction and operation of nuclear reactor plants. It is unthinkable that safety standards would be allowed to vary from state to state.

Over the years, the states have been given some say-so over nuclear plants, but on Tuesday, the Supreme Court of the United States handed down a decision which may be a watershed for the nation's nuclear industry. The states may ban construction of new nuclear power plants until the federal government devises a safe means of disposing of high-level radioactive wastes from existing reactors.

The high court, in effect, confirmed the right of nine states to place moratoriums on new power reactor construction. California, Connecticut, Maine, Massachusetts, Montana, Oregon, Iowa, New York and Wisconsin acted to prevent new reactors from being built.

In a sense, the ruling is a rebuke to the federal government, which, in these 44-odd years, has failed to provide a place for the disposal of the tons of nuclear fuel which have been burned in the nation's 80 utility-owned reactors. The spent-fuel assemblies have been piling up in large water tanks at the reactor sites, pending a federal government decision on where to put them eventually.

Congress, which has been the bottleneck in establishing a national repository for spent fuels, only last year agreed to a program to provide for disposal of these highly radioactive wastes by the 1990s. Sites are now being studied in four states for underground storage of the accumulated waste. South Carolina is not among the candidates.

The basis of the moratoriums was that nuclear plant shutdowns may be necessary some day because of the large accumulation of spent fuel; since the federal government has provided no place to store it permanently, the power plants would have to shut down to keep from generating any more radioactive wastes. That would also mean shutting down important electrical generating stations, and resulting serious economic problems.

The Supreme Court decision may be a bit late to have much impact since no new licenses for power reactors have been applied for in the past five years. There is also a dispute as to whether the ruling applies to the 57 reactors now under construction nationwide.

But the fact is, the nation's high court has expanded the states' rights in nuclear development decisions, possibly opening the door to even wider state participation in the future.

THE WALL STREET JOURNAL.

New York, N.Y., May 2, 1983

Both sides of the nuclear power debate have frequently been inattentive to economics. Pro-nuke factions have exhibited an enthusiasm that led them into power projects like the economic disaster in Washington state and technologies—like reprocessing and the breeder reactor—that are unlikely to prove economic any time soon. Anti-nuke factions have tended to demagogue the safety issue, trying to persuade the public any collection of nuclear material could easily turn into a mushroom cloud. Somehow it is little surprise to us that in retrospect the main lessons of the Three-Mile Island accident were that the public safety was not really endangered, but that there was an often unrecognized insurance risk in an industry in which you can lose $1 billion in half an hour.

We were reminded of all this recently when the Supreme Court upheld a California law that blocks nuclear power plant construction in the state until the U.S. "has approved a demonstrated technology or means for the permanent and terminal disposal of high-level wastes." On its face, this was an economic decision. The court said federal law did not preempt state regulation because states have traditionally had the right to make judgments about the *economic* feasibility of utility projects. And since California had decided that uncertainties in the nuclear fuel cycle make nuclear power an uneconomical and uncertain source of energy, it was entitled to put the chocks to further nuclear power development.

There is of course a certain sense to this. The full costs of nuclear power have to include waste disposal, just as the full costs of coal-fired power ought to include management of air pollution. And the court's decision could have some positive spinoff if it should happen to increase the pressure on federal and state authorities to make a decision on a nuclear waste disposal site.

Yet in another way, if the court thought the California law was merely a matter of economics, it was simply naive. Justice White, who wrote the opinion, failed to find any evidence that the state of California was trying to poach on the regulatory territory of the federal Nuclear Regulatory Commission, which has pre-empted the safety question. Surely, the state legislators who passed the bill must have had some inkling of the big fight that has been raging for years now over nuclear plant safety.

Back in the mid-'70's, we recall a visit from Barry Commoner, one of the leaders of the assault on nuclear power. He more or less boasted that his movement would succeed through harassment tactics that would delay nuclear plants, escalate their costs and make them uneconomical to build.

The anti-nuke lobby has had marvelous success with this strategy. Readers may recall the costly legal and regulatory battle Public Service Co. of New Hampshire and associated utilities have had to fight over the last decade to try to complete a large nuclear project at Seabrook, N.H. The fight against the project has had plenty of absurdities, but the worst was in late 1976, just after the election of Jimmy Carter, when a regional federal regulator in Boston blocked construction because of allegations that the plant, with its public investment at that time estimated at $2 billion, might damage some clam larvae.

The combination of long regulatory delays, costly and time-consuming litigation, pared-down projections of electric power demand and, of course, the disposal problem has indeed escalated the cost of nuclear power to utilities and their customers. But then the cost of coal-fired plants has risen too as a result of demand for expensive pollution control equipment. Pacific Gas & Electric, the plaintiff in the California case, still could see merit in contesting the California moratorium.

Now, however, the Supreme Court, probably without recognizing what it was doing, has handed the Commoner strategy a stunning new weapon. For years now, there has been technology for glassifying nuclear wastes so that they can be safely stored, preferably in some out-of-the-way place like a salt dome or cave. But the anti-nuke crowd has made potent politics against every attempt to legislate such a program, using scare tactics of the most outrageous sort to persuade the public that a bomb was being planted in its midst. Late last year, Congress finally passed a nuclear waste disposal act, setting a timetable for establishing a permanent underground repository. But Sen. Proxmire gave a veto right to any state selected as a site, subject to a congressional override.

So the anti-nuke forces will now turn their attention to preventing the timetable from being met, and we wouldn't want to underestimate their power. The court's decision is not likely to be a move toward better economics and more rationality in the nuclear debate. Rather, it will provide another economic foil for the notion that use of the atom to heat homes and run factories represents some horrendous crime against nature.

The Des Moines Register

Des Moines, Iowa, May 13, 1983

Two recent U.S. Supreme Court decisions confirm divided state and federal control over nuclear power. Several weeks ago, the court said that California can forbid the operation of nuclear plants until suitable facilities for permanent disposal of waste are available. Last week it ruled that Illinois cannot prevent interstate shipment of nuclear waste to a storage site within the state.

The latter was interpreted as a regulation of safety, which federal law reserves to the Nuclear Regulatory Commission. The former, although it also looks like a safety question, was held by the court to be a matter of economics, and states are permitted to regulate economic aspects of nuclear power.

The intent apparently was to allow utility commissions to decide the kinds of questions they have long decided about electricity supply: Is it adequate? Is it reasonably priced? etc. A lack of suitable disposal facilities could adversely affect electricity supply or price, although the California law seemingly was intended at least partly to curb nuclear plants on safety grounds.

The Illinois law, passed in 1980, prohibited out-of-state shipments of nuclear waste to a storage site at Morris, Ill., so the Nebraska Public Power District was unable to ship wastes from its Brownville nuclear station across Iowa by rail to the closest disposal site, Morris.

What it comes down to is that hardly anyone wants a nuclear-waste dump in the back yard, and those states that have them are not eager to take waste from other states. When this is argued on the basis of safety, though, the reluctant state loses, because the law says safety regulation is up to the NRC.

It should be. The federal government, more than any state, has the best technological know-how in this complicated area. It is better to have uniform safety standards. And when the question is one of interstate shipment of wastes (or interstate anything else), the Constitution gives control to the federal government.

Adequate waste-disposal facilities will not be found if every state, every community, is able to say, "Not here — take it somewhere else."

Part III: Nuclear Proliferation

Fred C. Ikle, then Director of the U.S. Arms Control and Disarmament Agency, warned in 1975 that "the spread of nuclear weapons capability is riding on the wave of peaceful uses of the atom." The transfer of peaceful nuclear technology, he said, was providing "not only the means, but also the cover" for a spread of nuclear weapons. This fear—that by enhancing other nations' technical capabilities through the sale of nuclear reactors or equipment, the United States might accelerate the global acquisition of nuclear weapons—is raised by many who oppose the domestic use of nuclear energy. It is beyond the scope of this book to explore the history of the non-proliferation stance of the U.S.—or how it relates to foreign policy decisions such as President Carter's sale of uranium to India or President Reagan's sale of heavy water to Argentina. Much of the general concern, however, is centered rather on two interrelated domestic energy issues: the development of breeder reactors and the reprocessing of spent nuclear fuel.

The concern arises because the fuel for American nuclear power plants, a mixture of two types of uranium, is partially converted into plutonium during the fission process. Some of the plutonium thus produced is burned by the reactor, and the remainder is contained in the spent fuel rods. This remaining plutonium can then be extracted through chemical reprocessing and used again either as fuel for the reactor or as a key ingredient in atomic weapons. When the United States' commercial energy program was initiated, it was assumed that the spent fuel from power plants would be reprocessed for use in breeder reactors. The "breeder" was so dubbed because it was designed not only to use plutonium as fuel but to produce more fuel than it burned. The U.S. prototype for commercial breeders, however, Tennessee's Clinch River reactor, has run into considerable political opposition. It now appears unlikely that it or others like it will operate for many years. The major commercial market for plutonium thus does not exist as envisioned. Although the Reagan Administration lifted the ban on commercial fuel reprocessing imposed by President Carter, there are as yet no domestic reprocessing plants in operation, in part because they lack customers. The only other major market for plutonium, the Government, maintains separate plutonium production reactors for use by the military. (When the goal is solely plutonium production rather than energy generation, these smaller reactors are a much more efficient source.)

It can be argued that breeder reactors and fuel reprocessing plants are domestic energy issues unrelated to the international proliferation issue. But because reprocessing is a key part of the breeder reactor cycle, and can be used to obtain weapons materials, many Americans feel that this technology should not be shared with those nations currently without nuclear arsenals. Smaller or undeveloped nations, particularly those without natural uranium supplies, are desirous of the breeder technology because it would provide them with the means to become energy-independent. Unfortunately, it is many of these same nations that are known or suspected to be trying to develop their own nuclear weapons. Even traditional nuclear reactors, of course, can be used to produce plutonium for weapons, although the process is much more inefficient and expensive than with the breeder reactors. Thus the Reagan Administration determination to become a "reliable" supplier of nuclear technology, fuel and equipment to other nations has heightened fears that the U.S. may be aiding the spread of atomic weapons despite international safeguards intended to prevent the use of nuclear reactors for military purposes. Other nations now also have the capability to export nuclear technology, however; it may be, as many claim, an empty precaution for the U.S. to forswear the domestic energy uses of nuclear technology in an attempt to prevent their misuse abroad.

Clinch River Breeder Reactor Project Appears Moribund

In the nuclear power plants now in operation in the United States, called light water reactors, water is used to cool the reactor core and to slow down the chain reaction of ricocheting neutrons which produces the heat that is eventually converted into energy. In the liquid metal fast breeder reactor, still in the developmental stage, the chain reaction is instead allowed to proceed much more quickly, and stray neutrons from the reaction are absorbed by a layer of nonfissionable uranium-238. During this process, the uranium-238, which is plentiful in nature but of little value, is converted into plutonium-239, which can then be removed and used as new fuel. The plutonium thus produced is also, however, a key ingredient in atomic bombs.

In 1972, Congress authorized funding for a project to demonstrate breeder technology on a large scale, to be built on 1,355 acres near Oak Ridge, Tennessee. The Clinch River plant, designed to produce more fuel than it consumed, was envisioned as an inexhaustible source of electrical power at a time when dwindling supplies of oil from the Middle East threatened the U.S. economy. The project was opposed from the beginning by antinuclear and environmental groups, in part because of the seemingly more catastrophic results of a potential core meltdown at a breeder reactor, and in part because of the enhanced opportunity it might present to terrorists who sought plutonium to manufacture nuclear weapons. President Carter also opposed the Clinch River project, calling repeatedly for cancellation of the breeder reactor because of the threat he felt it posed to an effective non-proliferation policy.

Congress remained on the side of the breeder until very recently. But in October 1983, only a year after workers began to clear the site on the Tennessee River, the Senate virtually killed the project by defeating an amendment to fund the completion of the Clinch River plant. Opposition in Congress had grown in the intervening years as the price of fuel for light water reactors dropped and the estimates of future electrical needs proved bloated. In 1982, the project had survived by only a one-vote margin in the Senate. Although a breeder reactor is less expensive to operate than the commercial reactors now in use, it is more expensive to build. Congress already had spent $1.5 billion on the reactor, and its final cost was estimated at $4 billion. Fiscal conservatives joined environmentalists in an unusual coalition to defeat the project's funding. Their concern over the federal deficit was part of the impetus for the project's defeat. Critics also contend that either nuclear fuel prices must skyrocket or the cost of constructing a breeder reactor must plummet before breeders become an economical form of electrical generation.

TULSA WORLD

Tulsa, Okla., March 12, 1981

PRESIDENT Reagan has decided to reverse Jimmy Carter's bad decision to stop work on the U. S. demonstration breeder reactor at Clinch River, Tenn., including $1.4 billion for the project in his 1982 budget.

The action will certainly renew the nuclear power controversy. But the breeder reactor — in the plans of U. S. atomic energy officials for 20 years or more — offers tremendous advantages while perhaps improving on some of operations usually opposed by anti-nuclear groups.

The breeder, so-called because it actually generates more plutonium than it consumes while simultaneously generating electricity, would, if put into widespread use, simply end the need for further uranium mining, thus eliminating much of the dangerous part of the nuclear fuel chain.

Additionally, the U. S. has in stockpile enough "tailings" from uranium operations to date to supply electricity for the U. S. for 200 years if used in breeder reactors.

The stockpile potentially contains five times as much energy as all the oil reserves in the Middle East.

Additionally, President Carter's halt of the breeder program for fear it would spur nuclear proliferation has proven to be meaningless. Other nations, including the Soviet Union, Great Britain, France, West Germany and Japan are pushing ahead full speed with breeder programs.

To keep the U. S. out of the breeder program would simply penalize the nation without having any real effect on the spread of nuclear fuel that can be used in the manufacture of weapons.

Reagan's decision, although certain to be opposed, is a sound one. We ought to get on with the breeder program.

The Houston Post

Houston, Texas, February 4, 1981

The fast-breeder nuclear reactor, kept in a state of suspended animation during the four years of the Carter administration, can apparently look forward to a new lease on life. President Reagan favors expansion of nuclear power with adequate safeguards. The Republican platform calls for more research funds for the breeder reactor. And at his confirmation hearing, Reagan's new energy secretary, James B. Edwards, reiterated his strong support for both the breeder and reprocessing of spent nuclear reactor fuel.

Revival of reprocessing and the breeder from their comatose state would mean a major shift from the nuclear non-proliferation policy pursued by President Carter. The purpose of this policy was to prevent the spread of nuclear material that could be made into weapons. Among the advantages of the breeder reactor is that, in addition to uranium, it uses plutonium as a fuel. By reprocessing spent uranium from conventional nuclear plants, plutonium can be extracted for use in breeders. And the breeder produces more fuel than it consumes. It could thus stretch the world's supply of uranium and curb the generation of nuclear waste. The problem of disposing of radioactive waste safely looms as one of the major barriers to expansion of commercial nuclear power.

But the plutonium produced by reprocessing and the breeder can also be converted to nuclear weapons. To thwart the spread of weapons-grade atomic material around the world, President Carter sought to discourage further development of reprocessing technology and used his first veto after entering office in 1977 against funds for the breeder reactor project at Clinch River, Tenn. Congressional supporters of the Clinch River breeder managed, however, to push through enough funding to keep it alive.

The Carter administration hoped that its policy would set an example the rest of the world would follow. But the rest of the world did not follow. Britain, France, the Soviet Union and other countries pursued development of the breeder reactor. And a two-year international study, initiated ironically by President Carter, concluded last year that breeder deployment and fuel reprocessing were probably inevitable. The administration was also inconsistent in following its policy. Last year, for instance, it supported the sale of uranium to India, which detonated a nuclear bomb in 1974 using American nuclear materials.

The International Atomic Energy Agency inspects many nuclear facilities around the world to insure that they are not being used for military purposes. But the IAEA does not have access to all such facilities, raising suspicions that there are dangerous leaks in the present system of nuclear safeguards. Urgent international efforts should be made to strengthen that system. At the same time, however, we must face the fact that the United States alone cannot prevent the spread of nuclear material and technology. Nor can it effectively discourage the reprocessing of spent reactor fuel or the building of breeder reactors. By adopting a policy of technological self-denial, we have handicapped ourselves. And we are in danger of falling behind in research and development on the breeder, which many experts see as the next generation of nuclear reactor. That policy must be changed if we are to avail ourselves of all the energy-production options open to us.

The Oregonian

Portland, Ore., January 14, 1981

In its final days, the Carter administration has seen the light. Departing Energy Secretary Charles W. Duncan Jr. announced, literally over his shoulder, "I think there is no time to lose in developing the breeder."

President Carter had opposed development of the nuclear power breeder reactor, particularly the Clinch River project in Tennessee, fearing it would hasten the spread of nuclear weapons. But this did not prevent France from moving ahead in an effort to capture this world market.

Duncan called for the nation to skip over the Clinch River prototype project and go directly to a commercial-size breeder like France's Super-Phoenix project. The Reagan team has sounded positive about the breeder option, a vital link in achieving energy independence by the end of the century. But it must act promptly if the U.S. nuclear energy program is to be rescued from the stagnation triggered by Three Mile Island.

The Detroit News

Detroit, Mich., March 23, 1981

The anti-nuclear people are aggrieved because President Reagan's proposed budget includes funding for the Clinch River, Tenn., breeder-reactor project.

We believe, with the Reagan administration, that Clinch River is an essential energy project.

Breeder reactors generate fuel as they work, and they deliver 60 times more energy from a pound of uranium than conventional hot-water reactors. Further, if breeder reactors were in use, their process would utilize mine tailings and other wastes. Thus, they could deliver the electric power required for the next 100 years without further mining.

Because breeder reactors utilize bomb-grade plutonium, they are opposed by those who fear a catastrophic nuclear accident. This is a recurring theme among romantics from the time of Mary Shelley: Man will destroy himself with the instruments of progress. Although understandable, this is not the kind of anxiety that should determine national policy. New technology always involves risks, but a country capable of developing such esoteric devices certainly has the wit to make them acceptably secure. And, indeed, extraordinary measures are taken to make the generation of nuclear power safe.

Does that mean 100 percent safe? No, that's not possible. And people who insist on an absolute guarantee are living on the wrong planet.

The sad truth is that other nations are striving much harder than the United States to achieve commercial breeder technology — as a defense against enormous oil bills that threaten to wreck their economies and expose their governments to blackmail.

France, Great Britain, and the Soviet Union now lead in the field, and Japan and West Germany are rapidly catching up.

There is no denying the high capital costs of breeder reactors. But their cost advantage in fuel consumption grows daily as the price of other fuels rise. More important, the breeder reactor would insure energy supplies.

The breeder will very likely become commercially viable in the not-too-distant future. Meanwhile, Clinch River scientists are striving to perfect the technology and refine its applications.

If they succeed, America will have plenty of electricity generated from domestic uranium. How can money spent to achieve such an objective be considered misspent?

THE SUN

Baltimore, Md., May 6, 1981

The breeder reactor, a nuclear power source that makes more fuel than it uses, was opposed by the Carter administration but may become a centerpiece of Reagan administration energy policy. The president wants increased funds for construction of a demonstration breeder reactor at Clinch River, Tenn. There is some opposition in the new Congress, but the chances seem reasonably good that it will support the administration's request.

Breeders are feared by many, not only because they are fueled by the radioactive element plutonium, but because they also *produce* plutonium (about 1.4 times more than they use). Nuclear bombs can be made from plutonium, and opponents say the breeder poses the threat of a "plutonium economy" in which shipments of this substance would be common and hijackings and conversion to bombs possible. Advocates retort that handling plutonium and producing bombs from it are so complex and difficult as to make these activities virtually impossible for terrorists.

The attraction of the breeder is its potential for stretching increasingly short supplies of uranium. A now nearly useless isotope, uranium-238, is abundant and can be converted in a breeder reactor to plutonium, which *is* useful in power plants. Waste uranium-238 now stored in Tennessee may hold as much as five times the energy of all the oil in the Middle East.

Opponents raise objections in addition to their fear of plutonium diversion—for instance, that the Clinch River breeder will be cooled by liquid sodium, a chemically highly reactive, and thus dangerous, substance. They add that the breeder is more expensive than other approaches, and represents an unjustified federal subsidy to electric utilities. Advocates reply that breeder safety questions will be thoroughly studied before actual operation—and that French calculations prove that over the lifetime of a breeder it will be cheaper than other techniques.

We believe all reasonable—and reasonably safe—options for energy production ought to be kept open. The Clinch River breeder qualifies and thus should be built. We wonder, though, about the Reagan administration decision to support a costly subsidy to the electric utility industry while making Draconian cuts in the budgets of other, equally promising approaches such as solar energy and conservation. Giving one technique a far more generous federal subsidy than another is not exactly letting the marketplace decide.

St. Louis Globe-Democrat

St. Louis, Mo., February 10, 1981

Sooner or later the American people are going to wake up to the fact that the Carter administration did them a terrible disservice by stopping development of the nuclear fast breeder reactor technology.

Why?

Because the fast breeder reactor offers a way of using nuclear fuel that is more than 100 times as efficient as present-day light water reactors.

Today's commercial light water reactors use U235 for fuel. But there are only seven pounds of U235 in every 1000 pounds of uranium ore. The balance is U238 and is of relatively little use in today's reactors.

The net result is that in today's nuclear fuel process there are 993 pounds of unburned U238 for every 1000 pounds of uranium fuel used in the reactor. This huge amount of U238 winds up in canisters as "tailings" from this process.

Breeder reactors can burn virtually all of the U238 fuel with very little waste, an incredible saving in uranium usage. If the United States had gone ahead and developed commercial fast breeder reactors as other industrial countries have, the nation's uranium supply would last for many centuries due to the ability of the breeder process to use practically all of the uranium fuel rather than only a small fraction.

Some idea of the potential of the breeder reactor can be gained by the fact that a pellet of breeder reactor fuel only about half the size of an eraser on a pencil can produce a year's supply of electricity for two people. This same pellet also can produce about the same amount of energy as 3 tons of coal, 12 barrels of oil, 500 gallons of gasoline or 75,000 cubic feet of natural gas.

If the United States should continue to stifle development of the breeder reactor — as it very probably won't now that Ronald Reagan is president — the inefficient use of uranium could result in virtually all of the available U.S. uranium reserves being mined and turned into "tailings" at nuclear storage centers.

Patently the United States should move ahead as rapidly as possible to develop the fast breeder reactor demonstration plant at Clinch River, Tenn., that was stalled by former President Carter. This new technology is badly needed to greatly increase the efficient use of the nation's uranium nuclear fuel.

The Des Moines Register

Des Moines, Iowa, May 18, 1981

Despite President Reagan's talk about eliminating wasteful government spending, his 1982 budget contained millions for one of the more wasteful and risky projects proposed in recent years: the Clinch River nuclear-breeder reactor.

The House Science and Technology Committee last week bucked the president and voted to kill the project. Representative Tom Harkin (Dem., Ia.) played a key role in the committee's action.

The breeder reactor was designed to demonstrate a form of nuclear technology that might produce virtually unlimited energy.

But the world energy situation has changed since then, and demand for electricity is slowing. Meanwhile, the estimated cost of the breeder has soared to well over $2 billion. Moreover, because the breeder produces plutonium, widespread use of this technology could encourage the spread of nuclear weapons.

Reagan's budget director, David Stockman, used these same arguments to help lead the fight against the breeder when he was a member of Congress. During a 1979 debate on the Clinch River program, Stockman told the House: "... despite the severity of the energy problem, we are not going to solve it by throwing globs of money at any project or any idea that comes down the pike."

But when the Reagan administration drew up its 1982 budget, Stockman's logic lost out to the administration's pro-nuc'ear tilt and to a desire to please Senate Majority Leader Howard Baker Jr. (Rep., Tenn.), in whose state the breeder would be located.

So Reagan's budget called for $254 million for the breeder, while it severely cut more promising energy programs such as conservation and solar energy.

The full Congress should support the committee's decision. At a time when Congress has cut back many federal programs, putting millions into the Clinch River project is hard to justify.

Democrat Chronicle

Rochester, N.Y., May 11, 1981

WHY, oh why, does the Clinch River nuclear breeder reactor keep popping up for a vote every year in Congress? The project seems to have more lives than a cat.

We hoped that it would have been put out of its misery by the Carter administration, which opposed it, but the Reagan team has given it new life. This despite the fact that the project violates the president's professed desire to let private enterprise shoulder more of the load in developing new energy sources.

The House Science and Technology Committee, in what is expected to be an extremely close vote, is considering spending $254 million to begin the breeder's long-delayed construction.

A breeder reactor of this type uses nuclear fuel that has been reprocessed from the spent fuel of conventional power reactors like RG&E's Ginna plant. The reprocessing process, however, also separates plutonium, which is usable in manufacturing atomic weapons.

The breeder also creates (breeds) more fuel as it generates power. That's a wonderful way to extend the world's supply of nuclear fuel, but it's also a way of expanding the plutonium supply.

The president hasn't yet articulated his strategy for curbing the spread of nuclear weapons to nations not now possessing them, but a "go" vote on Clinch River will make formulating such a policy more difficult.

And there are too many questions about the long-delayed Clinch River project to justify funding it: Is it by now obsolete? Are there better alternatives? Is it safe? The liquid sodium it must use as a coolant, for example, would react violently with water, should it escape the sealed coolant system.

THERE ARE other alternatives — even other breeders, such as a thorium-cycle breeder that does not produce plutonium. They should be studied carefully.

In 1978, the Rand think tank recommended that breeder research be broadened from the one which focuses only on the Clinch River type. The study said "the technical and economic superiority of the Liquid Metal Fast Breeder Reactor over other alternatives has not yet been demonstrated." There seems no reason to suspect these arguments are not still true.

Somehow, Clinch River has become a symbol of support for nuclear power in general, of new technology, of keeping up with Western European countries that are continuing their work with breeders. It doesn't merit that symbolism. And its supporters mustn't be dazzled — by the glitter of a system that promises to produce more fuel than it uses — into overlooking its faults.

We hope the thin thread of common sense prevails in Congress, and that Clinch River is put back in its coffin — permanently.

The Philadelphia Inquirer

Philadelphia, Pa., May 16, 1981

Using the Reagan administration's budget-cutting oratory as ammunition, opponents of the controversial Clinch River Breeder Reactor in Tennessee recently scored a major victory in a House committee, which voted to kill a $254 million appropriation for the project. It was a fitting tactic; it would be difficult to find a more wasteful and excessive project.

The experimental breeder reactor has been on the drawing boards since 1970, and is now estimated to cost $3 billion. Supporters argue that breeder reactors are necessary to provide a source of inexpensive fuel for the nation's commercial nuclear plants. Breeder reactors process plutonium in such a way that more fuel is produced, or "bred," than consumed, thereby generating an unlimited supply of fuel and energy.

Critics of the Clinch River project argue on a number of fronts, including serious safety questions which never have been addressed adequately. The excess plutonium generated in the breeder reactor could be used for nuclear weapons. Experts also claim the plant's design is outmoded, its technology far surpassed by that developed abroad. Also, they argue that the reactor would be nothing more than a massive federal subsidy to the nuclear industry, which claims it needs the reactor but has thus far refused to ante up any money.

The argument that carried the most weight with the House Science and Technology Committee, which voted 22 to 18 to reject the appropriation, was an adroit use of the administration's line on other budget matters.

Opponents used the words of the administration's budget-cutter, David Stockman, to win over undecided votes. In 1977, Mr. Stockman, then a member of Congress from Michigan, labeled the Clinch River project "totally incompatible with our free market approach to energy policy." The administration now, however, favors funding for Clinch River.

In addition to building the breeder reactor, the federal government would be required to subsidize its operation into the future, and that could easily cost billions of dollars over the next decade and a half.

With the economic evidence against proceeding with Clinch River, the administration's revival of the project can be viewed as nothing more than a political payoff to Senate Majority Leader Howard Baker of Tennessee. "This is the first vote but it certainly is not the final word," said Mr. Baker, after the appropriation was voted down.

The funding of Clinch River has evolved into a pro-nuclear, anti-nuclear dispute. That's a mistake, and Rep. Judd Gregg (R., N.H.) properly took note of the misconception during debate. "This is not a referendum on nuclear power. It's a referendum on economics," he told his colleagues.

In such a referendum, there can be only one vote on the Clinch River breeder reactor: a resounding no.

Chicago Sun-Times

Chicago, Ill., May 9, 1981

Thursday's vote doesn't cinch the matter, but the House Science and Technology Committee made a valuable statement with its 22-18 vote to kill the controversial and expensive Clinch River breeder reactor in Tennessee.

Opposing the plant were both Democrats who argue against the reactor on the ground that it represents a step toward wider plutonium proliferation and Republicans who see the $3 billion project as too costly an extravagance in these times of tight budgets.

The reactor would "breed" plutonium from uranium found in spent nuclear-plant fuel. And even *if* that is a good idea now—which has yet to be proved—there are good arguments that delays have already made the Clinch River plant's design obsolete.

President Jimmy Carter had tried to stop the Clinch River plant since 1977; his position was that fuel reprocessing was a necessary goal, but the Clinch plant wasn't the best way to reach it. In intervening years, funding has continued only at minimal levels, without significant progress toward plant completion.

The project's backers argue that ending the Clinch plan will cost hundreds of millions. True. But continuing it would still cost ten times the termination price—besides losing the chance to consider newer, better ways of reprocessing fuel.

The time now seems ripe for exploration of such alternatives.

SAN JOSE NEWS

San Jose, Calif., May 19, 1981

FOR nearly a decade, the Clinch River Breeder Reactor in Tennessee has been a kind of nuclear litmus test for politicians. If a congressman or senator supported the breeder, that meant he was pro-nuclear; if he voted against the breeder, he was anti.

In fact, though, there are all sorts of good reasons for even the most avid nuke-lover to oppose the Clinch River project. Which explains why pro-nuclear conservatives joined forces with anti-nuclear liberals to hand the project a critical House committee defeat May 7, and why we're glad they did.

The Clinch River project was begun in 1970; it has cost the taxpayers about $1.5 billion to date and would cost another $1.5 billion or so to finish. Spending that kind of money on this project at this time simply isn't a good investment.

Clinch River is known as a liquid metal fast breeder reactor — liquid metal because it uses liquid metal instead of water, as in conventional reactors, to transfer heat from the reactor to a steam generator; fast breeder because neutrons from its core irradiate a blanket of surrounding material and actually produce more fissionable fuel than the reactor consumes while it's running.

Nuclear foes criticize the breeder reactor for two main reasons. First, the potential for a calamitous accident is believed to be much greater than with a conventional reactor; unlike a conventional reactor, a breeder could actually undergo a small nuclear explosion. Second, the breeder produces plutonium, and plutonium can be turned into bombs; hence it's feared that breeder reactors would accelerate the spread of nuclear weapons.

But even if you discount these arguments, the Clinch River project still looks like a bad deal.

Since Clinch River was conceived, the state of the nuclear art has advanced considerably. It's now widely believed that the Clinch River design is obsolete and that the plant would be too small to be economically viable. Also, projected demand for nuclear energy has declined and estimates of world uranium reserves have increased, so the need for the breeder has become less urgent.

No less a pillar of the nuclear establishment than Edward Teller has said that Clinch River would be "obsolete before it started." No less a symbol of fiscal conservatism than David Stockman has denounced it (back in his congressman days) as a waste of tax dollars.

Stockman has changed his tune about Clinch River since becoming part of the Reagan administration, but he was right the first time. And the House Science and Technology Committee was right when it slashed $254 million for the breeder from the 1982 budget. We hope the full House and Senate will follow through and send this project to its timely and deserved demise.

The Morning News

Wilmington, Del., May 12, 1981

Tennessee's Sen. Howard Baker, in whose state the $3 billion project would be built, insists stoutly that the last word has not been heard. If the blow dealt to the Clinch River breeder reactor last week was not fatal, however, it sure came close.

The House Science and Technology Committee, with anti-nuclear Democrats combining with economy-minded Republicans, voted 22-18 Thursday to delete all construction funds for the project.

If the full House doesn't revive the project, or if the Senate doesn't oppose the deletion successfully, the committee action could mean the end of the proposal to build a 350-megawatt reactor that would produce more nuclear fuel than it uses up. There would be a cost of hundreds of millions of dollars, apparently, to terminate the project — settling with contractors and reimbursing some private investors — but these would be burial costs.

So would be interred another dream of bridging the gap between our present dependence on exhaustible — and fast diminishing — fossil fuels and the renewable energy sources of the future.

A couple of other bridges are looking better — if only slightly better.

The Massachusetts Institute of Technology, in its study "Coal: Bridge to the Future," is looking more hopefully at that resource. Coal now is used for only 18 percent of total U.S. energy, mainly for generation of electricity. The MIT study says coal will have to supply from one-half to two-thirds of all additional energy needed by the world between now and 2000. A byproduct of that additional exploitation will be that nearly 20 percent of all coal exports will come from the United States (as opposed to 6 percent now; we really might become what some have described as the "Saudi Arabia of coal." We're already exporting coal's companion product: acid rain.)

The National Coal Association estimates that exploitation will reach 980 million metric tons by 1985, compared to 703 million in 1979, which is all very well, assuming the environmental obstacles can be hurdled. Of course, not very much soft coal has been dug during the 49-day-old national miners strike.

Another bridge could be shale oil. Clifton C. Garvin Jr., president of Exxon Corp., says shale oil could provide half of a bridge between fossils and renewables and that such a bridge could last for 185 years. "That's quite a bridge," says Mr. Garvin in a current Fortune magazine survey entitled "Exxon gets serious about shale."

Well, if the surveys quoted are right, there are two trillion barrels of shale oil in some 16,000 square miles of rocks in Colorado, Utah and Wyoming. At our current rate of some 15 million barrels a day, that's plenty for Mr. Garvin's 165 years.

There could be some problems.

Water might be one. (It loomed large over the weekend in hearings over the Limerick nuclear plant in Pennsylvania and it could be even a larger problem in reference to shale.) Apparently, use of Exxon's process at its Colorado site requires from two to four barrels of water for each barrel of oil extracted. When the Colony shale plant goes on stream in 1985, producing some 47,000 barrels of shale oil a day, that would mean about 4 million gallons of water would be needed every day, in an area that isn't all that rich in water supplies. Exxon and the other companies interested in shale projects in the area seem confident, however, that they can solve the water problems.

All of which leaves us looking toward 1985 as the apparent start for both of these bridge construction projects. Meanwhile, we have our oil glut to wallow in. With a little bit of luck, and with OPEC's indulgence, maybe it will last long enough for us to start on these other dreams.

The San Diego Union

San Diego, Calif., May 12, 1981

The Clinch River breeder reactor has had so many lives that the vote against it last week in the House Science and Technology Committee is no occasion for an obituary. All the vote tells us is that there is still no consensus in Washington about future federal policy toward nuclear power.

For four years the Carter administration took the Clinch River project out of its energy budget, and Congress put it back in. Now, the Reagan administration is supporting the project, and the House committee, which always favored it in the past, has enough new members to come out with a 20-19 vote against it. So the project at Clinch River, Tenn., limps on toward an uncertain future.

We have supported the breeder reactor program all along. The logic for it was established years ago when commercial power reactors came into being. The spent fuel elements now being stored at power plant sites need to be reprocessed, and when they are, the plutonium they contain can best be disposed of by "burning" it in breeder reactors. This fuel cycle will vastly increase the energy to be derived from the original uranium fuel.

Completing the Clinch River breeder, then, is not a commitment to a new nuclear program, which is the impression that some of its opponents would give. Rather, it is a sensible step in the nuclear power program we already have. The $1 billion invested in the program so far should not be hostage to the questions still hanging over the long-range future of nuclear power.

The nuclear programs of Japan, Western Europe, and countries in the Soviet bloc all recognize the technological imperative of fuel reprocessing and breeder reactors. Sooner or later, U.S. technology will have to catch up. A majority in Congress will surely recognize this, even if Clinch River has lost its latest round in the House committee.

THE ARIZONA REPUBLIC

Phoenix, Ariz., July 28, 1981

THE U.S. House has voted to pour another $189 million down what some day will be the biggest rathole in the world, the Tennessee-Tombigbee waterway, familiarly known as Tenn-Tom.

On completion, Tenn-Tom will run 450 miles from the coal fields of southern Tennessee into the Gulf of Mexico through a series of dams, locks and canals.

Constructing it will necessitate the destruction of 100,000 acres of farmland, forests and wildlife habitats. The final cost is estimated at $3 billion.

The anticipated revenues never will meet the cost of maintenance and operation. Tenn-Tom will be a constant, endless drain on the Treasury.

This is probably the least defensible public-works project in history.

Tenn-Tom will provide no flood-control or hydroelectric benefits whatever. It merely will serve as a barge route to Mobile, Ala., for southern Tennessee coal operators.

The coal operators don't need such a route. They've been able to move all the coal they produce for decades by railroad and by a waterway, that didn't cost the government a penny, the Mississippi River.

In fact, Tenn-Tom has been called "a clone of the Mississippi."

Old Man River needs a clone like the Sahara needs sand.

When he was in the House, David Stockman could get positively eloquent in denunciation of the waste that Tenn-Tom represents.

Why then, as director of the Office of Management and Budget did he include the $189 million for Tenn-Tom in the administration's fiscal '82 budget?

The answer is that Senate Majority Leader Howard Baker Jr. of Tennessee demanded it.

Baker and several other powerful Southern senators want Tenn-Tom because constructing it has been providing employment for economically depressed areas in their states.

Also, a bonanza for contractors.

A majority of the members of the Senate and now of the House went along with them because the Golden Rule on Capitol Hill is, You vote for my water project and I'll vote for yours — no matter how ridiculous.

The Seattle Times

Seattle, Wash., July 23, 1981

A NUCLEAR-power project plagued by huge cost overruns, endless time delays, nagging management problems, unresolved safety and environmental questions, dubious contract procedures, possible fraud and bribery, skeptical legislative probes, and growing public frustration. In short, a "fiasco."

Sound familiar? No, it has nothing to do with WPPSS.

It's the Clinch River fast-breeder reactor in Tennessee, as described by a House investigations subcommittee in a preliminary study released this week.

Defenders of the project called the study shallow and amateurish, rife with errors and omissions.

Depending on whom one listens to, the breeder is either the hope of nuclear power's future or a horrendous white elephant. Thus far, it has bred a lot of controversy but no fuel or energy. Construction hasn't even begun; parts and supplies are sitting in a Tennessee warehouse.

President Reagan supports the breeder, while President Carter opposed it, mainly because he thought it would increase the spread of nuclear weapons. The breeder produces more fuel than it consumes to generate electricity. But the fuel it makes is plutonium, which can be used to build bombs.

However, France, Japan, and other developed nations are moving ahead on their own breeders, which seriously undermines the anti-proliferation argument.

The economic arguments are stronger. In a tight-budget year, with the administration touting a free-market energy policy, a federal subsidy for a costly, unproven nuclear technology seems inconsistent.

But about $1.2 billion already has been invested in the breeder, and termination costs probably would exceed $400 million. Moreover, 753 utilities nation-wide (including Seattle City Light) have contributed to the breeder program.

Despite its problems, the breeder project should go ahead. It's a promising technology that can increase the energy potential from uranium 70 to 100 times over conventional reactors.

Congress should approve continued funding for the breeder, but only on the condition that the administration find a tough director who can end the project's management problems, cost overruns, and contract irregularities.

Detroit Free Press

Detroit, Mich., July 30, 1981

IN THE MIDST of frenetic budget-cutting by House and Senate conferees, the House last week appropriated funds for two projects as useless as they are expensive: The Clinch River breeder reactor, the country's most pointless power project, and the Tennessee-Tombigbee, the most redundant waterway ever conceived.

The Clinch River breeder reactor was meant to demonstrate the feasibility of breeder reactors, which produce more nuclear fuel than they consume. It was originally supposed to cost $750 million. Construction has not even begun, but land acquisition and equipment costs have already topped $1 billion, and the final bill is estimated at $3.2 billion, not even allowing for inflation.

The Clinch River project was supposed to prove that breeder reactors can meet federal safety and environmental standards, but it probably cannot be completed without being absolved from them. The reactor may have problems getting licensed by the Nuclear Regulatory Commission on its present site.

Clinch River's electricity is not needed by its only customer, the Tennessee Valley Authority. The per-kilowatt cost will be so great that nobody, including the TVA, can afford it, so the reactor will require federal operating subsidies in perpetuity. The plant's design is already regarded as obsolete, and since Clinch River is obviously no argument for the commercial attractiveness of breeders — if anything, it's quite the opposite — when it is completed the Department of Energy plans to build another breeder reactor three times as big, to see if *that* one will work.

The Tennessee-Tombigbee is a failure on an even bigger scale. A waterway running roughly parallel to the Mississippi River, the Tenn-Tom will cut through the state of Mississippi from the Tennessee border to the Gulf of Mexico. It will cost more than $3 billion and when it is completed it will do no more than what the Mississippi River already does.

The tonnage figures projected for the Tenn-Tom have turned out to be largely fictional. There is little demand for it from shippers, who can use the Mississippi or existing railroads for transport. As with the Clinch River project, the only argument for completion of the Tenn-Tom is that it is already under way — that mistakes, if they are colossal enough, must roll on to completion.

Each year, however, supporters of the Clinch River and the Tenn-Tom have found it a little bit harder to line up the votes on their side. This year, a switch in 11 votes in the House would have defeated the appropriation for the Clinch River reactor. The Senate is yet to vote on the two projects, although the chances for defeating them are assumed to be slim, since Clinch River is dear to the heart of Majority Leader Howard Baker and Tenn-Tom has the support of President Reagan.

But simply scrapping the two projects would save $3 or $4 billion — more than the $1.5 billion cut from child nutrition programs when the budget conferees reduced them by a third, or the $1 billion to be cut from food stamps for the working poor, or the several hundred million dollars to be saved by eliminating the minimum Social Security benefit for three million elderly recipients. If Congress and the administration are sincere about cutting the budget, Clinch River and the Tenn-Tom offer them ample opportunity for maximum savings at minimal human cost.

The Providence Journal

Providence, R.I., July 23, 1981

The Clinch River fast breeder reactor has been a poor investment by the federal government from the beginning. But the charges of mismanagement and financial disaster made in a House subcommittee report are startling.

This prototype of a nuclear power plant that would produce more plutonium than it consumed is unnecessary, expensive and probably would be obsolete by the time it was completed. In that context, the assertion that it would entail a cost overrun of 450 percent is beside the point.

The contract with Westinghouse signed by the developers — including 800 utility companies as well as the federal government — is castigated as "unbelievably loose." The prime contractors are not required to turn in a good performance, according to the report of the House Energy and Commerce Subcommittee. The steam generators have thus far failed to work and their cost has gone up an incredible 1,000 percent.

That the Clinch River project is far behind schedule is not so much a management problem as a legislative one. The White House and Congress have been fighting over it for half a dozen years. President Ford first ordered a slowdown while a task force explored the possibility of safer ways to obtain as much nuclear power as offered by the breeder technology. President Carter persisted in opposing the project, but Congress tried to keep it alive.

Now the Reagan administration has left $250 million in its pared-down 1982 budget; and congressional opponents are trying to knock out even that amount. If that seems like a lot, it is only a drop in the bucket compared with the current estimate of $3.2 billion for the entire project.

Advocates — aside from those who support the project out of regional loyalty — contend that breeder reactors will be needed in the next two decades to meet energy demands of an expanding economy. They point out that other countries, including the Soviet Union, are building breeders and that the United States could find itself at the end of the parade in breeder technology.

But several studies — especially the Ford-Mitre report — have concluded that this country's energy needs are not so critical that breeders are needed, at least not before the next century. By waiting, and conducting only research on breeder technology, the United States could save money and, when there is a real need, build a more technically sophisticated reactor.

The reason for holding off is that breeders can be dangerous, not in themselves necessarily, but because they produce plutonium, the stuff of which nuclear bombs are made. The U.S. policy has been to avoid making plutonium except in its military program, to refrain from sending it abroad or from helping any foreign country produce plutonium. The aim is to avoid the risks of diversion to weapons use by unstable countries or by terrorist groups who might seize supplies by violent or clandestine means.

Nuclear countries that have a greater energy supply problem, such as France and Japan, are building breeders. But they are finding the plants exceedingly expensive, as well as hazardous from a weapons proliferation point of view. And they may well taper off in their programs.

Meanwhile, if the Clinch River project is shoe-horned through the budget process, it ought to be subject to all the safety requirements that other nuclear power plants must comply with. These requirements are laid down by the Nuclear Regulatory Commission. The Department of Energy reportedly is trying to get Clinch River removed from NRC jurisdiction. That would be a foolhardy move. What remains of the old Atomic Energy Commission, now a part of DOE, has always had an arrogant attitude toward safety controls.

Nothing could be less wise than to build a breeder reactor that is both uneconomically expensive and unresponsive to the safety measures that reactor accidents like Three Mile Island have shown the country to be necessary.

Arkansas Gazette.

Little Rock, Ark., November 7, 1981

The United States Senate has proved as accommodating for continued funding of the Clinch River Breeder Reactor project as it did earlier for more spending on the wasteful Tennessee-Tombigbee Waterway.

Although Senator Dale Bumpers of Arkansas went along with the additional funding for the Tenn-Tom project, he led an unsuccessful effort in the Senate (with the support of Arkansas's junior senator, David Pryor, also a Tenn-Tom supporter) over a two-day period to block a $180 million appropriation for the $3.2 billion Clinch River project. He was joined in the effort principally by a conservative Republican from New Hampshire, Senator Gordon Humphrey, who argued that the project is economically unsound. Liberal Senator Paul Tsongas tried but failed to cut the funding by half.

Both projects had the backing of President Reagan, but the implications of the Clinch River project went far beyond the fact that it represents wasteful federal spending at a time of budget austerity, or that it was the pet of Senate Majority Leader Howard Baker of Tennessee.

Former President Jimmy Carter slowed down the Clinch River project almost to a halt but not simply because it was and is a waste of money. A breeder produces more nuclear fuel than it uses and the plutonium derived from it can be utilized for nuclear weapons. Mr. Carter wanted to discourage the spread of nuclear weapons by all means at hand and the Clinch River plant would make the effort more difficult.

Other factors also have entered into the equation, but principal among them is the decreasing demand in growth for electricity, which would be the product of breeder reactors. There are increasing signs that nuclear power, even that produced with breeder reactors, cannot thrive except through massive federal aid that amounts to nuclear industry welfare.

Certainly the renewed funding for the Clinch River project is not surprising in light of the earlier House approval and the strong support of the administration, but this does not mean, not by a long way, that Clinch River is truly justified.

THE TENNESSEAN

Nashville, Tenn., December 15, 1981

THE Department of Energy has asked the Nuclear Regulatory Commission to exempt the Clinch River breeder reactor from standard licensing procedures to permit construction to begin sooner.

Rep. Morris Udall, D-Ariz., who chairs a House committee with jurisdiction over the NRC, and five members of the Senate Energy and Environment Committee have opposed the DOE's request. Mr. Udall contends that granting the exemption would establish an undesirable precedent for future licensing actions in the breeder reactor program.

He is right. Bypassing standard licensing procedures to speed up construction of the project gives an impression of careless haste which is not good in any aspect of the nuclear program.

The public needs to be assured at all times that decisions regarding the construction and operation of nuclear-energy facilities are being made in the most deliberate and careful way possible with all due consideration being given to questions of safety and proper scientific procedure. This cannot be done while the DOE is clamoring for exemptions to the standard methods in the rush to get started on a plant.

Some DOE officials claim that Congress intended some shortcuts be taken when it approved $154 million in first-year construction funds. There is serious doubt that Congress intended this. If it did it was not in step with the public.

No doubt the DOE believes it can bypass the standard licensing procedures at the Clinch River project in complete safety, and it may be right. But many people cringe when government officials speak casually about "shortcuts" in nuclear energy. The public wants no shortcuts in this area, and the NRC should proceed in the standard way to license the Clinch River reactor.

THE CHRISTIAN SCIENCE MONITOR

Boston, Mass., March 8, 1982

Some belated applause is in order for the US Nuclear Regulatory Commission. Late last week the NRC voted to deny a request from the US Department of Energy that the commission waive rules and allow construction work at the Clinch River Breeder Reactor site before required environmental studies were completed. In short, the DOE was asking that it be granted exemptions from rules that must be followed by other petitioners before the commission. The NRC's 3-to-2 vote correctly sent the DOE packing.

One would think that after everything that has happened concerning nuclear power in the past few years – from the Three Mile Island incident to questions about the effectiveness of the NRC itself – "going by the rules" would be the primary marching order for the nuclear industry. The Clinch River project has a formidable roster of backers, not least of all the Reagan administration and Congress. But no construction is yet underway.

That was what DOE wanted to do. Get ready for construction by putting in roads, building a cement plant, clearing land for a railroad. And, of course, all such non-nuclear construction work would make it more difficult to stop the project. Fortunately, the NRC saw through the ploy.

DAYTON DAILY NEWS

Dayton, Ohio, March 13, 1982

The Nuclear Regulatory Commission was right on line when it refused to grant the Department of Energy, the Tennessee Valley Authority and the Project Management Corporation any exemption involving the Clinch River Breeder Reactor.

The group was requesting permission to begin grading roads, building a wharf and a railroad spur and putting in a sewage system before any construction permits for the reactor were issued. The petition was based on cost savings. NRC, however, felt that delay costs wouldn't be great.

Unfortunately, the request can be resubmitted and probably will be after June when the term of one NRC member expires and President Reagan can appoint someone sympathetic to his feelings.

Mr. Reagan is firmly behind breeder reactors because they are designed to provide electricity with weapons-adaptable plutonium as a by-product. Large quantities of plutonium will be needed if Mr. Reagan is to get more nuclear warheads and atomic weapons in the U.S. defense stockpile.

But many experts argue, credibly, that the Clinch River breeder reactor concept already is outdated. And others fear, reasonably, that increased plutonium stocks will heighten chances of the deadly stuff falling into the wrong hands. There is also a real possibility of wholesale death-dealing should even the tiniest amount of plutonium escape into the atmosphere or a water supply.

If anything can kill the Clinch River reactor project once and for all, it is the present economics of that decade-old boondoggle. What started out in 1973 as costing $669 million already is up to $3.2 billion, and construction hasn't even started.

If hard facts and figures are needed, the NRC should ask for the details of the TVA board's recent decision to stop work on some of its nuclear reactors. Even the need for electricity couldn't justify the construction costs being passed on to its consumers. The same holds true for Clinch River, where neither the end product nor the byproduct justifes the mounting costs taxpayers are expected to pick up.

The Courier-Journal

Louisville, Ky., March 10, 1982

THE CLINCH RIVER nuclear breeder reactor has more lives than a houseful of cats. So there's no reason to suppose it was fatally wounded last week when the Nuclear Regulatory Commission rejected a Reagan administration request to speed the project up by a year. But there's also no law against hoping.

Critics of the project — which owes its continued existence mostly to the influence of Tennessee Senator Howard Baker — say the turndown might prompt Congress to discontinue its funds. Perhaps so, in a year in which Congress should be looking hard for ways to trim a ballooning deficit. But it seems more likely that the NRC will relent on the speed-up after the term of one of its five members, Peter A. Bradford, ends in two weeks. Another Reagan appointment will give the administration control of the board.

Actually, the Clinch River reactor never should have found its way into the current budget in the first place. Many of the one-time arguments for it have lost their snap. For example, a shortage of uranium to fuel conventional nuclear reactors once was forecast, and the breeder technology produces more atomic fuel, in the form of plutonium, than it consumes.

But no such shortage now is expected until well into the next century. The price of uranium, some authorities say, would have to be several times as high as at present to make the breeder equal, in economic terms, to the current variety of nuclear power plants.

Meanwhile, breeder technology is considered even more risky than the conventional technique. But, worst of all, widespread use of breeder technology would make weapons-grade plutonium an item of ordinary commerce, with all the security problems this would entail. And any of several technologies that are considered more benign are likely to be ready for use before there's any strong economic reason for breeder-reactor power.

About all that's left is the breeder's merit as an excuse for pork-barrel spending. So far, the project has bred nothing but cost-overruns. Its currently estimated $3.2 billion cost is four times the original estimate. Congress should not give this project the chance to multiply its costs again.

The Washington Post

Washington, D.C., May 20, 1982

JUST HOURS into his tenure as a member of the Nuclear Regulatory Commission, James K. Asselstine provided a nice lesson in public service. Standing up to intense pressure from the administration and the nuclear industry, Mr. Asselstine cast a key vote in which he placed his own "independence and objectivity, and that of the commission" above what he had clearly been appointed to accomplish.

Hanging in the balance was an ill-advised administration attempt to secure a special exemption from the NRC's licensing requirements for the troubled Clinch River Breeder Reactor. The case against wasting more money on this breeder demonstration program was becoming harder and harder to ignore, and the Department of Energy feared that unless it could show that construction was actually under way, Congress would finally kill the program this summer.

But there was a problem: the breeder had not yet passed the NRC's licensing examination, which meant that construction could not begin. So DOE asked for the exemption. In March, the NRC denied the request by a vote of 3 to 2. There was still hope though, since one commissioner who had voted "no" was about to leave. Administration officials and Chairman Nunzio Palladino vowed to find a right-minded replacement.

Under the NRC's rules, the decision could be reconsidered without new evidence until Monday, May 16. Mr. Asselstine's nomination was rushed through the Senate. He was confirmed last Thursday. On Friday, DOE formally requested reconsideration. Mr. Asselstine was sworn in Monday at 11 a.m., and Chairman Palladino called a meeting to reconsider the exemption request for 3 o'clock.

But here Mr. Asselstine's clear sense of how an ostensibly independent regulatory agency should operate upset the plan. Noting that DOE was requesting "extraordinary relief," he pointed out that his vote to approve could give "at least the *appearance* of a hasty and ill-considered judgment," and he voted "no."

DOE can still try to come up with new evidence and ask the commission to start the whole process over. Mr. Asselstine might then vote in favor—though we believe that is unlikely if he brings the same objectivity to consideration of the merits as he has to proper process. However it turns out, we salute his first move.

Newsday

Long Island, N.Y., July 15, 1982

If anything can give weak-kneed members of Congress the strength to resist throwing more money at the misbegotten Clinch River nuclear reactor, maybe it's a critical report by their own auditing agency.

At issue, for the umpteenth time, is the demonstration breeder reactor to be built at Oak Ridge in Tennessee, home state of the Senate's powerful majority leader, Howard Baker. In the 10 years since its conception, the cost of the project has grown fivefold. It's now up to $3.2 billion, and no one expects it to stop there.

The preliminary report by the General Accounting Office cautions:

"The continuing deterioration in the nuclear power industry and current information indicating that commercial breeder reactors are unlikely to be deployed for the next 40 to 50 years make it difficult to argue that developing the breeder reactor is an urgent task in the United States."

If anything, that's understating the case against the breeder. The arguments rest solidly on concerns over safety, expense, outdated technology, lack of need and the dangers of a plutonium-based fuel cycle.

The Reagan administration has nevertheless requested $233 million for the project in fiscal 1983. An effort to stop Clinch River fell only a few votes short in Congress last year. Now is the time to scrap it for good.

The Kansas City Times

Kansas City, Mo., July 9, 1982

President Reagan is proving just as persistent in trying to get the Clinch River breeder reactor plant built as President Carter was in seeking to have it killed. Just six weeks after the Nuclear Regulatory Commission rejected Mr. Reagan's request to begin preliminary construction at the Tennessee site before an NRC permit is issued, he is back seeking the same waiver. He might succeed this time, because in May a new Reagan appointee to the commission cast the decisive "no" vote so as not to appear a rubber-stamp member.

But in the meantime a surprisingly diverse coalition of environmental and conserative groups has formed in opposition to the 10-year-old project, which would demonstrate a nuclear power technology of generating more fuel than is consumed, thereby conserving uranium supplies. Ninety-one House members, including 36 Republicans, have asked the White House to withdraw a $252-million fiscal 1983 request for Clinch River, citing its cost escalation from $500 million in 1972 to $3.2 billion now, and claiming it would be obsolete when completed in 1988.

Mr. Carter opposed the project because it would produce plutonium, which also can be used in nuclear weapons. But powerful friends in Congress, including Republican Sen. Howard H. Baker Jr. of Tennessee, Senate majority leader, kept it alive. Mr. Reagan, on the other hand, while generally favoring minimal federal involvement in energy matters, has strongly backed nuclear power development.

The opponents believe current budget-cutting stresses will work against Clinch River, as will new critical reports on it by the General Accounting Office and a decline in annual electricity demand growth to 2 percent from the 7 percent projected in 1972 when Congress authorized it. An Energy Department study also has concluded that uranium prices would have to be eight times higher to make the breeder a serious option.

Clinch River, 10 years after its inception, may prove to be a project whose time never comes, despite the president's determined effort to keep it viable.

Los Angeles Times

Los Angeles, Calif., July 11, 1982

The General Accounting Office, which is in the business of giving expert evaluations of government programs at congressional request, has taken a long look at the fast breeder nuclear reactor project and given it a C-minus. Surely, in this era of tight federal budgets, that is a failing grade.

At issue is an experimental fast breeder reactor proposed for construction at Clinch River, Tenn. The Reagan Administration is requesting $253 million in fiscal 1983 for the project, the expected cost of which has ballooned to $3.2 billion—several times the original estimate.

Power-generating breeder reactors are designed both to produce plutonium and to burn plutonium that has been separated from spent fuel elements of ordinary uranium-fueled power reactors.

The idea is to make uranium reserves last longer and to provide a useful disposition of worn-out reactor fuel elements. But there are severe drawbacks.

Plutonium, once it is separated from burned-out uranium, can be stuffed into nuclear bombs as well as into power-producing reactors. And the International Atomic Energy Agency does not have inspection procedures that can be relied on to prevent such surreptitious diversions from occurring.

As a result, U.S. policy dating back to the Ford Administration has been to discourage other nations from nuclear power programs, such as breeder reactors, that depend on the separation of plutonium.

Obviously, U.S. construction of a breeder reactor would give other nations an excuse to go the same route, vastly increasing the danger of nuclear weapons proliferation among countries that do not now have them.

Even if this important factor is overlooked, breeders do not make economic sense.

Additional uranium reserves have been discovered. Projected demand for uranium reactor fuel has fallen precipitately as it has become more obvious that nuclear power reactors will not be built in the numbers anticipated because of safety considerations and skyrocketing costs.

As the GAO study noted, the result is that the country will not need breeder reactors before the year 2025. So it is hard to argue that developing the breeder reactor is an urgent task in the United States. The study affirms what was known already. The striking thing is the apparent determination of the Reagan Administration to push the project through anyway.

When it comes to synthetic fuels of greater economic practicality—gasoline made from oil shale or coal and coal gasification, for example—President Reagan takes the view that heavy government subsidies are unwise, that private enterprise will develop these alternative fuels when the market allows.

Nuclear power, it seems, is the one alternative energy source that is exempt from the test of the marketplace. The only explanation that makes even political sense is that Clinch River is in Tennessee—the home state of Senate Republican leader Howard H. Baker Jr.

That isn't a good enough reason to spend mountains of money on a project that, if it is ever needed at all, can be safely done without for another quarter-century.

Roanoke Times & World-News

Roanoke, Va., July 25, 1982

THE NUCLEAR Regulatory Commission has agreed to hear arguments again on accelerating construction of the Clinch River breeder reactor. By happy happenstance, along comes a new General Accounting Office report outlining all the reasons that the NRC ought to keep saying no.

GAO, Congress' auditing agency, doesn't contend that the $3.2 billion project should be ended outright. But it does strongly question giving Clinch River the high priority it has been assigned by the Reagan administration. In fact, says GAO, the nation probably will not need a fast breeder nuclear reactor before the year 2025 — if then.

The Reagan administration's romance with nuclear power is — like many affairs of the heart — hard to rationalize. It contradicts the administration's commitment to trimming needless federal spending. It flies in the face of economic and energy trends.

Most fundamentally, it ignores the problems besetting the nuclear industry. What need is there for a reactor to breed more nuclear fuel than it consumes, when the market for conventional reactors is dying and nuclear wastes are piling up?

At about the same time the NRC reopens discussions on granting Clinch River a construction permit, the House Subcommittee on Oversight and Investigations will hold hearings on the project. An effort to end financing for Clinch River barely failed in Congress last year, and Rep. John D. Dingell, D-Mich., subcommittee chairman, says: "This mild-mannered [GAO] report may be the stake through the heart that stops this nuclear Dracula from continuing to suck the taxpayers' blood."

Colorful imagery. But the Clinch River site isn't in Transylvania; it's in Tennessee, home of Senate Majority Leader Howard Baker. If the president can't do it, maybe Sen. Baker can explain why the country needs this welfare program for the nuclear industry.

THE WALL STREET JOURNAL.
New York, N.Y.,
September 27, 1982

Construction crews broke ground last Wednesday for the long-delayed Clinch River breeder reactor near Oak Ridge, Tenn. If some members of Congress get their way, however, that may be as far as the project will ever get.

Sens. Gordon Humphreys and Dale Bumpers plan to introduce an amendment tomorrow to the continuing budget resolution that would terminate funding for the nuclear reactor. Supporters of the amendment think they have a good chance at passage because the last crucial Senate test of the program last year won by only a two-vote margin after Majority Leader Howard Baker lobbied intensely for this big investment in his home state. Since then, several conservative senators, who are normally supporters of nuclear power, have apparently turned around on Clinch River.

One big reason is acceptance of the simple view that government should get out of the energy business. The success of oil deregulation attests to the fact that market forces can easily cope with our energy demands without government intrusion and no good purpose would be served by further massive federal subsidization of any energy project, even nuclear power. This was, in fact, the Reagan administration's initial view on Clinch River until the White House acquiesced to gain Sen. Baker's support in the budget and tax fights on Capitol Hill last year.

Another concern is money. Only last week the General Accounting Office concluded that the cost of the project could exceed $8 billion, more than twice the administration's current estimate. The cost totals of the 11-year-old program have already been revised six times, and critics now say the total could rise to more than $10 billion if the reactor encounters any construction delays and suffers escalation costs.

It's also becoming clearer to many that the only reason the federal government is so enmeshed in the project is that breeder technology isn't economical and the private energy sector doesn't want to waste its own money. At least one study projects that given the availability of relatively cheap uranium fuel, breeder reactors won't be economical until the year 2030 or beyond.

Besides these concerns, there is also the haunting worry about nuclear proliferation. Breeder technology provides an easy means of acquiring weapons grade plutonium. American support of such an uneconomic nuclear technology could encourage other countries to pursue their own breeder programs with more than just electricity production in mind.

It makes no sense, especially in light of current budgetary constraints, to sink billions of federal dollars into a nuclear project that won't be economical for at least 50 years. The Senate could do us all a favor by sinking the Clinch River reactor.

St. Petersburg Times
St. Petersburg, Fla., September 17, 1982

It was the late Sen. Everett McKinley Dirksen of Illinois who said of the federal budget, "A billion here, a billion there, and pretty soon you're talking about real money." In one of history's cute twists of fate, his son-in-law, Sen. Howard Baker, R-Tenn., is demonstrating just what Dirksen meant.

It's Baker's clout as Senate majority leader that has kept the Congress from saving some $6-billion by scuttling the Clinch River Breeder Reactor and the Tennessee-Tombigbee Waterway. The Senate came within two votes of killing both of them last year. The votes were nearly as close in the House. These are two of the most pointless, wasteful projects ever conceived. They survive for no better reason than that the money wasted is being lavished largely on three states — Tennessee, Mississippi and Alabama — that are well-fixed in Congress.

IN ITS TYPICAL fashion, Congress has blundered ahead with eyes closed and purse open, authorizing the projects on the basis of artifically low cost estimates and exaggerated benefit predictions, and keeping them going only on the notion that it's too late to stop.

What cannot be countenanced is the ruthless reduction of social programs for needy people while money continues to be lavished on such monstrosities as the Tennessee-Tombigbee Waterway and the Clinch River Breeder Reactor.

The Tennessee-Tombigbee, a 449-mile shortcut between the Tennessee River and the Gulf of Mexico to bypass the Mississippi River, was authorized in 1960 at $323-million. The latest estimates are now $4-billion (of which $1.2-billion already has been spent). Congress could save more than $3-billion by stopping it, selling right-of-way and using what has been built for recreational purposes and local navigation. Throwing more good money after bad is especially useless in this case. The volume of barge traffic that was supposed to justify the project is imaginary, according to both the General Accounting Office (GAO) and the Library of Congress. The Library's report also warned that it would cost more to ship coal on the canal than on the Mississippi, notwithstanding the canal's shorter length, because so few barges could be accommodated per tow.

THE CLINCH RIVER breeder reactor is indefensible on economic grounds, aside from the grave danger of manufacturing yet more plutonium. The reactor was planned as a joint public-private experiment at a time (1971) when nuclear power was popular and uranium shortages were forecast. (Breeder reactors will manufacture more fuel than they consume.) But expansion of the nuclear power industry has been paralyzed by safety questions and high construction costs, uranium is abundant and the annual growth of electrical consumption is less than half what had been forecast. According to the GAO, there will be no need for a commercial liquid metal breeder reactor (such as Clinch River) until well after the year 2020. To pursue Clinch River will cost at least $3.6-billion (the current official estimate) and possibly as much as $9-billion. Of that, the private nuclear power industry would have to pay only $257-million, the figure at which its contribution is frozen. Who said there is no golden goose? About $1.1-billion has been spent on Clinch River so far. It's high time to cut the losses.

Attempts soon will be made in both houses, when the annual public works bill comes up, to delete funding for both the Tennessee-Tombigbee and Clinch River. One danger is that their defenders will try, with the session running short, to perpetuate both of them under a continuing resolution not open to amendment. That would be a political swindle.

Once again, it will be argued that neither project costs much by federal standards — "only" $189-million this year in the case of the Tennessee-Tombigbee. But, as Dirksen noted, the sums do add up.

THE $6-BILLION that could be saved with a timely dose of common sense would be better spent balancing the budget or on public works projects that would put far more people to work and invigorate the economy in a lasting way, such as by repairing the nation's badly deteriorated highways, bridges, parks and public buildings — or by financing research and development of solar and renewable energy sources. What cannot be countenanced is the ruthless reduction of social programs for needy people while money continues to be lavished on such monstrosities as the waterway and the breeder reactor.

The paradox is glaring in Baker's case. As Senate majority leader, he has been the administration's point man for cutting social spending. He even defended the President's veto of the appropriation bill that meant continued employment for 52,400 nearly destitute older Americans. By overriding the veto, Baker's Senate colleagues spared both him and the President any further moral embarrassment on that issue. It's up to them now to stop the Tennessee-Tombigbee and the Clinch River projects as well.

THE COMMERCIAL APPEAL

Memphis, Tenn., December 23, 1982

WHEN THE HOUSE voted last week to cut off construction money for the Clinch River Breeder Reactor at Oak Ridge, Tenn., and the Senate saved it again by that one-vote margin provided by Majority Leader Howard Baker (R-Tenn.), the need for that nuclear energy project again was put into question.

The Clinch River reactor is to produce — or "breed" — more new fuel than it would consume in generating electricity. It would, therefore, assure the nation an inexhaustible supply of nuclear energy.

The need for this project and the desirability of it were widely recognized 10 years ago. Nuclear energy then was regarded as the wave of the future, a source of electrical power so cheap, some said, that there would be no need for meters to measure its consumption. Utility companies across the nation were building reactors and promising clean energy to replace the smokestacks of steam generation.

Then the Three Mile Island reactor failed, sending shock waves across the land. Licenses for new nuclear plants were held up while the interminable investigations were under way. New restrictions were written for such reactors and the utility companies hesitated to build. Nuclear energy had become extremely costly. President Carter halted the plans for the Clinch River breeder because he feared the project's plutonium fuel could be coverted into nuclear weapons.

Cost estimates continued to rise but bills providing money for it were passed in Congress 15 times. Finally, the government began clearing the 90-acre site for construction three months ago after the Senate saved the project's funding by a 49-48 vote. A Nuclear Regulatory Commission board still is reviewing the project and is expected to decide early next year whether construction should begin.

Critics have labeled Clinch River a pork-barrel boondoggle. Its proponents have insisted it was and still is essential to meet the nation's future power supplies.

The cost was estimated in 1972 at $700 million. The Department of Energy's latest estimate was $3.8 billion. The General Accounting Office says that if the interest that appropriated money could have earned is included in the calculations, the total cost to the taxpayers would be $8.5 billion. The Congressional Office of Technology Assessment estimates breeders will not be economically attractive until the mid-21st Century. France and West Germany, which like the Soviet Union have moved ahead with breeders, have cut back their programs. The virtual halt in nuclear generator construction, a slowdown in energy needs due to increased conservation and to the worldwide recession, and the discovery of new uranium reserves all have diminished the demand for the sort of fuel the breeder would produce and thus appear to make it less cost effective.

IT IS DIFFICULT to defend the breeder at a time when budget cuts are being demanded on all sides because the federal government is facing a $180 billion deficit this year. Critics have a point when they say that if this project is so important to the public utility companies of the nation they should shoulder a greater share of the cost. The plans called for a consortium of 753 utilities to pay only $340 million of the $8.5 billion — about 4 per cent. These are the same utilities which long have complained about the subsidization of the Tennessee Valley Authority's power production through the use of low-cost federal funds.

But it is impossible for anyone to predict the future energy needs of the nation. Every reasonable means of power production should be kept available. The fact that nuclear power presents formidable problems and costs now does not mean those problems and costs always will exist or that they will always be regarded as the impediments they are now. A cooperative arrangement between government and industry, fairly shared, now should be developed to keep alive the plans for the breeder at least until some better and more economical new source of energy can be demonstrated. The conference committee's decision to order the Department of Energy to spend $1 million of the new appropriation exploring ways for the utilities to share a larger part of the cost was sensible.

The News and Courier

Charleston, S.C., December 28, 1982

Despite James B. Edwards' plug for the Clinch River breeder reactor at a recent Charleston Rotary Club meeting, we believe that the widespread doubts about the project, reflected in Congress and in the media, are justified.

The first reaction of the House, expressed in a 217-197 vote to stop funding the breeder reactor, was right. Despite the approving quacking of the lame ducks, who will not have to account for their actions to the voters, the House managed to resist the temptation to pour more good money after bad.

As has happened time and time again the project was saved at the last moment by an agreement of the House-Senate conference committee to appropriate $181 million. But there was a difference this time. None of that money may be used for building the reactor or buying more components.

It makes sense to try and scale down the multi-billion project, which has already cost $1.3 billion, and transform it into an ongoing, but much more modest, research program. Breeder reactors are an energy option which should be kept open. They may be needed sooner than most experts believe, although they are unlikely to be economically viable before the year 2050. But there is no justification for steaming full-speed ahead on Clinch River when even foreign competitors, including France, are having doubts or, like Britain, are putting breeder reactor technology on hold.

Conservatives in Congress who formed an unlikely alliance with conservationists and anti-nuclear groups in voting against Clinch River were won over by one argument above all others. The estimated cost of the breeder reactor has risen from $700 million dollars in 1972 to $8.5 billion, according to the latest report of the General Accounting Office. Dr. Edwards, in his speech to Rotary, was, presumably, defending the project on the basis of the administration's own estimate of $3.6 billion. But it is questionable whether the breeder reactor is defensible on even that estimated cost increase. It is even more questionable to assert, as Dr. Edwards did, that the breeder reactor has been destroyed by "anti-technology groups." What has killed the breeder reactor has been escalating cost estimates and the nightmarish thought of a stream of taxpayers' dollars pouring into the Clinch River project and disappearing from sight.

Chicago Tribune

Chicago, Ill., December 13, 1982

With the federal deficit at such a huge level, Washington budget masters who've been trying to squeeze out a million here and a million there in savings are now turning to billion dollar items. There's one item of domestic spending that could save billions and produce relief, not pain, if cut. That's the $3.6 billion Clinch River Breeder Reactor project.

Concrete has yet to be poured for it, but it's already obsolete. It would expensively produce reactor fuel and plutonium for nuclear weapons in a world awash in cheap uranium, at a time when the troubled American nuclear power industry has substantially reduced its plans for expansion and its appetite for fuel. There may be a need for the Clinch reactor, but most likely in the next century.

The government hasn't the cash or even the credit to subsidize the future that way. Though $2.3 billion is what's being asked to finish it, the General Accounting Office estimates total completion costs could run to $9 billion.

Supporters argue that cancellation of the project would mean wasting the $1.3 billion that's already been spent on it. If you lose your shirt in a con game, you don't ask for two more shirts to keep trying.

The Clinch River project has advanced as far as it has because Sen. Howard Baker, whose home state is the recipient of this boondoggle, is majority leader of the Senate. If he'd like one day to be President, he'd do well to kill this thing and show that he's responsible enough to put the nation's interests ahead of his state's. He won't get far running for president of Tennessee.

THE MILWAUKEE JOURNAL

Milwaukee, Wisc., May 31, 1983

Rep. James Sensenbrenner (R-Wis.) has the right idea about the proposed Clinch River breeder, an unwise and unnecessary project that has been kept alive largely through the efforts of Senate Majority Leader Howard Baker.

"It would be better to admit it was a mistake to begin with and move forward on other fronts — ones less environmentally questionable and less expensive — to develop viable energy alternatives for the future," Sensenbrenner said in a recent newsletter to voters in the 9th District.

The breeder reactor was conceived under the assumption that the nuclear power industry would continue growing rapidly, requiring a large new source of fuel. Breeders produce more nuclear fuel than they use.

However, the spread of breeder technology worldwide could tempt many countries to produce nuclear weapons for the first time, increasing the danger of nuclear war. Thus, the breeder was a highly questionable notion even under the best of circumstances.

The project is less supportable than ever, now that the demand for electricity has leveled off and the power industry itself is de-emphasizing nuclear energy. Moreover, there are promising new technologies that may expand the usefulness of present uranium supplies without incurring the risks inherent in the costly breeder program.

President Carter sensibly halted the Clinch River project, but it was revived by the Reagan administration and kept alive in 1981 by the assiduous efforts of Baker. The project site is in Tennessee, Baker's home state.

A House subcommittee recently voted $1.5 billion to continue federal subsidies for the project through 1990, on the condition that the power industry pick up 40% of the costs in that period. A better answer, as Sensenbrenner points out, would be to call the whole thing off. It makes no sense to keep throwing good money after bad.

Wisconsin State Journal

Madison, Wisc., May 31, 1983

Last September, after 11 years of federal planning, bulldozers began clearing land near Oak Ridge, Tenn., for construction of the controversial Clinch River breeder reactor, a special kind of nuclear-power plant that once offered the dream of inexhaustible energy for the United States.

Now, however, the project appears to be running out of steam in Congress. The perennial question has come up again: Should the Clinch River project be killed?

The answer: Yes. Because of financial uncertainties about this plant, and the uncertainty of nuclear power in general in the United States, breedor reactors are a dream that — for now — will not come to pass.

Cost estimates have ballooned from $700 million to $3.7 billion for the Clinch reactor.

Taxpayers already have shelled out $1.5 billion; private industry has contributed $257 million. The General Accounting Office says it believes costs will climb as high as $8.8 billion.

The Department of Energy is trying to construct a financing package in which private investors will kick in more money to help complete the reactor. Yet, even under the best plan contemplated, taxpayers will be left holding the bag if the reactor fails, or if there is no market for its power.

The market is uncertain. The Tennessee Valley Authority, which promised to buy the power, now has no need for it. Nationwide, no new nuclear power plants have been ordered since 1978, and many plants on order have been cancelled.

The latest bad news for the nuclear industry comes from the state of Washington. Owing in part to overestimations on future electricity needs, two of five new nuclear plants under construction by the Washington Public Power Supply System have been scrapped.

The company is in danger of defaulting on municipal bonds worth $2.25 billion. The default could result in cancellation of the remaining three plants — on which $6.1 billion is owed — and the biggest municipal bankruptcy in U.S. history.

Added to these uncertainties at Clinch River are more concerning the environment and the reactor's production of plutonium — a key ingredient of atomic bombs. The case against the breeder reactor is formidable.

It's time for this expensive and questionable experiment to be put out of its misery. If the breeder reactor is as promising a piece of technology as its proponents claim, the nuclear-power industry should pick up where the taxpayer's left off.

Pittsburgh Post-Gazette

Pittsburgh, Pa., September 19, 1983

Energy Secretary Donald Hodel predicted Thursday that after years of rapid escalation, the costs for the Clinch River breeder reactor have finally leveled off. According to his sanguine assessment, the demonstration plant to provide the country with a virtually unlimited supply of uranium fuel can be constructed for a total cost of $3.6 billion.

Despite such long-range optimism, Congress at the moment seems more concerned with the fact that more than twice the project's original total estimated cost of $699 million has been expended on it before any construction has even begun. The federal government's demonstration project for a once-promising reactor technology can't shake the image of being — at any rate of inflation — an expensive boondoggle.

Mr. Hodel provided his assessment for the House Energy Conservation and Power subcommittee, which is reviewing alternative financing schemes to keep the Clinch River project alive after bruising setbacks in Congress earlier this summer. He had the embarrassing task of also acknowledging that an Energy Department official had withheld findings of about $300 million in additional estimated costs on Clinch River during congressional hearings last December.

But perhaps the most embarrassing aspect of Mr. Hodel's appearance has to do with explaining why the Reagan administration has decided to stake its political capital on a program that ought to be instead a number one target for federal cost-cutting.

The most plausible explanation continues to be President Reagan's desire to pay political tribute to Sen. Howard Baker of Tennessee, who has shepherded the project for his home state past all obstacles and through numerous resurrections in Congress for nearly 13 years. With Sen. Baker about to retire from the Senate, it's time for debate over the Clinch River reactor to be dominated less by pork barrel politics and more by hard economic analysis.

In response to a congressional call in June for significantly increased participation in financing of Clinch River by private industry, supporters are asking the government to guarantee about $1 billion in loans from private financers. The proposal not only would make taxpayers bear the risk of investments that Congress clearly intended private industry to share in more heavily; it would also remove appropriations for Clinch River from regular congressional review by placing them under a federal corporation in a so-called "off budget" category. Sen. Baker or no, Clinch River funds would be assured without a public debate for six years.

Those maneuvers prior to a crucial vote in the Senate before the end of this month are another misplaced effort to save a federal program that has been undermined by economic reality.

Clinch River was conceived originally as a principal source of fuel for light water nuclear reactors early in the next century. Steep downward revisions in the estimate of per capita electricity requirements, however, have combined with a sharp curtailment in light-water reactor construction. Consequently, analysts see a situation developing in which the hypothetical market for costly breeder-generated atomic fuels will be disrupted by sources already available at a much lower price.

Despite the administration's backing of a dubious new financing package, Congress ought to recognize that the energy industry itself fully appreciates the extreme financial risk of the breeder reactor. Until private investors are willing to assume a large portion of that risk directly, the Senate shouldn't be obliviously throwing tax revenues into the Clinch River.

Lincoln Journal

Lincoln, Neb., August 12, 1983

Informed critics of the still-alive, still-kicking Clinch River Breeder Reactor project offer a defensive case that the latest scheme by proponents is a financial shell game, with you-know-who really stuck for most of the additional $2.4 billion bill.

Where we are right now is that the House of Representatives — with absolutely no help from the three Nebraskans there — voted last December to deny additional funds for the Tennessee boondoggle, although the Senate subsequently OK'd, 49-to-48, continued funding. (Sen. J. J. Exon voted against, but Sen. Edward Zorinsky was for the expenditure.)

The Reagan administration backs the plutonium economy thesis behind the breeder reactor. A couple of days ago it submitted another plan for finishing the $4 billion experiment. Only on paper, and by refusing to look at details, would anyone say this most recent rescue would have the private nuclear power industry meeting its promised obligations.

It was a right-wing Republican, New Hampshire Sen. Gordon J. Humphrey, who initially protested the plan was a fraud. He contends that the private sector's contribution would stay very minimal, indeed.

The interesting array of fisically conservative, environmental, scientific and religious groups making up the Taxpayers Coalition Against Clinch River comes down even harder. The administration's Aug. 1 plan featuring bonds and federally guaranteed loans "will provide no new money from outside sources and does not ask the private sector to take any risk on investment . . . The government [actually] gives away to the private sector revenue and tax benefits it now possesses and then claims that this results in a new private sector contribution."

The only reason Clinch River has stayed alive so long is that Howard Baker, Senate majority leader, is its chief legislative champion. After the next election when Baker is gone, and Clinch River is finally abandoned, we'll all wonder what took us so long to reach the obviously logical decision.

The Chattanooga Times

Chattanooga, Tenn., September 1, 1983

The current issue of *breeder briefs*, published by the Clinch River Breeder Reactor Plant Project (CRBRP), carries a front-page endorsement of the breeder project by President Reagan: "I'm determined to have it built." Mr. Reagan had better have a deep pocket, for unless sentiment changes drastically on Capitol Hill, the 11-year-old project's days are numbered.

Its viability depends on a last-gasp effort called "alternative financing," which according to *Congressional Quarterly* is an effort to raise private funds for the project, thus diminishing the level of federal funding. But utility companies have so far contributed only about $257 million, a drop in the bucket compared to the project's expected total cost of $4 billion. When the CRBRP was conceived in 1972, it was to be equally financed by the federal government and private utilities. But as costs escalated, so did the federal share — about $1.5 billion so far. The Department of Energy estimates another $2.4 billion will be needed to complete the project.

Concern over the project's rising cost led Congress to order DOE to "vigorously explore proposals . . . that would reduce federal budget requirements" for the breeder. The House underlined that request by voting 388 to 1 on May 12 for a measure that could halt all further funding for the breeder on Oct. 1 unless Congress has approved a new financing plan. Breeder proponents hope to save the project with a plan that they say would require the private sector to pay 40 percent of the program's remaining cost.

It sounds like someone is cooking the numbers. DOE's first plan would have repaid private investors with a combination of project revenues, government guarantees and tax breaks, *CQ* reported. But the General Accounting Office argued that the ultimate cost to the government — which is to say, to the taxpayers — of raising money in this way would be as great or greater than that of simply borrowing from the Treasury.

If investors are given tax breaks, then technically the breeder would no longer be a part of the budget. But it would be receiving a backdoor subsidy. And if the government agreed to guarantee bonds used to build the reactor, that is not much different than issuing Treasury bonds to do the same thing. The upshot is that the plans suggested so far don't do much to increase the private investment in the project; if anything, it is the government's obligations that will increase. The more so because if, as some energy experts insist, it will be virtually impossible for the project to sell all the power it will produce, bondholders would simply line up at the Treasury's front door to redeem their investments.

So much for the so-called "magic of the marketplace," a concept to which the Reagan administration pays so much homage. The Clinch River bailout plans produced so far would draw some money from private sources but would force the government to assume all of the real costs, not to mention all of the risk. Congress has the distasteful duty of deciding whether to cancel the project and kiss $1.5 billion goodbye, or to keep the project going under a DOE funding plan that ultimately could cost taxpayers some $4 billion with no guarantee the expenditure will ever be recovered. It ought to keep in mind the incident in which a power system's construction project out West defaulted on bonds in part because unrealistic projections of power demands forced the cancellation of three nuclear power plants. Do we want the same at Clinch River?

The Atlanta Journal
THE ATLANTA CONSTITUTION
Atlanta, Ga., September 25, 1983

It is an article of faith with the Reagan administration that free-market forces should determine what happens in most areas of American life. If something isn't worthwhile enough to draw support and investment from the private sector, then it isn't worthwhile enough for the government to get involved in.

Curiously, though, the administration seems to forget this heartfelt principle when it comes to nuclear energy — particularly when it comes to the Clinch River Breeder Reactor in Tennessee. Private industry has all but lost interest in the project, recognizing it as the outdated, unneeded boondoggle it is. Yet the government continues to push for more millions to complete the facility.

Perhaps sensitive to this violation of the free-market dogma, the administration has lately come up with a plan by which the private sector would chip in another $1 billion toward the $2.5 billion needed to finish the reactor. On the face of it, industry willingness to invest that much money would seem to prove that there is an economic benefit to the project.

Aha! Now we find that there is a catch. Congressional budget officials have reported that the administration has set up such a package of tax credits, depreciation and interest deductions for these investors that the return for those who contribute could amount to an incredible 37 percent. With a deal rigged like this one, you could get the private sector to invest in a horse-drawn stagecoach line; the economic benefit of the project obviously doesn't enter into the industry's consideration at all.

The administration is clearly so desperate to get the appearance of private-sector support for this useless experiment that it will pay any price for it. The congressional budget analysis shows that the incentives will eventually cost the taxpayers more than would simply using federal funds alone to finish the facility.

The House voted last year to cancel this project, but the Senate kept it alive by one vote. Surely there is at least one senator who can read these numbers and realize that the time has come to ring down the curtain on the Clinch River pork barrel.

The Miami Herald
Miami, Fla., November 1, 1983

ONCE upon a time in the land of plenty, the lords who govern had a bright idea: Let's make a breeder reactor, they said. Give power to the people. Yes, said the lords of Congress, our fields of uranium are dying.

And so it was proclaimed. The man who would be President, known as Richard of Nixon, said: Go build this great reactor. Make it nuclear. Let it breed. Let it produce more plutonium than it consumes.

Plant it firmly in Tennessee at Oak Ridge. Near the river Clinch, in the great state of our patron Sen. Howard Baker, Republican extraordinaire. Let its beneficence flow throughout the Tennessee Valley Authority.

This is the way of the future, said the Senator Baker, this Clinch River Breeder Reactor. The people will be pleased.

But what will it cost? asked the people. A mere pittance, said the lords. A billion or two, give or take.

And the years passed in struggle, through loans, and bonds, and strife. But the great breeder reactor came a cropper. The fields of uranium did not die. The people had power aplenty.

The reactor did breed, all right, but t'was bills it produced — and the costs multiplied. So that in the year of 1983, the Houses of the lords of Representatives and the lords of the Senate said: Enough already! This is not boon, but doggle.

This great reactor has consumed $1.6 billion, and if completed would eat even more — say, $4.5 billion by modest estimate.

Thus did it come to pass. The lords of Senate, by a stunning 56-40 proclamation, killed the great breeder reactor. The Senator Baker was sad. "I sincerely regret the loss," he said. "One day we will regret not having an entry in this sweepstakes in this developing field."

But the people should not be saddened by this timely death. To the contrary, they should be pleased. For it wasn't plutonium that fueled this great reactor; it was money. And the people gladly would keep that for themselves.

THE PLAIN DEALER
Cleveland, Ohio, September 30, 1983

See if you remember the following phrases: "energy autarky," "too cheap to meter," "philosopher's stone," "hypothetical core-disruptive accident," "large-scale alchemy," "energetic disassembly," and "rat hole." Those are all phrases either coined or rejuvenated in the debate over nuclear energy, specifically the sodium-cooled fast breeder proposed to be built on the Clinch River in Tennessee.

Our favorite is "hypothetical core disruptive accident," which is techno-speak for a potentially murderous core meltdown. Another favorite is "energetic disassembly," which was coined a few years back as a euphemism for explosion. All of which is evidence that the propaganda war surrounding the Clinch River reactor has contained some of the worst writing as well as the worst thinking in modern times.

The breeder reactor was borne of technological altruism by the early nuclear pioneers. But 40 years and billions of dollars later, it is still little more than a dangerous concept. Of the three taxpayer-financed breeders built in this country, two have suffered at least partial core meltdowns and the third, a research facility, was designed to test fuels rather than breed plutonium. Further, the energy forecasts that supported the initial breeder concept have changed so radically that construction of a commercial version simply doesn't make sense now. Uranium-reserve estimates have changed as well, obviating the need for the breeder's fuel-production capacities.

Proponents of the breeder therefore have changed their strategy, largely by making fallacious appeals of philosophy. Hence lobbyists invoke the idea that technology is salvation, regardless of its form (a popular thought in Hiroshima), and contending that both the Russians and the French have a scary lead in breeder development.

They fail to point out, however, that progress, not technology, is salvation, and that part of progress is knowing when to quit. They further fail to mention the massive problems faced abroad in breeder research and development. They also scoff at legitimate fears that breeders, which can "breed" weapons-grade plutonium, will contribute to proliferation.

But let's not split hairs. The plain truth is that, given current technology, a breeder reactor that would produce "energy autarky," a "plutonium economy," and electricity "too cheap to meter," is still little more than "large-scale alchemy"—the fabled "philosopher's stone." It is a myth, the pursuit of which has cost American taxpayers billions of dollars and threatens to cost billions more.

Given all the above, you are right to wonder why the government is still seeking ways to fund the breeder program on anything more than a limited research level. One reason is that the Reagan administration favors it. Perhaps more to the point, though, is that Sen. Howard Baker, R.-Tenn., is a powerful man, and because his state stands to gain so much from the Clinch River reactor, he has worked hard in its support.

That is why the complicated and flawed financial scheme cooked up by the atomic lobbies and the government bypassed the House, where it surely would have lost, and went straight to the Senate, where there is a slim chance that it may pass. The reasoning is that Baker might try to insert it into a continuing budget resolution, which would keep the House from voting on the project's financing independently.

But remember the last phrase, which was "rat hole," as in "money down a ———." And that is what the breeder reactor is: A giant, misconceived, overstated rat hole. It is magic we can live without—a myth the Senate would be wise to kill.

ST. LOUIS POST-DISPATCH

St. Louis, Mo., November 7, 1983

The Clinch River breeder reactor project has been pronounced dead on a number of occasions, only to rise again. Therefore, it is hard to believe that the project is actually dead. Yet after a decisive 56-40 vote in the Senate, none other than Sen. Howard Baker of Tennessee has acknowledged as much.

It was largely Sen. Baker's political muscle that kept the Clinch River project in Tennessee alive long after the program's economic justifications had withered away.

But there are limits to what even a powerful lobby in league with the Republican majority leader of the Senate can accomplish — particularly now that the senator has announced his intention to retire. As a result, the public, which has already been bled for $1.5 billion for the breeder, will not be forced to provide more money to keep the project alive.

The credit for that goes to a bipartisan coalition that formed against the breeder, which included liberals and conservatives, anti-nuclear power forces and those who support nuclear power but felt the breeder to be both wasteful and unwise.

As Republican Sen. Gordon Humphrey of New Hampshire, a breeder critic, put it, "Each time it came up, we got through to a few more senators who had been embarrassed to support something they didn't really favor, out of respect for Howard Baker." Recently, a Senate majority finally recognized the breeder for what it had become and took action.

When it came time to drive a stake through the heart of the monster, Missouri's Democratic Sen. Thomas Eagleton was there, once again. But GOP Sen. John Danforth remained loyal to the breeder and to Howard Baker to the end, perhaps hoping that the project would once again show signs of life. Better, we say, to let the Clinch River breeder, at long last, rest in peace.

The Washington Times

Washington, D.C., September 26, 1983

The Clinch River breeder reactor was started 13 years ago as a demonstration project, and it has since demonstrated itself to be worthy of termination. Congress knows this, and should, with all due dignity, punch its lights out.

The Congressional Budget Office, in an analysis of a cost-sharing scheme designed to save the staggering behemoth, blasted a hole between its eyes. The plan would raise the government's ante, not lower it, and that's talking big money.

In 1970, when Clinch was born, its estimated cost was $400 million. Three years later, it jumped to $700 million. As of today, some $1.5 billion has been spent, and it's still not finished. The most optimistic completion projection is now $4 billion, with some estimates going to twice that.

Noting this progress, we should also take a look at the way the nation's energy picture has changed since Clinch River was conceived. Back then, it was assumed that the 7 percent annual electric power growth rate seen in the 1960s would continue. Instead, it has dropped below 3 percent. Energy planners also expected over 1,000 new nuclear plants to be on line by the year 2000, making breeders a good idea since fuel supplies would have become low, and breeders can draw more energy from uranium than conventional reactors. But nuclear plant start-ups are as common today as new primary colors. There's plenty of uranium around.

Clinch backers are trying to slide the plan through the Senate on a continuing resolution. To block this, more than 100 House members want to attach an amendment to the resolution requiring that the project be financed under separate legislation. That would effectively remove it from the continuing resolution, allowing the cost-sharing plan to be voted on in committee on its own merits. Having none, the plan would certainly fail, thereby killing Clinch River once and for all. The breeder's backers know that the only way this project can pass is through the back door. And if the Senate goes along with the strategy, it is abandoning its responsibility to the American taxpayer.

The end has been long in coming. Last year, Congress demanded that a cost-sharing plan be devised. This year, we have one. Figure it's Clinch River's best shot. And since this supposed cost-sharing plan would raise the total cost $250 million more than if the government paid the whole ticket, consider it a dud. As we've said many times over, if Clinch is such a hot idea, the private sector would be lining up with its own fists full of dollars to get a piece of the action, instead of queuing up for government handouts.

The Hartford Courant

Hartford, Conn., November 5, 1983

Did the Senate succeed this time in driving a stake through the heart of the Clinch River Breeder Reactor?

The experimental project with pork barrel appeal has been rescued many times, but the 56-40 vote against additional funds sent a message. U.S. Energy Secretary Donald P. Hodel said he would begin the "orderly termination of the project."

The reactor, authorized in 1970, would produce more fissionable fuel than it would consume. The technology was seen as aiding a conversion to nuclear-generated electric power to reduce dependence on foreign oil.

But a Senate majority has come to see it as too expensive and probably unnecessary, given nuclear power's diminished importance in meeting future power demand. Also undermining support for the breeder was the threat it posed, as a producer of nuclear fuel, to the proliferation of nuclear weapons.

It took some courage to cut off funds after spending $1.7 billion for preparing the site on the Clinch River in Tennessee. It is better to swallow that loss, however, than spend the additional $3 billion to $7 billion it would cost to complete the project. Research on breeder technology can continue without it.

There is another benefit to burying the Clinch River Breeder Reactor once and for all: It is symbolic of pork barrel public works projects that have proved almost irresistible in the past. Having broken ranks on this one, maybe Congress will have the nerve to judge other expensive, unnecessary projects on their merits, rather than on political appeal.

Denver, Colo., November 6, 1983

THE Clinch River nuclear breeder reactor seems finally to have been killed by Congress after years of controversy.

It succumbed to several factors: the government's unwillingness to finance the skyrocketing costs, the private nuclear industry's reluctance to assume a larger share of the financing, the slowing in the growth of demand for electricity, the opposition of environmentalists to nuclear power, and fears that breeder reactors, which produce plutonium, will contribute to nuclear weapons proliferation.

But cost was the main objection. The estimated price of the experimental project was $800 million in 1970. The latest price tag was put at $4.2 billion by supporters and twice that by opponents.

The government already has spent $1.7 billion on design, engineering and excavation. It may cost as much as $300 million more to close out the project.

It seems a shame to have spent that much money with so little to show for it. Had the power industry, which had contributed only $300 million, been willing to assume more of the risk, the project might have been saved.

The important feature of breeder reactors is that they produce more fuel than they consume. Breeder plants are in operation in France and the Soviet Union and are under construction in Japan and West Germany.

The object of building the plant on the Clinch River in Tennessee was to develop the technology necessary for general use by the electrical power industry in the United States.

It's true that the ever escalating costs of the project ate up enthusiasm for pouring in even more dollars, especially at a time when federal budget deficits were soaring.

But as oil and gas reserves become depleted, as they eventually will, and as controversy grows over the effects on clean air of using more coal, the nation could come to regret that the Clinch River project was killed. Certainly it will not get less expensive as time passes.

Reagan Reverses Carter Stance on Nuclear Fuel Reprocessing

Although vaguely worded, President Reagan's first definitive statement on national energy policies in July 1981 signaled a significant change from the nuclear energy policies of the Carter Administration. President Carter had stressed U.S. restrictions on exports of nuclear material and equipment that could be used to build nuclear weapons; the United States should set an example to other nations by restricting the export of technology or equipment for plutonium reprocessing, he said in 1977. In line with a determination to "defer indefinitely" the use of plutonium as a fuel in commercial nuclear plants, Carter said federal funding for reprocessing plants and breeder reactors would be reduced. Reagan's statement, by contrast, stressed the importance of improving the reputation of the United States as a supplier of nuclear technology to other countries. In this way, the U.S. could regain some of the international influence it had lost, he said, and would be in a better position to enforce non-proliferation guidelines. Reagan reaffirmed U.S. opposition to the spread of nuclear weapons, saying that "further proliferation would pose a severe threat to international peace, regional and global stability." The statement promised, however, that the U.S. would once again be a "predictable and reliable partner" in nuclear energy technology sales abroad. Reagan supported the development of reprocessing and breeder technology, and said the Administration would not inhibit the development of these technologies "in nations with advanced nuclear power programs where it does not constitute a proliferation risk."

In October of the same year, President Reagan announced that the federal government was lifting the domestic ban on commercial reprocessing of spent nuclear fuel that had been in place since 1977. None of the three U.S. commercial reprocessing centers—at Barnwell, S.C., Morris, Ill. or West Valley, N.Y.—was in operation at the time of the Reagan pronouncement; and the Clinch River breeder reactor project, in Tennessee, had been delayed indefinitely by cost overruns and the opposition of the Carter Administration. (See pp. 130-143.)The Administration took pains to separate Reagan's policy statement on domestic nuclear power from his July policy statement on the non-proliferation of nuclear weapons. Energy Secretary James B. Edward called the new domestic policy "a logical step to development of advanced nuclear reactors," and said the lifting of the commercial reprocessing ban "did not go into the question of proliferation abroad."

The News and Courier

Charleston, S.C., February 24, 1981

Any time The Wall Street Journal bad mouths a segment of the private energy industry it is noteworthy. When the object of the Journal's censure is the Allied General nuclear fuel reprocessing plant at Barnwell it becomes doubly interesting to South Carolinians.

The Wall Street Journal asserted in an editorial that nuclear fuel processing is uneconomical, and the "breeder" reactor which would use the plutonium produced so costly, that the French, who lead the world in breeder technology, have been unable to persuade their frugal utilities to even buy the breeder.

We would have supposed that, since the world's supply of uranium is finite and all of its energy potential is not extracted by the conventional nuclear reactor, the Journal would look kindly upon the salvage of spent fuel rods. That would have been the easier for the Journal, since its business offices are located some distance from Barnwell and the Reagan administration is proposing to turn South Carolina, not New York, into the globe's nuclear garbage dump. But here we have the Journal lining up with former President Carter and the angels in opposition to nuclear fuel reprocessing and breeder reactors. Strange, but welcome.

Our opposition all along to the Barnwell operation has been based on more personal grounds. It has never appeared to us wise to locate such a facility in South Carolina's earthquake zone, very near the epicenter of the state's last major quake, in fact. Over and above that, there is sufficient empirical evidence to convince a reasonable person that nuclear poison spewed into the atmosphere from reactors — let alone fuel reprocessing — constitutes a serious health hazard to the surrounding area.

The Oregonian

Portland, Ore., December 29, 1980

President-elect Ronald Reagan's choice for secretary of energy, former South Carolina Gov. James B. Edwards, is sending cold shivers up the spines of nuclear critics. Edwards is the quintessential nuclear man: the staunchest supporter of unconditional nuclear power development ever to hold a top government energy post.

That should be good news for the struggling nuclear industry that has not reported a reactor order or sale in the United States since 1978 and only two orders since 1977. Yet, Edwards may be almost too vehement a nuclear proponent to be helpful — like getting a carnival barker when what is needed is a quiet, dogged diplomat.

A dentist by trade, Edwards pushed nuclear energy alternatives in his home state of South Carolina from 1975 to 1979, including promotion of the $1 billion federal nuclear waste reprocessing plant at Barnwell. Yet, so adoring of the atom was he that his nuclear advocacy was not very effective.

Should he be confirmed by the Senate, Edwards should not try to move too quickly in trying to rescue the sagging nuclear industry from the Three Mile Island doldrums.

The majority of citizens in the United States, according to several recent national polls, are more hesitant about nuclear power than they once were. They still favor the nuclear option, but they want assurances that the reactors will be safe, that they will be built to critical stress and earthquake standards and that the nuclear waste storage problem will be solved.

In facing these issues, Edwards should take a tip from his own profession: Don't pull a tooth until convinced that the nerves are deadened and that any ensuing pain and suffering by the patient will be only temporary.

ST. LOUIS POST-DISPATCH

St. Louis, Mo., March 13, 1981

Overall, the Reagan budget approaches federal subsidies for nuclear power in a way that seems generally consistent with the ideological thrust of the new administration. For example, the president proposes to reduce the subsidy for the production of enriched uranium for fuel, and his budget cutters completely eliminated a $600 million plan to put an unfinished South Carolina reprocessing plant into operation. Nevertheless, spokesmen for the president insist, the administration favors reprocessing and is "determined to restore the nuclear industry in this country."

"Given all these signals," Raymond Romatowski, acting Undersecretary of Energy, told reporters, "there's a good chance the private sector will reassert itself." One signal that clearly went beyond the expression of good intentions is the $222 million allocated for the controversial Clinch River breeder reactor program. Another — with perhaps more far-reaching significance — is the proposal to allocate $27 million toward the clean-up of the crippled nuclear power plant at Three Mile Island.

A major drive is currently underway to get the federal government to pick up much or even all of the clean-up cost, over and above the $300 million in insurance held by the utility that owns the facility. Two-thirds of the insurance money has already been spent, and the last $100 million is being doled out largely to contain the after-effects of the accident.

The ultimate cost of the clean-up is estimated to be in excess of $1 billion. So $27 million is but a drop in the bucket. But it could be the drop that signals a flood of federal dollars. As such it must be an encouraging sign to the nuclear industry, which has been trying for two years to minimize the significance of TMI and find a way to pay for the clean-up that will not effectively undermine economic viability of nuclear power.

Long Island, N.Y., February 15, 1981

President Reagan's energy secretary wants a big increase in federal spending for nuclear power. He shouldn't get it—especially if the public isn't sure it can trust nuclear power's regulators.

After spending more than a year investigating the 1979 nuclear accident at Three Mile Island, investigators for the House Interior and Insular Affairs Committee concluded last week that utility company officers had lied about it.

The investigators said that Metropolitan Edison had "presented state and federal officials misleading statements . . . that conveyed the impression the accident was substantially less severe and the situation more under control than what the managers themselves believed and what was in fact the case."

This contrasts with the Nuclear Regulatory Commission's conclusion a week earlier that—despite a "clear failure" by Met Ed to collect, analyze and release information about what was going on in the plant reactor as the accident developed—there was no evidence that anyone at Met Ed had lied about the situation.

"Met Ed was not entirely forthcoming," the NRC said, but "information was not intentionally withheld." The NRC imposed no penalty for this lack of candor.

Considering the enormous amount of support the government has given the nuclear industry, it's no wonder some people think the regulators are more interested in promoting nuclear energy than in making sure it's as safe as humanly possible.

Now, with budget cuts threatened in virtually every energy field, Energy Secretary James Edwards thinks nuclear power should get more rather than less.

This is already the most heavily subsidized industry in the United States. In fact, if the government hadn't pushed the utilities so, hard and helped them so much, they might not be overcommitted to nuclear power now.

Breeder reactors and reprocessing of spent radioactive fuels rank high on the new energy secretary's list of priorities. They should rank a lot lower until it's clear that the safety of conventional nuclear power is being adequately regulated.

THE WALL STREET JOURNAL.

New York, N.Y., February 13, 1981

Nearly four years ago, President Carter announced the curtailment of nuclear-energy programs designed to use high-grade plutonium to fuel power plants. He scrapped a plutonium reprocessing plant at Barnwell, S.C., and eventually cut off funds for a demonstration breeder reactor at Clinch River, Tenn. Mr. Carter took the decision in hopes of slowing the international spread of bomb-making nuclear material. We agreed with the President's decision then; whatever the foreign policy arguments, these programs made no economic sense.

Energy Secretary James Edwards, saying "We have got to get nuclear on the move again," interrupted budget cutting Tuesday to say that the administration will seek "substantial" funding to revive the projects. Just as the programs were uneconomical then, so they remain today.

Several recent studies, done both in the U.S. and abroad, conclude that conventional nuclear power generation is far less expensive than either reprocessing or breeder technology. According to one study, uranium prices would have to double before reprocessing spent fuel would break even, and then reprocessing would only save about 2.5% on the cost of generating electricity. While the sources of new uranium supplies remain plentiful—and no one can foresee when they'll run out—reprocessing spent fuel is simply a waste of money.

For the breeder reactor to become commercially feasible, uranium prices would have to rise about sevenfold, though breeder technology would provide a larger cost savings than using plutonium in conventional light-water reactors. France, probably the world leader in the breeder, cannot even get its own utilities to buy the breeder because of the high expense. The main virtue of the American breeder seems to be that it is located in Senate Majority Leader Baker's home state.

We agree with Mr. Edwards that the administration should give fresh impetus to nuclear power, especially in solving the technically easy but politically difficult problem of waste disposal. But plutonium energy technology won't become economically feasible until the year 2030 or so; we can wait ten or twenty years to see if uranium starts to run out. There is no need and no excuse for new subsidies for its development in the midst of a budget emergency.

The Boston Herald American

Boston, Mass., April 29, 1981

A decade ago, reprocessing of spent fuel from nuclear power plants seemed like a good idea. Today it is still a good idea, but it has fallen into disrepute.

The Reagan administration should rescue the privately owned reprocessing plant which stands nearly complete and idle at Barnwell, S.C. It shoudl be operated by the government as part of the nation's nuclear energy program.

Each ton of spent fuel contains enough uranium and plutonium to generate, when reprocessed into new nuclear fuel elements, the energy equivalent of nearly 200,000 barrels of oil.

And reprocessing would reduce the waste disposal problem from nuclear energy plants.

The federal government persuaded Allied Chemical Co. and General Atomic Co. to build the Barnwell plant. A permit was applied for in 1968. Construction started in 1971. The partners each invested nearly $200 million of private money in the venture. The government invested nothing, but did help make a site available.

But in 1976 President Gerald Ford, in a surprise announcement, created a task force to study the wisdom of nuclear fuel reprocessing. And the next year President Jimmy Carter suspended hearings and declared that there would be no reprocessing.

Allied Chemical and General Atomic were left with no return on their investment and no way to recover the investment. They want the government to buy the plant and operate it.

The government has operated several reprocessing plants for spent fuel elements from the government's own military reactors.

A similar reprocessing plant for civilian fuel rods is justified by economics. France and Britain have big ones. There are 17 all told around the world on this side of the Iron Curtain.

The U.S. plant was shut down primarily because of fears that the spread of reprocessing technology around the world would increase the dangers of nuclear proliferation. But the technology is spreading, whether we like it or not. And we will pay a heavy economic penalty if we don't use it.

Other nations operate their reprocessing plants as government enterprises. We should too. Government regulations and political policy changes make it impossible for private enterprise to do the job in this era of public controversy over every aspect of the nuclear power industry.

It may go against the grain of policy-makers in the Reagan administration to take over a private enterprise. But the government has an obligation to Allied Chemical and General Atomic, based upon the fact that they built the plant at the government's request. And a takeover would be in the national interest.

THE SACRAMENTO BEE

Sacramento, Calif., July 18, 1981

Since the first atomic bomb was exploded, every American president has expressed his determined opposition to nuclear proliferation, and now Ronald Reagan has, too. But neither American policy pronouncements nor the international treaty the United States sponsored has prevented five additional nations from exploding nuclear weapons, nine nations from gaining the capacity to build a bomb today, and 16 other nations — including such bellicose regimes as those in Libya, Iraq, Taiwan, South Korea and Argentina — from obtaining the skills and materials to build nuclear weapons in this decade.

Clearly, the non-proliferation effort is in need of more than good wishes, and next week may be, as Sen. John Glenn has written, "the watershed moment" for action. With world attention still focused on the problem because of the recent Israeli attack on Iraq's nuclear reactor, and with the leading non-Communist industrial nations — the world's major suppliers of nuclear fuel and technology — about to begin their annual economic conference in Ottawa, an "extraordinary opportunity" is at hand.

President Reagan's attempt to seize that opportunity is commendable, but the policy statement he has issued on nuclear proliferation leaves much to be desired. His "goal" of strengthening the International Atomic Energy Agency is wise, but it will take more than talk to put some teeth into an agency which now cannot even make full inspections of the nuclear facilities for which it is responsible and can impose no sanctions in case of violations.

Reagan's pledge to "discourage the transfer of sensitive nuclear material, equipment and technology" was all but invalidated by his announcement that he was, at the same time, lifting the Carter administration's ban on the export of breeder reactors. These nuclear power plants produce more fuel than they consume, continually generating the raw materials for nuclear weapons development. And with a breeder reactor, a small nation would become virtually free of even the limited inspections and prohibitions that nuclear fuel suppliers can now impose.

A much harder line is necessary if the United States is to convince its allies to join in strengthening the current, ineffectual international system for controlling proliferation. We certainly cannot count on the French (who are planning to resupply Iraq) or the Germans and Swiss (who jointly supply Argentina and Brazil) to take the initiative in this matter.

Difficult as it may be to get their cooperation, the supplier nations must finally agree to ban the spread of breeder reactors, to refuse nuclear help to nations that do not abide by the Nuclear Proliferation Treaty (something the United States itself was unwilling to do when push came to shove in India last year), and to export only fuels that have been doctored to make processing for use in weapons more difficult. These are the only remaining chances non-proliferation has.

President Reagan said a major part of his non-proliferation strategy would be to supply near-nuclear nations with conventional arms instead and to use American diplomatic and military power to stabilize regions that might otherwise feel a need for nuclear weapons. This strategy is extremely risky: As likely as not, it would provoke hostilities and aggravate regional arms races. Indeed, as long as stepped-up weapons sales and regional policing are American policy, it would be foolhardy not to at least accompany them with the most stringent non-proliferation effort that can be mustered.

Beginning tomorrow, the president will be in Ottawa face to face with the French, German and Canadian leaders who can make or break his non-proliferation hopes. There is still a chance to make more of this opportunity than he has so far.

Columbia, S.C., March 19, 1981

THAT WAS certainly a strange presentation Secretary of Energy James B. Edwards made before the Senate Armed Services subcommittee on strategic and theater nuclear forces last Friday.

We don't have a transcript and are relying on press reports, but it may be the kind of situation which will land him in big trouble at the White House. Maybe we should say "bitter trouble," in view of recent reports of his "disfavor" in the Administration.

Dr. Edwards arrived in Washington with pro-nuclear flags flying. He and President Reagan were in agreement on the necessity of reviving the ailing nuclear power industry in the United States. It is integral to the nation's energy policy.

A key to turning on the nation's nuclear power program is the Barnwell Nuclear Fuels Plant, which was built by Allied General Nuclear Services. It was designed to reprocess spent fuel from utilities' reactors, a reasonably cheap means of maintaining a nuclear fuel supply for years to come.

A bit of recent history is in order. BNFP has never operated, although it is the only reprocessing plant for commercial fuel in the nation. President Carter suspended reprocessing before it started because he feared some byproducts might find their way into foreign hands and become nuclear bombs. International atomic agency studies have since endorsed reprocessing, however.

Recent Democratic congresses have provided research funds for BNFP to keep it open, to keep the staff together, in the event the national policy swings to reprocessing — which Secretary Edwards thinks it should. And we agree.

The Administration apparently is not ready to make a decision on reprocessing right now. And although Dr. Edwards has thumped the drums for funds to keep BNFP in business for another year (1982), the White House won't even go that far.

That is the awkward position Dr. Edwards was in last Friday. He had his orders from the White House to oppose a $13 million appropriation agreed to by a House subcommittee to keep BNFP open, essentially to protect the reprocessing option. He did as he was ordered, but grumbled that he's never found himself working for someone else before.

The secretary then proceeded, with some prodding from senators, to argue forcefully for the reprocessing option. He said the federal government has a "responsibility" in Barnwell because it encouraged Allied General to build in the first place. He said the President wants to "move on this" and that "I feel very strongly this is one of the things we have to do."

To all of that, Sen. John Warner of Virginia commented, "The Reagan administration's commitment to reprocessing is refreshing."

Secretary Edwards' performance was more transparent than artful. But he was in a no-win situation. He was at the same time the President's good soldier and a prisoner of his own convictions. And as the Administration's mouthpiece, he appeared to be talking out of both sides.

Oregon Journal

Portland, Ore., July 18, 1981

A total of 17 "minor powers" either operate nuclear reactors or have them under construction (Argentina, Brazil, Cuba, India, Iran, Mexico, Pakistan, Philippines, South Africa, South Korea and Taiwan) or operate research nuclear reactors (Chile, Colombia, Egypt, Iraq, Israel, Libya and some members of the nuclear power club). Who regulates or watches these potential merchants of nuclear disaster?

The job falls to the International Atomic Energy Agency. The IAEA tries to convince countries to sign the Nuclear Non-Proliferation Treaty of 1968. The United States, Soviet Union and Britain signed that treaty. France did not but said it would fulfill its provisions.

While 114 nations have signed the treaty, Western countries frequently agree to provide nuclear plant technology to poor, oil-importing countries. The price includes a promise that the poor country won't use nuclear technology to build bombs, a promise that may become the Third World's biggest joke on the superpowers in the 20th century.

The IAEA employs 150 inspectors who supposedly make sure all nuclear facilities will be used for peaceful purposes. But a country can choose its inspectors. Iraq, aligned with the Soviet block, permitted only Russian inspectors from the IAEA to look at the facility bombed by Israel.

In addition, Third World countries often make it difficult for IAEA inspectors to obtain visas and move their inspection equipment through customs.

Some U. S. intelligence officials believe South Africa, Taiwan and Israel are cooperating to build nuclear bombs.

Israel showed what it thought of IAEA inspectors when it bombed Iraq's nuclear reactor. But what will the rest of the world do if one of the Third World countries announces that it has built an atom bomb? A nuclear madness has overtaken the world. Unable to keep a corner on the spread of nuclear technology, the major powers had better have a plan for dealing with the first nuclear renegade that turns such technology into a weapon of war. Assure us, somebody, that such a plan exists.

DESERET NEWS

Salt Lake City, Utah, July 20, 1981

In early 1977, less than three months after taking office, President Carter announced a strict new U.S. policy designed to help curb the spread of nuclear weapons.

By contrast, it has taken the Reagan administration more than twice as long before it finally got around to announcing its own non-proliferation policy.

To the critics of the new administration, this delay reflects what they maintain is President Reagan's lack of interest in the issue, and they cite some of his campaign statements to back up this claim.

To Reagan supporters, however, the delay merely reflects the fact that nuclear know-how has spread so far that there is little the U.S. can do on its own to keep other nations from acquiring atomic weapons.

So the new administration is planning to rely far more than President Carter did on international safeguards to pursue nuclear nonproliferation, though details are sketchy. There is talk about encouraging regional and global stability to reduce the impetus behind nuclear arms races. And Reagan wants to reverse the Carter resistance to the development of the fast breeder nuclear reactor, which produces the plutonium that can be used in nuclear weapons.

So far, international non-proliferation efforts have been directed largely at regulating the trade in nuclear materials and technology. But there are large gaps in such efforts. Moreover, the international system does not address the reasons why countries without them want to develop nuclear weapons.

At the heart of this system is the Nuclear Non-proliferation Treaty of 1968 signed by 115 nations but not several important ones — including China, India, France, Israel, Pakistan, South Africa, Argentina, and Brazil.

Under the treaty, nations without nuclear weapons pledged not to try to obtain either nuclear weapons themselves or the capability of building them.

In return, nations with nuclear weapons promised to supply the "have-not" countries with nuclear technology for peaceful purposes. Moreover, the nuclear nations — especially the U.S. and Russia — promised to "pursue negotiations in good faith" toward nuclear disarmament.

Though most of the nuclear nations have done a reasonable job of making good on the first promise, they have made little or no progress toward actual disarmament. This situation weakens whatever incentive the non-nuclear nations have to keep their part of the bargain.

As India and Iraq have demonstrated, it's relatively easy under the international control system to divert nuclear materials from peaceful to military purposes despite periodic inspections.

Even in the United States, which is supposed to have strict controls on nuclear materials, there is a potential for abuse. For example, there has never been a public explanation of what happened to 200 pounds of weapons-grade uranium that disappeared from a Pennsylvania plant in the mid-1960's. Some observers charge the uranium ended up in Israel.

Nor does the international system, as it now stands, come to grips with the feeling of insecurity and the desire for prestige at home and abroad that prompts various nations to seek nuclear weapons. But, then how can it?

No wonder the Reagan administration has taken its time about formulating a new non-proliferation policy of its own. The gap between ideal and actual practice, always substantial in international affairs, is particularly acute when it comes to nuclear weapons. Maybe the best the world can realistically expect is just to keep trying to muddle through.

THE SUN

Baltimore, Md., July 26, 1981

Harry Truman shut the barn door.
Dwight Eisenhower opened it.
Jimmy Carter closed it again.
Now Ronald Reagan has reopened it.

That, encapsulated, is the story of American nuclear policy, at least that part of it concerned with the release of nuclear knowledge and equipment—and the concommitant problem of nuclear weapons proliferation. It is a frustrating story because the United States, as the world's first nuclear power, was never destined to be the only nuclear power. The hunger of nations for for energy, prowess and prestige would not permit it.

In a series of articles starting today, *Sun* reporter Robert Ruby makes the point that nuclear proliferation has already happened; with ten countries now possessing a nuclear weapons capability and, 10 to 20 more on the threshold. "The problem for the 1980s," he writes, "is not so much preventing the spread of nuclear weapons as managing it, making it survivable."

President Truman did not appreciate that the frontiers of science easily penetrate international frontiers. President Eisenhower engaged in the wishful thinking epitomized by the slogan, "Atoms for Peace." President Carter presumed, wrongly, that the United States could use domestic-style regulatory methods to control how other nations deal with basic energy, military and commercial objectives. Now President Reagan, in the name of "realism," has junked Carter export restrictions and, instead, wants the United States to be a "predictable and reliable partner for peaceful nuclear cooperation." Otherwise, "other nations will tend to go their own ways and our influence will diminish."

Mr. Reagan's course is indeed more realistic, but that in no way guarantees its success or frees it from from awkward choices and contradictions. To slow the spread of nuclear weapons, Mr. Reagan is prepared to increase the flow of conventional arms to nations beset with regional security fears. To strengthen the Non-Proliferation Treaty and the International Atomic Energy Agency, he will make the U.S. a more dependable nuclear supplier to cooperating nations.

So long as the president can speak in generalities, his position is persuasive. But policy has a way of getting down to specifics—of reconciling U.S. conventional arms for Pakistan, for example, with that country's aggressive pursuit of nuclear bomb-making capability. While Mr. Reagan's America will continue to oppose the transfer of sensitive material to countries seeking to join the nuclear weapons club, he offers only "grave concern" as the U.S. response to the proliferators. The United States—and the world—has much managing and surviving to do.

Pittsburgh Post-Gazette

Pittsburgh, Pa., July 25, 1981

The Reagan administration's recently unveiled policy to help prevent nuclear proliferation is long on idealism but rather short of specifics.

Its thrust is that the United States, while remaining a "clearly reliable and credible" supplier of nuclear technology for peaceful purposes, will commit itself to stopping the spread of nuclear weapons abroad.

How it plans to achieve this goal is not spelled out in very great detail. Moreover, while it is not surprising that the United States will continue to sell conventional weapons, there is a question of whether nations receiving such technology will be satisfied ultimately with anything but nuclear weaponry. Recent events in the Middle East hint at, if they do not irrefutably confirm, the likelihood of this eventuality.

This makes all the more important the question of how the Reagan administration intends to enforce its commitment to the sale of nuclear technology for only "peaceful purposes." In this regard, the new policy properly declares support for the Treaty of Tlatelolco, which establishes a nuclear-free zone in South America, and the 1970 Non-Proliferation Treaty, which promotes peaceful use of nuclear energy while trying to prevent the spread of nuclear weapons.

It also is on target in its call for support of the International Atomic Energy Agency (including an endorsement for greater funding) and a "high level of intelligence activities" to detect weapon-making activities in nations with nuclear power. And, finally, it shows proper recognition of the weakness to date of international efforts to control non-proliferation.

But what is missing is an extension of the policy that actually outlines concrete solutions to these shortcomings. One possibility is stricter adherence to the Nuclear Non-Proliferation Act passed by Congress in 1978. The key here is a prohibition of nuclear exports to any nations that don't accept safeguards at their facilities. (The Reagan administration has asked for modifications in the act to accommodate sales to Pakistan, which hardly offers encouragement on this score.)

Another is a push for more power for the International Atomic Energy Agency, which currently has no authority to prevent a country from diverting nuclear fuel for weapons use. Yet another matter to be considered is greater pressure on those nations that neglect international safeguards. France and Italy, for example, have long been accused of selling nuclear power without proper non-proliferation safeguards. Other nations, like Israel, China and Pakistan, also have virtually ignored the non-proliferation treaty.

By whatever means it is pursued, the Reagan administration's policy will stand apart from other diplomatic drivel only if tangible guidelines for non-proliferation and specific methods of dealing with those who violate them are developed.

The San Diego Union

San Diego, Calif., July 25, 1981

The United States must assume a more positive role in international efforts to control the spread of nuclear weapons. But how?

President Carter was as dedicated to nuclear nonproliferation as any president could be, but his policy initiatives in that area got nowhere.

Now President Reagan has issued a policy statement on nonproliferation and it appears to correct the major mistake in the approach of Mr. Carter. The latter attacked the problem by limiting the export of U.S. nuclear technology and materials and urging other nuclear-exporting countries to do the same. The failure of this initiative was evident even before the June 7 Israeli raid on an Iraqi reactor brought the non-proliferation issue to the fore.

Mr. Reagan's statement restores an earlier perspective to American policy — one that acknowledges that the United States can best contribute to safeguards and controls by being a "reliable supplier" of technology and fuel for nuclear power programs. By going back on promises to share our nuclear technology with other countries, we only encourage them to try to develop that technology on their own — and with it the capability of making bombs.

While reaffirming U.S. support of the Nuclear Non-Proliferation Treaty and other international safeguard programs, Mr. Reagan puts new emphasis on creating conditions that would reduce the motivation for countries to go their own way with nuclear development. This is a broad and ambitious goal, but it recognizes that insecurity about future energy supplies is what motivates countries to try to develop independent nuclear power programs, and that military insecurity motivates some of them to arm themselves with nuclear weapons.

The Reagan policy puts the United States back on a more promising track. It is said to have been greeted with approval at the Ottawa summit. It remains, however, a statement of guidelines and objectives rather than a summary of how the United States will proceed on specific issues on the non-proliferation agenda.

The statement promises to bring pressure on countries whose nuclear programs make them a "proliferation risk," but administration officials have declined to identify any countries now held to be in that category. It calls for strengthening the inspection and safeguard program of the International Atomic Energy Agency, but does not point to the weaknesses in that program that need to be addressed.

The world is flirting with catastrophe if the proliferation of nuclear weapons continues unchecked. There are now six nations known to possess atomic bombs. Nine others are believed to have the capability of producing them. Another 16 may join that category by 1990. Mr. Reagan says he has told the State Department to give "priority attention" to the proliferation problem, and well it should.

The Times-Picayune
The States-Item

New Orleans, La., July 20, 1981

In an apparent reaction to criticism from former President Jimmy Carter and others that it has not been concerned enough about the spread of nuclear weapons technology, the Reagan administration has come out unequivocally in opposition to nuclear weapons proliferation.

Opposition to the spread of nuclear weapons is a "motherhood issue" of the first order. Still, the Reagan administration seems to have been genuinely impressed by the perils of proliferation as a result of Israel's bombing of the Iraqi nuclear reactor near Baghdad. And President Reagan, in direct reference to the "ominous events in the Middle East," has proclaimed that opposition to the spread of nuclear weapons is a "fundamental national security and foreign policy objective" of his administration.

To this end, the administration calls for strict adherence to nuclear safeguards and treaties by all nations, in particular the 1970 Nuclear Nonproliferation Treaty. The problem here is that not all nations are parties to the treaty, and several, over which the administration has little or no influence, appear eager to join the nuclear club.

But if the administration is serious about making nonproliferation an important element of its foreign policy, it can pressure its allies to help stem the flow of nuclear weapons technology. France was the primary supplier of the technology for the Iraqi reactor, and Saudi Arabia, with which the United States enjoys good relations, has vowed to finance the rebuilding of the Iraqi reactor. In both cases, the Reagan administration might exercise influence. Similarly, the administration might be able to dissuade Pakistan, to which it has committed a large increase in military aid, to forego its ambition to build nuclear weapons.

The administration's proposal to strengthen the International Atomic Energy Agency, which is charged with inspecting nuclear facilities to see that Nonproliferation Treaty signees are playing by the rules, is in order. The agency's inspectors failed to detect anything amiss with the Iraqi reactor, opening it to charges of ineptitude, incompetence or — at the extreme — even collusion.

A major potential weakness in Mr. Reagan's nonproliferation stance is his willingness to accept breeder-reactor development in nations with advanced nuclear power programs, provided there is no proliferation risk. While some governments insist breeder-reactor development is essential to development of nuclear energy for peaceful purposes, the reactors breed more fuel (plutonium)than they use, and plutonium can be used to make nuclear bombs. Seeing that the extra fuel is not used for weapons development could be a major challenge of its own.

The Reagan policy statement is an encouraging development, but one that can be judged better for its effectiveness after it has been in practice for some time.

The Dispatch

Columbus, Ohio, July 27, 1981

THE REAGAN administration's nuclear arms control policy announced recently represents a sober-minded reappraisal of the president's earlier views on the issue and is a positive step towards limiting the world's nuclear arsenal.

The new policy:

- Reaffirms U.S. support for international efforts to control nuclear arms.
- Vows to work towards regional and world stability as a way of reducing the need for nuclear arms.
- Offers to expand the U.S. nuclear protection "umbrella" to nations that shun nuclear arms.
- Promises to make the United States a "predictable and reliable" trading partner for those countries seeking nuclear capability for peaceful purposes.
- Increases federal checks on exported nuclear material to ensure it cannot be used for unauthorized military purposes.

The president, in announcing the policy, has come a long way from his statement, made in the heat of last year's campaign, that he didn't think "it's any of our business" whether other nations want to develop nuclear weapons. Although he subsequently retracted the statement, it was enough to raise doubts at home and abroad about his perception of the problems inherent with the spread of nuclear arms.

There can be no doubts now. The president says nuclear controls stand as a "fundamental national security and foreign policy objective" that will generate considerable interest and effort from the administration.

The policy lays the groundwork for additional action. U.S. Sen. John Glenn, D-Ohio, wants the president to plan for a world nuclear energy policy conference that would, among other things, strengthen international nuclear controls. An aide to the president says the administration shares Glenn's concerns and is considering his suggestion.

The Reagan policy might also encourage the newly elected Mitterrand government in France to reject previous French policy and adopt a rigid controls stance. Since France is a leading world supplier of nuclear material and equipment, such a change would go a very long way towards limiting the nuclear threat.

These changes in attitudes and in governments hold the promise of significant advances in an area of grave concern. It is to be hoped the promise will soon become reality.

DAYTON DAILY NEWS

Dayton, Ohio, July 18, 1981

"The challenge of the 1970s was whether we could limit the number and spread of nuclear weapons," said New York Sen. Dan Moynihan. "The answer was that we could not. The challenge of the 1980s is whether we can prevent their use."

Glenn

The challenge is getting very dangerous. President Reagan, months ago indifferent to nuclear proliferation, now is echoing the warning. The Reagan administration is preparing new non-proliferation guidelines. Preventing the spread of nuclear weapons probably will be discussed by leaders of the industrialized free world meeting for an economic summit next week in Ottawa.

One of Congress' most vocal experts on this danger, Ohio's Sen. John Glenn, notes that current international sanctions and safeguards for nuclear development aren't enough. Too many of the nations that can build a nuclear bomb are embroiled in regional tensions — Israel, Pakistan, South Africa. Other erratic nations such as Libya and Iraq could build a weapon within 10 years.

Safeguards can be tightened with stricter controls on certain radioactive materials, international provisions for spent fuel, stricter support for the non-proliferation treaty. But too many deadly elements are loose in the world to guarantee they won't be picked up.

The trends are not reassuring. One State Department official said world opinion on this danger won't be galvanized until dictator-led Pakistan explodes its bomb. Even that guess is optimistic. Such a widespread affection for superkiller bombs may be so deep and myopic that humankind may not get its consciousness raised enough to do anything dramatic about it until a few countries start incinerating each other.

No longer is the world racing with its optimism but against its pessimism.

The Hartford Courant

Hartford, Conn., July 25, 1981

Even Ronald Reagan, who was saying not very long ago that it is none of this nation's business if other countries develop nuclear weapons, has come around to recognize the extraordinary danger that the spread of such weapons presents.

The policy guidelines he announced last week to discourage nuclear proliferation are largely consistent with those of former President Carter, who considered the issue of paramount importance.

The guidelines veer from former policy mainly by recommitting the nation to become a "reliable nuclear supplier" of nuclear technology to nations that do "not constitute a proliferation risk."

Mr. Reagan probably meant the nations of Western Europe and Japan, which he seems to assume will use technology and materials strictly for peaceful purposes.

The administration does, on the other hand, give considerable stress to the need for cooperation among exporting countries to prevent the spread of such technology and materials elsewhere. Mr. Reagan called particular attention to the need for curbing the spread of nuclear arms in the Middle East.

Although the guidelines lack specifics, the Reagan administration already has gone so far as to ask the Turkish government to stop shipping equipment to Pakistan that could be used to help build an atomic bomb.

Despite their protestations of innocence, the Pakistanis seem to be preparing to test an atomic bomb. Such preparations tempt attack by both the Israelis, who wiped out a nuclear reactor in Iraq recently, and India, a long-time foe of Pakistan that already has the bomb.

The administration might have a greater immediate impact first by reviving former President Carter's policy of halting aid to Pakistan until it submits its nuclear development program to international inspection.

The administration has, on the contrary, offered a $3 billion economic and military aid package for Pakistan, despite that country's refusal to sign the international nuclear nonproliferation treaty.

Further, the administration should consider including among its guidelines a recent suggestion from the Ripon Society, a think tank tied with the Republican Party.

The society suggested that the United States impose economic sanctions on any country that ships weapons-grade nuclear materials or facilities that could be used to fabricate nuclear weapons.

Turkey might recognize an economic threat, even if it does not recognize the common threat posed each time another nation arms itself with nuclear weapons.

We hope that recognition of danger, which the Reagan administration seems to have come to in its general guidelines, will soon be followed by specific and effective proposals to stop it.

SAN JOSE NEWS

San Jose, Calif., September 25, 1981

A peculiar ideological schizophrenia afflicts the Reagan administration. While advocates of solar power and conservation get lectures about the virtues of the free marketplace, nuclear power continues to receive billions of dollars in federal subsidies.

The latest case in point is President Reagan's plan to revitalize the nuclear industry. The plan has not been officially unveiled yet, but a draft of it indicates that Reagan wants the government to subsidize development of the "breeder" reactor and to pick up part of the cost of disposing of radioactive wastes, as well as to prune some of the red tape away from the reactor licensing process.

Some pruning probably is in order: It now takes 10 to 14 years, on the average, to build and license a nuclear power plant in the United States, and paper work and litigation add significantly to each plant's cost. But regulatory reform shouldn't become a pretext for tearing down safety standards. Judging by the Nuclear Regulatory Commission's recent report that 21 of the nation's 72 operating reactors are "below average" in safety and other areas, we need more effective oversight, not less.

Nor is Reagan's draft statement telling the whole story when it blames the nuclear industry's troubles on "a morass of federal obstacles including unnecessary regulations . . ." Regulatory hassles are a factor, but utilities have other good reasons for not building new nuclear plants: the high cost of money and the slowdown in the growth of demand for electricity.

Donald Winston of the Atomic Industrial Forum, a nuclear industry trade association, was quoted in a recent Los Angeles Times article as saying, "The problem of nuclear is the problem of the utility industry. They're not building anything. What's happening to the nuclear industry is happening everywhere."

A high-cost, extended-payback investment like a nuclear plant just doesn't make economic sense to utilities at a time when interest rates are soaring and power demand curves are sagging. But the administration seems determined to swim against the tide and pump more billions into the development of nuclear technology, including the ill-advised Clinch River breeder reactor project.

While subsidizing breeder reactor R&D is, at worst, foolish, another of Reagan's nuclear ideas is downright dangerous.

The Carter administration imposed a ban on commercial reprocessing of spent reactor fuel because a byproduct of reprocessing is plutonium, which could fall into the hands of a foreign country or a terrorist group itching for its own nuclear bomb. Reagan not only would lift the ban but actually wants the government to promote commercial reprocessing by helping with the disposal of highly radioactive wastes and by buying plutonium to use in the breeder reactor program.

All in all, the Reagan program looks to us like an unwise, unwarranted and frighteningly unaffordable government intervention in energy matters that are better left to the private sector. The administration ought to take its own advice to heart and butt out of the nuclear marketplace.

CHARLESTON EVENING POST

Charleston, S.C., November 10, 1981

Both partners in the Barnwell Nuclear Fuel Plant have said they want out. What then is $10 million for research at Barnwell in fiscal 1982 still doing in the federal budget?

A spokesman for General Atomic was quoted last week as saying it is commercially impracticable to proceed with the reprocessing of spent fuel. Therefore, "we want out...We want to divest ourselves," Creighton Galloway, General Atomic's executive vice president, told a reporter. Several weeks ago, Allied Corp. announced it was making provision to write off BNFP and eventually shut down the operation. Meanwhile, Congress has given final approval of $10 million for research at the plant. Now that Allied and General Atomic have said they have no plans to invest further in the plant, when will the politicans get the message?

The Providence Journal

Providence, R.I., August 18, 1981

The Reagan administration, in a draft of a new policy, seems to be going counter to its own interests in the control of nuclear power plants and nuclear materials.

In reaching for a policy that would govern U.S. actions around the world, the administration's science advisor sounds as though he would weaken regulations put in place by the Nuclear Regulatory Commission since the 1979 accident at Three Mile Island. There can be no mistaking the impetus to develop the fast-breeder reactor that past administrations had tried to halt. Similarly, the draft ignores past warnings against commercial reprocessing of spent reactor fuel.

The administration understandably wants speedier licensing of nuclear power plants already completed or under construction. Most people burned by high oil prices agree, as do many who fear that coal-burning plants will cause intolerable air pollution and health problems.

Yet undue haste in licensing new plants could create new dangers, already foreshadowed by numerous incidents of flawed equipment and human error. Certainly, new plants should not be permitted to start up while serious safety questions remain about components that have caused trouble in older ones.

The administration wants to cut down the elapsed time between start and completion of nuclear plants. The nuclear industry blames the NRC for throwing obstacles in its path. But the NRC itself has recently streamlined hearing procedures.

Laymen find it difficult to sift out the facts, because licensing is a maze of hearings and regulations, federal and local. Some critics charge that the NRC is bogged down in paperwork. Others, including NRC staff, complain about a shortage of personnel.

Moreover, there is growing evidence that the almost complete lack of new starts on nuclear plants stems from the reluctance of the investment community to risk billions of dollars on plants that are becoming so expensive as to be economic question marks.

The breeder reactor is an example. The Clinch River breeder project cannot be built without the huge federal outlay proposed in congressional bills. Previous U.S. administrations considered the breeder dangerous because it would produce plutonium, the material used in nuclear bombs. More to the point, it was considered unnecessary to U.S. needs in the next 20 years.

Reprocessing of spent reactor fuel has been banned for several years to prevent production and distribution of plutonium, which might be seized by terrorists or diverted to weapons use by other countries. A ban on export of plutonium or reprocessing equipment is regarded by many as futile, yet it can at least slow down the spread of nuclear weapons.

Whether protests against relaxing U.S. policy will have any effect in changing the draft statement, Americans ought to be aware of the hazards involved in the current proposals. No part of the country needs nuclear power more than New England, but no one in this region wants to risk another major accident. Other countries are less concerned about the spread of nuclear weapons capability, yet the United States, as still the main supplier of nuclear materials, has an obligation to continue its efforts to make controls work.

The Washington Post

Washington, D.C., September 7, 1981

A DRAFT of a soon-to-be-released presidential policy statement on nuclear power has been circulating for several weeks. It bypasses the industry's pressing economic woes, focuses on developing an exceedingly expensive and more dangerous new technology, and ignores nuclear power's single greatest problem, the loss of public confidence. If eventually adopted, it will make matters worse for an industry already in dismal shape.

Not a single new reactor order has been placed in four years, while there have been many cancellations. The four companies that build reactors have limped along by relying on exports and supplying parts required by new regulations imposed after Three Mile Island. They cannot continue in this way very much longer.

Utilities are discovering that it is far harder to operate reactors properly and much more expensive to build them than they had thought. Tenfold increases between estimates and actual costs are now commonplace. State power commissions are ordering investigations of lower-cost alternatives. Delays caused by citizen opposition are lengthening. These are generally not emotional 1960s-type sit-ins, but determined challenges over cost, siting, evacuation plans, waste disposal and other matters that were once the quiet province of engineers and accountants.

The administration chooses to interpret all this as being due to financial ill-health on the part of the utilities and overly cumbersome federal regulation. Each is part of the problem—but only a small part. No matter how financially strong a utility is, it will not turn to nuclear power if its directors or state regulators feel that this energy source is too costly or technologically uncertain.

Similarly, while no one would deny that the Nuclear Regulatory Commission can and should speed up its licensing process and reduce duplication and delay, such changes are not likely to make much of a difference in the 10 years or so it now takes to get a nuclear plant from the drawing board to power production. Much of the delay commonly blamed on the NRC is a result of faulty or overdue parts delivery, of labor problems, of construction errors and of local citizen opposition.

When it comes to breeders and reprocessing, the administration throws its otherwise strict reliance on the marketplace to the four winds. There is no reasonable prospect that breeders will ever be economically competitive. Yet the outmoded Clinch River breeder reactor is to be completed despite the fact that, after the expenditure of $2 billion in public funds, private industry is only willing to put up less than 10 percent of the billion or two more needed to complete it.

•

There are three commercial reprocessing plants in this country. One is incomplete and will only be finished if the government spends several hundred million dollars. A second was finished but never run. The third reprocessed fuel from 1968-1972, was a technical and commercial failure and was abandonned by its owner, Nuclear Fuel Services. The federal government and New York state have agreed to pick up the tab for cleaning up the contaminated site. As though none of this had ever happened, the draft policy pictures reprocessing—which also has severe proliferation risks—as necessary and financially attractive. Yet it implicitly recognizes the lack of private interest by pledging to buy the resulting plutonium "to provide a stable market."

The lack of a waste disposal policy, after 30 years, is probably the greatest single cause of public antagonism to nuclear energy. The administration's plan says only that the secretary of energy, "working closely with industry," will "proceed swiftly" to build a repository for nuclear waste. This completely misses the point: what is holding up the construction of a waste site is state and local opposition to having one and a serious debate over what technical criteria such a facility should meet to be licensed by the NRC. The plan to brush all this aside by quickly building an unlicensed facility is destined to fail. Public concern—rightly or wrongly—is too broad and too deep.

Nuclear power needs help and should get it—it is an important part of the country's energy supply—but not the kind of help the administration appears ready to give. Things would be improved by spending research funds on designing a safer and more efficient version of current reactors, a design that could be standardized, then quickly licensed. This could be done for a fraction of what is being squandered on the breeder. No public funds should be spent directly, or indirectly, on reprocessing: nuclear power is a mature enough technology to stand on its own. Finally, the administration needs to spend the time—frustrating though it will certainly be—developing a publicly acceptable waste disposal policy. If these things are done, nuclear power's future will brighten.

The Honolulu Advertiser

Honolulu, Ha., October 19, 1981

President Reagan's decision to lift the ban on domestic reprocessing of spent nuclear fuel is a mixed blessing. There are some practical pluses, but until companion policy regarding overseas reprocessing is ready, disturbing questions will remain.

Reprocessing recycles used fuel to produce plutonium which, if it is of high enough quality, can be made into nuclear weapons.

That fear led Presidents Ford and Carter to cut back, and then eventually halt, U.S. reprocessing. Carter also imposed controls on foreign countries which use American-supplied fuel.

THE REAGAN approach, which will encourage commercial reprocessing, may have a number of effects on domestic and international issues. They include:

- Reduced pressure to find more storage sites for the spent fuel. Since it is recycled, it stays in the "system" longer. This positive aspect, however, may be overshadowed if steps to better ensure plant security are not undertaken.
- As to possible international impact: If the administration (as some suspect) allows industrially advanced nations with good U.S. relations to reprocess their spent fuel, there may be a reduced need to find out-of-country storage sites. The possibility of Japan using Palmyra for such a purpose comes to mind.

A MORE COMPLETE assessment must await the administration's foreign reprocessing policy. But while that is being drawn up, a larger debate may be getting underway in this country.

Reagan's proposal, and others aimed at reducing the time to license nuclear power plants, assumes nuclear power will play an important role in American energy development. This is no surprise, since the president promised as much during the campaign.

Still, given the safety questions involved and the often emotional nature of the nuclear debate, it could become a heated issue. In that sense, real concerns about reprocessing may become secondary as larger implications of Reagan's boosting of nuclear power become clear.

Fort Worth Star-Telegram

Fort Worth, Texas, September 17, 1981

A plan whereby nuclear fuel would be recycled for use by the government — including, presumably, for use in weapons — has enough merit to warrant serious consideration by the Department of Energy.

The proposal apparently has been around for several years in one form or another, but it is currently being discussed with renewed interest by energy experts within the Reagan administration.

The plan would entail reprocessing by the government of spent nuclear fuel from commercial power plants, with the resultant fuel products to be used as the government saw fit. Use of the reprocessed fuel in the production of nuclear weapons is one possiblilty being considered.

Critics of nuclear weapons have voiced expected hostility to the proposal, and their arguments certainly should be heard before any final decisions are made as to the dcsirability of including this plan in the nation's official energy policy.

But two other aspects that add a certain logical appeal to the proposal should be taken into consideration also:

- The reprocessed fuel would provide the government with a new and reliable source of plutonium, an element that some experts fear will be in short supply before long.
- The reprocessing would relieve private power facilities of the problems of storage and disposal of the spent fuel.

The latter factor alone should help make the utilities amenable to the plan, although the industry has generally favored completely separate private and governmental nuclear programs in the past. Disposal of nuclear waste material is a major problem now and promises to be an even greater one in the future, as more nuclear power plants go on line.

It well may be that, in the final analysis, the plan's drawbacks would outweigh the advantages, but it certainly merits careful, objective study with an eye toward including its provisions in the administration's formal nuclear policy that is expected to emerge in final form over the next several weeks.

Roanoke Times & World-News

Roanoke, Va., October 12, 1981

PRESIDENT Reagan's decision to resume domestic reprocessing of spent nuclear fuel touched a nerve. Knee-jerk critics of nuclear power and of weapons policy immediately responded, conjuring up images of insidious health hazards and of A-bomb-toting terrorists on every street corner.

Emotional overreaction should not be allowed to blur the facts. There is not enough justification now for the United States to resume commercial spent-fuel reprocessing; a strong case — in terms of economics and international safety — can be made against it.

The U.S. government has its own facilities for reprocessing spent fuel from nuclear reactors. From this it extracts plutonium, which can be burned as nuclear fuel or made into atomic bombs. (This man-made element also is extremely toxic, but — short of self-ingestion — to get it into the human body is rather difficult.)

Apparently the nation doesn't have enough of this ghastly stuff; nuclear weapons production has been increasing and Uncle Sam reportedly is running short of plutonium. Revoking an order by his predecessor, Ronald Reagan has reopened the door for the private sector to get back into reprocessing.

The Reagan approach fits in with his July 16 statement taking a more relaxed view of spent-fuel reprocessing abroad. There are dangers in that, but it's true there are limits to what the United States can do to discourage other nations from producing, storing and selling plutonium.

Still, it does not follow that the United States should add to the global supply of plutonium. Reprocessing plants, says Victor Gilinsky of the U.S. Nuclear Regulatory Commission, are "potential bomb factories." About 22 pounds of plutonium are enough for a crude implosion-type device, and the technology is no longer much of a secret. Terrorists can't just walk into a plant, federal or commercial, and carry out the product; but ways of obtaining it surely can be found, and may not be detected. The U.S. plutonium inventory is short by enough matter to make several bombs, and nobody knows where the material is.

Does the country need plutonium as fuel? Gilinsky says no; the horizons for nuclear power have shrunk in recent years, and "the present generation of uranium-fueled power reactors can continue to satisfy the needs of electric power for many years to come." Unreprocessed spent fuel adds to the growing problem of radioactive waste disposal, but safe, practical solutions exist; it's basically a matter of persuading a state or region to act as a repository.

Some see the Reagan order as an attempt to revive the troubled nuclear power industry, which already has benefited by billions of dollars in federal research and investment. Getting the commercial sector back into reprocessing could require billions more.

Three such plants exist in the United States. All three had technical and financial difficulties. Only one ever operated: the Nuclear Fuel Services plant in West Valley, N.Y., which was shut down in 1972 after it was found that radioactive leaks were contaminating workers and the environment. Owners declined to invest the estimated $800 million needed to make the facility safe.

Similar problems plagued a $65 million plant in Illinois; it never did start up, and General Electric abandoned its investment.

Finally, there is the $250 million plant that Allied General Nuclear Services was building near Barnwell, S.C. Before Jimmy Carter called a halt, the company was asking the government to put up hundreds of millions more dollars to complete construction.

If more plutonium is needed — a highly debatable thesis — the government should add to its own facilities. No commercial plant can be built or operated without massive federal subsidy. That would be a strange route for a cost-cutting, free-market-minded administration to take.

The Boston Globe

Boston, Mass., December 24, 1981

The President's nuclear power development plan revealed in October called for a go-ahead on the Clinch River breeder reactor and an end to the ban on commercial reprocessing of spent reactor fuel. The rationale for reprocessing is to extract plutonium from fuel rods for use as fuel in the breeder reactor.

This dual course is unwise, not merely because the breeder reactor is expensive technology that is not needed, but because if the nation embarks on a plutonium path, it will vastly complicate the problem of nuclear weapons proliferation.

The President wants to provide moral support for the nuclear power industry. A meaningful pronuclear policy would focus on rebuilding the tattered confidence of the public and the investment community after Three Mile Island and Diablo Canyon. The main stress would be on effective regulation and on subsidization of certain safety programs. As the industry's track record improves, the public's perceptions will follow.

It would be unwise in the extreme, however, to open the door to the plutonium era. Once freed from irradiated fuel rods, plutonium is easy-to-handle bomb material. If a trigger mechanism is available – and it is not difficult to create one – a crude nuclear "device" can be fashioned overnight.

At present there is relatively little plutonium in the world, and what exists is under tight military controls. Within three or four years, under the plan Reagan encourages, there will be vast quantities of it and the control system seems likely to be overwhelmed.

The breeder reactor was initially designed when it was thought that world uranium supplies were limited and that there would be 1000 US commercial reactors in operation by the year 2000. The breeder, which runs on plutonium rather than uranium and produces more than it consumes, was seen as a solution for the impending dearth of fuel.

Since then, however, added uranium supplies have been found, and new reactors have been built at only about one fifth the rate initially expected. The sole economic rationale for unlocking plutonium has, therefore, vanished. Commercial power can be produced until at least the middle of the next century with the "throwaway" fuel cycle used to date. True, some potential energy is left in the spent rods – but the plutonium stays locked safely within as well.

The Administration plan ignores the practical reality that abundant plutonium is likely to be synonymous with nuclear proliferation. A single plant can process 1500 tons of spent fuel per year, producing 33,000 pounds of plutonium. The tightest inspection procedures are commonly assumed to lose track of at least 1 percent of the material being controlled. In this case, that would amount to 330 pounds. Only 10 or 20 pounds are adequate for a Hiroshima-sized nuclear explosion.

As a function of its availability, then, it seems a possibility that some plutonium will be purposefully diverted. As time passes, chances will multiply that some government or political group that has built a bomb or bomblet will cross its threshold of desperation and touch one off.

Maybe the most calming aspect of the Administration's push for reprocessing is that the owners of the three partially completed commercial reprocessing plants in this country are plainly not interested. That is partly because they know the breeder reactor has nothing to do with the problems of the industry, which are financial. The plant owners may also be more sensitive than the President to the liabilities of operating "commercial" reprocessing facilities, which will be branded bomb factories.

However, if the domestic nuclear power industry may spurn Reagan's plutonium plan, some other nations, such as France, are close to crossing the same Rubicon. They should be stayed from that course, not encouraged by official American policy to pursue it. It will be a historic error if, in its wish to help the commercial nuclear power program, the Reagan Administration needlessly hastens the world past the point of no return on plutonium. That would undercut this country's ability to seek effective controls on nuclear proliferation and leave an indelible stain on the future.

The Houston Post

Houston, Texas, September 27, 1981

The Reagan administration is chewing on an old and controversial idea — using waste fuel from commercial nuclear power plants to make atomic weapons. Even reports that it was being considered again drew sharp criticism from anti-nuclear organizations. And the electric power industry may oppose it.

Opponents of government conversion of commercial reactor waste to nuclear arms production argue that it would be contrary to what we have long practiced and preached both at home and abroad: Atoms for peace should be kept separate from atoms for war. To erase that barrier, they contend, would undercut our nuclear non-proliferation policy. The nuclear power industry, for its part, is hesitant about being linked to the military applications of nuclear energy by having its spent uranium fuel turned into weapons-grade material.

Yet government use of spent commercial reactor fuel for our nuclear arsenal offers at least a partial solution to the problem of disposing of high-level radioactive wastes piling up at nuclear plant sites around the country. It would also relieve what military planners warn is a looming shortage of plutonium, a key ingredient in our nuclear weapons. To extract the plutonium from the spent fuel, however, would require reprocessing. President Carter ordered a halt to commercial reprocessing of nuclear wastes in 1977 in an effort to curb the global spread of nuclear weapons and the technology to make them.

The Reagan administration has said it plans to lift the reprocessing ban as part of its program to revive the nuclear power industry. But Energy Secretary James B. Edwards stressed that the idea of using commercial reactor wastes is still in the "conversation phase." Another Energy official observed, however, that the dividing line between the peaceful and military applications of nuclear energy "may be more psychological than real." Still, no decision to blur that distinction should be made without a determination at the highest level that it is necessary to national security.

The Miami Herald

Miami, Fla., October 15, 1981

PLUTONIUM is the critical element in nuclear weapons. Nuclear power reactors, such as Florida Power & Light's at Turkey Point, generate plutonium, but it is inherently mixed together with highly radioactive material in the reactors' spent fuel. In that state, it can't be used in a nuclear weapon.

President Reagan has decided that industry should be allowed to build facilities that chemically separate plutonium out of spent reactor fuel. If separated, the plutonium can be used anew as reactor fuel — or to make nuclear weapons.

Presidents Ford and Carter both thought that permitting huge supplies of plutonium to be produced commercially raised high risks that nuclear weapons might proliferate throughout the world. They both also recognized that terrorists might steal enough plutonium to make crude but deadly nuclear weapons. In 1977, President Carter banned commercial reprocessing of spent nuclear fuel.

President Reagan, according to a Government spokesman, didn't even consider whether his decision to end that ban might increase the danger of nuclear-weapons proliferation. All experts concede that proliferation is the central peril posed by nuclear-fuel reprocessing. Therefore, Mr. Reagan's failure even to consider it before lifting his predecessor's ban is simply irresponsible. And irresponsibility on this grave matter is intolerable.

The danger cannot be overstated. Constructing nuclear weapons ordinarily would be beyond the capacity of, say, a terrorist group such as the PLO. If they had reprocessed plutonium, however, skilled technicians could build a PLO nuclear bomb in a few months. Or a Khadafy bomb. Or a Khomeini bomb. Should such material be an item of commerce, bought, sold, shipped, stolen, as if it were just another product?

The weapons threat, while the most serious problem with commercial reprocessing of nuclear fuel, is hardly the only one. Nuclear wastes become much more difficult to handle, according to an authoritative 1977 Ford Foundation study. Conventional spent reactor fuel is relatively easy to store. Reprocessed fuel produces high-level radioactive wastes, acidic liquid wastes, trash contaminated by plutonium, and more. The United States still has no answer on where and how to store its nuclear wastes permanently. Yet Mr. Reagan has just complicated that problem further.

The supposed advantage of reprocessing is that it expands the fuel supply available for nuclear power. Yet the Ford Foundation study concluded that at best it would expand the supply by 20 per cent. That study said this would not be necessary in this century, and probably never. Any economic benefits arising from reprocessing were "very small even under optimistic assumptions," it added.

What's more, it is unlikely that any commercial facility can survive economically without heavy Government subsidy. Before they were banned by President Carter, the two that were attempted in the United States both failed. None operates abroad without subsidy.

Viewed from any angle, Mr. Reagan's decision appears indefensible. Congress should overrule his decision, and fast.

THE MILWAUKEE JOURNAL

Milwaukee, Wisc., October 24, 1981

Apparently it will take more than presidential pumping to breathe new life into the nation's moribund nuclear power industry. President Reagan's recent support for commercial nuclear-fuel reprocessing has collided with some tough economic considerations.

In announcing several steps to bolster nuclear power development, Reagan lifted the prudent ban that Presidents Jimmy Carter and Gerald Ford had imposed on fuel reprocessing. Carter and Ford had feared that reprocessing, which produces plutonium, would increase the risk of nuclear-weapon proliferation worldwide. Reagan's decision to lift the ban flew in the face of his own promise, announced earlier this year, to maintain efforts to prevent the spread of nuclear weapons as "a fundamental national-security and foreign-policy objective."

Yet, despite Reagan's new stance, the owners of the nation's primary fuel reprocessing plant announced the other day that it would be closed as a commercial failure by 1983 if no buyer offered to take it over.

Allied Corp., which owns the plant at Barnwell, S. C., said that, in the current economy, utilities were not willing to buy reprocessed fuel. A spokesman for the company cited the expense of environmental studies necessary for the transportation and use of plutonium. And he predicted that no one would be willing to go into the reprocessing business unless the government promised to buy reprocessed fuel. (Neither of the nation's other two reprocessing plants is operating.)

This reprocessing issue is just one more indication that the future of nuclear power is now dependent on the government's willingness to bail out the industry by easing environmental regulations and providing other subsidies, such as a guaranteed market for reprocessed fuel. And while Reagan so far has balked at offering substantial subsidies, he has broadly proposed cutbacks in the regulatory process.

We think Reagan is right to resist any financial rescue of the industry, particularly when the expected demand for electricity has fallen sharply in recent years and the cost of constructing nuclear power plants has risen greatly. But the president is taking unacceptable risks in his go-ahead for commercial fuel reprocessing and his proposals for easing the government's regulatory oversight.

While U.S. Stalls, Other Nations Develop Nuclear Energy

As the United States' embattled nuclear industry struggles for survival under the burdens of increased costs, public opposition and decreased demand for electrical power, the rest of the world is moving ahead with their nuclear energy programs. France has progressed furthest among the large industrial nations, currently producing nearly one half of its electricity with nuclear reactors. The first commercial breeder reactor is scheduled to begin functioning in France this year. France's rapid nuclear development particularly rankles U.S. industry officials because the standardized French reactors are copied from U.S. reactor designs. Great Britain has projected a rise from 16% to 20% of its electricity generated by fission by the end of 1984. In Japan, a relative newcomer to the nuclear energy field, the atom generates 19% of the nation's electricity, as compared with 13% in the United States. As these and other nations plow ahead at varying paces with their nuclear power programs, proponents of nuclear energy in the U.S. worry that this nation will fall behind and be forced, in the future, to buy its reactors from abroad.

Houston Chronicle

Houston, Texas, June 11, 1980

The "lessons" of the Three Mile Island nuclear plant problems are interpreted differently in Paris, London, Moscow and other seats of power abroad than they are in Washington. While the U.S. nuclear industry remains in a state of inertia, the rest of the world is pushing full steam ahead with the development of nuclear power.

The French have been the most ambitious, with a commitment totaling some $8 billion annually for the development of a breeder reactor. On completion of their "Super Phoenix" project in 1982, it is predicted that French scientists will have a 10-year technological lead on their American counterparts. By 1985, 55 percent of France's electric power will come from atoms. The British government concluded its own independent investigation of Three Mile Island with a decision to accelerate nuclear power development. Following a favorable public referendum last March, the Swedes plan to double nuclear output. The story is the same throughout the rest of Europe and Japan.

The East Bloc nations are pushing ahead with similar speed. Comecon, the Soviet bloc's economic alliance, has a target of producing 25 percent of its electric power with nuclear reactors by 1990, up from 5 percent.

The effects of the Three Mile Island incident on the U.S. nuclear industry could hardly be more devastating. In 1979 there were no new orders by U.S. utilities for reactors and 11 cancellations of earlier orders. More importantly, the burden of proof for safety was shifted squarely onto the nuclear industry's shoulders; this even though no lives were lost and the industry's safety record is unparalleled.

For all of the self-righteousness of the "no nukes" protests that have come into vogue, the hard facts are that nuclear power development is not extravagance, but necessity. Indeed, exploitation of *all* energy sources will be needed if there is to be economic growth in the next 20 years.

For all their potential, the energy "exotics" such as solar, wind and geothermal cannot possibly meet the surging demand for power, which is expected to grow at the rate of about 4 percent per year. That leaves nuclear power, along with coal and natural gas, for power to keep America's factories running and her houses heated and cooled.

Yes, there were lessons to be learned from Three Mile Island. But they're the ones being applied by the rest of the world, not by crusaders who would, quite literally, leave us jobless, powerless and in the dark.

THE INDIANAPOLIS NEWS

Indianapolis, Ind., May 16, 1980

Quickly now, what country uses more nuclear energy than any other country?

Britain? The Soviet Union? The United States? France? No. Belgium.

One-fourth of the country's electricity comes from atomic plants. That is 6 to 7 percent of its total energy needs. In eight years, Belgium plans to get 53 percent of its electricity from nuclear plants or better than 16 percent of its total energy requirements.

The Belgians are embracing nuclear plants, according to the *Christian Science Monitor*, to help diversify their energy supplies. The country is heavily dependent on imported oil, primarily from OPEC members, and is trying to obtain long-term contracts with suppliers while cutting back on imports. Other energy sources are natural gas from the Netherlands, Algeria and Nigeria and coal from West Germany, the United States and South Africa (to make up for its declining coal production).

The Belgians, whose per capita use is greater than the West Germans, the British and the French and only slightly less than the Americans, are facing up to the necessity of using nuclear power to fulfill their energy needs. The United States should make the same sure commitment — more sure than it is already — to developing safe nuclear power.

FORT WORTH STAR-TELEGRAM

Fort Worth, Texas, July 16, 1980

Anyone with "nuclear plant neurosis," as one observer describes it, would do well to consider the example of the Japanese.

In Japan today, there are 16 atomic energy plants in operation and plans have been announced for 19 more.

That is in spite of the fact that the Japanese have had a firsthand, horrible experience — in the atomization of Hiroshima and Nagasaki — of what nuclear power could do in its earliest, crudest form as a weapon.

You'd think the Japanese would have developed an anti-nuclear nurosis. They certainly couldn't be blamed for it if they had.

Instead the Japanese plan to build more nuclear plants.

The answer is simple. The Japanese know that the risks of atomic power are small when compared with the risks of an energy-short Japan.

Total energy demand in the United States will increase between 40 percent and 80 percent by the end of the century, the Electric Power Research Institute estimates.

"Even the low end of the range suggests an enormous need for added power facilities," the *Morgan Guaranty Survey* says.

"Clearly, failure to move ahead with nuclear energy risks power shortages in this country in the 1990s and beyond" and, with all that, "higher inflation, slower economic growth, continued energy dependence, and national insecurity."

The current anti-nuclear activity in the United States isn't common to the world. Nations abroad are pushing ahead with ambitious nuclear power programs, spurred by precarious oil supplies and rising prices, the *Morgan Guaranty Survey* reports.

France even proceeds with a "breeder" reactor program. Germany, Spain and Italy plan major nuclear expansion. Belgium, Switzerland and Sweden plan to produce 40 percent to 50 percent of their electricity by nuclear in 1990.

Some nations, Japan being among them, probably would risk the nuclear option out of self-preservation even if nuclear power generation were riskier than it is.

The wonder of nuclear power neurosis is the exaggerated concern. The Rasmussen Report estimates the likelihood of a nuclear accident claiming 100 or more fatalities to be one in 100,000 years, and the likelihood of an accident claiming 1,000 lives is once in a million years.

Those who point to the Three Mile Island accident can be comforted in the fact that the Three Mile Island safeguards worked. The Rasmussen assessment remains intact. Nuclear is about as risky as being hit by a falling meteorite.

DESERET NEWS

Salt Lake City, Utah, July 6, 1981

Despite the recent announcement by France's new Socialist government that nuclear power plans for a proposed 5,200-megawatt plant have been placed on a back burner, the Paris government in no way is abandoning nuclear power development.

The new president of France, Francois Mitterrand, has already agreed that all nuclear power plants under construction shall be completed. But before any new plants are started, he wants a government review of the nation's energy policy. The matter will be debated next fall by the recently-elected National Assembly.

For the past 10 years, France has been developing nuclear power at a fast pace. The country has no oil production and very little coal and natural gas. Nuclear power was seen as the best alternative to the importation of high-priced oil.

France now has 24 nuclear power plants in operation. That many more are under construction and at least a dozen are on the drawing boards. From a fourth to a third of the country's power comes from nuclear plants. By 1985, nuclear plants will produce more than half the country's electricity.

In France, nuclear-generated power costs only a third as much as that produced by oil-fired plants and half as much as the output of coal-fired plants. For this reason, Mitterrand should find it hard to drop plans for nuclear expansion no matter how much he may want to.

Even with the rapid construction of nuclear plants, France has had to import increasingly expensive quantities of oil. The cost of these imports went from 14 billion francs in 1973, to about 110 billion francs in 1980. Opposition to the nuclear program in France is minor, the smallest in any industrial nation. After a decade of education by the government, the general public is pretty well sold on the need for nuclear development.

Without abandoning this development, the Mitterrand government wants to review the situation, emphasize conservation, develop other sources of energy, and research alternate fuels and resources. This makes sense provided the program is not studied and debated to death.

If France continues the energy policy it has followed for the past 10 years, it could soon outclass the U.S. in developing nuclear power. This should give some pause to those who have insisted that the U.S. back off from such technology.

THE DENVER POST

Denver, Colo., April 3, 1981

WHILE THE Carter administration was allowing U.S. nuclear power policies to drift, other nations moved to take over leadership of this industry pioneered by Americans in the 1950s.

The world's first nuclear power stations were adapted from U.S. atomic-powered submarines. The first breeder reactor was not built in France or Russia, but in Idaho. U.S. nuclear power set the early pace.

Our nuclear technology is still excellent. But it is mired in an inheritance of red tape left by anti-nuclear policies of the Carter administration. Another factor was the Three Mile Island accident two years ago. Although costly in monetary terms, the event caused no injuries. Yet it created questions in the public mind that helped dampen U.S. nuclear construction.

The extent of the trend is shown by a new survey by the Atomic Industrial Forum, a trade group. It suggests that while the United States wallows in nuclear indecision the rest of the world moves ahead. Since 1979 not a single U.S. nuclear plant has been ordered; other nations have plans for 14. Our 76 operating plants now compare with 180 stations elsewhere.

France and Japan are moving up rapidly. With its new breeder reactor technology added to 70 conventional, water-cooled plants, France expects to produce 56 percent of its power from the atom in 1990. Japan will surpass France in total megawatts by the year 2000. Last December, the Soviets started up a 600-megawatt breeder reactor, the world's largest.

Invoking national pride to suggest a U.S. policy of "catch up" isn't proper. But what is significant is that nuclear power economics is proving itself worldwide. Such proof is available here as well. Illinois' Commonwealth Edison Co. says its six newest coal-fired plants produce electricity for 32.6 mills per kilowatt-hour compared with 17.3 mills for the firm's six nuclear plants. The gap has widened every year since 1974.

Fortunately, the Reagan administration is reversing the Carter tendency to surrender U.S. electrical options to the "soft energy" lobby and its excessive emphasis on conservation and diffuse energy sources. There is nothing wrong with conservation and alternative energy sources, but there is a present need for nuclear power in large, efficient central generating stations.

Reagan's decision to build the Clinch River experimental breeder reactor in Tennessee is a sign that the administration intends to move nuclear's share of the U.S. electric power mix from 12 to 25 percent by the 1990s.

There are encouraging local developments. The St. Vrain plant of Public Service Company of Colorado, near Platteville, has been kept at a limited operating level because of its pioneering design. Now the Reagan administration has approved speeding the plant up to its full capacity of 330 megawatts. It is a welcome decision: St. Vrain is potentially an extremely efficient and safe design. It must, however, prove itself in this trial period.

The nation also needs to standardize and shorten permit procedures. It presently may take more than 10 years to complete a plant after groundbreaking. Such delay increases cost to companies. Additionally, the Nuclear Regulatory Commission must improve performance. After the Three Mile Island accident was caused by valve failure, it was disclosed that a similar defect in another plant had been reported earlier. NRC neither circulated the report nor asked a check of other valves. This does little for NRC credibility.

The industry is helping its own cause by maintaining a good safety record. Despite a tiny minority of dedicated opponents, public opinion supports nuclear energy. For example, a recent poll by National Geographic showed that despite the magazine's strong support of "soft" energies, 52 percent of its readers believe nuclear is a "desirable" form of energy.

What citizens are saying is that the nuclear trade-off is acceptable. It is a technology requiring scrupulous safety procedures in return for low-cost power production without smog or acid rain. And because nuclear's fuel source — uranium — is applicable almost exclusively to power production, it can economically supply world electrical needs for hundreds of years if it is developed wisely.

The Detroit News

Detroit, Mich., April 17, 1981

A bumper sticker proclaims: "Nuclear Plants Are Built Better Than Jane Fonda."

Unfortunately for the nation, Ms. Fonda seems to be having the last laugh. The release of her film, *The China Syndrome*, coincided with the Three Mile Island mishap, and that may be part of the reason America's production of electricity from nuclear power declined the past two years from 12.5 to 11 percent.

While an average of 200 nuclear reactors have provided electricity to the free world for a decade, the United States remains hesitant. The Nuclear Regulatory Commission has reduced the production of U.S. plants to 55 percent of capacity.

In contrast, Belgium generates 23 percent of its electricity from nuclear power; Sweden, 22; Switzerland, 17; the United Kingdom, 14; and France, 13 percent. By 1983, the French will triple their nuclear capacity and 12 other nations, including Russia, will have doubled theirs.

In one year, nuclear energy could replace one million barrels of oil burned daily in the United States to produce electricity. This would reduce import costs by nearly $15 billion a year. Nuclear critics, for their part, assert the cost is too high because safe waste disposal is impossible and atomic-fuel supplies are inadequate.

But safe disposal techniques were documented by the American Physical Society in 1978, and Sweden adopted these methods with minor modifications to the satisfaction of a national plebiscite in 1980. Energy from fission is available for the foreseeable future. Improved uranium-mining techniques, the likelihood of a fission-fusion hybrid, and improved design of reactors will provide fuel for thousands of years.

The issue of nuclear energy is being addressed by other nations while the industry in the United States staggers amid the whims of four federal regulatory agencies. Whatever technological edge the U.S. industry once possessed is lost. President Reagan favors regaining the initiative, and his call for action has provoked the familiar alarms.

Hysteria, of course, makes for grand drama and blockbuster movies. But it serves poorly as an underpinning of policy.

THE BLADE

Toledo, Ohio, June 15, 1981

THE United States continues to penalize itself by falling behind other industrial nations in the construction of nuclear power plants, a dismal fact reported in a new study by the Atomic Industrial Forum, an association of industry, utility, and university representatives.

Even considering that western European nations and Japan have more incentive to go nuclear because of their heavy dependence on imported fuels, the figures are devastating. As described in an article by Blade Science Editor Michael Woods, the number of foreign orders for atomic plants in 1980 was more than double the number in 1979, while in the United States no new orders at all were placed by utilities either in 1979 or 1980.

There are, of course, many reasons for the lack of progress in this country, among them overreaction to the Three Mile Island accident, increased costs, bureaucratic delays, and the presence of alternative sources of energy including hydroelectric power, oil, and coal. But even so there is a need for a more substantial nuclear-power capability in the mix than now exists — at present only 11 per cent of U.S. power is generated by nuclear plants, less than any other industrial nation. Hydroelectric power is limited by geography, oil is a dwindling nonrenewable resource despite the current glut, and coal still presents pollution problems, even though these are gradually being overcome.

In addition to lagging behind such countries as France, West Germany, Great Britain, Japan, and even the Soviet Union in nuclear power-plant construction, the United States also is dropping behind in fast-breeder technology, the techniques of reprocessing atomic fuel, and the disposal of fuel wastes.

Not only did domestic utilities not place any new orders for nuclear plants in 1979 or 1980, but they also canceled plans for 16 reactors and delayed construction of 69 others. And that is a direct reflection of the oppressive regulatory factors which have spelled both inordinately long periods from planning to completion and the high costs that go along with such delays.

Other nations do not impose such penalties on their own industries, a fact reflected in their impressive rates of installation of nuclear facilities. It is time the Federal Government and the Nuclear Regulatory Commission in particular began to encourage rather than discourage the planning and construction of more of these plants in the United States.

THE ARIZONA REPUBLIC

Phoenix, Ariz., December 14, 1981

AS the owner of the world's fourth largest supply of crude oil and gas, Mexico could conveniently fuel a new industrial revolution with its new-found domestic energy supplies for years to come.

But Mexico wisely recognizes that even its 72 billion barrels of proven petroleum reserves are a finite source, to be used only as necessary.

Hence, Mexico has embarked on a crash program to develop nuclear energy as a principal source of power.

In today's dollars, at least $25 billion has been earmarked by the government to building a string of 20 nuclear reactors. Total output of the new plants will be 20,000 megawatts — which will therefore double Mexico's present total electrical power production of 17,000 megawatts.

Mexico's nuclear commitment is in sharp contrast to the chaos north of the border, where nuclear power is held hostage to legal assaults and bureaucratic red tape.

The scent of a major nuclear reactor building program has lured companies from throughout the world to bid for the work.

The United States, Canada, Britain, France, West Germany and Sweden are the major contenders.

Among all the nations seeking Mexico's nod to build the plants, the United States may be in the weakest position.

The nuclear system being promoted by U.S. companies involves the use of enriched uranium reactors, instead of heavy water reactors that use crude uranium, of which Mexico has ample reserves.

In 1978, then-President Carter held up a supply of enriched uranium for a small trainer reactor in Mexico, insisting the United States retain the right to supervise uranium in Mexico.

Some Mexican officials fear, therefore, that Mexican nuclear power could be held continuous hostage in the future to U.S. enriched uranium supplies if it buys the U.S. technology.

Moreover, Mexicans also remember the Carter administration's death blow to a deal for Mexican natural gas when Carter would not agree to the price Mexico wanted.

At home and abroad, the U.S. seems to be all thumbs in handling energy affairs.

The Oregonian

Portland, Ore., June 23, 1982

Failure of the United States for the past four years to proceed with an orderly nuclear power plant construction program will slow the rate of the nation's economic recovery and likely will increase the nation's energy dependency.

These are the conclusions of an article published in the current issue of Harvard Business Review — a well-documented report that suggests that the Reagan administration cannot possibly hope to quicken the nation's economic pulse unless the nuclear industry is able to break out of its slow- or no-growth pattern.

Evidence of U.S. energy dependence on foreign sources, usually quantified in terms of barrels of oil pouring in from OPEC nations, can be gathered for nuclear energy with the little-noted announcement in the Wall Street Journal that the United States, for the first time, will begin importing nuclear power from Canada this summer.

Canada's National Energy Board in late April gave its approval for the export to utilities in Maine and Massachusetts of up to 205 megawatts of power (one-fourth the output of a Trojan nuclear plant).

The specter of importing nuclear power from another nation, while a disturbing trend that could continue if the U.S. economy heats up suddenly, could be accommodated as an interim necessity while the nuclear industry is revived to health. The gloomy prospects forecast in the Harvard Business Review article, however, are another matter.

Peter Navarro, a researcher at Harvard University's Energy and Environmental Policy Center, has analyzed the decline in this nation's electricity reserves. He projects impending brownouts on the basis that the reduced economic growth rate in the nation is a superficial variable that has created erroneous load-growth assumptions for the future. Can anyone be certain this is not also true of the Pacific Northwest?

Navarro claims that the nation's electric utilities, in aggregate, must spend more than $300 billion for new power plants in the next decade "just to keep the lights on and the economy growing." If he is right, where will the risk capital for these multibillion-dollar ventures come from? Utility executives in the nation already are pursuing strategies of minimizing capital spending, while recent discussions on Wall Street have revealed that nuclear plant bond buyers are becoming increasingly timid as both risk and uncertainties surrounding nuclear power increase.

The best sign of some relief for the beleaguered nuclear industry is the news that Congress, at long last, may resolve the 25-year struggle to mandate a national policy for managing and storing nuclear waste.

Legislation to accomplish a nuclear waste management program, with schedules set for the operation of interim storage facilities and permanent geologic respositories, appears closer to reality than at any time in the past. Once such a program becomes law, many nuclear projects, placed in limbo by states awaiting resolution of the nuclear waste question, may proceed, no doubt allaying much of Wall Street's anxiety.

The Union Leader

Manchester, N.H., September 5, 1983

"Today we find ourselves in the ironic position of watching Europe and Asia move more aggressively toward greater use of nuclear power while in the U.S. plant orders have come to a halt, and a majority of the public believes we should build no more nuclear plants."

Thus spoke Harold B. Finger, president of the U.S. Committee for Energy Awareness, before a recent meeting of the Uranium Institute in London. What is really ironic, however, is that domestic foes of nuclear power and those they have deceived seem not at all concerned by that reality, if indeed they are aware of it at all.

The facts are alarming. Today, we have 85 licensed reactors. The rest of the world has 207. Last year about 13 percent of our total electricity came from nuclear; ***nine*** other countries surpassed that!

Finger underscored the irony of our predicament:

"We debate endlessly over an experimental breeder reactor while other countries embrace breeder technology. We have mothballed our only reprocessing plant while other countries possessing fewer energy sources move toward fuel conservation through reprocessing. Only now have we begun addressing the *political* issues of waste disposal while many of you have long recognized the *technological* feasibility and a few of you have moved ahead with disposal programs."

One the basic reasons for America's role reversal, from leader to follower, Finger believes, is that the public, with little perception of the problem faced by utilities that must plan more than a decade ahead for new electric generating plants, misunderstands the variety of energy, electricity, and nuclear power issues.

According to a USCEA poll of public opinion:

- A large part of the public no longer believes that additional electric capacity will be needed, or believes that if it is needed, it will be easy to provide it.
- An increasing percentage opposes further building of nuclear plants, more because they think it isn't needed than because they're concerned about safety.
- In somewhat a contradictory position, almost half believe nuclear energy will be generating close to half of our electricity by the year 2000.
- About half think nuclear plants can explode like an atomic bomb!
- About two-thirds think that no method has yet been developed for safe and permanent disposal of nuclear wastes.
- Fully 82 percent are very worried or somewhat worried about safety with regard to radioactive waste disposal.

Against these misperceptions, Finger arrayed some key facts:

Since the 1973 Arab Oil Embargo the United States has cut back on the direct use of all energy forms *other than* electricity by nearly 15 percent. However, the demand for electric power has ***increased*** more than 20 percent in the past ten years—the only energy form that increased in demand, representing 90 percent of new energy supplies over this same period.

The blunt truth that foes of nuclear power generally, and the Seabrook Nuclear Power Plant specifically, refuse to face is that although we ***now*** have an abundant supply of electricity, we can look forward to increased need for electrical capacity in the decades ahead. This new electricity will ***not*** be provided by solar power or renewable sources such as wind and geothermal, or from sources that are simply too valuable and costly to burn in power plants. At least in the foreseeable future, it will come from ***coal and uranium***.

Yet, despite the fact that nuclear power is estimated to have saved American consumers **$30 billion to $40 billion** since 1974, despite estimates that the phase-out of the 85 plants now licensed to operate would cause costs of electricity to jump about **$120 billion** between now and the year 2000 (*$220 billion* if the plants now under construction were stopped), the public — thanks in large measure to the preference of the news media to alarm rather than to inform — remains in a state of near-panic. This despite the fact that 730 reactor years of U.S. commercial reactor experience have produce ***not one*** injury to the public — and that includes Three Mile Island — to say nothing of the 2,100 reactor years in the rest of the world.

The anti-nukes are running the biggest con game in town.

THE DAILY OKLAHOMAN

Oklahoma City, Okla., April 11, 1983

COMMERCIAL nuclear power may be stalled in the United States, land of its origin, but it's developing like gangbusters elsewhere in the world.

That's evident in the annual survey by the Atomic Industrial Forum of nuclear power plants outside the United States for 1982. Electricity generated from uranium fission is on the upswing in 40 countries, despite the global recession.

Among the striking developments reported in the AIF survey are the first reactor order by the People's Republic of China for two generating stations near Shanghai by 1988, start-up of Brazil's first plant, major moves toward nuclear self-sufficiency by India and Japan, which began commercial operation of spent-fuel reprocessing plants, and South Korea's launching of a breeder reactor development program.

AIF tallied five new reactor orders in 1982, and eight new plants went into service. By comparison, while six new plants were licensed in this country, no new reactor orders were recorded. And none of those new orders elsewhere went to American firms.

France added to its lead as the nation most dependent on nuclear energy for electricity; atomic stations now provide more than 40 percent of that nation's power. France remains on track toward its goal of having essentially all of its base-load power requirements from nuclear plants by the end of the century.

Japan also is forging steadily ahead on all nuclear fronts, increasing its installed nuclear capacity more than 10 percent last year with the addition of two reactors and bringing the total to 25 currently in operation.

The Geneva-based International Atomic Energy Agency reports that nuclear power plants now account for 9 percent of the world's electricity — up a full percentage point from 1981.

Quite obviously, the anti-nuke minority hasn't made much headway anywhere except in this country, where public apprehension has been exploited in the wake of the Three Mile Island incident (which harmed nobody) and by such biased movie thrillers as "The China Syndrome."

What the public apparently doesn't realize is that the U.S. lead in nuclear technology is slipping away, as other countries move ahead with development while we stand pat.

For example, those five new reactor orders last year were all for Light Water Reactors, the type developed in this country that represent more than 60 percent of all foreign nuclear power commitments. Yet none of those orders went to American firms.

There is no rational explanation of why foreign countries like the Republic of China on Taiwan can put a nuclear plant on line in five years, while it takes more than twice that long in the United States. At this rate nuclear power could join the list of other American inventions and technologies whose potentials were perceived more accurately abroad than at home.

Part IV: Economic Issues

All of the controversies over nuclear plant safety and nuclear wastes will become largely irrelevant if the nuclear power industry fails to survive financially. Even before the Three Mile Island debacle in 1979, the optimistic 1960's industry slogan describing nuclear-generated electricity as "too cheap to meter" had been turned into an effective taunt by industry critics. Lengthening plant construction schedules, a growing body of costly regulatory requirements, an electrical demand that fell far short of predictions, public concern over such issues as the possibility of plutonium thefts from nuclear plants: all of these have combined to cause serious setbacks and skyrocketing costs. The list of cancelled plants continues to grow longer and longer as utilities decide that their ventures into nuclear power are too costly to complete.

From the standpoint of the consumer, the completion of these increasingly expensive plants is often an unwelcome prospect. Most plants currently nearing completion will replace coal-burning plants, and will cost area ratepayers far more than they are now paying for electricity. Although industry officials claim that over their useful lifetimes the plants will actually save customers money, this actually depends on such hard-to-predict factors as the future price of oil and national demand for electricity. Until the plants are turned on, utilities usually cannot pass on construction costs to consumers; the sharp initial jump in price when a plant enters commercial operation goes mostly to repay accumulated construction costs, plus interest, to the plant manufacturers or other loan source. Cases have been reported in which new plants will increase the cost of electricity to ratepayers more than threefold. In some cases, the new plants will provide substantially more energy than the area requires. It is this increasing cost-per-kilowatt-hour, many analysts say, that may spell the death of the nuclear power industry.

The alternative to completing an expensive nuclear plant, however, can be equally unattractive to consumers. Cancelling a plant, particularly if construction is well under way, can itself cost billions of dollars. The costs of abandonment are usually absorbed by the Government (because of tax laws), and by shareholders and ratepayers. With the continued rise in construction prices and the decreasing costs of alternative fuels, more utilities than ever before are making the decision to cancel nuclear plants when they are already more than half-completed. Since the cancellation costs of a nuclear plant rise in proportion to the amount that has been invested in it, these late cancellations entail greater contributions by ratepayers. A new dimension in cancellations was reached with the 1983 default of the Washington Public Power Supply System on $2.25 billion in municipal bonds. The impact of the default was felt on Wall Street, as well as by the economy of the Northwest. Although an atypical case, the "WHOOPS" default has dramatized the potentially grave consequences of multi-billion dollar plant cancellations.

In a separate area of nuclear power economics, that of plant accident liability, the ratepayer has also been involved. The ballooning cost of the cleanup at Pennsylvania's Three Mile Island-2 plant has been and will continue to be borne in part by electricity users in the area. Widespread apprehension about the low limit on the financial liability of the industry in the event of an "extraordinary nuclear occurrence," set at $560 million by the Price-Anderson Act of 1957, has made this aspect of nuclear energy economics a highly controversial one. Numerous attempts have been made in Congress to establish a higher threshold of financial responsibility for the plants and their owners, and thereby reduce the risk that injured victims of an accident would receive little or no compensation.

Many Factors Contribute to Skyrocketing Construction Costs

Everyone is agreed that the cost of building a nuclear power plant has skyrocketed in the last ten or twelve years; by some estimates, the average total construction cost of a plant completed in 1983 has increased tenfold over one completed in the early 1970's. Some of the worst examples of cost overruns have repeatedly made newspaper headlines: New York's Shoreham plant, started in 1965 at an estimated final cost of $241 million, still not on line and expected to cost at least $4 billion; Michigan's Midland plant, launched in 1969 at an estimated cost of $267 million and now about 85% complete, having cost $3.4 billion so far; New Hampshire's two-reactor Seabrook plant, estimated originally to cost less than $1 billion, but now put at $5.8 billion, with Unit 1 slated to go on line in 1985 but the future of Unit 2 uncertain, etc.

Part of the explanation for such tremendous cost overruns can be traced to unanticipated increases in inflation and in interest rates. Beyond this point, however, nuclear industry adherents and critics are in complete disagreement about the reasons for the cost overruns. Proponents say that massive delays in construction have been caused by public hearings concerning the safety of proposed plants and by the frequent design modifications required by the Nuclear Regulatory Commission. Most of the new regulations, they say, are unnecessary to plant safety. Critics, on the other hand, maintain that most of the new regulations have been instituted because of poor management and lax quality control by the industry, pointing to such incidents as the flaws found in California's Diablo Canyon plant blueprints. (See pp. 56-60.) A resounding blow was dealt to the industry in January 1984 when the Nuclear Regulatory Commission denied an operating license for the Byron Nuclear Power Station near Rockford, Ill. Construction of the $3.35 billion Byron plant, owned by the Commonwealth Edison Co., had been nearing completion. The NRC said they had "no confidence" in the utility's monitoring of contractors' work on the site. (See also pp. 224-235.)

One controversial measure sought by the industry to alleviate financial distress during the construction of nuclear plants is to allow electric utilities to charge customers for plants still under construction. If the money spent on power plants were included in the rate base of a utility's prices, the industry argues, customers would have to pay higher rates but would avoid the huge increase that occurs when a new plant goes on line. They also maintain that consumers would pay lower electric bills over the life of the plant. The objection to the "CWIP" (construction work in progress) idea from the consumers' viewpoint is not just that there would be an immediate hike in rates. Many ratepayers, such as the elderly, would end up paying for nuclear plants that would never benefit them, consumer groups argue, and in some cases customers might pay for the construction of a plant that was eventually abandoned.

The Union Leader

Manchester, N.H., May 22, 1979

With all of the well-publicized difficulties experienced by the nuclear industry in recent months, it is noteworthy that the positive side of the nuclear industry is seldom reported upon — let alone emphasized— by the news media.

Yet, according to the 1978 economic survey of 43 of the 48 nuclear plants that responded throughout the nation, the cost of generating electricity from nuclear plants has remained stable, for the third year in succession, while the cost of generating electricity from coal and oil has risen. The cost comparisons compiled each year by the Atomic Industrial Forum have produced some interesting statistics:

A nuclear kilowatt-hour of electricity generated by nuclear plants cost, on the average, 1.5 cents in 1978, the same as in 1976 and 1977.

A coal-generated kilowatt-hour of electricity cost 2.3 cents —up from 2 cents in 1977 and 1.8 cents in 1976.

The cost of an oil-produced kilowatt-hour of electricity rose to 4 cents from 3.9 cents in 1977 and 3.5 cents in 1976.

According to a recent news release from the Atomic Industrial Forum, nuclear power's contribution offset consumption of the equivalent of 135,000,000 tons of coal, or 2.9 trillion cubic feet of natural gas, or 470,000,000 barrels of oil. "Importing that much oil, at the new, higher OPEC prices," AIF President Carl Walske pointed out, "would add more than $6 billion to our trade deficit, thereby aggravating our national security and inflation problems."

As a sample of the direct savings to consumers, which totaled $3 billion when comparing weighted cost averages of nuclear with oil and coal, Green Mountain Power in Vermont calculated that a residential customer would have had to pay $106.44 more for electricity last year if the utility's nuclear power had been generated by fossil fuels. Central Maine Power's customers would have had to pay $56 more.

In view of these facts and statistics, one wonders how some consumers can fool themselves into believing that the attempt to halt nuclear power plant construction in this country is a matter of little concern to them.

THE KANSAS CITY STAR

Kansas City, Mo., April 23, 1979

Much has been made of how the accident scare at Three Mile Island in Pennsylvania may blight if not terminate the future of nuclear power generation as a result of public and official alarm over the safety of this energy source. But the even tougher fact is that nuclear energy was in serious trouble long before the mishandled valves at the Harrisburg reactor threatened a fuel core meltdown.

The steeply escalating cost of nuclear power plants, the many years now required to bring one from the drawing boards to completion and the nagging uncertainties of legal resistance by nuclear opponents and of how radioactive wastes will eventually be disposed of — all these have cooled the electric utilities' enthusiasm for nuclear.

There are presently 72 units in operation, providing nearly 13 percent of the nation's electricity. But whereas 34 new reactors were ordered in 1973, fewer than half a dozen were ordered between 1974 and 1979. Costs approaching $2 billion a plant and the 12 years' average time needed to design, license and build it are causing utility decision-makers to shrink from such a commitment, particularly when the difficulty of winning rate increases from state regulatory agencies makes it harder for the power companies to borrow expansion capital. And even more particularly when the electric demand growth rate has slowed to the point where the need for these huge new units is put in doubt.

In most parts of the country nuclear still is a less expensive means of producing kilowatts than a coal-fired plant, but construction costs are going up much faster on the nukes. A New York consultant for Congress's General Accounting Office reports that the cost of building nuclear plants went from $335 a kilowatt in 1972 to $850 by the end of 1977. In the same period the per-kilowatt cost on coal-fired plants increased from $322 to $535, even allowing for the added expense of flue gas scrubbers to remove air pollutants.

It seems certain that the federal Nuclear Regulatory Commission, in an effort to restore public acceptance of this power mode, will come down hard on safety requirements and procedures, which would increase the time that reactors will be shut down for safety inspections or modifications. If nuclear energy, with all its newly magnified problems of dependability and possible health hazards, is also going to lose its cost advantage over fossil-fueled generators, then the projections of future nuclear power growth will have to be sharply curtailed.

CHARLESTON EVENING POST

Charleston, S.C., July 12, 1979

While a House Interior subcommittee was hearing testimony explaining how much a ban on nuclear power would cost American consumers — about $119 billion through the year 2000 — Duke Power Co. was announcing how much nuclear power was saving customers. The utility asked the N.C. Utilities Commission for permission to lower its rates to save residential customers about $1.40 each month.

During a test period from November through April, Duke had to buy less coal than it expected. The reason for that was the efficiency of its Oconee nuclear power plant. Thus, the saving could be passed to consumers.

The Oconee reactors are the "sisters" of the Three Mile Island reactor which was involved in an accident earlier this year. This prompted the subcommittee hearings, chaired by Rep. Morris K. Udall, who warned that the nation ignores the Three Mile Island incident at its peril.

We contend that if Americans ignore the advantages of nuclear power, it also will be at their peril, and the consequences could be much more severe than most nuclear protestors seem willing to acknowledge.

The Courier-Journal

Louisville, Ky., June 12, 1979

SAFETY, or the lack of it, gets most of the attention in the nuclear energy debate. Rightly so, because no other industry has the potential of causing thousands of casualties with a single accident, or of laying waste huge sections of a nation.

But the safety issue isn't what's ails the nuclear industry most, except indirectly. It's becoming clearer that nuclear power is falling short in plain old dollars and cents.

That's really nothing new. The loss of the billion-dollar Three Mile Island nuclear plant at a single stroke was a violent reminder of the risks of nuclear power. But the handwriting was on the wall long before the Pennsylvania fiasco. The cost of constructing nuclear plants, one prominent energy consultant estimates, has been increasing at triple the rate of inflation and half again the rate for coal-fired plants. No wonder no nuclear plants have been added this year or last to the 126 already built or on order.

The judgments that led to that result were made long before the Three Mile Island episode. They reflect findings such as those of investment banker Saunders Miller, who told a University of Louisville symposium last weekend that coal-generated electricity, for plants completed this year, already is 10 percent cheaper in this part of the nation than nuclear power. There's every indication that the comparison will become much less favorable to nuclear power in the future, he asserted.

That seems true even if the outgrowth of Three Mile Island isn't, as seems likely, a radical and expensive increase in safety requirements. Nor does it take into account all the undetermined costs, such as the unsolved problems of waste disposal and of dismantling or otherwise making nuclear plants safe after their relatively short lifespan of 30 years or so.

In fact, the complete cost of nuclear power hasn't been calculated. As Mr. Saunders pointed out, the real costs involve not only construction and operation, but a multitude of things that hit the taxpayer rather than the utility ratepayer. Those include, besides waste disposal and dismantling, tax breaks, accelerated depreciation, cheap enrichment of uranium fuel at government plants and limits on insurance liability far below actual risk.

In view of this analysis, the proposal of a nuclear expert that Public Service Indiana's Marble Hill site eventually have six huge nuclear reactors rather than the presently planned two seems an exercise in fantasy. The plan, in the current issue of *Science* magazine, perhaps would make sense if the nation were committed to all-out pursuit of the nuclear option.

But as matters stand, it would seem that Public Service Indiana's ratepayers should be vigorously questioning the *two* nuclear reactors planned for Marble Hill, rather than worring about six. After all, it's the ratepayers, rather than PSI, who are likely to pay the steepest price if the nuclear critics are right and PSI is wrong.

Long Island, N.Y., August 5, 1981

The times are not propitious for building nuclear power plants. Huge cost overruns and long delays plague the industry. Some utilities have even canceled plants already under construction.

But once a plant is begun, how is a utility to know when it's throwing good money after bad—especially when estimates of a plant's cost at completion vary widely?

In the case of Nine Mile Point 2, a 1,080-megawatt plant now under construction near Oswego, N.Y., the difference between the high and low cost estimates is nearly $2 billion. The low figure, $3.7 billion, comes from Niagara Mohawk, which is building the plant in cooperation with LILCO and three other utilities. The worst-case estimate, $5.6 billion, appears in a draft report done by consultants for the Public Service Commission, although the final version of the report omits the figure and merely notes that the increase could be "significant."

Several interested parties, including the State Consumer Protection Board and the state attorney general, have petitioned the PSC to hold hearings on the economics of Nine Mile Point and the need for it. Utility executives say this would be repetitious and serve no particular purpose.

We're not so sure. The Nine Mile Point plant is about one-third finished, but probably a good deal less than a third of the money has been spent. As LILCO has found at Shoreham, costs tend to rise steeply as a nuclear reactor approaches completion.

LILCO has also just announced a $280-million cost increase and a four-month delay for Shoreham, which is about 85 per cent finished. It seems to us that a financially extended utility should welcome attempts to find out whether completing Nine Mile Point makes economic sense.

LILCO's own latest estimate for its 18 per cent share of Nine Mile Point is close to $1 billion. So far it has spent about $250 million on construction and interest payments; that leaves $750 million to go. In our view, hearings are worth holding if they can help clarify whether Nine Mile Point warrants the additional investment or whether LILCO and the other utilities would be better off to scrap the project and take the resulting losses.

DAYTON DAILY NEWS

Dayton, Ohio, July 5, 1979

When Exxon talks on energy, people listen. Exxon is saying nuclear power is too expensive.

The June 19 *Esquire* quotes an internal Exxon study made in late 1977 and recently updated that concludes that nuclear power is not a good investment opportunity for Exxon. The plants, it says, are not cheaper than coal-fired ones, but actually more expensive. And the thriftiness of operation has always been the main reason for building nuclear plants, which are more expensive to start with.

Add into the equation the potential costs of compensating employes and the public from overdoses of radiation and the cost of developing safe storage methods for nuclear wastes and the cost of tearing down and disposing safely of an old nuclear plant — and you've got a whopping bill.

The bill, as usual, is footed by the customers.

Three Mile Island may not be what kills nuclear power or slows it to a crawl. It may be the balance sheets. They don't demonstrate and get arrested, but they grab the attention of executives quicker than any picket sign ever will.

ST. LOUIS POST-DISPATCH

St. Louis, Mo., May 31, 1981

Repeated leaks at a Japanese nuclear power plant, some of which were initially concealed from Japanese nuclear authorities, have led to the resignation and indictment of some of the firm's officers and to a punishment that goes beyond the relatively small fines that might have been imposed under Japanese law. In what might be termed nuclear capital punishment, Japanese authorities decided to penalize the utility owners by ordering the plant shut down for a period of time, thereby depriving them of the income it might have produced.

A similar discipline has been used by Pennsylvania utility regulators, who have removed the crippled plant at Three Mile Island from the rate base. As a result — and because of the burden of cleaning up the disabled plant — TMI's owners are seeking a federal bailout to avoid bankruptcy.

If TMI has utility investors worried, so should a recent study by Charles Komanoff, an independent energy consultant. He says the economics of nuclear power are getting worse, not better. He found that the cost of building nuclear power plants rose 142 percent between 1971 and 1978, even after adjusting for inflation, and that nuclear plants will continue to be significantly more costly than coal-fired plants. He says the difference in capital costs is already sufficient to overcome the lower operating costs of nuclear plants, and that coal's advantage will grow in the years ahead.

Although his findings are questioned by the nuclear industry, Mr. Komanoff has a track record that adds to his credibility. Five years ago, he predicted that the newer, larger nuclear power plants then coming on line would operate at only about 55 percent of their rated capacity — and not at the 70 to 80 percent predicted by the industry. Subsequently, nuclear plants have run at between 50 and 62 percent of capacity.

Small wonder investors are growing cautious about nuclear power. It's one thing to take risks, quite another to invite capital punishment.

The Burlington Free Press

Burlington, Vt., August 20, 1981

Gov. Richard A. Snelling asked for and got a tart response from anti-nuclear groups after he threw down the gauntlet last week by urging officials of the Central Vermont Public Service Corp. to go ahead with plans to build either a nuclear or coal-fired power plant in the state.

While he qualified his support for a nuclear plant by saying it is impractical until the waste disposal problem is solved, he indicated his confidence that the federal government would come up with the answer.

He said the power companies have been "intimidated" by a public that wants cheap electricity but rejects nuclear and coal-fired plants, fossil fuel facilities and transmission lines.

What perhaps irked nuclear power plant opponents the most was his neglect to mention alternatives, including energy conservation; wind, water and wood-fired generation; and use of solar energy.

And they expressed fear that his stance on a Vermont nuclear power plant in tandem with President Reagan's pledge to push aside obstacles to nuclear reactor construction could be ominous for the future.

"People in government have got to realize what's involved here and crack down on their governments," said Esther Poneck, clerk of the New England Coalition Against Nuclear Pollution. She charged that Snelling was interested in the approval of "some powerful and influential people whom he would like to please."

Jeanne Keller of the Vermont Public Interest Research Group objected to Snelling's suggestion that the federal government has the responsibility for solving the waste disposal problem, saying the nuclear industry already has been subsidized "to the tune of $20 billion." She said the government should consider support of solar development or conservation on that scale.

Winooski Mayor Dominique Casavant said construction of a new plant was unnecessary and urged greater stress be placed on conservation.

Having been soundly criticized for his remarks, Snelling this week declared that those who objected to his endorsement of a new power plant were "enemies of thoughtful public policy."

Aside from the yet-unsolved waste disposal problem is another practical consideration: Construction costs for nuclear power plants are so high that it is almost impossible for them to undersell the electricity that is derived from other sources. Several firms already have cancelled plans to build nuclear generators on that basis. While some have said regulations have forced the costs up, increases in prices of material and labor have affected their plans almost as much as federal rules.

Before the public can be expected to favor the need for another nuclear power plant, it must be convinced that there is no alternative which would provide less expensive, and perhaps less dangerous generation of electrical energy.

WORCESTER TELEGRAM.

Worcester, Mass., January 4, 1981

The price of oil and the price of money may be doing more than all the protests ever held to fulfill the dreams of anti-nuke demonstrators.

High oil prices are a large part of any electric bill because of this region's dependence on foreign oil. Customers of utilities in this area are noticing a sudden jump in the fuel-charge part of their bills because of rising oil prices.

But big electric bills, along with a slowdown in the region's economic growth rate since the 1973 Arab oil embargo, have cut demand for electricity far below what the forecasters were expecting.

The Western Massachusetts Electric Co., which started planning a $2.5 billion nuclear power plant in Montague in 1974, has scrapped those plans.

At the time, demand for power was increasing at an annual rate of between 6 and 8 percent. But the 1973 Arab oil embargo and conservation cut demand far below the utility's forecast, the company says.

There was no increase in demand for the company's electricity last year, a spokesman said. Conservation is cutting demand. The company suspended plans to build the 2,300-megawatt plant in 1978 but held off on shelving them completely. The plant originally was expected to begin generating power between 1988 and 1990.

The company wanted to watch demand for another few years, "but it's clear now that this plant would be a mismatch for demand," a spokesman said.

Western Massachusetts Electric has not ruled out nuclear power as the best answer to new power plant construction questions. The company will turn again to nuclear power if demand starts to rise.

But for now, those people in Western Massachusetts who marched on the planned construction site can breathe a sigh of relief. Conservation is doing for them what protest marches could not.

Western Massachusetts Electric is listening to the economics of nuclear power, if not to the protestors' voices.

BUFFALO EVENING NEWS

Buffalo, N.Y., September 13, 1981

In voting unanimously to order a special hearing on the Nine Mile Point 2 nuclear power plant now under construction near Oswego, the state Public Service Commission has moved responsibly to resolve questions raised in the wake of escalating costs and construction delays.

The hearing will focus on an independent consultant's audit and the PSC's own staff report, which examined the costs and timetable for completing the plant in comparison with alternative means of generating additional energy.

The plant, still only about one-third complete, was budgeted at about $380 million when construction began in 1975. The Niagara Mohawk Power Corp. and the four other private utilities associated with it in the project currently estimate the cost of the plant at $3.7 billion.

While the independent audit found that the 1986 target date for beginning commercial operation is feasible, it warned that interest costs, regulatory changes and other factors could increase the final cost significantly above the $3.7 billion estimate. Meanwhile, the PSC's own staff report put the probable figure at close to $5 billion.

Critics of the project have conceded that if construction is halted, the utilities probably would be allowed to recover the $1.3 billion already invested in the plant through a PSC-approved increase in their electricity rates — a matter of obvious importance to Western New York customers of Niagara Mohawk as the major investor in the plant.

Contrary to claims of the plant's opponents, both the PSC study and the utilities agree that there will be a clear need for the electricity generated by the plant in the late 1980s and beyond. Thus in the view of NMP officials, the key issue awaiting the PSC resolution is whether any substitute for the Nine Mile 2 nuclear plant could better serve residential and industrial consumers if it were abandoned.

According to the utility spokesmen, completing the plant would be less costly in the long run than either converting it to coal — if this were technologically feasible — or building a separate new coal facility from scratch, with all the protracted delay in licensing procedures this would entail.

In addition, as evidence of the need for a reasonable proportion of nuclear-based power in the total generating mix, the utilities cite the state's present excessive reliance on costly imported oil, the growing obsolescence of existing power plants, the dangers of acid rain from coal-fired plants, escalating coal prices and problems in rail haulage of coal.

The Nine Mile 2 plant is by no means unique in experiencing construction delays and cost overruns. These have plagued the entire nuclear power industry NMP officials attribute the problems in part to financial difficulties that utilities have encountered since the mid-1970s and to stiffened federal safety regulations in the wake of the Three Mile Island crisis. The PSC's independent audit emphasized the need for improving management efficiency at the site.

Whatever the PSC's conclusion after reviewing the issues in dispute, it should keep its pledge to issue its decision before the end of the year. There have already been ample studies, and if completion of the project proves to be the best option, it is imperative that the construction schedule move ahead expeditiously without still further costly delays.

The Dallas Morning News

Dallas, Texas, October 31, 1981

LET us face it — economically speaking, the times are discombobulated. It no more surprises that the Comanche Peak nuclear plant should be facing more delays and more cost increases than that interest rates still hover somewhere in the stratosphere.

Naturally the news about Comanche Peak, delivered this week by its owner, the Texas Utilities Company, is not what one would like to hear. The nuclear plant's Unit 1 was to have started operating last year, its Unit 2 next year, at a total cost of $779 million.

The newest target dates are 1984 and 1985, respectively. The projected cost? We pause for a good shudder: $3.44 billion.

The truly graphic statistic concerns interest costs, which now are estimated at $780 million — almost exactly what TU originally thought it would take to build the whole plant.

The statistic reminds us that the economic turmoil of the '70s will exact a high price for several years to come. How relatively serene were the '70s when they began, with inflation less than 5 percent. Then came the energy crisis and the explosion of federal spending, with double-digit inflation and interest rates following in their train. Thanks to inflation of 130 percent since 1972, it would take $1.8 billion in 1981 dollars to buy what $779 million purchased when Comanche Peak was planned.

In the nuclear industry, Comanche Peak's problems are far from unique. Nuclear plants are large, complicated things to build, and they get more so all the time. The state of the art is perpetually changing, necessitating alterations in design. Similarly the licensing process keeps changing.

At that, the new cost estimates make Comanche Peak no more expensive than many nuclear plants now abuilding, and considerably cheaper than others.

None of this would console us were it not for other statistics that Texas Utilities advances. Cost run-overs notwithstanding, Comanche Peak is expected to produce cheaper power in the 1990s than all other conventional sources.

The company projects that by 1990 the per-kilowatt hour cost of natural gas and crude oil will reach 19.5 cents; for coal from a new lignite unit the cost will be 7.8 cents, and for out-of-state coal 13.6 cents. For fuel from Comanche Peak the cost is expected to be 6.5 cents.

Let's hope Comanche Peak's costs don't rise any more — though they could. At the same time, let's be glad we are as far along as we are with so sorely needed a project. For our part we see no reason to suppose that Texas Utilities hasn't done the best job that could be done under trying circumstances.

Dallas, Texas, November 9, 1981

Anybody want to buy part of a nuclear plant — an unbuilt one yet? Austin's share of the South Texas Nuclear Project is for sale, local voters having decided they can do without nuclear power for the foreseeable future.

Austin has from the start lived on uneasy terms with the nuclear project. Anti-nuclear convictions are widespread in this university city; moreover, as with many another nuclear installation, costs have escalated, even as the completion date has receded. The other day, in a referendum on what to do about the plant, Austin said, enough already.

One anti-nuke ad admonished voters that "Your grandchildren will be paying off your debt" for the nuclear plant. In fact the grandkids are likelier to be paying through the nose for the expensive fossil fuels that will furnish all their power, the nuclear option having been foreclosed. However much nuclear fuel costs, it will seem cheap in the '90s, compared to oil and natural gas — unless all the studies we know of prove to be wildly wrong.

It's Austin's right to gamble on the future. But what a gamble — one we're glad that Dallas doesn't appear ready to contemplate.

The Atlanta Journal AND THE ATLANTA CONSTITUTION

Atlanta, Ga., September 27, 1981

REMEMBER years ago, when the visionaries told us that one day nuclear power would make electricity "too cheap to meter"? Well, it hasn't exactly turned out that way. Construction of nuclear power plants is getting so expensive that the electricity they produce is not much cheaper than that from coal plants, and the gap is still narrowing.

With all the other questions there are about nuclear power — plant safety, reliability of power production and radioactive waste disposal, for instance — the question of cost becomes more and more important. If we are to accept the uncertainties, then we should be assured that we are getting a real bargain in return.

Which brings us to the issue of Georgia Power's Plant Vogtle, now under construction near Waynesboro. The utility reported the other day that the estimated total cost of the plant has gone up again, by another $1.9 billion, to $5.5 billion. The share of that which Georgia ratepayers will shoulder when the plant is put into the rate base sometime in the future (it is not part of the current rate hike case) will be more than $2.6 billion, unless the utility can sell another 16 percent of the plant to some Florida power companies. And the latest escalation in costs has that proposition in some doubt.

The problem is that the plant is going to produce electricity that Georgia simply doesn't need, and won't need for some years to come. That situation arises because the plant was planned back when electricity consumption was rising steeply every year; in the last couple of years, however, the annual increase has slowed considerably as people began to conserve energy of all kinds. Georgia Power can't be criticized too harshly for that miscalculation, but it — and the state Public Service Commission and the ratepayers — are stuck with the decision of what to do with the half-finished plant.

There is logic to Georgia Power's contention that it makes no sense to cancel the plant now, wasting much of what has been spent already, since one day we will need its electricity. But there is also logic to Consumers' Utility Counsel Vic Baird's observation that as costs continue to skyrocket, that tradeoff may change and the wiser decision may be a different one. He wants the PSC to conduct an independent study of the issue to see just what is the proper course; the PSC has indicated it wants a study, too, but can't afford it.

We believe there should be an outside look at the problem to see if the tremendous increase in the cost of the plant warrants a shutdown of the project, or some other decision. The study would cost $200,000 or more, but if it shows Georgia citizens the way to save millions down the road — either by canceling the plant or by continuing it — it will be money well spent.

Chicago Tribune

Chicago, Ill., January 20, 1981

Two unsolved problems, one in Pennsylvania and one in Washington, illustrate why the original optimism about nuclear power plants has faded considerably. At Middletown, Pa., the Three Mile Island plant still stands unused and unusable, with radioactive water more than eight feet deep in the containment building almost two years after the spectacular accident there. In the northwest, the price tag on the Washington Public Power Supply System's five nuclear generators, which started at $4 billion 10 years ago, is now $17 billion and still rising.

The hangup at Three Mile Island is that though millions have been spent on a "demineralizer system" to filter the radioactive water, the Nuclear Regulatory Commission [NRC] has not yet approved using the system. No one knows quite how to handle the extremely concentrated radioactive stuff that filtering would produce. The NRC would like the Department of Energy, which normally handles only military wastes, to take charge of the material, but the D.O.E. says congressional clearance would be required for that. So no one knows when, whether, or at what cost the Three Mile Island plant will return to service, if ever.

Builders of the five plants in Washington are having a variety of troubles. The power system says that regulatory requirements account for much of its cost overruns. The NRC replies that the defects it finds in construction are not of its doing. An NRC inspector quoted himself as having said, "How can you guys inspect this and not weld it right? How the hell can we trust you to do everything right?"

Besides, the "fast track" construction on the plants has led to almost endless changes in specifications during construction, leading to much dismantling of work already done. The massive project has already borrowed so much that some lenders are up against their limits on investments in one place. The project can hardly issue $160 million in bonds every two months forever.

And there still is no national policy about ultimate disposition of radioactive waste from nuclear power plants.

We're going to have to face these problems because we're going to have to depend on nuclear power, at least for several decades. But it can no longer be assumed that nuclear power will prove as much of an economic bargain as once expected.

SYRACUSE HERALD-JOURNAL

Syracuse, N.Y., January 21, 1982

The Public Service Commission is leaning toward giving the five utilities headed by Niagara Mohawk a "go-ahead" to finish Nine Mile 2 nuclear power plant near Oswego.

That attitude is reflected in its search for some way to contain serious cost overruns on which, as rates are figured, utilities can still make profits. The PSC, for instance, is considering an overall cost ceiling.

Is that workable?

If the plant was brought in under a PSC-set ceiling, the utilities would enjoy a windfall, pretty much of their own making, according to some PSC thinking..

If the plant was built at a higher-than-ceiling cost, the utilities (probably investors) would have to absorb the extra expense.

We don't know what's feasible or what's legal.

We do appreciate the PSC's positive approach; namely, to carry out its staff recommendation and order completion of the plant.

If the PSC and the consortium of utilities can agree on a system of cost containment, so much the better for consumers and investors.

That's worth striving for.

THE DENVER POST

Denver, Colo., January 16, 1982

WITH THE PUBLIC already sensitized to such "disaster" code names as Diablo Canyon and Three Mile Island, one might conclude that nuclear power generation is a fading technology as inflation drives up construction costs and recession reduces electrical demand. Some partisans argue this case articulately.

Indeed, the current nuclear outlook is gloomy, with blueprints for several new atomic plants having recently been shelved. But this trend is far from uniform, suggesting the nuclear downturn may not be permanent.

For the other side of the picture, one might turn to Commonwealth Edison, the Chicago-based utility which for a decade has served 8 million people in northern Illinois with electricity from a half coal-fired, half nuclear system.

Despite the "ban nuclear" movement, Commonwealth Edison is going ahead with its atomic construction program. In the face of delays and complex construction requirements, the utility insists nuclear fission is a cheaper source of power than high-sulfur Illinois coal or low-sulfur coal shipped from Montana or Wyoming.

The firm now operates six nuclear units at three locations and plans to double that number by 1986, raising to 72 percent the total kilowatts produced by splitting atoms. Critics would say this is wrong because nuclear plants cost at least a third more than a coal-fired plant of comparable size.

But do they? There are situations — such as in Washington state — where overbuilding and delays have produced ruinous nuclear construction costs. Yet, in the Chicago area, no such disaster in construction costs is apparent.

Take, for example, Commonwealth's 2,200-megawatt Braidwood nuclear plant, now half finished. Its total cost is $2.74 billion. A 1,000 MW coal-fired plant proposed at Las Animas, Colo., by Public Service Company is estimated at $1.5 billion. On the basis of output, the Commonwealth plant is less costly. But the real bargain, comparatively, is in nuclear fuel.

"The fuel costs are way lower," says James Toscas, a spokesman for Commonwealth Edison. "We accept regulatory delays and uncertainty because we know how to deal with them and we know that in 1980, alone, we saved $500 million by running a half-nuclear system instead of all coal. It's simple economics. If it weren't cheaper to do what we're doing, we wouldn't do it."

By 1988, three-fourths of Commonwealth's customers will be illuminating their houses, cooking meals and powering their appliances with electricity generated by light-water nuclear reactors. The firm assumes, of course, that the federal government will make early decisions on disposal of nuclear "ashes" presently being stored for re-use and permanent burial. Reagan policy makers have promised such decisions, and there's no technical reason why they should fail.

The public's psychological fears of nuclear mishap — towering out of all proportion to any damage caused so far — obscures an important fact: the nuclear option hasn't been discredited in all parts of the country and, when the recession ends, that option may rise again as an important contender. A nuclear system, as Commonwealth Edison is proving, can produce impressive results in safety and on the bottom line.

The Oregonian

Portland, Ore., January 5, 1982

Cost alone is not the best way to judge the relative merits of thermal power plants using competing fuel sources. On the basis of cost-per-kilowatt-hour, nuclear power still rates as a better buy than plants fired y coal, oil or gas.

Other variables to examine are performance, reliability and plant capacity. One would expect controversial nuclear energy to do worse than fossil-fuel-fired plants in performance and reliability, if only because of strict operating procedures required for safety. Yet, this is not so.

A recent survey of utilities that own and operate plants fired by atoms, coal and oil reveals that nuclear energy was not only a cheaper electricity source than coal and oil, but also:

— That the 56 nuclear plants in the survey produced more kilowatts relative to their capacity than did 36 coal and 21 oil units. The oil-fired plants in the survey were especially poor performers, producing only 39.5 percent of the electric energy promised by their boilers, as compared to 59.3 percent for the nuclear plants.

— That nuclear plants in the survey were forced off-line by component failures or other causes less often than were the coal plants.

The only category in which nuclear energy was markedly inferior to coal and oil was in plants' availability to produce power, which reflects, more than anything else, the governmental red tape in shutting down or starting up a nuclear facility.

Based on these results, it is difficult to imagine why U.S. utilities have placed no new orders for nuclear power stations in the past three years. Cost or performance is not the reason. Uncertainty. That's the reason.

Oregon Journal

Portland, Ore., January 18, 1982

Watching attempts to mothball Washington Public Power Supply System nuclear plants Nos. 4 and 5 blurs perspective. WPPSS isn't alone. Nuclear power is in serious economic trouble nationally, not just in the Pacific Northwest. Several other nuclear plants under construction are in as much trouble as WPPSS.

The Tennessee Valley Authority has mothballed five of eight reactors that it is constructing. A TVA staff report says TVA is overbuilding and doesn't need the electricity that would be produced by the other three plants. Staff's conclusion: Perhaps the other three reactors should be mothballed.

In New England, the Seabrook nuclear plant has survived environmental demonstrations by the Clamshell Alliance, but the plant is beset by economic difficulties.

Public Service Co. of New Hampshire, the lead agency in building Seabrook, began two years ago reducing its ownership from 50 percent to 35 percent in the plant's two reactors. But the New Hampshire Public Utilities Commission now says it should reduce its share of Seabrook further, to 28 percent, or the utility might even go bankrupt. Seabrook's Unit 2 could be mothballed.

Underlying the TVA and Seabrook financial problems are predictions that the power generated by the plants may not be needed. Even if Seabrook's Unit 2 isn't built, the New England Power Pool says the region's generating capacity seems adequate until 1991. Boston Edison Co. says it thinks its customers will have adequate capacity through the end of the century.

Utility companies made predictions in the early 1970s about the need for electricity that were based on post-World War II experiences. Now the realization is coming that they overbuilt.

That's why this region needs better forecasts of future energy needs. Utilities have been making forecasts since 1948. The Bonneville Power Administration is making its first extensive energy forecast, and it is due within a month. The Regional Power Council by congressional directive must make a forecast within 15 months.

So it's not difficult to see why no utilities have stepped forward to offer publicly to buy WPPSS Nos. 4 and 5.

It is difficult to understand why Puget Sound Power & Light Co. applied in December to the Nuclear Regulatory Commission to move its proposed Skagit nuclear plant to the Hanford reservation and why Portland General Electric is continuing efforts to license its proposed Pebble Springs nuclear plants.

These are hard times for any utility group to try to start nuclear plants. The Pacific Northwest should go at slow-bell speed in preparing for expensive new generating capacity. Let's see first what the regional and national forecasts say. Some research groups, such as Carnegie-Mellon's Energy Productivity Center, expect the national demand for electricity to decline slightly in the 1980s. The cost of overbuilding can be devastating to a region, perhaps even leading to financial insolvency for utilities.

The Knickerbocker News

Albany, N.Y., January 19, 1982

Until recently, the controversy surrounding the Nine Mile 2 nuclear power plant in Oswego, although filled with confusing and often contradictory facts and figures, came down to a simple choice: Complete the plant, at a cost of about $5 billion, or abandon it and write off the $1.3 billion already invested. Either way, consumers would pick up the tab in the form of higher rates.

But now the Public Service Commission is examining some alternative solutions that might save consumers money. And although Karen Burstein of the state Consumer Protection Board wonders if these alternatives might not be an attempt by the PSC to get itself off the horns of a dilemma, she also observes that the proposals are innnovative and "better than nothing." On that basis alone, the PSC should be encouraged.

One alternative the PSC is studying is whether to put a ceiling on the cost of a project. Utilities would then have to absorb any costs above that mark but could pocket the difference if the project was completed with fewer dollars.

No utility would argue with the latter part of this plan, but the idea of having to absorb cost overruns is a different matter. John W. Keib, senior systems attorney for Niagara Mohawk, one of the five utilities with an interest in Nine Mile 2, warns of opposition to any "up front determination" of what the company can include in the rate base.

There are, of course, legitimate concerns raised by the PSC plan. Would it encourage inflated estimates of future projects in order to avoid any chance of overruns? Would it encourage shortcuts, possibly at the expense of safety, to keep within the ceiling?

Despite these reservations, however, the PSC initiative is welcome because it would remove the profit guarantee that utilities enjoy even if they make grave errors in the cost estimates.

True, the utilities can argue that passing along risk to stockholders will only discourage investment, but that position ignores two persuasive counter-arguments. One is that the 1981 tax act provides for tax-free investing in utility stocks and should help offset any investor defection. The second, and sounder, argument is that utilities just might be a lot sounder investment if their managers' feet, while not exactly being held to the fire, were at least within warming distance of it.

The Evening Gazette

Worcester, Mass., March 31, 1983

The Seabrook nuclear power plant is a troubled project and its troubles are made worse by the announcement that the New England Electric System wants out of Seabrook II. NEES says it still will need the power produced by Seabrook I, 70 percent complete and scheduled to go on line by 1985, but it wants to sell its 10 percent share in Seabrook II to some other utility.

NEES says that if it can't sell its share in Seabrook II, it will recommend that the project, 17 percent finished, be put in mothballs. The problem with that is that Seabrook I, standing by itself, probably cannot produce electricity at competitive rates. Even with both units on line, the power may cost more than 12 cents a kilowatt-hour. If a single unit had to carry all the capital costs, the power might cost up to 20 cents.

The NEES stand is a gamble. It is a gamble that oil prices will stay depressed and that demand for power will stay essentially flat. NEES hasn't built a new generating facility in more than 10 years. It is relying heavily on conservation and coal to get it through the next 20 years. It has converted several of its oil-burning plants to coal and is building a collier to haul the coal from Virginia.

Other utilities are not so well situated. They will need that Seabrook II power by 1990, they say. Otherwise, they fear, they will have to buy even more expensive power from other utilities. Some of them are still heavily dependent on foreign oil.

Massachusetts Municipal Wholesale Electric Company, which owns 10 percent of Seabrook and which serves a number of towns in Central Massachusetts, is doing its own study. But MMWEC officials have always maintained that they will need the power from both Seabrook I and II to get through the 1990s.

The ideal solution to the problem posed by NEES would be a willing buyer. However, other utilities have been trying to sell their shares of Seabrook without success. As the contract is written, the Public Service Company of New Hampshire, builder and prime owner of Seabrook, has the power to block any proposal to suspend work on Seabrook II.

But PSNH is on shaky financial ground, and no one knows what is going to happen. The only sure thing is that the price of electricity is going up, whether Seabrook II is built or not.

The Hartford Courant

Hartford, Conn., April 1, 1983

A recurrent phrase in discussions among members of the General Assembly's Energy and Public Utilities Committee is that regulating utilities is more of a theology than a science.

Certainly faith must be the primary motive behind the committee's approval of a bill that would prohibit utilities from charging ratepayers for a power plant's construction before it begins producing electricity. There's no science to justify it.

The bill to ban charges for construction work in progress — CWIP — does have some common-sense appeal: Why should people have to pay for something, such as the Millstone III nuclear power plant in Watertown or the Seabrook I plant in New Hampshire, before they derive any benefit from it?

Why? Because if they don't, they could end up paying even greater and more abrupt increases when the plants go on line. If CWIP is prohibited, the cost of financing a nuclear plant could be driven up and, once the plant kicks into service, consumers would be suddenly faced with even more whopping electric rate hikes to pay for it.

CWIP should not be employed for every project. But the state Department of Public Utility Control, which already prohibits CWIP charges for electric utilities, should not be prevented by law from changing its mind. If the evidence is convincing that the imposition of CWIP charges in a specific case would be in the long-term interest of the ratepayer, the DPUC should be able to act.

In fact, the Federal Energy Regulatory Commission, which governs wholesale electric contracts, recently ruled that customers in many cases should pay CWIP charges, because to exclude them could result in the loss of needed investments.

State utilities commissions, which set the retail rates, should also be free to invoke CWIP when it is advantageous to customers. In fact, more than 30 states already do allow such charges.

CWIP is not a new concept even for Connecticut. Although their situation was not exactly parallel to that of the electric utilities, water companies in the state already have been allowed to charge customers for the construction of facilities before they were put into use.

CWIP need not be a one-way street, either. If construction charges of a power plant are allowed to be phased in before the plant is on line, rate increases after the plant is built might be better phased in to smoothen the impact on consumers.

Another interesting possibility that has been bandied about is a requirement that a public representative serve on the boards of private utilities that are granted CWIP charges. If the public is to be made a hostage investor in a utility project, it ought to have some say in that utility's decisions.

In the meantime, however, a law should not be passed that would prevent the DPUC from taking advantage of a potentially useful regulatory tool, merely because it contravenes the theology of some legislators.

The Morning News

Wilmington, Del., May 10, 1983

IF THE owners of the nuclear generating station a few miles from downtown Salem, N.J., are fined $850,000 as the Nuclear Regulatory Commission proposes to fine them, *someone* must pay that sum. Who?

Spokesmen for the utilities told the Associated Press that their stockholders may ultimately pay the fine, but customers won't.

"It would come out of company earnings," said Edward Anderson, a spokesman for Public Service Electric and Gas Co.

PSE&G and Philadelphia Electric Co. own a total of 85.2 percent of the plant. Delmarva Power and Light and Atlantic City Electric each own 7.4 percent.

"As far as the treatment of any fines, when this is over with, it wouldn't be billed to customers," added Mr. Anderson.

All four companies may feel that they really wish to shoulder the expense of what would, if not reduced or eliminated upon appeal, be the largest NRC fine ever imposed for a failure to observe federal safety regulations. It could be useful public relations to assure customers that the financial costs of any error in such an operation would come from "earnings." That's extremely difficult.

The rules that govern the utility industry admittedly can at times benefit the ratepayer.

These rules, however, determine that such earnings, by and large, come in the form of money paid to the utility by its customers. If the customers were not going to pay the fine, "ultimately," nobody could.

That, in general, is what makes public utilities tick. It is a system that was designed to "guarantee" certain levels of income to investors in exchange for their willingness to allow themselves to be regulated.

When the price of a federal fine, a barrel of oil or bucket of coal is not accounted for in the utility's rates, there simply are no earnings.

That is the system. Society may need to change it — very, very gradually — but basically it's been working a long time.

THE BLADE

Toledo, Ohio, May 10, 1983

IT is seldom these days that one hears anything good about nuclear power. But Toledo Edison's announcement last week that it expects to cut electric rates up to 14 per cent for six months is good news, indeed, for Edison customers. And it is due in large measure to the operation of the Davis-Besse nuclear power station near Oak Harbor.

Residents of this area are well aware of the spotty and discouraging operating record that Davis-Besse has chalked up since its inception in August, 1977. It often has seemed that the generating plant was out of service more than it was in, for a variety of reasons. Consumers also are aware of the costs entailed in that kind of on-and-off operation.

But now that the facility has been operating more efficiently — at a reported 92 per cent of the time since last September — the promise that nuclear power has held for electricity consumers in this country for years is becoming more evident in northwestern Ohio. The result is that Davis-Besse is meeting nearly half of all the electric power requirements of the region and is doing so at roughly one-fifth of the cost of most of Toledo Edison's conventional coal-fired plants.

Adding to the overall saving of about 50 per cent of the fuel-cost portion of customers' bills has been the operation of the Bay Shore coal station, which is regarded by Edison officials as one of the most efficient of its kind in the nation. Together, these plants have enabled the utility to schedule the 14 per cent rate reduction for next August. The reduction still must be approved by the Public Utilities Commission of Ohio and should reduce the average residential bill by about $5 per month.

That may not seem like a great deal of money — and Toledo Edison even now has rate-increase requests in the works with the PUCO dealing with the non-fuel portions of bills — but the reduction will amount to some $33 million and it is far better than none at all in these times of constantly rising energy costs.

Moreover, in view of the bad press that nuclear power has been getting from various segments of the population in recent years, the Davis-Besse record of late is highly encouraging. It demonstrates even to the most skeptical observer that even though nuclear-power construction has been drastically curtailed in this country, largely because of economic factors, this method of power generation has a place in the energy-producing sector of the United States for a long time to come.

A reduction of 14 per cent in electricity bills over a period of six months is solid evidence of that reality.

WORCESTER TELEGRAM.

Worcester, Mass., May 8, 1983

The Massachusetts state government may want to take a look at a new law the Connecticut Assembly passed last week. Gov. William A. O'Neill says he will sign it.

The law is an attempt to ease the difficult road ahead for Connecticut electricity consumers when Millstone III, the state's fourth nuclear power plant comes on line in 1986. The cost of the plant — now put at $3.54 billion — will increase electricity bills by as much as 25 percent, according to some analysts.

The law does two things: It allows Northeast Utilities, which is building the plant, to increase electricity bills now to help with construction costs. But it will also require the utility to phase in the steep rate increases that will be needed after 1986. In other words, it will stretch out rate increases over the next few years so customers can cope better.

Some Massachusetts customers are going to be in for sharply higher electricity bills when Millstone III and Seabrook come on line. Several Massachusetts utilities own parts of both.

Since both plants lie outside the state, the Massachusetts Legislature cannot act unilaterally as the Connecticut Assembly has done. If a compromise is to be reached, it will have to be done in cooperation with Connecticut, New Hampshire, and the utilities involved.

However, it's time to start thinking about the problem. Seabrook I is scheduled to go on line late next year, a deadline it probably won't meet. But unless costs at the project can be kept in line, many Massachusetts electricity bills are going to zoom. If so, a gradual increase might be preferable to the alternative.

TULSA WORLD

Tulsa, Okla., December 14, 1983

IT IS widely assumed that the opponents of nuclear power have won their campaign to portray that source of energy as too dangerous to be used.

Nevertheless, by 1990, the U. S. will rely on nuclear power for 19 percent of its electricity and the rest of the industrialized world will generate 23 percent of its power by nuclear means.

Question: If nuclear power is as dangerous as those who fought it so hard contend, won't the 120 or so plants in operation by the end of the decade bring terrible risks?

The answer, of course, is no, giving the lie to the doomsayers who oppose nuclear power largely on emotion and fiction.

Under present economics, new nuclear power production would be more expensive than energy produced by oil and gas and even coal, in some cases. But that is largely due to needless governmental regulation and delay and a lack of commitment by the government to developing "package" plants that conform to the most rigorous safety standards while expediting construction.

Countries like France, Japan and Great Britain, which have committed to safe nuclear generation have cut years off the construction schedule, enabling nuclear power to retain its economic advantage over coal.

Once the current nuclear plants under construction are completed by 1990, the U. S., if it is to use nuclear power, must put the scare of the 1970s behind it and build more reactor/generators.

The disadvantages of the alternative, coal-fired plants, are becoming more apparent. The costs to the environment from wholesale coal mining, air pollution and terrific transportation charges are making nuclear plants look comparatively safer and cheaper every day.

The Boston Herald

Boston, Mass., September 11, 1983

THERE DOES not seem to be much reason or urgency to build a power generating plant whose final construction costs might be five times greater than first estimated — especially if the need for it has, for several reasons, decreased.

Thus it appears that the Public Service Co. of New Hampshire, principal owner of Seabrook 1 and Seabrook 2, acted wisely in deciding to suspend building of the second reactor until December 1984, when New England's power needs and the dollar resources of the consortium underwriting the controversial project will be clearer.

Seabrook has now been under construction for several years. There's no dispute here about the benefits of nuclear power and its value in helping us to be energy-independent. But since the several years in the '70s when boycotts and OPEC price hikes made imported oil a cause of impoverishment as well as our principal source of energy the situation has changed drastically — and for the better.

Seabrook 1 is expected to begin producing power by mid or late 1985, and is rated to make an additional 1150 megawatts of electricity available to New England. Seabrook 2 has the same rating.

There is now no Mideast oil boycott or threat of one; supplies of oil are plentiful and OPEC has been pretty much de-fanged as a price gouger. There is increasing use of coal and of water power in the generation of electricity. If natural gas is ever properly deregulated, and solar is both exploited and made less expensive, they could contribute even more to energy independence. Conservation has already proven its value.

But what has been largely forgotten is the pact under which the Province of Quebec will, in October of 1986, begin supplying New England with potentially nearly as much power as both Seabrooks combined. There is already a firm agreement on Phase 1 of that agreement, under which 690 megawatts of "excess" hydroelectricity will be available to us in just about three years. Phase 2, still being negotiated, would, if implemented, add another 1300 megawatts to that by the end of the decade. In all, 1990 megawatts of hydroelectricity would be added to our power resources by the end of the decade — and there is no guarantee that work on Seabrook 2 would either be resumed or completed by then.

All of which, it would seem, could make Seabrook 2 a luxury rather than a necessity. It's worth watching, and tracking, because if Seabrook 1, plus hydro and all the other alternatives New England has resorted to these past 10 years do the job, we will have moved a long and encouraging way toward the goal of energy independence.

St. Petersburg Times

St. Petersburg, Fla., July 26, 1983

The morning of March 28, 1979 will not be forgotten for years to come. The most frightening commercial nuclear accident in U.S. history began at 4 a.m. on that day. As word of the radiation leak spread, people who lived near the Three Mile Island (TMI) nuclear plant at Harrisburg, Pa. began leaving work, frantically picking up children at school, loading their cars and driving away. Fear was the most common response. A mass evacuation was not ordered, but Gov. Dick Thornburgh urged all pregnant women and preschool children living within five miles of the power plant to leave.

Since then, the nuclear power industry has been haunted by memories of the TMI accident. It has become the focal point for those who have doubts about the safety of atomic power plants and those who question whether new facilities should be built.

AFTER THE TMI accident, the Committee for Energy Awareness (CEA), an industry trade association, was created to soothe the public's fear. With that goal in mind, the CEA is planning to spend $25-million during the next 12 months to promote nuclear power. Soon, Americans will be bombarded with TV, radio and newspaper ads in an effort to alter what CEA believes are misconceptions about nuclear power. The organization wants to convince the public that nuclear power is safe and efficient, according to Pat Wheeler, CEA projects manager.

If the nuclear industry wants to spend millions on a public-relations campaign, that's the industry's business. Investors should pay for it. The problem is that utility customers will be picking up $21-million of the tab — and most of them don't even know they're paying for the advertising blitz. Florida Power Corp. customers already have donated $36,500 to the cause.

Even more astonishing, Florida's Public Service Commission (PSC) approved charging ratepayers for the pro-nuclear power campaign last year when it considered Florida Power's request for a rate increase. The PSC allows utilities to charge customers for membership dues in trade organizations if the money is spent in ways that will benefit customers, such as research into better conservation measures or more efficient methods of operating power plants. That is a legitimate charge to ratepayers.

FLORIDA POWER contends that the ad campaign will be "informational," rather than promotional. FPC spokesman Larry Shriner argues that it is important for the public to get the message that nuclear power is safe and efficient.

Florida Power's 800,000 customers so far have paid a little less than a nickel each for the national advertising campaign. The money is not the real issue. The principle is. Utility investors should pay for image-building promotions, not customers.

Arkansas Gazette.

Little Rock, Ark., June 4, 1983

Arkansas Power and Light Company has pledged $500,000 to a national public relations campaign promoting nuclear energy as economical and safe, and it hasn't decided whether it wants customers or company stockholders to pay for the contribution.

This is one expense the state Public Service Commission should not allow AP and L to place in its rate base so that the utility's customers will have to dig into their own pockets. The utility is entitled to its views on the efficacy and desirability of nuclear power, but as a public utility it has no rightful calling to charge its customers to promote this view.

It may be argued that AP and L already passes along contributions to the Edison Electric Institute, which lobbies Congress in favor of nuclear power as well as other matters. But if two wrongs can make a right, the channeling of $500,000 in ratepayer money into a public relations campaign in favor of nuclear power coupled with the contribution to Edison would not be one of those occasions.

It was just last month that the United States Supreme Court unanimously upheld California's law prohibiting construction of new nuclear power plants until methods are developed for safely storing or disposing of radioactive wastes produced in generating electric power. Seven other states have similar laws. Electric utilities by 1982 had accumulated 8,000 metric tons of radioactive wastes. Since the accident at Three Mile Island in 1979, no federal permits for new construction have been issued. Congress last year provided for permanent underground storage facilities, but the first one cannot be opened until 1998 at the earliest.

Against this background is the power industry's position, backed by AP and L, that electric power production at nuclear plants is safe and economical. The industry offers such an assertion as fact, but given the circumstances, the assertion reflects an opinion of a directly interested party that stands to gain or lose a great deal by public acceptance or rejection of nuclear power.

There are no public plans at the moment for construction of additional nuclear plants in Arkansas, although AP and L continues to operate two units near Russellville. The utility naturally has a strong interest in seeing public acceptance of its view that nuclear power is safe and economical. It does not follow, however, that the "captive" ratepayer of a public utility should be required in effect to endorse and promote that same view by the simple act of paying his monthly utility bill.

WORCESTER TELEGRAM

Worcester, Mass., September 11, 1983

The utilities that are struggling with the ever-higher cost of the Seabrook nuclear power station in New Hampshire have done the right thing. They have agreed to slow construction of the second nuclear unit until Seabrook I is complete and operating.

From this point forward, Seabrook II work will progress at a pace sufficient only to keep the project alive. That means planning and construction resources can be concentrated on the first nuclear unit, more than 80 percent complete.

New Hampshire residents are worried that even one nuclear power station will cause huge rate hikes for them. Completion of the first unit will certainly mean higher electric rates for those utilities involved — Massachusetts Electric and Mass Municipal Wholesale Electric Co. among them — but the start-up of Seabrook I will have some beneficial effects on those companies, too.

Start-up and the sale of power from Seabrook I will reverse the flow of funds for utilities that have borrowed money to finance the construction of the plant. Instead of spending borrowed money on the project, they will start to see a positive cash flow. Whether start-up occurs in December of 1984, as Public Service of New Hampshire estimates, in late 1985, as the utilities commission says, or in mid-1986, as some of the member utilities estimate, it will at least clarify the financial picture for PSNH.

The slowdown allows extra time for utility planners to assure themselves and try to convince regulators that Seabrook II is really needed. That will be a hard sales job indeed. When the two units were planned, demand for electricity was rising sharply. Since construction began, demand has fallen until it is nearly flat and shows almost no sign of picking up substantially.

When Seabrook I starts producing power, some of its customer utilities will be able to reevaluate their own needs for new generating capacity.

Builders of the Seabrook station have been claiming for years that theirs was a wise investment in cheap power for the future. When all the charts showed oil prices rising, power at Seabrook prices was a good deal. The price of oil is proving them wrong in 1983, but their claim that oil prices could shoot up again is a strong enough one to give pause.

Slowing work on Seabrook II may set the stage for eventual cancellation. At the least, it will delay the day of reckoning and reduce — at least for now — the exposure of New England's electric customers.

Unless fuel prices change markedly, there is no way that Seabrook II can be justified. By slowing the project to a crawl, the companies have bought time to adjust to the idea that Seabrook II ought to be scrapped — and they have bought time to work on ways to do that as efficiently as possible.

Post-Tribune

Guarding Your Interests Daily

Gary, Ind., November 15, 1983

The cost of electric power is a common puzzle across America. In several areas, the outlook for nuclear power plants is bleak. A dozen or more electric utilities, including Indiana's largest, Public Service Indiana, face formidable economic problems. The new nuclear facilities are draining the utilities' resources.

The trouble is rooted not in nuclear power itself, but in the cost of building plants. That is what largely forced cancellation of the Northern Indiana Public Service Company's Bailly nuclear project in Porter County. Construction had barely begun, but a chunk of the costs has to be borne by the customers.

In some areas, the words "financial ruin" are used to describe the problems. A private study by a Cambridge, Mass. group says that many of the new reactors will produce electricity at double or triple the price of OPEC oil.

The report added that "American ratepayers and utility company shareholders and creditors have yet to learn, let alone pay, the full price for nuclear electricity."

The crises all seem to be linked to timing and the long lag time between start and finish of nuclear plants. Rising costs are deadly. Predicted rate increases of 30-80 percent indicate how severe the squeeze is.

Hearings on the Public Service Indiana's Marble Hill Plan in southern Indiana are going on in various parts of the state.

The pessimistic picture of the industry suggests that the dream of cheap energy from nuclear power won't be achieved on a wide scale. The study, which nuclear industry officials have not commented on, says the share of electricity coming from nuclear power may not rise above 20 percent. What cancellation of plants, and the burden of tremendous costs, will do to the utilities and to their investors may be a more important consideration.

These fears may be exaggerated, but they warn of a unique American tragedy that will bring severe economic hardship to thousands. Managers of the troubled utilities may be blamed for lacking vision, but that is too simple.

Charleston, W. Va., September 10, 1983

COAL has nothing to fear from nuclear energy for the foreseeable future, if Caroline J.C. Hellman and Richard Hellman, writing in *The Wall Street Journal*, know whereof they speak. Caroline Hellman is a research assistant professor in energy economics at the University of Rhode Island and Richard Hellman is a professor of economics at the same school and has worked for a number of federal agencies, including the Federal Power Commission. Together they wrote *The Competitive Economics of Nuclear and Coal Power*.

"The technological and economic design of nuclear power has been the biggest failure of any civilian industry in history," say the Hellmans. "No private risk-taker would ever have built today's nuclear-power industry without unlimited cost-plus provisions and government subsidies."

Electricity produced by a nuclear plant if ordered today would be at a minimum twice as expensive as electricity produced from a coal plant. Reason for the huge discrepancy in price is that the technology for nuclear-generated electricity was rushed into premature mass commercialization.

"The (nuclear) industry," says the Hellmans, "has been its own worst enemy in refusing to recognize that nuclear power is uneconomic and getting worse. To this day, the Tellers, the Weinbergs, the Atomic Industrial Forum and the Edison Electric Institute are leading the industry down a path of fool's gold."

In addition, the industry isn't learning from its mistakes. The authors made 163 comparisons of nearly identical younger and older nuclear units and used four learning criteria. In 39 percent of the comparisons performance improved significantly, improved not a bit in 23 percent of comparisons and "deteriorated significantly" in 38 percent of the comparisons.

Nuclear power exposed to valid economic analysis can't justify the expense and isn't in the same ballpark with coal power. One reason is "sufflation" or the escalation of original capital-cost estimates attributable to technological problems as distinct from price rises attributable to the usual inflationary factors.

Nuclear plants for a variety of reasons aren't very likely to last more than a quarter of a century. Indeed, some utility companies — American Electric Power, for one — recognize this. Coal plants are good for 30 years.

Eventually, nuclear power may supplant coal power, but it won't soon happen.

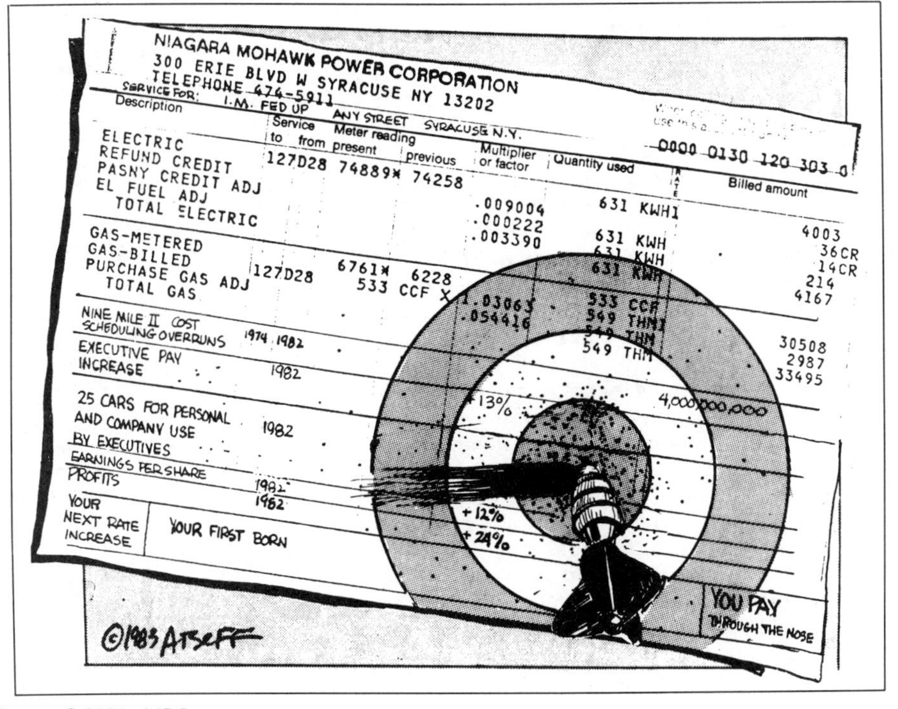

Roanoke Times & World-News

Roanoke, Va., November 19, 1983

NUCLEAR POWER in the United States is an endangered species. Unless the situation changes, its days are numbered, and the country will lose an important energy source.

Nukes themselves will be around for several years. There are 84 nuclear generators with operating licenses; 56 more have construction permits, and there are four reactors on order. But no new order has been placed since 1978.

Partly that's because demand for electricity, no matter how it's generated, is rising much more slowly than a decade ago. But the bottom's dropped out of the nuclear-power market because 1) it takes a long time to plan for and build one, 2) costs of such plants have risen astronomically, and 3) making reactors safe, reliable and politically acceptable has turned out to be much more difficult than anticipated.

There's no single solution. But all of those problems could be eased if the industry could develop two or three standard reactor designs. That's one suggestion of a recent study by the nuclear engineering department of the Massachusetts Institute of Technology.

The usual course is to seek economies of scale: If you're going to put all that money into building a nuclear reactor, might as well make it big. Instead, MIT's people like the idea of smaller, more efficient reactors that could become the model for the industry in the 1990s.

Fragmentation in the industry also complicates matters. Victor Gilinsky, a federal nuclear regulatory commissioner, notes that 58 utilities own nukes, and this makes regulation difficult: "It is only a slight exaggeration to say that we get nearly 60 different solutions to every safety problem." He would prefer that utilities let a few large operating companies run their nuclear plants; such companies could hire the experts that this field requires.

These aren't all of nuclear power's problems. Others include finding safe, orderly methods for radioactive-waste disposal and mothballing obsolete plants. Given the political will, however, those are soluble. Nuclear technology can be unforgiving of mistakes, but that ought not deter us from trying to make it more safe and usable.

The MIT study sees a possible turning point in the fact that the breeder reactor — a project that absorbed most of the time, money and attention in nuclear R&D for more than a decade — finally is being phased out. The next step, says the study, is up to the electrical utilities: They'll have to supply the pressure to achieve new directions in nuclear-power research.

That could be asking a lot of the utilities. They tend to be conservative if not unimaginative. Many of them, burned by public reaction during the energy crunch, also like to keep a low profile. But it's utilities that will decide, by their orders, whether nuclear power has a future and, if so, what shape it will take. The peaceful atom's potential has never been realized, but it shouldn't be written off yet.

The Houston Post

Houston, Texas, December 14, 1983

What happened to the promise of cheap, abundant nuclear-generated electric power? A decade ago some experts were predicting that by the year 2000 half of the world's electricity would be produced by nuclear reactors. Current estimates have cut that projection in half amid growing evidence that the bloom is off nuclear power, especially in the United States. In the past 10 years, 102 unfinished nuclear plants have been scrapped in this country.

A recent study by the Worldwatch Institute concludes that the cost of electricity produced by nuclear reactors in the 1980s will be higher than electricity from coal- or oil-burning plants. A major reason for the higher cost of nuclear power, says the study, is huge cost overruns produced by poor management, high interest rates and haphazard government regulation.

Worldwatch, a Washington-based research organization that studies global trends, used utility industry data on some 30 nuclear power plants to make its calculations. It projected that the "average lifetime generating cost" of a new plant would be 10 to 12 cents a kilowatt hour, more than 65 percent higher than a new coal-fired generating plant and 25 percent higher than a new oil-fired plant.

The Worldwatch study cited other factors that have depressed the nuclear power industry: Energy conservation spurred by OPEC's steep oil price increases, a severe global recession and the Three Mile Island incident that resulted in tougher federal regulation of the nuclear power industry. But stiffer government safety rules can hardly be faulted. Their laxity in the past has contributed to many of the industry's current problems, including a persistent, vociferous anti-nuke movement.

Still, the Worldwatch study's criticism that the nuclear industry has, in effect, shot itself in the foot by its own inefficiency is especially distressing. First, the present oil glut could quickly evaporate if the Persian Gulf supply were disrupted. Other industrial countries that depend more heavily than we do on Mideast oil are already engaged in ambitious nuclear expansion programs to protect themselves against a cutoff and higher prices.

Secondly, even if there are no more energy crises, the day will come when the world will run short of oil. At that point we will need nuclear power in substantial quantities. When that time comes, the public should be satisfied that it is safe and that its price has not been grossly inflated by mismanagement and misregulation.

Plant Cancellations Increase: The Costs of Abandonment

Over one hundred nuclear plants have been cancelled since 1972. The largest number of cancellations in one year since the birth of the nuclear power industry was in 1982; in that year, 18 plants were cancelled, seven of which were already under construction. This growing phenomenon, as the electric utilities are caught between rising construction costs and falling electrical demand, has brought with it a host of legal and economic problems. While the investment in a plant not yet under construction usually only runs to tens of million of dollars, when a plant already under construction is scrapped, the costs of abandonment can run into billions. Although the first inclination of many observers is to say that the utilities should pay for their own mistakes, they are in fact often not able to do so. Absorbing the total cost of a cancelled plant, particularly if is close to being completed, could bankrupt most utilities. It is not only the utilities' shareholders, therefore, who are hurt by plant cancellations. The apportionment of cancellation costs is usually decided by the state public utility commission, and in the great majority of cases the ratepayers are forced to pick up part of the cost. This is usually done by allowing the utility to include amortized portions of the uncompleted plant in the company's expenses, which are paid by customers in the form of increased rates over a five-to ten-year period. Even the threat of a utility bankruptcy is threatening to ratepayers, since a bankrupt electrical company is usually exempt from the ordinary limits on rate increases set by state regulators. A small minority of states, however, have refused to allow utilities to collect significant portions of abandonment costs from ratepayers; among them are Oregon, Ohio and Arizona.

One of the two most costly cancellations to date, the abandonment of four plants in 1982 by the Tennessee Valley Authority, cannot readily be applied to other situations because the TVA is a huge government-run utility. (The other was the cancellation of a series of plants owned by the Washington Public Power Supply System—see pp. 180-187.) Other major cancellations discussed in the articles that follow include the Bailly plant in Indiana, representing an investment of over $200 million; the Pilgrim 2 plant in Massachusetts, at an investment of nearly $300 million; and two Black Fox plants in Oklahoma, in which over $350 million had been invested. The cost of decommissioning a nuclear plant, mentioned in some of the editorials, is routinely passed on to ratepayers and covers the dismantling of a plant after its useful life is over.

The Charlotte Observer

Charlotte, N.C., June 26, 1979

...The final referee is the almighty dollar....

—David Springer, president of the High Rock Lake Association

The High Rock Lake Association, a citizens' group that fought to prevent construction of the Perkins nuclear generating station on the Yadkin River in Davie County, is jubilant following Duke Power Co.'s decision to postpone the project. (The river feeds High Rock Lake at a point common to Davie, Rowan and Davidson counties.)

Duke announced last week that, along with Perkins, it would postpone construction of a third nuclear unit at its Cherokee plant (near Gaffney, S.C.) and delay the start-up of two other units there until the late 1980s.

While Duke officials say their faith in the safety of nuclear plants is undiminished, their action acknowledges that nuclear power is becoming a risky investment. Their act of fiscal prudence pleased Wall Street as well as anti-nuclear activists.

Mr. Springer is right about money. Producing electricity is a business. To attract investors, power companies must not put money into unsound investments. The decision about the soundness of investments must take into account all the factors: public safety, demand for electricity, technological efficiency, the political climate. With all those things in mind, Duke has considered nuclear power an efficient, economical way of producing electricity — not as efficient or economical as once predicted, but nevertheless more so than coal. So Duke has promoted nuclear power.

The Three Mile Island accident changed things. Nuclear procedures and designs are being reevaluated. Nuclear licensing standards are likely to become tougher. Shutdowns are likely to occur more frequently. Citizen protests against nuclear plant construction are likely to spread. Concern about radioactive wastes will intensify, as it has here in response to Duke's proposal to haul spent fuel through the county.

The political climate, as much as the complex technological debate over safety, makes the future of nuclear power in the United States at best uncertain — particularly since the growth rate in demand for electricity is declining. The investors who can put up the millions of dollars needed for nuclear plant construction don't like uncertainty.

Duke's decision is therefore a sound one. Forced to turn their attention to alternate fuels, Duke and other power companies can no doubt develop cheaper, more efficient ways to use them — including coal, wind and solar power.

But those sources, valuable as they may be, are not the immediate keys to a reliable energy supply in the Piedmont. In the short run, the best insurance is to make the wisest possible use of the energy existing plants produce. Conservation will be vital, and its importance must be reflected in the rate structure as well as in the voluntary habits of Piedmont consumers.

ST. LOUIS POST-DISPATCH

St. Louis, Mo., May 13, 1979

The economic fallout from Three Mile Island is beginning to be felt on Wall Street, where investors have drastically hiked the cost of borrowing money for nuclear power plants. The most dramatic example is found in the recent $150 million debt issue from the North Carolina Municipal Power Agency. The NCMPA is purchasing a 75 percent interest in Duke Power Co.'s Catawba nuclear power plant, and will sell electricity to 19 North Carolina cities. The agency is totally dependent on atomic power, according to *The Wall Street Journal*, and its tax exempt revenue bonds are the first all-nuclear offering since the Pennsylvania accident.

Last November, when the North Carolina agency borrowed $400 million, it paid only a small premium over other non-nuclear utility bonds. But the agency had to pay a premium six times as large — nearly a half a percentage point higher — to sell the recent, smaller issue. The 7.325 percent yield the agency was forced to accept is comparable to 14.75 percent return on a taxable bond. That yield, in turn, is between 30 and 40 percent higher than the return commanded by routine public utility bond offerings. If Wall Street sees the risk of nuclear investment, Main Street investors will not be far behind.

Portland Press Herald

Portland, Maine, May 17, 1979

Electric utility spokesmen are expecting too much in arguing that consumers shouldn't grumble about footing the additional cost of power while Maine Yankee's nuclear plant is temporarily out of service.

Utility officials told the Legislature this week that the nuclear plant has saved consumers $200 million during its six years of operation. Having saved customers all that money, they shouldn't complain now when electricity bills are temporarily higher, utility spokesmen reasoned.

That argument won't impress most Mainers. It's a little like suggesting that Americans shouldn't complain about the high cost of Mideast oil because the sheiks sold it to us so cheap for so long.

That said, however, let's get down to the core issues involved in legislative proposals to bar utilities from charging customers for the higher cost of alternative energy while Maine Yankee is sidelined.

The first is this: The burden of the cost of the more expensive alternative energy can't be permanently shifted off the shoulders of the consumer. In the long run the customer is going to pay for it because the user is the only—repeat only—source of utility income. The burden can be shifted from present to future customers but in the long run *some* customers must pay.

Equally fundamental, there's absolutely no reason for lawmakers to be tinkering with the nuts and bolts of public utility regulation in Maine. Those decisions, as Public Utilities Commission Chairman Ralph Gelder rightly observed, "are for the PUC to decide after hearings on all aspects of the case."

Granted, it may be politically tempting for lawmakers to get involved in utility rates. But the temptation ought to be resisted; it's bound to result in poor utility regulation.

EVENING EXPRESS

Portland, Maine, June 27, 1979

The Maine Yankee atomic power plant at Wiscasset will be shut down again in September.

That's not good news. It will reduce the supply of electricity and increase the intensity of controversy.

Presumably, as in the past, Central Maine Power Co. will have to make up for Maine Yankee's lost power by buying electricity or stepping up production at conventional plants, or both.

This up and down business with the nuclear plants does nothing to take the steam out of the heated dispute over nuclear power.

This shutdown will accommodate another directive from the Nuclear Regulatory Commission which wants a check made for hairline fractures in the pipes of certain steam generating systems. The NRC requires the inspection be made within 90 days. But the Wiscasset plant was scheduled to go down in September anyhow "to replace a refurbished transformer."

It's unfortunate that the transformer replacement and the inspection the NRC now requires could not have been made while the plant was down earlier this year.

Closed in March while the NRC assured itself that the plant wouldn't pop its pipes in the event of an earthquake, the Wiscasset installation survived an earthquake practically centered outside its gates. It also convinced the NRC. The plant went back on line only last month. With only four months of operation it will be down again.

Maine consumers are still sputtering about paying for the cost of replacement fuel during the last shutdown. The September shutdown is expected to last only a couple of weeks but the replacement cost probably will be added onto the customers' bills. It's going to add to the dissatisfaction and it's going to give the anti-nuclear forces another wedge.

With the nuclear plants going up and down like yo-yos, the confidence in nuclear power, and the cost reduction it represents, is weakened.

The Courier-Journal

Louisville, Ky., August 4, 1979

THE EXECUTIVES of Public Service Indiana should be ready, if they're not, to bewail the day the slick salesmen of the nuclear industry turned them from the relative certainties of coal-fired power to the awesome imponderables of nuclear generation. Continuing bad news from the Marble Hill project magnifies points that should have been evident long ago.

A skeptic never would have accepted the nuclear industry's rosy cost comparisons of nuclear generation versus coal, even discounting entirely the risks of human and environmental disaster associated with the atom. The case for constructing a billion-dollar-plus nuclear plant in a region with easy access to coal and good transportation never was convincing.

It is even less convincing today, with much of the Three Mile Island plant in Pennsylvania in ruins, and the need for added safety precautions confirmed by federal investigators. The cost of constructing nuclear generating plants was soaring far faster that that of coal-fired plants even before the Three Mile Island fiasco. It's sure to escalate after the final report on that affair is written.

This outlook is made even bleaker by PSI's difficulties in having concrete poured for the Marble Hill plant's containment shell, a matter so serious the Justice Department has been asked to consider prosecution. The company's explanations are further evidence that PSI wasn't quite aware of what it was getting into.

Even as company officials issue soothing assurances that all will be made right, PSI continues to insist that the matter is all cosmetic anyway — just a case of a few blemishes that might disturb the public. (A cheerful communique reports that many of the 170 improperly repaired voids in the concrete were less than a foot in diameter, but neglects to mention one void that was 15 feet wide.)

None of this indicates any safety threat, we are assured. But if PSI can't even get concrete poured to exacting standards, a layman may well wonder how it will perform when faced with the far more difficult tasks of building and operating the actual generating and safety systems.

Continuing evidence that PSI is making a hash of the job emphasizes the need for a searching look by the government at whether the company is capable of building and operating a safe nuclear plant. But an even better question is whether PSI and its customers and neighbors wouldn't all be better off if the utility cut its losses right now and abandoned a project that was ill-advised in the first place.

The Detroit News

Detroit, Mich., October 22, 1979

As if there weren't controversies aplenty surrounding nuclear power plants, "decommissioning" has been added to the list.

When the Palisades and Big Rock nuclear plants, which are owned by Consumers Power Co., wear out 27 years from now, they cannot be simply closed, abandoned, or torn down. They need to be decommissioned, meaning that their dangerous radioactive material must be removed or contained.

Consumers Power has already figured such decommission charges into utility bills, but inflation and Nuclear Regulatory Commission (NRC) rule changes have raised anticipated costs. The company is asking the Public Service Commission (PSC) to create a fund to handle the future expense.

The PSC staff has recommended that the fund be created by raising utility customer's bills. They reason that present customers are underpaying at the expense of future customers.

The future cost estimated by PSC and Consumers staffers, based on a relatively rosy 6 to 8 percent annual inflation rate, would equal $526 million in the dollars of 2006. That would mean the collection of $19 million extra from ratepayers each year, or about 40 cents per customer per month.

But PSC Commissioner Willa Mae King, Atty. Gen. Frank Kelley, the Michigan Citizens Lobby, and the Public Interest Research Group (PIRGIM) are demanding a hearing on the whole decommissioning issue before the PSC agrees to a rate increase.

That sounds like a pretty good idea.

Even the sanguine PSC staff must admit that the $526 million estimate involves so many assumptions that a crystal ball would serve almost as well as technical analyses.

Consider the problem of method. Presently, the decommissioning process may require the removal of radioactive material and then guarding the structure forever, or sealing up the plant with concrete and steel, or dismantling the plant and removing it to a federal dump.

The PSC is assuming dismantling, a more expensive option but one the staff sees as more likely to be required by the unknown and unknowable NRC rules of the year 2006. The state agency is also assuming that the same technology of decommissioning will be used 27 years from now and that its cost will reflect an inflation rate of 6 to 8 percent annually — two leaps of faith that are at least as arguable as they are well-intentioned.

How is the fund to be collected and handled? If the money is collected from ratepayers, should the fund be managed by a trustee or given over to Consumers Power to invest at its pleasure? The PSC staff recommends that the utility collect and keep the decommissioning funds, thus providing the company with three decades of involuntary investment by ratepayers.

Concern about the security of such a utility-controlled fund is justified. As Mrs. King reasonably notes, the company could go bankrupt before the 27-year deadline (as, for example, in the wake of a nuclear accident. It's extremely unlikely, but who knows?).

Citizens would then wind up paying three times — for the fund, for the federal taxes on the money collected for the fund, and for the still-required decommissioning costs of the disabled plant.

So the question: Is there any way around the corporate taxes that must be paid on income collected specifically to defray the decommissioning costs, taxes that double the expense to the consumer?

And, finally, what's the rush? The delay caused by a hearing might be useful in clearing the air without exacerbating a problem that won't pop up until 2006 A.D.

THE MILWAUKEE JOURNAL

Milwaukee, Wisc., October 3, 1979

The announcement by Wisconsin Electric Power Co. that it is considering spending $30 million to $40 million to alleviate problems with steam generator tubes at the Point Beach Nuclear Power Plant poses serious economic and safety questions. The State Public Service Commission, which had planned to examine the issue, now needs more than ever to probe the implications.

One big issue, which the utility should not be allowed to settle on its own, is whether Point Beach Unit 1 should be reopened after its current shutdown for refueling or kept out of service pending generator replacement. Related issues include the cost of (1) replacing the generators, (2) operating Unit 1 safely if it is restarted before new generators are ready to be installed, and (3) buying replacement power during shutdown for generator replacement.

The PSC had been under pressure to probe the steam generator tube problems anyway. Wisconsin's Environmental Decade, an environmental group, and three state senators had been prodding the commission to examine both the safety and economic issues involved.

The PSC has been reluctant to jump into the safety issues, since it may lack legal jurisdiction (nuclear plant safety being chiefly in the baliwick of the federal Nuclear Regulatory Commission). However, there seems no way to divorce the safety issue from nuclear plant economics, over which the PSC clearly does have jurisdiction.

So, we hope the PSC examination will deal thoroughly with the safety questions as well. And if it does not, then certainly the NRC should, before Point Beach Unit 1 is returned to service.

Sentinel Star

Orlando, Fla., August 4, 1979

FLORIDA Power Corp. has closed its Crystal River nuclear generating plant down — again. This time the problem is a loose seal and leaky pipes.

And, once again, Florida Power officials are telling their ratepayers they are going to have to pay for the problem. It is a refrain Florida Power customers, already paying among the highest rates in the state, have heard before.

Of course, if there were any question of safety, the plant had to be closed down. Safety must always come first. The question is why the ratepayers are always the ones to pay for somebody's mistake.

Allowing a utility to make a fair profit — as is done with monopolies regulated by the Public Service Commission — should not mean that stockholders are insulated from loss.

If Florida Power stockholders had to pay occasionally for the mistakes of its management — such as in the $8 million oil daisy chain rip-off — then maybe stockholders would demand better management.

If that happened maybe Florida Power, which is beginning to look like the original Hard Luck Kid, would stop having so many unfortunate occurrences in its operations.

SYRACUSE HERALD-JOURNAL

Syracuse, N.Y., June 28, 1979

The New York Public Service Commission intends to scuttle construction of two nuclear power plants in the Town of New Haven, Oswego County.

The PSC moved after Ecology Action, an Oswego County opposition organization, recommended that the plants, which were scheduled to generate electricity in the 1990s, not be built.

The PSC acted capriciously.

The evidence is found in the reason, cited by its own publicity spokesman, that the PSC couldn't figure out whether New York State Electric & Gas Co. or Long Island Lighting Co., partners in the two ventures, owned a controlling interest in either one or both.

* * *

And, of course, the PSC must expect us to believe it hasn't the power or the will to find out.

That's an excuse for a Ripley "Believe It or Not" cartoon in The Herald-American comics section.

The PSC is also fearful we may produce too much power in New York State before 1995.

It holds little faith in our growth.

It also lacks vision.

We cannot visualize the PSC accepting the contamination of coal-fired plants if oil becomes so dear that no utility can afford to keep oil-fired plants in continuous use.

The only substitute we can turn to in this century is nuclear.

We can visualize if gasoline prices follow oil prices to the stratosphere that hundreds, even thousands of electric cars will appear on our roads and streets as if by magic.

Where will their batteries be recharged? At home and in parking lots all over the state.

Will cleanly generated power be there in 1995?

* * *

The PSC is asking the state's siting board to rescind its approval of the proposed, 1,250-megawatt plants.

We see no reason for the siting board to duplicate the PSC's hysterical playing to the gallery of the moment.

The Providence Journal

Providence, R.I., December 14, 1979

New England Power Company's decision not to build a nuclear power plant in Charlestown can be lived with. But if that is only the first of a series of such retrenchments, as some might hope, New England may well become the Appalachia of the 1990s.

When NEPCO originally decided to build a nuclear plant, it did so on the basis of projected demand and because it has had two decades' experience in producing nuclear power safely at half the cost of conventional fuels. But these perceptions and considerations have been altered by changing circumstances. Since then nuclear power has become an environmental and political football; OPEC has raised the world price of oil by a factor of ten; and the nuclear accident at Three Mile Island in Pennsylvania has heightened public fears about nuclear safety.

Those who have fought the Charlestown project, for the most part, are motivated by such fears. They see nuclear power as a threat to the natural environment and to human health. There are other considerations — doubts that nuclear power's economic advantages still exist, doubts about NEPCO's projections of future energy demand. These may not be the primary issues for the opponents of nuclear power, but they are central to the issue. And to NEPCO's future plans and intentions.

NEPCO has a new scenario both for future electric power demand and how to meet it. Demand projections have been significantly scaled down. Critics will say "I told you so," but it is a question whether NEPCO was overoptimistic or whether the nuclear oppostion has scared away potential new customers. Whatever the case, while the 15-year "NEESPlan" pointedly leaves out the Charlestown plant, it includes several new nukes — Seabrook, Millstone and Pilgrim — all of which are under heavy attack by environmentalists.

NEESPlan also bows to critics by heavily emphasizing conservation. It looks to alternative energy sources, such as coal (though many authorities consider that more of a health and ecological threat than nukes). It proposes some experimental projects, such as burning solid wastes for power production. But when all is said, nuclear power remains a key element of NEESPlan.

The question now is whether the nuclear opposition is willing to compromise, or if it intends to fight to the death. And if the latter, the death of what — New England's economic future?

New England is, has long been, a depressed region heavily dependent on imported fuel. It has no defense against extortionate fuel prices or cutoffs. In recent decades, it has relied on one great resource to keep industry here and to attract new jobs — a highly skilled work force and a highly educated academic cadre. These assets have outweighed the disadvantage of high-cost energy. So far.

But in the post-OPEC energy era, the equation has changed. Energy availability and price are now an overriding consideration. Ignore that reality and not only will outside capital shun New England, but local capital will flow out even faster than it always has. That leaves us with our work force. In a society as mobile as this, can anyone believe that the region's best brains and ablest workers will not follow the sun?

That is one alternative to nuclear power that we have not seen discussed in the anti-nuclear literature.

TULSA WORLD

Tulsa, Okla., August 27, 1981

THE BLACK Fox nuclear power plant, which would have started operations next year under the original construction schedule, is dead, the victim of misguided anti-nuclear protesters, Government red tape and inflation.

An analysis of the project by the prestigious firm of Touche Ross & Co. tells the story. Instead of the original 1973 estimate of $1.5 billion, the plant could cost as much as $10.1 billion, and that only if the regulatory logjam were broken.

Other securities analysts agree the escalating costs of nuclear plants render them virtually an impossible option for power companies. The same analysts say the nuclear fuel option is "absolutely necessary."

While Touche Ross recommends conversion of Black Fox to a coal-fired facility, that option would require starting almost from scratch. The firm estimated the completion of Black Fox as a coal-fired plant at between $5 billion and $6 billion.

The same groups which successfully killed Black Fox as a nuclear facility have vowed to fight it as a coal-fired project. Their contention is that no additional electrical generating capacity is needed.

Touche Ross disagrees. Additional capacity will be needed if Tulsa and Oklahoma are to benefit from the obvious growth that is already occurring and is certain to accelerate.

By 1988, additional power will be needed. Work would have to start almost immediately if power plants are to be built and ready to deliver power by that date.

The strategy of the anti-nuclear, anti-coal, anti-growth forces has succeeded in the case of Black Fox. Delay, protest and force ever increasing Government regulations and let inflation and high interest rates do the rest.

Sadly, the same tactic might work if the decision is made to convert to coal.

The result will be that sometime around 1990 the public will awaken to the fact that lack of electricity is hampering economic development.

When that happens, you can be sure the same protesters and "consumer" groups will blame the utilities for not building plants.

Lost in that debate will be the simple fact that when we had the opportunity to build them, we didn't.

THE WALL STREET JOURNAL.

New York, N.Y., July 20, 1981

All right, so Harold Robbins we're not. But we think we have an interesting tale to tell.

Some time ago, Cleveland Electric spent about $60 million doing preliminary spadework on a nuclear power plant. In January, 1980, it decided to drop the project because of political and regulatory hang-ups. It's not unheard of for power projects to be abandoned in their early stages—though usually at $100 million to $200 million a pop—because of changing capacity needs, or financial constraints or bureaucratic foul play which has nothing to do with the utility.

Until it decided to drop the project, the expense was carried on Cleveland Electric's balance sheet as construction work in progress; it did not, however, figure into the rate of return calculation made by the Public Utilities Commission of Ohio. After the project was scrubbed, Cleveland Electric asked the PUC to allow the expense to be amortized over 10 years and to include it in rate-base calculations.

This is also pretty standard. Usually PUCs will let utilities recover the spent dollars, but not the interest payments the utility owes on the debt borrowed for the project.

All fine and dandy. Then a little number called the Office of Consumer's Counsel, an Ohio state agency, complained in court that the ratepayers should not be burdened with the abandoned project's costs because it was not a cost of doing business. The Ohio Supreme Court agreed, arguing that the whole thing was a policy matter, not an accounting one; thus the state legislature should make the determination.

Well, whatever you think of the plot so far, you're probably not terribly distressed. After all, in the natural order of things utility managers and investors should pay for mistakes and ratepayers should be spared. Yes; but just let us complicate matters a bit by spinning out a hypothetical ending.

Cleveland Electric takes the loss in its next quarter. Its earnings drop, ah, rather steeply. So steeply, in fact, that its bond ratings drop. Well, a utility closed out of the debt market is like a summer without baseball. No place to go, nothing to do. The utility will just have to go back to the PUC and ask for a rate increase and the PUC will just have to grant one: the long way around to the same solution, right? But, actually, this is a little worse than a boomerang; not only do ratepayers and investors pay, the industry pays as well.

The cost to the industry will be a loss of flexibility. Cleveland Electric's protracted turmoil will have a chilling effect on those utilities thinking, for good reasons, of dropping projects. They won't. And it will freeze those utilities thinking, for good reasons, of expanding. They won't either.

There is another moral to the story. Forcing shutdowns of nuclear power projects may seem like good clean fun to the people engaged in the activity, but someone has to pay. And however much that Ohio "consumer" agency pretends to be doing for the electric ratepayers, economic reality dictates that the main burden will fall on them.

THE INDIANAPOLIS STAR

Indianapolis, Ind., August 28, 1981

The decision to cancel the Bailly nuclear power plant project was a difficult one which ought to be greeted with relief but no real joy.

That huge hole in the ground in Porter County at Lake Michigan now is one of the symbols of our time and perhaps of our nation: It is a high technology dynamo that is not to be, a contemplated elbowing-aside of nature that proved cumbersome beyond tolerance, another bit of energy independence gone sour, an investment of a kind that once seemed gilt-edged but in today's time of shrunken ambitions came more to resemble a plunge into a murky swamp.

The site was a poor one, events and closer examination showed.

It was lawful all along, though. The federal government, which may soon adopt tougher siting criteria that would rule out Bailly's earlier approval, has some responsibility for what took place haltingly over many years.

The Three Mile Island accident is not now a cause for panic. But it remains a sign of the need for great care and wisdom in nuclear plant design and siting. Three Mile Island also stands as a huge investment out of production indefinitely. There is no assurance of who will absorb the clean-up costs or the lost revenues, how much they will be, or when the plant may operate again.

Northern Indiana Public Service Co. made the good faith decision to go nuclear a decade ago, based on projected needs and costs. It has alternatives, not excluding a possible nuclear site in a more remote area of North Central Indiana.

But the possibilities for gas-fired generation in the high demand area of Lake and Porter counties should not be overlooked as federal policy on gas changes. And careful revision of energy need projections clearly is in order. Some areas of the nation have had a 10 percent reduction in energy use over the last three years, with slowed economic growth a big factor.

It may be that a portion of the $200 million now wasted at Bailly could be counted legitimately in the cost of a replacement plant if one is built. But we don't think the company's rate payers should carry all that freight.

The stockholders should not, either. Some sharing is in order. The extensive federal involvement in nuclear generation development includes limiting of liability and assumption of some of it, in the case of accidents. Surely the same policy approach can be used to allow reasonable federal tax treatment of an investment loss in which the government's site approval was crucial.

The Bailly story also is a wry commentary on the very questionable public policy that CWIP (construction work in progress, an approach of charging today's utility customers for productive capacity that is months or years from use) would involve if adopted by Indiana.

Under CWIP in its full version there would be no argument about who bears the cost of this failed dream. The customers of NIPSCO would already have paid it.

THE COMMERCIAL APPEAL

Memphis, Tenn., September 3, 1981

WHO GENERATES delays in building nuclear power plants? Government regulators and "environmental extremists" most often get the blame. Utilities get their share, too, for overestimating electric demands and for running into financing problems.

Yet, none of these parties is really "at fault." Delays, it seems, are built into the system.

Consider the case of the Northern Indiana Public Service Co. It announced last week that it was canceling the Bailly nuclear plant it's been planning for a decade. The reason: Costs originally estimated at $187 million in 1970 would be $2.13 billion by time the plant came on line.

What or who is responsible for that increase? Inflation, for one, and high interest rates to pay for the plant. Both problems could have been avoided had construction begun when the plant was announced, but it didn't. There were objections to the site, only 35 miles from the Chicago Loop and abutting the Indiana Dunes National Lakeshore. There also were licensing procedures to be satisfied as both plans and construction went forward.

Do such factors lead to longer-than-necessary delays?

As Peter A. Bradford, a member of the Nuclear Regulatory Commission, pointed out in a letter to The Wall Street Journal, the United States could have faster construction if that were its only goal:

"In France, for example, reactor sites are often selected and approved before any decision is made to build a plant. Therefore, the site approval process does not have an impact on construction time. Furthermore, there is only one utility (state-owned) and it may buy only one reactor design from only one vendor. Since that design is fully familiar to licensing authorities and since the site is previously approved, the regulatory process is more simple."

THE UNITED STATES could do business that way, but it hasn't chosen to. It sees the benefits of having a multitude of independent utilities, which can decide for themselves whether a plant should be built and, if so, how. Yet, because the product involves splitting atoms — albeit for peaceful purposes, companies must show that their plants will be safe as well as sound investments.

That is the cost of being part of the U.S. nuclear power industry — a cost, it should be noted, that isn't all that inefficient, given that more nuclear plant licenses have been granted here than in the rest of the free world combined. Given the benefits, is the cost too high?

Chicago Tribune

Chicago, Ill., August 29, 1981

Northern Indiana Public Service Corp. announced Wednesday that it has abandoned its bitterly controversial plan to build the Bailly nuclear power plant on the shore of Lake Michigan between Portage, Ind., and the Indiana Dunes National Lakeshore.

The company's decision is good news for the many, including The Tribune, who have opposed the Bailly site. But the victory must not be misinterpreted. It was not a victory over the construction of nuclear power plants in general; it was a victory only for the public's right to a voice through legal channels in the location of these plants. Though the company's critics reflected all kinds of opinions, the only argument that won judicial support was that the Atomic Energy Commission (now the Nuclear Regulatory Commission) had violated its own rules in August, 1974, when it approved the Bailly site, which is less than two miles from the Portage city limits.

This was the chief argument raised by Business and Professional People for the Public Interest, the most persistent and influential of Bailly's opponents.

The site was objectionable also because it posed a potential threat to thousands of visitors to the Dunes and an esthetic threat that the Department of the Interior denounced as "appalling."

But it was the violation of the two-mile limit that guided a Federal Court of Appeals in rejecting the AEC's approval. This decision was later sent back by the Supreme Court for further study. But accidents like the one at Three Mile Island in 1979 focussed new attention on the potential danger of having a nuclear plant too near a population center and strengthened the determination of opponents. This finally persuaded the company to give up.

The public are losers as well as winners in the NIPSCO decision. The losers include not only the company, which sank more than $100 million into the project, but also its many Indiana customers who are now denied, for the foreseeable future, the advantages of nuclear power from the plant. These advantages include cleanliness and an abundant supply of fuel at relatively stable prices — something that has helped keep Chicago's electricity prices well below New York's. Anybody who interprets NIPSCO's surrender as an argument against nuclear power in general is misreading the legal history of the case and blinding himself to the need for more domestic sources of energy.

There are no villains in this story. At the time NIPSCO made its plans a decade ago, nuclear power was being hailed as the symbol of the future. Having made so large an investment, the company was understandably reluctant to abandon its plans even when the euphoria over nuclear power wore off. In retrospect, it should have recognized earlier that its adversaries consisted not only of anti-nuclear zealots but also of reasonable people able to tell a good site for a plant from a bad site. If there is a lesson in the whole unfortunate affair, it is that the importance of developing more nuclear power must be tempered with an insistence that the plants be built in relatively uninhabited areas When adversaries like BPI have a good case, they don't just go away. And the cost of transmitting power a few more miles is trivial compared with the real and psychological costs of a nuclear accident, even a minor one, in a populated area

The Boston Globe

Boston, Mass., September 26, 1981

When nuclear power plants suffer malfunctions, the system may go through a sudden shutdown known to the trade as a "scram." The purpose, as might be expected, is to protect the reactor and the rest of the world from serious damage.

Boston Edison went through a money "scram" Thursday, when it decided to cut its losses by cancelling commitment to its Pilgrim II nuclear power plant. The decision was wise.

When Pilgrim II was conceived 10 years ago, assumptions were different from those confronting Edison and other electric utilities today. Most conspicuous among the changes is the financial environment. In 1971, utilities could borrow money for less than 8 percent. Today they are lucky to find long-term financing at twice that rate.

That change, plus the fact that utilities underestimated the costs of constructing nuclear plants that could comply with stringent safety requirements, has made nuclear power plants expensive.

When the project began, it was expected to cost about $400 million and to be in operation by today. Today, had Edison persisted with the project, its projected cost was about $4 billion by the time the plant would have ultimately been completed in 1990. Some students of electric utilities believe that that cost would have driven Edison into bankruptcy.

It is striking that if interest rates were higher by only 2 percent than Edison's assumption, the cost would have risen to $5 billion. Conversely a 2 percent reduction would have lowered the cost to $3 billion.

To its credit, in killing Pilgrim II, Edison has kept its objective of reducing oil dependence for making electricity to only 20 percent by 1990 compared with 70 percent today. Doing this will not be easy. Key elements in this project are purchase of lower-cost nuclear and hydropower from Canada, conversions of oil-fired Edison plants to coal, and possible investment in other nuclear projects in New England, at New Hampshire's Seabrook or Millstone in Connecticut.

The toughest unresolved question in the wake of the Pilgrim decision is the treatment of Edison's 59 percent share of the approximately $331 million being sacrificed to end the project, plus a possible rate of return on the capital.

Edison will ask the Department of Public Utilities to allow it to charge customers the approximately $195 milion over a 10-year period, which it says works out to about $1.70 a month for a residential customer.

The task for the DPU is not simple. It must decide whether in the long run it will be more costly to consumers to impose the full costs of the Pilgrim decision on stockholders or on utility customers. Edison contends that if it absorbs all of the capital costs of Pilgrim, it will be unable to pursue other capital projects, especially coal conversions, that would eventually lower fuel bills to consumers.

The DPU does have a middle road, allowing recovery of the capital but denying application for a rate of return on it.

In any event, the crucial issue for the DPU is to make a decision that will provide the lowest long-term costs for customers.

It is premature for anyone to judge that question in the light of the Pilgrim decision. The DPU should make sure the issues are clearly understood by everyone, especially the public, before rendering its decision.

The Register

Santa Ana, Calif., September 1, 1981

Judging from TV footage and new photos, members of several environmental groups were delirious with joy Friday after plans were abandoned to build a $1.8 billion nuclear power plant on the shore of Lake Michigan near Hammond, Ind.

An AP photo in this newspaper showed about 25 members of the "Bailly Alliance Citizens Group" whooping it up with cheers and raised fists.

And why not? Between the Bailly Alliance and such other organizations as the "Business and Professional People for the Public Interest" (a creation of a Chicago-based public interest law center), the ever-vigilant Indiana and Illinois greenies had finally done what they had set out to do ten years earlier:

- Caused the Northern Indiana Public Service Co. (NIPSCO) permanently to abandon plans to bring abundant, low-cost electricity to a large section of Indiana.
- Put the utility and its investors through 10 years of frustration and legal and engineering expenses totaling $205 million.
- Put so many legal and governmental roadblocks in front of NIPSCO that original cost estimates of $187 million 10 years ago had risen to a prohibitive $1.8 billion.

Indeed, the environmental causies had once more done their jobs. What might have been progress in the effort to expand national energy capabilities had been stopped in its tracks. Indian Dunes National Lakeshore, a sort of sandy oddity located near the proposed site of the Bailly reactor, had been "saved." The greedy scum that operate utility companies had been firmly put in their place.

Nineteenth-century lifestyle, here we come.

There is so much environmental quackery rampant in the U.S. today — hysteria over imagined catastrophes piled atop mountains of "what ifs" and "suppose thats" — it is truly a wonder that a single nuclear facility has been constructed in the United States since useful purposes for atomic energy were first conceived nearly 40 years ago.

Closer to home, we are presently seeing the problem first-hand at San Onofre and at the proposed San Diablo Canyon reactor site where protestors regularly roost, spewing forth fact mixed with nonsense, gullibly picked up in an unquestioning manner by newspeople who ought to know better.

Finally seizing advantage of the situation are the politicians, whose ears are naturally always open wide to the advice of environmentalists. Gov. Brown did not become an expert in Medfly eradication overnight. It took help.

It is extremely doubtful if the mass of people living along Lake Michigan are as happy as the environmentalists today. Among them must be a few realists who clearly understand that few of us are willing to read by kerosene lamps, cook on wood-burning stoves or go to work on horseback. Maybe others can see beyond the decision to abandon the project to realize the $205 million already spent is going to be eventually retrieved from NIPSCO customers.

In short, it was an expensive victory. Demonstrably safe nuclear-generated energy was traded away, and for what? Windmills? Coal-fired generators? Solar panels? When the lights begin dimming in Hammond, maybe we'll find out.

The Evening Gazette

Worcester, Mass., January 22, 1982

It's beginning to look like the end of the line for nuclear power growth in New England, and maybe the country.

The New Hampshire Public Utilities Commission says in a few weeks it will order an end to construction work on the second nuclear power plant at Seabrook. Boston Edison has canceled its plans to build Pilgrim II and is threatened with a huge fine by the U.S. Nuclear Regulatory Commission for the negligent way it has been running Pilgrim I. In Connecticut, Millstone III is still under construction, but is coming under increasing fire for its mounting costs. And in the far West, the Washington Public Power Supply System voted last week to pull the plug on two of the five nuclear plants it has been building, despite the $2.25 billion already spent on them.

As far as New England goes, it may get Seabrook I and Millstone III on line, but that will be it. Other sources of electricity must be found for the 1990s.

Two years ago, the prospect of an end to nuclear power plant construction would have been frightening. New England utilities saw nuclear power as the antidote to the region's reliance on foreign oil. Almost one-third of the electricity used here is produced by nuclear plants. They helped save New England industry.

But today there are other possibilities. Oil-fired generating plants are converting to coal. More hydropower is being developed. Wood-burning power plants are on line in Maine and Vermont. Proposals to generate electricity by burning trash are being considered.

For New England, however, the two most important energy developments have to do with Canada. New England utilities are collaborating on a project to bring electricity from the huge hydroplants in northern Quebec down through New Hampshire to connect with the New England grid. They hope to have 600 megawatts on line by 1988 or sooner. And a $600-million 36-inch gas pipeline from Quebec to Rhode Island will convey 300 million cubic feet of Alberta natural gas a day, one-third of it for New England.

Not long ago it seemed that New England was at the end of the energy line. But now, it appears, it may be at the beginning of the energy line from Canada. We can thank our lucky stars that Canada has plenty of electricity and natural gas to export to its nearest neighbor.

WORCESTER TELEGRAM.

Worcester, Mass., March 17, 1982

The staff of the Nuclear Regulatory Commission has either some bad news or some good news, depending how you look at it.

It reported to Congress last week that 19 nuclear power plants now under construction — including the second one at Seabrook, N.H. — probably will be canceled because of flat demand for power, high interest rates and rising costs.

The report was prepared for NRC member John Ahearne before March 4. On that date, the Tennessee Valley Authority announced it would stop work on two reactors in Tennessee and one in Mississippi, laying off 4,800 workers. Those three units and 16 others were on the NRC list of likely casualties. Three of the 16 are in South Carolina, two in Tennessee, and one each in Illinois, Mississippi, North Carolina, Pennsylvania, Indiana, New York, Virginia, Louisiana, New Hampshire, Texas and Georgia.

That news will give heart to the groups and individuals who oppose nuclear power as a threat to health. But it should give pause to the investment community and to the people who purchase their power from utility companies involved in the projects, particularly here in New England.

Boston Edison Co., which canceled its plans for a second nuclear unit in Plymouth late last year, is trying to get rate-payers, as well as stockholders, to share the costs. If all 19 nuclear power projects are canceled, as the NRC report suggests, investments that were to be worth billions will be wiped out.

The plants, some more than a third finished, were to begin operating at various times from 1986 into the next decade. They represent nearly a fourth of the 71 nuclear plants that have been granted construction permits.

Few electric utilities can afford to write off the sort of investment that a nuclear power plant project requires. The utilities will argue before the regulators that their shareholders should not shoulder all the burden. They will argue in some cases that absorbing the whole cost will drive them into bankruptcy.

In Massachusetts, customers of Boston Edison and those served by municipal light companies are already getting higher bills because of the recent shutdown of Pilgrim I. They will not be happy with further rate boosts caused by what looks like a mistaken decision to build Pilgrim II. They will not be impressed by arguments that Pilgrim I has saved them money over the years. The same holds for the customers of municipal light companies that owned a piece of Seabrook II, if that project is canceled.

The worrisome thing for Massachusetts and New England is that the loss of Pilgrim II, especially if it is coupled with the loss of Seabrook II, will leave the region still perilously dependent on foreign oil. Although some plants are converting to coal, oil will still produce more than half of our electricity for at least the next five or six years.

That is the long-range problem that the region's planners should address as soon as possible.

The Kansas City Times

Kansas City, Mo., April 3, 1982

The woes of the nuclear power industry continue to mount. Scheduled new plants, some well along in construction with many millions of dollars invested, are being abandoned or at least deferred. And some of the 72 existing reactors, pressurized water type models, are suffering corrosion problems in their miles of steam-generator tubing. When one springs a leak and spills some mildly radioactive water, the public reaction usually is out of all proportion to the potential health hazard involved. But confidence in this energy source has been further eroded nonetheless.

A Nuclear Regulatory Commission study reports tubing problems in at least 40 operating plants, accounting for 23 percent of the shutdowns not related to refueling. A spokesman for the Atomic Industrial Forum, the major nuclear trade association, says they are spending $50 million researching the tube degradation problem. The Ralph Nader Critical Mass group calls nuclear power "potentially the largest product failure in the history of American business." Twenty-two reactors in nine foreign countries have tube problems also.

The power plants to replace the present ones as they complete their service life or develop insoluble maintenance problems just aren't being built. Since January 1979 a total of 33 proposed reactors have been dropped and no new ones have been ordered. Last month the NRC predicted that 19 reactors now under construction will be canceled and within days the Tennessee Valley Authority dropped three projects on that list.

Safety concerns are only a part of the difficulty. Heavy and escalating construction costs — compounded by high interest rates — and power demand growth that last year amounted to a bare 0.3 percent are chilling the electric industry's will to stay the course on these burdensome enterprises. And if somehow nuclear power should survive all these rigors, one expert told the Atomic Industrial Forum recently that the uranium will run out by the year 2025 — unless of course this country converts to the breeder reactor or the fusion process is perfected. The dream of a boundless energy bonanza a generation ago has soured, and the hum of those nuclear generators has the sound of a low moan.

The Oregonian

Portland, Ore., December 13, 1982

Not too surprisingly, the bottom has dropped out of the nuclear power plant business worldwide. For the first time in the relatively short 30-year history of the nuclear industry, not a single nuclear reactor order was placed in 1982 — a global shutout.

Industry officials have attributed this setback to worldwide recession and an accompanying decline in the growth rate of use of electricity, yet most are confident that given reasonably swift economic recovery, the international market for nuclear reactor sales will become bullish again. That view is too optimistic.

Taiwan, Mexico, Spain and South Korea postponed or canceled new reactor orders in 1981 and 1982 and likely will proceed toward a more diversified energy program in the future. U.S. vendors, meanwhile, are discovering growing foreign competition in the nuclear reactor business from Japan, France, West Germany, Italy, Canada and Sweden.

What is developing, then, is a shrinking nuclear reactor sales pie to be sliced up to serve more vendors.

These dismal future prospects spell even more trouble for those vendors that have not had any reactor orders since the mid-1970s and consequently have lost skilled manpower, including design, engineering and project management teams.

Assuming economic recovery takes hold, prospects call for orders of only two to four nuclear units a year worldwide, compared to a high of 15 orders in 1977 and 1979. Clearly, there are no anticipated boom days ahead for nuclear power salesmen.

Unless major shifts occur in the negative world climate of opinion toward nuclear energy — with nations like Austria and Denmark, for example, reversing national moratoria against construction of new nuclear power stations — prospects for a nuclear reactor sales recovery in the 1990s are dismal, indeed.

Continued shrinking of this international market for nuclear power would mean for the United States a decline in domestic nuclear reactor manufacturers, thus a potential for deterioration in product services that competition spurs.

An adverse impact on U.S. balance of payments and possibly even greater reliance on foreign sources of energy also are likely consequences of a continuing, depressed market for the peaceful atom.

THE DENVER POST

Denver, Colo., March 11, 1982

RECENT construction halts at several large U.S. nuclear power plants is due only in part to high cost — and regulatory delays — involved in building atomic reactors.

It is also clear that recession is cutting into U.S. electrical demand. From 7 percent annual growth a few years ago (which meant demand doubled every 10 years) U.S. electrical use last year flattened out. Small increases in the West and South were offset by reduced demand in the Midwest. Public Service Company of Colorado is growing by less than 2 percent.

Current trends thus bear careful watching. So far in 1982, we've had abandonment of two big nuclear plants in the Pacific Northwest, with heavy losses suffered by the Washington Public Power Supply System. Besides mismanagement, the system overestimated its power markets disastrously.

Last week the Tennessee Valley Authority ordered a shutdown of three nuclear projects because of high costs. In the words of a TVA official, "we simply don't need all the plants we have under construction." But TVA is building five other nuclear plants authorized in the days of 7 percent growth. Will slackening demand doom them, too?

Unfinished nuclear plants are the first shelved because, while they are cheaper to run than oil or gas plants (and sometimes even coal), they are more costly to build. But coal plants are vulnerable, too: Utah Power & Light Co., last week canceled plans for a fourth coal unit in Emery County, Utah. The firm said industrial electrical growth in its territory — which includes southwestern Wyoming — isn't up to expectations.

Some closures — like the Washington state fiasco — are ordered in desperation. Others are prompted by fear of citizen outrage if bill-payers are asked to pay for new plants that might not be needed for some time to come.

The danger in such short-range judgments is that we may be short of power when demand rises with economic recovery. But the planners do face difficult problems. For example, when a Tokyo factory makes a car which supplants a sale formerly made by Detroit, U.S. power demand is also displaced. How much shrinkage is involved, and for how long? An answer is difficult.

And how lasting are conservation efforts? Insulation is relatively permanent, but some things that look like conservation may be only consumer resistance. The next national economic spurt may bring a new wave of purchases of electrical gadgetry.

"We really won't know the real causes of reduced demand until the economy gets back to normal," says Ray Schuster, an official of the Electric Power Research Institute, Palo Alto, Calif. "We believe the long-term trend toward greater electrical use is unimpaired but we can't say just which uses will recover and which will not."

It's vital we do know. Mothballing a plant today may seem to be ultimate wisdom. But the picture would change swiftly if a sharp economic recovery were to put us into brownouts and rolling blackouts a few years hence.

DAYTON DAILY NEWS

Dayton, Ohio, March 24, 1982

Credit the Tennessee Valley Authority for taking a hard look at its vision of a 17-reactor nuclear power system and admitting that the price of the dream is too high.

TVA will stop construction on three reactors and will scrap plans for any others. Five reactors presently are on line producing power for 2.3 million customers in five states.

Economics was the sole reason for terminating the project. When the push for nuclear power began in the 1960s and early '70s, interest rates were 4 percent and demand for more electricity was increasing at an annual 7 percent clip.

In the late '70s came a sagging U.S. economy, escalating interest rates, a string of eye-opening minor mishaps in on-line reactors and rollover rate hikes for utilities. Doubts began to cloud the dream.

TVA's electric rates doubled within five years while the demand for electricity dropped. To finish the three reactors under construction would take $3.2 billion on top of the $2.1 billion already in them. Reality replaced the dream.

In making its decision, TVA board members noted that if the need for electricity picks up, the unfinished plants can be turned into coal-fired generators at a lesser cost.

As tough as the decision was, it was the correct one.

Now, if the Nuclear Regulatory Commission can be convinced that the Clinch River Breeder Reactor proposed for Tennessee isn't needed either, America's taxpayers could be saved some $3 billion.

Newsday

Long Island, N.Y., June 27, 1983

Gov. Mario Cuomo's task force on Shoreham is confronted by a bewildering diversity of opinion about the costs of abandoning a nearly completed nuclear power plant without ever allowing it to open. So the group needs first to make a credible assessment of abandonment's impact on Long Island Lighting Co. customers and shareholders, Suffolk taxpayers, the economy of Long Island as a whole and the state's future power supply.

The economic data produced to date must be viewed with a critical eye: It's in the interest of LILCO to overstate the cost of mothballing the plant, and it's in the interest of Suffolk County Executive Peter Cohalan and the county legislature to understate it.

Beyond that, however, the consultants hired by the county, the Energy Systems Research Group of Boston, evidently elected to perform their calculations on the assumption that scrapping Shoreham should cost LILCO's customers nothing. But shifting the cost of abandonment to the shareholders doesn't make it go away.

It's true that the shareholders should and would shoulder some of the exorbitant cost of building Shoreham if it were established that mismanagement by LILCO were partly to blame. But since they'd have to do that whether the plant opened or not, the possibility doesn't belong in a cost comparison.

Further confusion results from the different time spans assumed by LILCO and the county's consultants. LILCO used the presumed 40-year life of the plant as a guide; ESRG picked 20 years.

It seems doubtful that Shoreham will be producing electricity 40 years from now, if for no other reason than advances in the technology of power generation. But starting this year and using 20 years as the base produces even greater distortion, chiefly because ESRG calculates that two coal plants to replace Shoreham's output would not be needed until 1998 and 2000. That means the considerable costs of those plants show up in the calculations for only five and three years, respectively.

Still more confusion has arisen from mixing 1983 dollars and inflated future dollars. A 6 per cent inflation rate makes an immense difference over 30 years.

The task force could make comparisons between the Shoreham-in and Shoreham-out scenarios more realistic and easier to understand if it used the same basic assumptions in both. That means calculating the cost of each without trying to divide the loss between customers and investors, and using the same time span — 30 years would seem a reasonable compromise — and same-value dollars throughout.

The Houston Post

Houston, Texas, January 18, 1984

The federal Nuclear Regulatory Commission may have been sending a message to the nuclear power industry when it denied an operating license for a virtually completed nuclear power plant in Illinois, citing quality-control failures in its construction. The NRC has had, in the words of its chairman, Nunzio Palladino, "deep concern with quality assurance lapses at nuclear construction sites." But this latest action, unprecedented in the 25-year history of the industry, raises troubling questions about the NRC's own policies and procedures.

The Commonwealth Edison Company of Chicago began construction of its $3.35 billion Byron Nuclear Power Station near Rockford, Ill., eight years ago. It had planned to start loading fuel before July of this year and to begin generating electricity before the end of the year. But the NRC's atomic safety and licensing board voted unanimously against granting an operating license on safety grounds. The tough language of the commission's denial charges "fraudulent" performance by the plant's electrical contractor and unsatisfactory quality control by other contractors.

The NRC further charges that Commonwealth Edison, the nation's largest nuclear utility with eight reactors in operation and three others under construction besides the Byron plant, "has a very long record of non-compliance with NRC requirements." If that is the case, where was the commission when the utility was compiling that "very long" record? Does "very long" include 1975 when the NRC granted a building permit for the Byron nuclear plant? Why did the agency wait until the plant was nearly complete to unconditionally deny the operating license, contending that even reinspection could not assure that work on the plant is satisfactory?

Is the law under which the NRC operates at fault, perhaps barring regulators from taking more timely action? If so, it should be changed, mainly for the sake of utility consumers, who ultimately pay for debacles in reactor construction.

Many of the problems plaguing the nuclear industry are self made, stemming in all too many instances from failure to follow federal regulations. But the government must also bear a heavy share of the responsibility for the short-circuiting of nuclear power development when it lets a multibillion-dollar generating plant go beyond the point of no return before saying, in effect, "Scrap it."

Detroit Free Press

Detroit, Mich., January 18, 1984

THE CRASHING noise you hear around the country this week is the sound of the prospects of the nuclear industry collapsing like dominoes. In the last five days the industry has suffered two unprecedented blows: First, a Nuclear Regulatory Commission panel refused an operating license to the $3.5 billion Byron Nuclear Power Station in Illinois because of inadequate quality control during construction. Second, an Indiana utility has announced the abandonment of the half-finished Marble Hill plant, into which $2.5 billion has already been sunk.

The latest news comes while the dust is still settling from the default of the Washington State Public Power Supply system — aptly nicknamed Whoops — after the cancellation of two of its five projected nuclear plants. And the roll call of other troubled projects tolls on, including Shoreham on Long Island, Seabrook in New Hampshire, Zimmer outside Cincinnati — and Midland in Michigan.

The economics of nuclear power have become abysmal. Huge cost overruns in construction and the down-the-line costs of mothballing the plants after a relatively short productive life mean nuclear power is about the most expensive way imaginable to heat your house or toast your muffins. The cost escalation is only partly due to regulatory delays and design revisions belatedly mandated by the NRC. More often it has been due to construction failures, lapses of quality control, mismanagement and the inherent complexity of nuclear power.

At Midland, such factors have pushed up the cost of completing the two nuclear generation units to about $4.4 billion, according to the company. The Public Service Commission estimates the cost at closer to $6 billion. It is insupportable that Michigan consumers, residential and industrial, should be forced to shoulder such a burden.

Midland has gone through a series of mishaps too long to chronicle here. Each time, Consumers promised the problem could be corrected and that it would be the last. Each time, the utility dug itself deeper in debt and the costs of completion have escalated. It is time for the cycle to stop.

Barring some miracle to be wrought by the company and the PSC, cancellation of the Midland plant now has to be considered seriously. To go on would be to saddle an already troubled utility with even more massive debt, in the fragile hope that the burden could eventually be shifted to its customers.

But the events of the past week will make lenders more skittish than ever. And forcing Consumers' customers to bear the entire cost of Midland would put Michigan industry at an even greater cost disadvantage compared to industries in other states and also damage Michigan's efforts to attract manufacturing jobs. Consumers has a 10 percent rate increase pending, and has said it would seek another 27.7 percent increase when Midland comes on line. Critics of the company say the necessary rate increase would be as much as 50 percent or more.

Abandonment is no less risky a course. It, too, will cost Consumers' customers something. But as events in Illinois and Indiana have shown, it is no longer unthinkable. What is unthinkable is that we should end up with a $6 billion plant whose prospects of winning a license from the NRC are dimming, and whose output would be unaffordable if it did.

A successful solution at Midland has to do two things: Spare the state the costs and chaos of having Consumers Power slide into bankruptcy, and spare Consumers' customers from having to swallow the entire $3.5 billion already sunk in the project, much less the additional billions that would be required to complete it. There is no precise, readily available formula for achieving that, although both the utility and its contractors ought to be held to account for the costs of mismanagement and sloppy work. There appears to be even less likelihood, though, of finding a formula that would permit completion of the plant at a cost any of the parties can afford.

THE SAGINAW NEWS

Saginaw, Mich., January 19, 1984

As goes Indiana, so goes Michigan? In the wake of the scrapping of the Marble Hill nuclear power plant by Public Service Co. of Indiana, some, including Attorney General Frank J. Kelley, think it certainly should.

The Indiana utility abandoned construction of Marble Hill, a twin-reactor plant on the Ohio River, saying it could not afford to finish the project. Kelley and intervenors such as Mary Sinclair insist that Consumers Power Co. and its ratepayers are in the same fix with the Midland plant.

There are some unsettling parallels. Marble Hill also suffered long delays and huge cost overruns. It threatened the utility's ability to survive financially. It, too, was well along toward completion; one reactor was almost 60 percent built.

But the parallels are not conclusive — and Michigan's conclusions should not be too hasty.

With all its troubles, the Midland plant still is expected to come in at about $6 billion, according to state Public Service Commission estimates. It's a huge sum. Yet Marble Hill's final cost was projected at an incredible $7.7 billion. The Indiana utility had spent $2.8 billion, while Consumers has put about $3.5 billion into Midland, 83 percent complete and holding.

There is a great difference as well in the relative capacities of the utilities. Consumers is four times bigger than Public Service. It has years of experience operating other nuclear plants, while Marble Hill was the Indiana firm's first such venture.

That doesn't mean that further delays may not cripple the completion of Midland, or that safety problems may not prevent its operation.

But assuming Midland goes ahead, the Marble Hill stunner does hold the lesson for Michigan that no party to a mistake should be exempt from its effects. While Consumers seems to expect ratepayers to foot the whole bill for Midland, no matter what the cost, the Indiana utility has cut its dividend by 65 percent. And while Kelley wants to sock stockholders exclusively, Indiana is considering an emergency $105 million rate boost to keep its utility afloat.

The other lesson from Marble Hill may be of some small — very small — comfort. No matter how bad things get — someone probably has it worse.

THE INDIANAPOLIS NEWS

Indianapolis, Ind., January 19, 1984

FOR SALE: Partially completed Marble Hill nuclear power plant located on a scenic 960-acre Ohio Valley site near Madison. Property includes two reactors, one 60 percent complete and the second 30 percent complete. Reactors built to withstand demolition. Solid construction. One construction worker claims, "A wrecking ball would bounce right off of it."

Asking price: $2.5 billion (but willing to negotiate).

Contact Public Service Indiana and the Wabash Valley Power Association.

In a brief announcement earlier this week, Public Service Indiana Chairman Hugh Barker pulled the plug on PSI's ill-fated Marble Hill nuclear plant. Barker said PSI's board of directors believes the Marble Hill station will be needed for future power requirements, but added, "... given the realities of today's political environment and financial markets, PSI clearly cannot continue to be part of the project.

Wabash Valley, which is made up of 22 rural electric cooperatives — mostly in Northern Indiana — has a 17 percent share in the Marble Hill project. It is presently weighing its options, which include the possible conversion of Marble Hill into a conventional coal-fired plant.

The risks of fallout from a completed nuclear power plant are problematical. The risks of fallout from a nuclear power plant terminated in the throes of construction are absolute. In the case of Marble Hill, the fallout includes:

- A drop in PSI's stock from 24⅞ per share on Dec. 31, 1982 to 10 immediately following the termination announcement. The total value of PSI common stock has dropped about $750 million and PSI's preferred stocks and bonds have dropped more than $250 million.
- PSI quarterly dividends have dropped from 72 cents to 25 cents a share. More than 14,000 Hoosier PSI stockholders will suffer.
- The loss of about 8,000 construction jobs.
- A proposed rate hike affecting about 540,000 PSI customers to be filed with the Indiana Public Service Commission seeking about $105 million in emergency rate relief. Wabash Valley officials say they will probably seek a 30 percent emergency rate increase.
- Further PSI rate increases can be expected in the future both to recoup losses on Marble Hill and as electrical use increases and PSI's existing coal-fired generators wear out.

The reasons for the failure of Marble Hill are many. PSI has been widely criticized for having "bitten off more than it could chew" in the ambitious project. Certainly its oversight and management left a lot to be desired. Many of the construction firms also deserve criticism for faulty workmanship and, in some instances, bid rigging.

The federal government should shoulder some of the blame for failure to have consistent policies for construction of nuclear power plants. Government-caused delays have not only doomed Marble Hill, but have also been a factor in the termination or massive cost overruns on many other projects.

Marble Hill might have been salvaged but for a lack of political courage on the part of legislators and Gov. Robert Orr, who refused to consider trended rates or Construction Work in Progress rates which would have reduced the overall cost of Marble Hill.

As for that piece of real estate on the Ohio River, it appears to be a bit of a white elephant. But with the increasing cost of coal-fired electrical generating plants, anticipated demand for more electricity as the economy improves, the likelihood of stringent EPA regulations on coal-fired generators to reduce acid rain and growing concern about the effects of the buildup of carbon dioxide in the atmosphere, Marble Hill just might turn out to be a good speculative investment after all.

The Hartford Courant

Hartford, Conn., January 20, 1984

It's still January, and 1984 already is shaping up as a harsh year for the nuclear power industry.

Last week, the federal Nuclear Regulatory Commission refused an operating license for a nearly completed, $3.35 billion nuclear power plant in Illinois. This week, an Indiana utility decided to abandon a half-completed nuclear power plant on which $2.5 billion had been spent.

Other nuclear projects, including the Zimmer plant in Ohio and the Midland plants in Michigan, are teetering on cancellation.

These latest chapters in the beleaguered industry's history should prompt serious stock-taking where it can still do some good — as it can in New Hampshire.

The Seabrook nuclear power project bears many of the same earmarks for economic disaster as the abandoned Marble Hill station in Indiana, but its situation is even more precarious. The principal stockholder, Public Service Co. of New Hampshire, is a small, weak company without the resources to handle the project's huge cost overruns. The company originally estimated that the two units being built would cost less than $1 billion; the official estimate is now $5.8 billion, but some analysts say the cost will go as high as $9 billion.

Many of the 14 out-of-state utilities with shares in Seabrook, including Connecticut's Northeast Utilities and the United Illuminating Co. of New Haven, have been putting pressure on the New Hampshire company to halt construction on unit 2 as one means of capping the project's cost.

The PSC has agreed to delay construction on unit 2, but insists that both plants will be completed.

The recent abandonment of the Indiana plant and the deepening threat to other plants suggest it would be smarter to terminate Seabrook's unit 2. That would be one way to help prevent further runaway costs and possible loss of the entire project.

THE ARIZONA REPUBLIC

Phoenix, Ariz., January 23, 1984

THOSE under 30 years of age cannot remember what it was like when electric power was produced in plants that belched black coal smoke or acrid natural gas fumes.

That was before nuclear generation, before the Environmental Protection Agency began requiring anti-pollution equipment in conventional generating plants and, not coincidentally, before monthly electricity bills became major household budget problems.

Those who paid electric bills back then might even say those were the "good old days."

Factors that were not present 30 years ago now plague the electric utility industry, and either drive up costs to consumers or threaten utilities' ability to meet demand.

In the past week or so, three nuclear plants [in Indiana, Illinois and Ohio] were abandoned before going into operation.

The closings brought to 90 the number of nuclear plants in the United States mothballed, cancelled or converted to other fuels since 1975, leaving 77 nuclear plants operating and another 61 under construction.

No other industries in the nation — not even mining or petroleum — have had to endure the oppressive costs of interminable litigation [much of it frivolous], delays brought on by bureaucratic squabbling, scare tactics of anti-technology and anti-growth groups, required government regulatory reporting and purchase of exotic safety and anti-pollution systems. The net result has been a dramatic increase in costs never before associated with electric generation that ultimately must be passed on to consumers.

But then comes the most frustrating Catch-22 of all — the refusal or reluctance of rate-setting utility commissions to allow rate increases to recover these costs.

Reduced utility revenues mean lower utility bond ratings, which require higher interest rates [also paid by the consumer] on future borrowing.

However, if the public wants the power industry subjected to these regulatory and litigation rigors, then it will have to accept higher costs.

Arizona's first nuclear generating complex, Palo Verde, is enduring much of the same.

However, unlike plants mothballed or cancelled in other areas of the country where economic growth either was slow or non-existent, Palo Verde is sited in an area where annual growth rates of 2 percent to 4.5 percent have been common for more than a decade, recessions notwithstanding.

Thus far, organized opposition to Palo Verde has been represented by small groups that have invoked unsuccessfully, scare tactics involving imagined water shortages, nuclear radiation dangers and contractor fraud to impede Palo Verde.

If central Arizona ever lacked sufficient electric power, and therefore the resource to maintain economic stability, whatever costs are associated with Palo Verde now would seem insignificant when placed alongside the crippling costs of unemployment and economic stagnation.

Washington Power System Declares Default on Municipal Bonds

Plants No. 4 and No. 5 of the Washington Public Power Supply System were two of five being built to meet the energy needs of the Pacific Northwest. In an unusual arrangement, all five were to be financed through the sale of tax-exempt municipal bonds. Some 88 public utilities in Washington, Oregon and Idaho had an ownership interest in No. 4, near Richland and No. 5, near Satsop. No. 4 was about 23% complete and No. 5 about 14% complete when the board of directors of WPPSS voted in January 1982 to stop construction. The plants were to have been built at a cost of $2 billion each, but by 1981 that estimate had risen to about $6 billion each. The board had originally planned to mothball the plants until late 1983, but that plan fell apart when several of the participating utilities backed out of an agreement to pay their share of a $150 million fund needed for orderly suspension of the projects. Antinuclear groups and utility customers, who were facing substantial rate increases, had made it clear at public hearings that they were opposed to the mothball plan and wanted the projects scrapped. WPPSS had little choice but to abandon the two plants. Their termination would ultimately cost an estimated $531 million. A $2.25 billion bond debt, with an additional $5.15 billion in interest, would still have to be repaid over a 30-year period.

In July 1983, WPPSS oficially said it would be unable to repay the bond debt, resulting in the largest municipal default in U.S. history. WPPSS filed a document stating its inability to meet debt payments after the Supreme Court refused to reconsider a June 1983 decision freeing Washington utilities from their contracts to pay for the cancelled plants.

Of the three other plants, administered and financed separately with backing from the federal Bonneville Power Administration, construction was proceeding only on Plant No. 2, which was 97% complete. The No. 3 plant was mothballed for three years in July 1983. WPPSS had had trouble raising funds for the No. 3 plant, which was 75% complete and required about $963 million for completion, because of the severe financial troubles of Nos. 4 and 5. No. 1 had been mothballed in April 1982, when it was about 60% complete. One of the possibilities for repaying WPPSS bondholders involved drawing from the accounts of Nos. 1, 2 and 3, which could be done only if WPPSS decided to declare bankruptcy.

Oregon Journal

Portland, Ore., November 12, 1981

The Washington Legislature, in its final act before adjournment, did something that could affect Oregonians' electric bills. A bill passed the Washington Legislature requiring a $1.5 million feasibility study of the last two nuclear plants being built by the Washington Public Power Supply System.

It may be that it will be wise for WPPSS to stop construction on Washington Nuclear Plants 4 and 5. WPPSS has sold slightly more than $2 billion in bonds to construct the plants, the last two of a five-plant construction program.

When WPPSS decided to build the plants, the estimated cost was .2 billion. But now it's $8.2 billion and perhaps will cost more, meaning that WPPSS must sell more bonds in the future than it anticipated when it approved the plants.

WPPSS already has issued more than half of the nation's outstanding tax-free municipal bonds, leaving Wall Street investment houses with a buy or no-buy option and a potential veto over WNP 4 and 5.

Even if Wall Street accepts the next bond offering, the new Northwest Power Planning Council could deal WNP 4 and 5 a fatal blow if it refuses to make the Bonneville Power Administration a guarantor of the bonds for the two plants. Bonds for the first three WPPSS plants have such a guarantee from BPA or utilities.

Washington state Sen. King Lysen, D-Seattle and a WPPSS critic, has predicted a "day of reckoning" when ratepayers will have to repay these billions of dollars in bonds through higher electric bills. With the advent of the regional power act, that means electric customers in Oregon, Idaho and Western Montana as well.

The Washington Legislature passed three bills aimed at WPPSS. In addition to the feasibility study, the Legislature permitted WPPSS to negotiate the sale of construction bonds. The latter measure may persuade bond buyers to refuse to buy bonds until they can negotiate a higher interest rate.

A third bill will expand the WPPSS executive board from seven to 11 members, the four new members coming from outside the supply system.

But it may be too late for WPPSS to gain control over soaring construction costs, making the stop-build alternative on WNP 4 and 5 more attractive. The Washington feasibility study could find that the two nuclear plants won't be cost efficient and should be halted, at least temporarily.

The Oregonian

Portland, Ore., May 5, 1981

That a group of Washington state fruit farmers, having no experience in building nuclear power generating plants, undertook to construct five complex atomic giants and were able to borrow billions of dollars in municipal, tax-free bonds is, belatedly, causing red faces on Wall Street.

The Washington Public Power Supply System, suffering from inexperienced management on its rural-dominated boards of directors and from other problems, managed to come in some 500 percent over its budget and six years behind schedule. This is less amazing than the failure of the municipal bond market in New York to assess the underlying risk of lending a record amount of money. It was a sum that dwarfed the bonded debts of New York City that brought national handwringing in 1975.

Like other elements in the nation, Wall Street assumed government would bail it out. The financing of the first three plants seemed secured by the Bonneville Power Administration through a complex net billing plan, but the Internal Revenue Service pulled the plug on that by ruling that future net billings would not be tax-exempt.

Plants 4 and 5, now in mothballs and only 23 and 17 percent complete, cannot count on BPA to bail them out. Plants 1, 2 and 3 will require at least $4.2 billion in more borrowing to be finished. Under a new Washington state law, the voters may have the final say on new WPPSS (Whoops) bond issues if the courts uphold the November-passed measure. If all plants are built, the total cost will come to $23.7 billion.

The costs of these plants, compared with other gigantic projects, such as the $7 billion for the Alaskan oil pipeline, helped depress the bond market and have dumbfounded investors, once they seriously considered the magnitude of the enterprise.

The economic impact can hardly be overemphasized, whether to the Pacific Northwest, where much of the historic low-cost hydroelectric power advantage has been blown away, or to the portfolios of investors (mostly institutions), whose huge losses range up to 60 percent of the value of their holdings.

The security, of course, is the ability of the supply system's consortium to pass on its losses to ratepayers, who face paying the costs even if no power (highly unlikely) is ever produced to pay back the borrowed billions.

Wall Street may lose some money, but it can't lose all so long as ratepayers can be charged among the 88 utilities in the supply system. And don't discount Wall Street's power to get a future federal bailout. It even has the luxury of blaming the disaster on "everybody," as one Wall Street analyst did recently after another said the problems began snowballing when only one essential element was missing — skilled management.

But what has not been satisfactorily answered is how sophisticated investors in the municipal bond market would lend billions of dollars without looking harder at the qualifications of the managers who would spend it.

The danger now is that all of us in the Northwest will be lumped by Wall Street in a poor-risk category.

The Idaho STATESMAN

Boise, Idaho, January 20, 1982

The imminent abandonment of two uncompleted nuclear power plants in Washington — which will leave 88 Northwest utilities obligated to pay $2.25 billion for absolutely nothing — is a warning to consumers. The warning is: Beware of politicians and bureaucrats who want to make you pay for grandiose projects that should be financed by private investors.

The complex history of the Washington Public Power Supply System's five nuclear plants, two of which almost certainly will not be finished, boils down to one omnipresent ogre — construction mismanagement allowed by WPPSS. Such mismanagement would have been much less likely under a privately owned company.

No company accountable to stockholders would ever have set out to build five nuclear power plants at once. Nor would the executives of a privately owned company have stood for the waste that allowed cost overruns of 600 percent.

Like so many financial catastrophes, this one started out with a dream. Bonneville Power Administration bureaucrats, the administrators of publicly owned utilities and Northwest politicians foresaw a future in which nuclear plants would provide the region with cheap electric power the way hydro projects had in the past.

Unfortunately, they ignored economic reality in a way no investor-owned company could afford to do. The bureaucrats and politicians even admitted as much, saying private capital would not develop nuclear energy alone because of the technical difficulties and the opposition to nuclear. The government, they said, had to step in to prove that nuclear's problems could be overcome. They had to do it, they said, to save the country from a catastrophic energy shortage in the future.

So, the BPA agreed to underwrite the first of those nuclear plants, assuming the financial risk for construction of WPPSS plants Nos. 1, 2, and 3. In short order, along came WPPSS Nos. 4 and 5, the two that appear doomed. Representatives of the 88 utilities that agreed to sponsor Nos. 4 and 5 now say they signed on with the understanding that BPA would assume the risk for Nos. 4 and 5 as it had for Nos. 1, 2, and 3. Thus, billions of dollars were committed by decision-makers who were under the impression that, if worse came to worst, all of BPA's ratepayers would pay the bills.

Worse, of course, did come to worst. Over the years, the estimated cost of the five plants soared from $4 billion to $24 billion.

Now, after an incredible waste of resources, nearly everybody in the Northwest is going to get stuck. The hardest hit will be the ratepayers in those 88 utilities. They will have to pay off the uncompleted WPPSS Nos. 4 and 5, which the BPA was unable to bail out. The rest of us will pay as the bloated costs of Nos. 1, 2 and 3 are distributed through BPA's rates.

All of which is so obviously awful that you would think the federal government had learned its lesson. Unfortunately, that's not so.

Our own Sen. James McClure recently shepherded through Congress a bill that would force ratepayers to assume the risk for another massive project, a $43 billion gas pipeline from Alaska to the lower 48. Again, it's the same argument: We must have the energy, and private capital won't do the job alone.

Also, there are movements afoot at the Federal Energy Regulatory Commission and in Congress to allow even investor-owned utilities to make their ratepayers foot the bill for construction work in progress.

Such goings-on make us fear that the country will not learn from the history of WPPSS and that, as the saying goes, we will be doomed to repeat its mistakes.

The Seattle Times

Seattle, Wash., January 21, 1982

IN FRANCE this week, anti-nuclear activists fired rockets into a nuclear-power-plant construction site.

In Washington State, where nuclear plants self-destruct, such drastic action isn't necessary.

The Washington Public Power Supply's board of directors will vote tomorrow on a "controlled termination" plan for Nuclear Projects 4 and 5, which now appear doomed never to be completed. A much-preferable "mothballing" plan for the two plants unfortunately collapsed last week.

The big question now is, can termination be kept under control? Arranging loans from WPPSS' 88 participants, settling contractors' claims, selling or salvaging the plants' assets, and paying off the $2.25 billion in debt already incurred will be extremely tricky.

The process could rapidly deteriorate into "uncontrolled termination," in which WPPSS and the other sponsors would simply abandon the plants. The result would be chaos — massive lawsuits; default on the bonds; plummeting credit ratings; negative repercussions on the state, counties, cities, school districts and other public agencies; and possible receivership in which a federal court would take over much of the Northwest power-supply system.

This, admittedly, is a "worst-case" scenario that everyone hopes won't play out. But some close observers of WPPSS remain skeptical that termination can be kept under control.

In sorting out the WPPSS mess, things will get worse before they get better. A colleague of ours refers to it as "the Pacific Northwest's Vietnam" — which is scarcely an exaggeration.

We're the first to admit we don't have any magical solutions to WPPSS' problems, but here are a few steps that seem to make sense:

WPPSS management must be restructured — by the Legislature, if necessary — so it's less unwieldy and ineffectual. The seven-member executive board clearly needs broader decision-making power. The board's "virtually unworkable legal structure," in the words of Edward Carlson, chairman of UAL, Inc., led to the regrettable — but inevitable — resignations of Carlson and two other outside board members last week.

WPPSS should hire a chief financial officer who really knows the business and has the authority to act. Managing Director Robert Ferguson has made progress, but his strength is construction, not finance.

Finally, WPPSS must concentrate on finishing Project 2 at Hanford — now about 80 per cent complete — and thus regain some badly needed credibility. Encouragingly, the U.S. Nuclear Regulatory Commission this week said that WPPSS has the organizational and technical competence and experience necessary to ***operate*** the plant. The N.R.C. didn't comment, however, on WPPSS' ability to ***finish*** the plant — which now must be given top priority.

THE CHRISTIAN SCIENCE MONITOR

Boston, Mass., June 7, 1983

One can hardly disagree with the observation of a congressional aide that the US government has "a gigantic problem here." The problem is the possible default any day now of the Washington Public Power Supply System (WPPSS). WPPSS is a utility put together to build five nuclear power projects in Washington State. It is also the largest single issuer of tax-exempt bonds in the nation. If WPPSS – dubbed "Whoops" – defaults, the ramifications would be far-reaching, for many of the bonds are backed by the federal government as well as 88 utilities in the Pacific Northwest.

Also involved in the legal tangle are scores of banks, industrial customers, law offices, local and state governments – and Secretary of Energy Donald Hodel, who as former head of the federal Bonneville Power Administration (BPA) helped get WPPSS under way.

No matter from what angle one looks at WPPSS, the conclusion seems inescapable that the utility made enormous policy misjudgments. It also illustrates the danger of seeking solutions to national problems through crash construction projects. Because Northwest officials expected a sharp increase in future demand for electricity, they promoted the building of the five nuclear power plants. The project, initially supposed to cost $4.1 billion, has now ballooned to $23 billion. Of the five reactors, two have been canceled; one was shut down a year ago for lack of financing, although 60 percent completed; one was closed in late May for lack of financing, although it is 74 percent completed. That leaves one plant, which could begin operations next year.

Should the federal government bail out WPPSS, and pay off some $2.25 billion in debt? Legislation to that end has been introduced by Idaho congressman George V. Hansen, who argues that whether Uncle Sam steps in directly or not with financial assistance, taxpayers will eventually be saddled with the debt since the BPA was a catalyst for the project.

The better approach, as urged by the Reagan administration, would be to avoid a federal bailout – and the bad precedent that would be set. Why should all US taxpayers have to bear the cost of mistakes made by a consortium of regional officials and utilities?

The honorable course would be for Northwest political and utility officials to eliminate the unnecessary reactors, raise electricity rates to obtain additional revenues, and then restructure WPPSS's debt.

THE SACRAMENTO BEE

Sacramento, Calif., June 23, 1983

If they had come about 10 years earlier, the recent decisions of the Supreme Court of Washington and an appellate court in Oregon would have been entirely laudable. Both courts invalidated the contracts signed by a group of public utilities, under which the utilities had agreed to pay off the nuclear power plant construction bonds of the Washington Public Power Supply System (known as "Whoops"), even if the plants were never built. The Oregon court held that, under Oregon law, a public utility can't make such a commitment without the approval of the voters. And the Washington court ruled that, under Washington law, publicly owned utilities aren't allowed under any circumstances to sign a contract that guarantees another utility's construction bonds — much less to burden their own ratepayers with a promise to pay for a power plant that might never deliver any power.

They were marvelous, pro-consumer decisions, and if the same position had been taken 10 years earlier by the courts or by any regulatory body or, better yet, by the utility companies themselves, it would have saved everybody a great deal of trouble.

But the courts' decisions came instead *after* the collapse of the Whoops nuclear project. And that made their impact much different. They didn't *prevent* an unsound public investment; they invalidated a promise that thousands of private investors had already relied upon in buying Whoops bonds. It was an incredible precedent to set, suggesting that all sorts of public agencies in the Northwest — if not elsewhere — can be bailed out of all sorts of revenue bond obligations by after-the-fact court findings that they never should have agreed to them.

Chemical Bank, which has represented all the bondholders in these legal proceedings, is now suing the utilities' officials and attorneys for fraud and misrepresentation. It was these same people who first promised investors that the contracts were binding — and then, when they wanted to get out of them, argued in court that the contracts had never been any good.

The utility officials, of course, are also busy suing Whoops officials for misleading them. But as the two court opinions make clear, if they had been less free with their customers' money, the utilities need not have been taken for so much. Indeed, the best use that can now be made of the two court decisions might be as evidence against the utilities in the forthcoming trials. Some Whoops bondholders might even consider it worth the price the courts imposed on them.

Minneapolis Star and Tribune

Minneapolis, Minn., June 24, 1983

"Whoops" becomes a more accurate nickname for the Washington Public Power Supply System with each new twist in the system's financial troubles. Last week, the Washington Supreme Court ruled that the state's public utility districts need not pay their two-thirds share of $2.25 billion spent on two canceled nuclear plants. Although the ruling may accurately reflect Washington law, it dishonors the utilities and tarnishes the state.

Whoops originally planned five nuclear plants. Public utility districts in Washington purchased two-thirds of plants 4 and 5. They expected to pay for the plants by selling their share of its electrical output. But the contracts specified that the utilities would pay whether or not the plants were completed. Whoops used those contracts to sell revenue bonds to finance construction.

Obviously, the utilities did not expect what eventually happened: Power demand fell while the cost of the Whoops plants increased. Plants 4 and 5 were canceled, but not before Whoops spent $2.25 billion. However, when Whoops sought to collect, the utilities sued, claiming that they did not have the authority to sign the contracts in the first place. The Washington Supreme Court agreed. Because the issue involves only state law, the decision cannot be appealed to the federal courts. (Lower courts in Oregon and Idaho have issued similar rulings. Public utilities in those states had smaller shares of the canceled plants.) As a result, Whoops almost certainly will default on more than $2 billion in bonds.

By their suit, the Washington utilities demonstrated extraordinary bad faith. They voluntarily, even eagerly, signed contracts with Whoops. They looked forward to enjoying the fruits of the two nuclear plants. Their bond counsels assured underwriters that the contracts were valid, binding agreements under Washington law. Those assurances were used to satisfy potential bond buyers. Then, when things went sour, the utilities sought, and found, a technical reason to renounce their pledges.

If the utilities are not responsible for their debts, Washington state should be. The utility districts are political subdivisions of the state, governed by state laws. The districts concluded contracts that violated those laws, but the state let them do it. If Washington, through its courts, could declare the contracts invalid after the money has been spent and the project has come a cropper, then the state had a responsibility to examine and approve the contracts before the damage was done.

Whoops likely will require years and many lawsuits to untangle. But one lesson already seems clear: Washington failed to properly protect the individuals and institutions that invested billions in the state's energy system. Buyers must accept a certain risk when they invest in revenue bonds. But they should reasonably expect that public entities will live up to the terms of their contracts. By its acquiescence to the Whoops financial arrangements, Washington encouraged that expectation. Unless the state now finds a way to help make good on the debt, the bond market would be justified in shunning future offerings from Washington.

The San Diego Union

San Diego, Calif., June 24, 1983

Investors who purchased $2.25 million of tax free municipal bonds issued by the Washington Public Power Supply System (WPPSS) to finance construction of two nuclear power plants had good reason to believe their money safe. A score of the state's municipal power companies had signed contracts guaranteeing the bonds.

But the Washington state supreme court last week sent shock waves through the nation's financial centers in ruling that the contracts are invalid. By irresponsibly flouting the principles of contract law, the ruling clears the way for what could be the largest default in the history of the municipal bond market.

In voluntarily signing the contracts, the utilities guaranteed payment of the bonds whether the nuclear plants are built or not. But when construction was halted due to cost overruns and decreasing demand for electricity, the utilities sought relief from their contractual obligations.

If the court had based its decision on judicial equity, it would have ordered the utilities to pay. Long-honored principles of contract law safeguard commercial relations by requiring non-fraudulent contracts to be fulfilled. There is good reason, however, to believe the court acted to satisfy the state's consumers who do not want electricity rates increased.

Nothing else reasonably explains why the court decided that the utilities had no right to guarantee payment of the bonds and that the utilities could not have been expected to know demand for power would decrease.

Some of the best bond attorneys in the nation had thoroughly studied the contracts before they were signed and could find no flaws. As for knowledge of future power demand, nobody has ever been able to guarantee the accuracy of economic and financial forecasts. If knowledge of the future were a requirement, no contract would ever be executed.

The Washington ruling could damage more than the bank accounts of the unfortunate bond purchasers. The fear of default is bound to make many investors think twice before putting their money into any type of municipal bond, or possibly state bonds, without the prospect of higher interest rates for their risk. This would mean higher costs for taxpayers who bear ultimate responsibility for general obligation bond issues.

The Washington court could do everyone a favor by reconsidering its ruling and ordering the utilities to pay their legal debts.

The Washington Post

Times Herald

Washington, D.C., July 14, 1983

WHAT HAPPENS when a public agency defaults on $2.25 billion worth of municipal revenue bonds? The country is apparently about to find out. The Washington Public Power Supply System—Whoops, as it is universally known—has for months been sliding toward the brink where it now perilously teeters.

The reasons for this disaster have very little to do with the recession and the larger economic troubles of recent years. It is a case of a coalition of small utilities that undertook a gigantic plan to build five large nuclear reactors—a venture for which they had neither the managerial competence nor the technical experience. The project rapidly developed huge overruns of cost as, simultaneously, the planners began to realize that their original estimates of power demand were greatly exaggerated. They were building machines at too high a price to generate electricity for which there would be no customers.

Of the five reactors in the Whoops plan, one is now nearing completion. Two have been mothballed, about two-thirds completed. Two have been cancelled, and it is the bonds floated for those two that are the immediate issue.

The bonds were originally issued on the strength of contracts under which local utilities promised to stand behind them. But last month the Washington State Supreme Court held that the utilities didn't have the legal authority to make that kind of commitment. The costs of a default on this grand scale are likely to be widely distributed. Investors have been reminded that their risks are real. In response, risk premiums will rise. That will not be helpful to the national economy, since interest rates are already rising for other, unrelated reasons.

But among all the gloom and losses, some of the results of this fiasco will be very much to the good. It is the financial counterpart of the Three Mile Island accident, another lesson that nuclear power is an unforgiving technology. This country is going to need more reactors because, properly run, they provide power that is both safe and clean. But, as Whoops demonstrates, it is unwise to leave the management of nuclear systems to amateurs and people whose previous experience is limited to operating dams. As for the bond holders, it seems that the securities industry had got rather casual about handing out large sums of money regardless of borrowers' qualifications and performance, as long as there was a piece of paper somewhere saying that someone would pay for the mistakes. Starting right now, the lenders are going to begin using their considerable influence to enforce more careful standards on the builders of large power projects—and that is not at all a bad thing.

THE  SUN

Baltimore, Md., July 1, 1983

The financial disaster the Washington Public Power Supply System brought on itself by overbuilding nuclear power plants in Washington state is at least as serious as the technological disaster that hit the investor-owned Three Mile Island nuclear plant in Pennsylvania three years earlier. That evens the score between privately and publicly-owned utilities: One big boo-boo each.

Joseph C. Swidler, a respected utility analyst, says "The experience with the WPPSS bonds, billions of dollars of which are threatened with default, is evidence that the capital costs of public systems can be heavy and that their managements . . . not necessarily inspired." But he adds that "as in the case of the private systems, some are better than others." That's certainly so. The investor-owned Baltimore Gas and Electric Co. generally performs well. The same is true of the publicly-owned Tennessee Valley Authority.

WPPSS (or Whoops!, as wags call it), a wholesale power supplier for municipal and other public utilities in the Pacific Northwest, made two immense mistakes: It based its nuclear power-plant building program on grossly inflated estimates of electric demand, and it even more grossly underestimated plant costs (a situation critics charge was aggravated by sweetheart deals with contractors). As a result, it had to halt construction on two uncompleted nuclear plants on which it had already spent $2.25 billion from tax-exempt bonds it had issued. Now the Washington Supreme Court says the local utilities that had agreed to purchase power from the cancelled plants are not responsible for the debt. If the decision stands, it means bankruptcy for WPPSS and default on the bonds — the largest in bond market history.

Fortunately, there is no general municipal bond market panic, because investors know how unique the WPPSS situation is. But bankruptcy and default would mean, among other unpleasant things, higher premiums for policyholders of insurance companies that invested in WPPSS bonds. If, on the other hand, bankruptcy were not allowed and the local utilities had to pay, the traditionally low electric rates in the Pacific Northwest would go up only to about Baltimore's current level of about 5.6 cents a kilowatt hour.

The Washington Supreme Court treated WPPSS as an independent entity. But there is a strong case to be made that it is a creature of the local utilities and that they ought to pay. This, surely, is an issue that will require a U.S. Supreme Court judgment.

THE ARIZONA REPUBLIC

Phoenix, Ariz., July 27, 1983

THE record $2.25 billion municipal bond default of the Washington Public Power Supply System reaches far beyond the start of its awesome financial shakeout.

The consortium — known as "Whoops" — includes not only the 23 Washington public utilities and 88 other Northwest utilities involved in two terminated nuclear power plants, but other aspects — the nuclear industry itself, the bond market, nuclear regulation issues, the obligation of ratepayers, and many other legal ramifications.

The utilities and industrial customers joined 25 years ago to build five nuclear power plants in Washington state to meet large expected demands for power.

The default immediately involves only the two terminated reactors. However, Chemical Bank of New York, trustee for the bondholders, may attach the three partially completed plants and all other assets of the consortium.

Of the three plants, two have been shelved although they were two-thirds completed, and one nears completion.

Costs skyrocketed from first estimates of $4.1 billion to $24 billion amid charges of gross mismanagement and technical incompetence. Estimates of power needs also were exaggerated.

Chemical Bank called the default when "Whoops" was unable to provide $32 million in missed interest payments on bonds.

The default underscores that municipalities and other government bonds are not risk-free. And this should raise questions about the bond market itself.

Questions should include how the securities system checks out the qualifications of borrowers and their prospects for success.

They should specifically include whether individuals with little knowledge of highly technical nuclear and other industries should be launching big projects involving massive funding.

Stricter lending rules are necessary if strong confidence in the bond markets is to be retained.

Following so many construction delays and foul-ups, regulations regarding the construction of nuclear plants should again be reviewed to determine their effects on events at "Whoops."

Some Northwest congressional members want the federal government to pick up much of the debt "Whoops" has incurred.

Such a request is premature to say the least.

The cost of default could soar to $7 billion, including all interest obligations.

There's no reason why taxpayers should come to the rescue of the bondholders and stockholders.

The default may take years to shake out, but most responsibility clearly lies with the lenders and borrowers.

THE KANSAS CITY STAR

Kansas City, Mo., July 27, 1983

It may lack the high visual drama of the 1979 Three Mile Island accident, but the $2.25-billion municipal bond default by the Washington Public Power Supply System could have at least equally traumatic effects on the nuclear power industry. More than that, this largest such default ever could have disastrous impact on the entire municipal bond market itself.

The bonds were issued by the WPPSS system, known irreverently as "Whoops," to finance its No. 4 and 5 nuclear reactors, being built for 88 Pacific Northwest electric utilities. These two unfinished projects now have become the most spectacular casualties of a nuclear power industry beset by high construction and borrowing costs, slowed electric demand and a widespread loss of public support. Chemical Bank, trustee for the bondholders, precipitated the default by going to court to force the system to admit it was out of money, except for $25.6 million which WPPSS turned over to the bank.

Because this is the first municipal bond default of such magnitude, it is unclear what will happen next. Chemical Bank might try to obtain money from the system's three other nuclear units, funded by $6.05 billion in other bonds, but WPPSS opposes that. The 88 utilities involved have won a decision by the highest court in the state of Washington that they do not have to pay off the debt. This leaves one other possibility, a federal bail-out, but that is unlikely unless the effect on the municipal bond market is catastrophic.

The one certainty is that resolution of this unprecedented situation is likely to be months if not years in the courts. Whatever the outcome, it is another doleful milestone for a nuclear power industry that misread the future and built too much too soon. It will be particularly unfortunate if the resultant shock waves damage others outside that industry through a decline in confidence in municipal bonds generally.

The Hartford Courant

Hartford, Conn., July 28, 1983

The magnitude of the Washington Public Power Supply System's grand plan for nuclear power development in the Northwest made its fall hard enough to be felt across the country.

Maybe that's just what the rest of the country needs to be jolted into keeping energy projects within manageable proportions.

The big crash came this week when WPPSS — commonly and appropriately called "Whoops" — announced that it couldn't pay back the $2.25 billion it borrowed to pay for two nuclear power plants that were never finished.

The public-power agency thus brought on the biggest municipal bond default in the nation's history, an event with still undetermined effects on the tax-free municipal bond market.

The default, however, is only the latest event in a long string of disasters that have beset WPPSS's original plan for construction of five nuclear power plants in the Pacific Northwest. Further construction on plants 4 and 5 had to be canceled; construction of plants 1 and 3 has been suspended. Only plant 2 is now being completed.

What happened in the Northwest is on a gargantuan scale, but some of the elements of the debacle are familiar. WPPSS utilities started with an unquestioning faith in nuclear power, despite the fact that the region once had the lowest electric rates in the country because of its hydropower resources.

The utilities grossly overestimated electric power needs and grossly underestimated the costs of nuclear construction. From an original estimate of $4.1 billion to build the five plants, the projected costs escalated to $23.8 billion.

That is a staggering increase, but proportionately less than the cost hike of the Millstone III nuclear power plant in Waterford from the original estimate of $400 million to $3.54 billion.

The cost escalation at Seabrook Station in New Hampshire recently was enough to move a minority of power companies with equity shares in the project to ask that its second reactor be canceled or delayed while work continues on the first.

The series of miscalculations that got WPPSS into trouble were, thus, not unique, and they weren't made in a vacuum.

Among those who should bear some responsibility for the disaster are those New York bond counselors and financial advisers who persuaded the citizen-board members of the Northwest public utilities to agree to take on colossal debts. The trustee for the bondholders who put the $2.25 billion into the canceled plants is the Chemical Bank of New York, which, one expects, would have had a better eye for a bad investment.

Litigation resulting from the WPPSS's tangled affairs is likely to go on for years, and expect to hear a call — hopefully, a call that will go unheeded — for some kind of federal bailout.

The cold fact is that electric consumers in the Pacific Northwest will probably end up paying for the mistakes that were made by WPPSS, and amplified by the sheer size of the project, for generations. It is better, sometimes, to think small.

TULSA WORLD

Tulsa, Okla., July 27, 1983

THE WASHINGTON Public Power Supply System defaulted Monday on municipal bonds sold to raise money for nuclear power plants in the Pacific Northwest. Although the default was certainly no surprise, the shock waves have yet to be measured.

WPPSS bond default is the largest in U.S. history. The system skipped payment on $2.25 billion worth of bonds Monday, but the total loss could top $7 billion.

WPPSS sold the bonds to build five nuclear power plants. But WPPSS supervisors weren't experienced with construction projects of that size and problems arose. These were compounded by rising building costs and the delaying tactics of the anti-nuclear movement. The estimated cost of the five plants rose from $4 billion in the mid-1970s to $24 billion by 1981.

The major question in the WPPSS default is who is going to pay for this fiasco? There are essentially two choices: Pacific Northwest utility ratepayers or the federal government.

If Uncle Sam is talked into a bailout a la Lockheed and Chrysler, you and I are going to foot a share of the bill. Such a plan would create severe inequities. According to the National Law Journal, even if the ratepayers have to pay for the default, their utility bills would still be less than half the national average. What justification is there for asking taxpayers in New England, for example, to help pay for the WPPSS default when if the burden were placed on WPPSS ratepayers they would still have lower utility bills than New Englanders?

Congress has seen fit in several instances to help financially troubled businesses. Such action is rarely justified. The WPPSS default is not one of those rare occasions.

THE BLADE

Toledo, Ohio, July 29, 1983

THAT complex default case involving the Washington Public Power Supply System has been used by some critics of nuclear power as an example of the inherent waste and high cost of that mode of power production. Nothing could be further from the truth.

The WPPSS, which represents a consortium of utility companies in Washington state and other parts of the Pacific Northwest, undertook in the late 1970s to build five nuclear plants. These, it was thought, would perpetuate that region's supply of cheap electricity hitherto provided mainly by huge federal hydroelectric power projects.

But as the cost ballooned from an original figure of $4.1 billion when the project was originally conceived to an estimated $24 billion by last year, the supply system abandoned plants 4 and 5. Plant No. 2 is 90 per cent completed and is supposed to begin generating power early next year. The other two plants are mothballed because of a lack of funds to continue construction.

The default occurred with respect to the two terminated plants. The power consortium says it does not have enough money on hand to pay its debts on those two plants which, it can be presumed, will never be finished.

A long court wrangle is in prospect, but the immediate effect on the financial markets was not noticeable because the default long has been anticipated. However, utility companies may find it more difficult to market their bonds. Many of them have been having a hard time doing so even before the default.

The cause of the default, of course, was not the fact that the Washington supply system decided to supplement its generating capacity with nuclear power. Nuclear power is efficient and clean, and it could be economical if government agencies and nuclear protesters were not doing their best to make it otherwise.

The problem in Washington state was not nuclear power, but grandiose expansion plans that far overestimated the region's electric needs. There is no conceivable demand for five additional nuclear plants in a region still sparsely populated even by midwestern standards.

There is not much sentiment for a bailout in Congress, partly because, unlike the Chrysler case, there is not much chance of a financial turnaround anytime soon. But the time may come when some — though presumably not all — of those nuclear plants will be needed. It will be much more costly to build them at that time.

Default or not, consumers of the utilities involved in the WPPSS system will have to pay for the cost of the plants, whether completed, partly completed, or terminated. The long and complicated legal battle over default will generate no additional power. It merely will enrich the lawyers fighting over the corporate and physical remains of the public power supply system, including the plants that are likely to be completed.

The proper approach would be to form a new entity to take to take a fresh and realistic look at the region's power needs and, one way or another, to complete the plants required to fill that demand.

The Times-Picayune
The States-Item

New Orleans, La., July 28, 1983

The Washington Public Power Supply System — WPPSS for short, "Whoops" in popular speech — has defaulted on $2.25 billion, the largest municipal bond default in U.S. history, and neither the precedent nor the path to it is pleasant to contemplate.

WPPSS supplies power to utilities in Washington State, and in the early 1970s it embarked on what has been called one of the largest nuclear power projects ever undertaken — five nuclear plants. Projected demand growth was predicted to outstrip production within a decade, and nuclear power was determined to be the most economical source for the long run. With a triple-A bond rating, WPPSS soon became the country's largest borrower by means of tax-free municipal bonds.

Then things began to go sour, and there are two sets of proposed reasons why. The matter has already become a tangle of lawsuits, and more are coming. One set of reasons offers fraud, negligence, breach of contracts and violations of security laws. Another cites poor planning, mismanagement, bad advice, cost overruns, inflation, rising interest rates, a slack economy's lower demands and costly design changes ordered by the Nuclear Regulatory Commission.

Whatever the reasons, the five plants have shrunk to one probable plant, scheduled to begin operating in February, seven years late and $2 billion over budget. Work on two others has been indefinitely suspended, and two more have been cancelled as unneeded.

These last two were the straws that broke WPPSS' financial back. Three plants were guaranteed by the federal Bonneville Power Administration, but plants 4 and 5 were financed by 88 utilities that were to buy their power. That procedure was disputed, and when the state Supreme Court ruled that the utilities had no authority to make such an arrangement and no obligation to pay back the bonds, a WPPSS default became virtually inevitable.

These events in a far corner of the country have resonances in our corner. Middle South Utilities, too, made a basic decision that nuclear power rather than coal was better for our general area. The cost of Louisiana Power & Light Co.'s Waterford 3 plant has been vastly run up by inflation and regulatory change-orders. There is a controversy over the cost-sharing arrangement of regional utilities, New Orleans Public Service Inc. included, for the Grand Gulf nuclear plant.

No one down here seems in danger of financial collapse, but some of the problems that beset WPPSS exist here — and, apparently, in most places that started up nuclear power projects at about the same time.

The Orlando Sentinel

Orlando, Fla., July 29, 1983

With a group of Washington state electric utilities now in default on an almost unbelievable $2.25 billion in bonds, the obvious question is who gets stuck with the loss. This time, making good on the Washington Public Power Supply System's obligations is not a duty of the American taxpayer. The ratepayers, the bondholders and, in some cases, their insurance companies will have to work this one out.

It's clear that low rates for electricity in the Pacific Northwest certainly leave ample room for paying up, and meeting one's obligations still is the right thing to do. After all, this disaster was spawned in a dream world of forever low rates. Even with recent increases, rates in Washington average about 3.5 cents per kilowatt hour, compared with Florida Power Corp.'s 7.3 cents in Central Florida.

But there is another question that may be even more important. Could it happen again? It can.

Utilities everywhere are finding that as prices go up, consumers cut back. It's called a "death spiral." The same thing happened with big cars when OPEC sent oil prices skyward. And you better believe it is going to happen to telephone service if bills take the leap that phone companies are predicting.

The problems of Whoops, as this group of utilities is popularly known, didn't come up suddenly. The seed was planted in the 1940s when huge dams, built with tax money to blunt the Great Depression, came on line with dirt-cheap hydroelectric power. It produced a false economy. Homes, farm irrigation and industry all were designed as though kilowatts could be flaunted. And as more power-intensive industries moved in, the capacity of the big dams was soaked up. Nuclear plants were ordered to fill the gap.

That's when trouble came to a head: The nuclear projects dragged on for a decade and the costs multiplied by six. Meanwhile, a recession shut down industrial customers. Then, as rates went up, the nation's most wasteful users of electricity cut back. Suddenly revenues fell and there was no money to pay off the bonds.

Incredibly, it is estimated that bond obligations could be met if rates were raised at most by 1 cent per kilowatt hour. That would leave Washington consumers well below the national average of about 7 cents. Yet they dodge through a cop-out loophole designed by their state Supreme Court, believing they have a guarantee on that 40-year dream.

What the Pacific Northwest must do now is raise those rates and pay that debt. For all of us, there is the lesson that dreams don't last forever.

The Oregonian

Portland, Ore., August 7, 1983

Oppenheimer & Co., the New York brokerage house, deservedly has egg on its face for trying to wreak revenge on the taxpayers of Washington state for the default of the Washington Public Power Supply System. The WPPSS fiasco does not mean that the state's taxpayers will be unwilling to pay off general obligation bonds, especially when those bonds are backed by the ability of the state to levy a property tax to back the bonds.

Washington, needing to sell $149.9 million in general obligation bonds to finance capital construction projects, will venture into the bond market this week. The state already has an agreement with Salomon Brothers Inc. to head a group of underwriters on the bonds. The interest rate will be negotiated early this week.

Interestingly, Oppenheimer, which earlier urged that no one buy Washington's bonds, has agreed to participate as a member of the underwriting group. Oppenheimer last week retracted its don't-buy-Washington-bonds advice, saying that the firm believes the state's credit and taxing power are "sufficient to recommend purchase by Oppenheimer clients."

That is not to say that the interest rate will not be high or that Washington taxpayers will not have to pay some penalty because of WPPSS.

Last week, the bond market's interest rates were heading up. Washington state, burdened by the Legislature's ease in issuing bonds in the past, has a credit rating of single-A, one to two notches below Oregon's, depending on the rating service.

But financial experts talk of Wall Street's revenge in terms of a concentric circle theory, with public utility districts at the center. PUDs participating in abandoned WPPSS plants 4 and 5 will be unwelcome on Wall Street for some time. The city where the PUD is located can expect to pay a lesser penalty, and the county and state even less.

By coincidence, Cleveland last week sold its first bonds since its 1978 default. The issue sold out on the first day the bonds were offered, at a 10⅛ percent interest rate.

Rapid City Journal

Reno, Nev., August 1, 1983

Whoops!

That is what some people say when a dish slips through their hands.

It is also what more than 80 northwestern public utility districts and four cities are saying to people who invested in their Washington Public Power Supply System, affectionately known as Whoops.

Because of inflation, cost overruns and bad management, WPPSS found that it could no longer afford to build two nuclear power plants for which it had issued bonds. It also decided it could not afford to pay back $2.25 billion it owes on those publicly held bonds.

So the utilities went to court, seeking to avoid payment. The Washington Supreme Court listened sympathetically to their heartrending tale and ruled that the utility districts and four cities of WPPSS had no authority under state law to underwrite the financing — even though this same state law has been used for years to do just that. So, said the court, the districts and cities naturally had no obligation to repay their trusting investors — who, of course, had invested mainly because they had been told their money was backed by public agencies.

Whoops!

And tough luck, gang.

Tough luck to Paul Bonseigneur of Bellevue, Wash., whose retirement income slipped through his hands just like that broken dish.

Tough luck to 63-year-old widow Pearl Brickner of New York City, who invested $36,000.

And tough luck to all those other folks who lost money by foolishly having faith in public agencies.

The Wall Street Journal notes that the utilities could have met their debts by raising rates only 14 percent — which would still have left their customers paying only 57 percent of the national average in the power-rich northwest. But we're not talking about honor. We're talking of corporate greed and spoiled consumers. And how much easier it is to walk away from responsibility than to meet contractual obligations.

All this should bother you even if you are not one of the unfortunate holders of worthless WPPSS paper. It should bother you because the WPPSS default will make it more difficult for municipalities everywhere to sell bonds. It should bother you because the default is typical of the growing trend for Americans to skip out on debts through misuse of overly easy bankruptcy laws. But, most of all, it should bother you because this is one more example of the untrustworthiness of public agencies, which sometimes operate more like con men than upholders of the people's trust.

The public nature of this debt must be emphasized. It is one thing for people to knowingly risk their money while investing; it is quite another to invest money thinking there is no risk, then be undone by legal maneuvering around the fine print.

Fortunately (just like in the ancient Greek plays) retribution is at work. The utilities and cities, including Seattle, will be tied up in expensive and lengthy lawsuits. Chemical Bank of New York, which represents the bond holders, plans to attach the assets of several cities and utilities. Every municipality in the northwest will find it exceedingly difficult to raise funds through bonds; if they sell bonds at all, buyers will demand appropriately exorbitant rates. Even schools and school districts will be hampered. There will be fewer new public buildings, elementary schools, bridges and public works projects of all kinds. And those that do get built will cost taxpayers much, much more.

Consumers won't get off easy in the long run, either. Poor financial ratings for the utilities will eventually mean higher power rates, despite their slithering out from under their present obligation.

Seeing all this, perhaps the state governments of Washington, Idaho and Oregon will step in to cover the debt, or to find some other way to assure repayment (besides a federal bailout, which would be just another way of avoiding responsibility, and isn't likely anyway).

Such responsible action would be most wise from the long-range financial standpoint, as well as from the standpoint of honor.

But if not —

Whoops!

SYRACUSE
HERALD-JOURNAL
Syracuse, N.Y., July 27, 1983

As soon as Washington Public Power Supply System (WPPSS) officially defaulted on $2.25 billion it borrowed to build nuclear plants, talk began about a possible federal bailout.

It is the largest default in municipal bond history, and will hurt thousands of small investors. Although exact numbers are unavailable, individuals account for between 50 and 75 percent of the bondholders and, consequently, the losses. For some, default will mean loss of what they considered safe retirement income or money needed to put children through college.

Despite potentially devastating effects on some people, the federal government should not bail out WPPSS (appropriately dubbed "Whoops").

▽ ▽

The government shouldn't set the precedent, especially since there are other ways to recover some or all of the bondholders' investments.

WPPSS was created in 1957 to meet a rapidly increasing need for electric power in the Pacific Northwest. The area has tremendous hydroelectric generating capabilities but the Bonneville Power Administration, a U.S. agency that sells hydroelectric power to utilities in the region, said it wouldn't be able to meet the demand by the early 1980s. WPPSS decided the cheapest way to meet the increased demand was to build nuclear plants. It planned five, at an estimated cost of $4 billion.

▽ ▽

Bonneville underwrote bonding for the first three, and 88 utilities in seven states took responsibility for the final two. In 1974, these utilities directed WPPSS to build the two plants and said they would be responsible for payments even if the projects were not completed.

By 1981, it was clear projected electrical demands were excessive. The sky fell in for a variety of reasons — increasing concern about nuclear safety, soaring costs (last estimate for the five plants was $24 billion, six times the original estimate), lower electric demand because of depression in the forest-products industry, poor planning and mismanagement.

Plans for units four and five — the ones backed by the utilities — were canceled even though $2.25 billion in bonds to cover them had been spent.

▽ ▽

Possibility of default has mounted since the plant orders were withdrawn. For all practical purposes, default became inevitable on June 15, when the Washington Supreme Court ruled local utilities didn't have to stand by their pledge to pay back bonds on the unbuilt plants.

That decision can't be allowed to stand, although the Washington court recently reaffirmed it. The 88 utilities were involved in a business deal. They made a pledge and should be held to it.

▽ ▽

The utilities argue that making good on bonds will cause electrical rates to soar. That's true but rates in the Northwest are among the lowest in the nation — about 3 cents a kilowatt hour for residential customers (compared with Niagara Mohawk's 7.2 cents per kilowatt hour in June and rates as high as 13 or 14 cents for some downstate utility customers).

The 88 utilities can afford to pay. The WPPSS default is a sad case — one that will mean hardship to individual investors unless it's resolved. But it is not a situation that merits a federal bailout.

The Des Moines Register
Des Moines, Iowa, July 28, 1983

Back when the Washington Public Power Supply System acquired the nickname Whoops — from its initials, WPPSS — surely few realized how sadly appropriate that name would become. This week Whoops committed the largest municipal-bond default in history, confessing inability to pay $32 million interest on $2.25 billion in bonds.

The cause of the default is easier to trace than the consequences. WPPSS made the same kind of miscalculation that several Iowa utilities did a decade ago in estimating future demand for electricity — only Whoops did it on a much grander scale.

WPPSS was organized in 1957 to build and manage electric-power plants for 29 publicly owned utilities in the Northwest. In 1970 it obtained a contract under which the giant federal Bonneville Power Administration agreed to buy electricity from three nuclear plants and to raise rates if necessary to pay for them.

If things had stopped there, it might have worked out all right. But Whoops let itself be carried away by a proposal of 88 municipal utilities that it build two more nuclear plants. The 88 agreed to pay for the plants even if they were not completed. It soon became apparent, however, that there would be no market in the foreseeable future for all of the output of those new plants.

To make matters worse, Whoops got caught up in the construction-cost spiral that has afflicted the nuclear industry. Estimates on one plant alone skyrocketed from $1 billion to $5.5 billion; but Whoops blithely issued more bonds to cover the cost overruns. Two years ago, the money ran out again, but investors had stopped buying WPPSS bonds. Early in 1982, when the last two plants were only 14 percent completed, they were mothballed.

The 88 utilities saw that they would have to raise their rates through the ceiling to pay off the bonds. Since they would be more or less buying a dead horse, they refused. Now the Washington State Supreme Court has found a legal loophole to get them off the hook. And that was the end for Whoops. It told the Chemical Bank of New York, trustee for the bondholders, that it couldn't pay, now or in the future.

There is muted talk of a federal bailout, and the bank is trying to decide whom to sue. The outcome probably won't be known for several years. Meanwhile, municipal-utility bonds have suffered a very visible black eye and probably will be harder to sell for a while.

Eventually, if the bond gurus and investors decide that Whoops was a special case, no permanent damage may have been done — except to the unfortunate electricity users of the Northwest, who used to have some of the lowest rates in the country and now face some of the highest.

WORCESTER TELEGRAM.
Worcester, Mass., July 27, 1983

The financial woes of a public power authority in Washington state have been giving the utility industry shivers for months. The worst fears were realized this week when the authority defaulted on part of its $2.25 billion obligation to people who invested in its tax free bonds.

The Washington Public Power Supply System is called Whoops by its detractors. Years ago it contracted to build five nuclear power plants based on contracts with member utilities. Those contracts were ruled invalid this year. The effect was to free WPPSS members from obligations to back the WPPSS bonds with their revenue. Chemical Bank in New York called for repayment and WPPSS said Friday that it cannot pay.

Massachusetts has reason to be concerned. The default is the largest ever in the history of municipal bonds. Worse, perhaps, is the fact that 30 Massachusetts communities are, like towns in Washington state, members of a public power authority that is involved in an ever more costly series of nuclear power projects.

Massachusetts Municipal Wholesale Electric Co., and its member towns, are already struggling to stay even with huge cost run-ups at the Seabrook nuclear power station in New Hampshire. The New York Times reports that the company has postponed a bond sale because of higher-than-expected interest rates. MMWEC has already published a special edition of its newsletter designed to answer investors' questions about how the company differs from WPPSS and how its bond offerings are guaranteed by member towns.

The key to the WPPSS default was a lack of specific legal authority for member towns to back up WPPSS contracts with power purchase agreements. Massachusetts law contains such permission, and that part of the law may protect MMWEC. But clearly, Massachusetts members of the Seabrook construction project face an era of ever higher financing costs — partly because their borrowings will be affected by news of the WPPSS default.

Those rising costs will hit the member towns and the utility customers in those towns. The electricity departments in Shrewsbury, Holden and other communities already have announced sharply increased rates. And that's only the beginning. When high-cost nuclear power comes on line in two or three years, ratepayers in the MMWEC towns will be in for a double shock.

Three Mile Island and Plant Accident Liability: Who Pays?

The first estimates on the cost of decontaminating and repairing the Unit 2 reactor at Three Mile Island after its 1979 accident hovered around $400 to $500 million. (General Public Utilities Corp., the owner of the plant, was covered by insurance for $300 million.) The latest estimates on cleanup costs have risen to over $1 billion. Meanwhile, the nearly one million gallons of radioactive water that spilled into the reactor building and into the building adjoining it have been decontaminated and removed. But the hardest part of the cleanup remains—decontaminating the reactor building and removing the destroyed core—and the question of who is to pay for the bulk of the cleanup remains unresolved.

In November 1981, the Reagan Administration pledged to ask Congress for $123 million in federal aid over three years for the cleanup of the Unit 2 TMI reactor, and gave its support to a plan by Pennsylvania Gov. Richard L. Thornburgh under which the costs of the cleanup would be shared by the federal government ($190 million), Pennsylvania ($30 million) and New Jersey ($15 million), a pool of utilities ($190 million), GPU ($335 million) and GPU's insurers ($300 million). (The commitment made by President Reagan was less than the federal contribution Thornburgh had proposed.) In 1982, part of Thornburgh's plan, in the form of a bill requiring utilities to chip in toward the cleanup costs for six years, was passed by the Senate Energy Committee. The legislation foundered, however, because of opposition to requiring ratepayers and shareholders of all nuclear utilities to share in the cost. But Thornburgh's plan did not die. In January 1983, Edison Electric Institute, a trade association of almost 200 private utilities that together provided about 80% of the nation's electric power, approved a plan under which its members would contribute a total of $150 million over six years. As of January 1983, the federal government had committed up to $70 million to the cleanup, Pennsylvania had appropriated $5 million for the effort, GPU's customers were paying $36 million a year in higher rates, and the insurers had paid about $300 million.

The Salt Lake Tribune

Salt Lake City, Utah, April 30, 1979

Shouldn't the owners and operators of nuclear power plants be required to pay all the damages caused by an atomic accident, not just part of them?

For more than 20 years, Congress has set a sharp limit on such liability.

But that limit clearly ought to be reappraised in light of the accident at the Three Mile Island nuclear power plant near Harrisburg, Pa., the nation's worst atomic mishap.

Actually, it should not have taken this mishap to prod the House Subcommittee on Energy and the Environment to undertake its current review of nuclear safety in general and the 1957 Price-Anderson Act in particular.

This law limits a company's total liability for a nuclear accident to $560 million even though actual damages to persons and property could be much larger.

Yet in the first two weeks after the mishap at Three Mile Island, at least four class action suits were filed involving claims well beyond $560 million.

It's easy to imagine worse nuclear accidents with much more extensive claims. Yet the Price-Anderson law is vague about what injuries must be compensated . . . never envisioned that large claims could result from small releases of radiation such as those at Three Mile Island . . . and its indemnity limit has not been revised to account for inflation.

By freeing firms of worry about huge damage claims from a nuclear accident, this limit was designed to encourage the atomic industry to grow.

Up to a point, this principle is understandable. Because a private insurer could easily be wiped out by the huge claims from a serious nuclear power plant mishap, there are sharp limits to which insurers are willing to extend their coverage. Likewise, nuclear power firms themselves have to be concerned about the possibility of being bankrupted by a nuclear mishap.

But there's also room for wondering how safety-conscious any industry will be when there are definite limits on how much it must pay for its own mistakes.

Moreover, if the risk of catastrophe is really as slight as the nuclear power industry still says it is even after the Three Miles Island incident, why won't the industry assume full liability?

Rep. Ted Weiss of New York has what sounds like a sensible solution. He has introduced a bill that would lift the present $560 million ceiling and make nuclear facilities liable up to the total amount of their assets. At the same time, atomic plants would be enabled to obtain increased coverage not just from individual insurers but from a pool of private insurance firms.

In any event, the Three Mile Island mishap has clearly demonstrated the inadequacy of the present limit on nuclear liability. The more damage claims this mishap produces, the more seriously the Price-Anderson law is open to question.

Pittsburgh Post-Gazette

Pittsburgh, Pa., April 18, 1979

In conducting its own "postmortem" on the costs of replacement electricity for the crippled Unit 2 at Three Mile Island, the state Public Utility Commission must decide how to assign one of the largest and most tangible penalties for the nation's most severe and frightening nuclear accident.

As a matter of principle, the penalty — an estimated $1 million a day — should be assigned to those who were most responsible for the human and mechanical errors precipitating the event. And if federal and state investigations of the accident have not yet produced the full and complete accounting President Carter and others have called for, it is nevertheless possible to reach several firm preliminary judgments to guide the PUC's solomonic deliberations.

First among the PUC's findings ought to be recognition that the 346,000 customers served from Three Mile Island by the Metropolitan Edison Co. were *not* to blame for the incident. In fact, some (though not many) suffered disruptions, emotional distress and higher than normal (if not truly unsafe) levels of radiation.

To charge the utility's customers would be unfair in every sense and if it elects to follow Met Ed's recommendation for just such a resolution, the PUC will need its own special "containment" procedures to protect itself against adverse public reaction.

But if customers are not to pay, then either the utility or the government must be assigned the cost. There is a case to be made for government-backed pooling of such risks, but such arguments also lead in the direction of the nationalization of the utilities and just do not suit the circumstances of this accident. So by fault and default, the PUC should see that the costs of replacement electricity from the Three Mile Island accident are charged to the some 177,000 stockholders of General Public Utilities, Met Ed's parent company.

Those investors have in fact been insured against the extreme costs of cleanup for the $780 million power plant and against the largest liability claims in the 22-year history of nuclear accident insurance underwriting. Still, Met Ed officials fear that if stockholders are made to bear uninsured damage to the utility's rate structure, they will not again invest in the company. The fear is legitimate, but, in soliciting its investors, the utility invited them to share the risk of the nuclear power venture.

Stockholders may now complain that they share no blame for the irresponsible deactivation of Unit 2's backup cooling system, the human error which allowed an otherwise containable pumping system breakdown to go completely amok. Such complaints, however, should be directed to the utility's managers, not to its customers.

The Virginian-Pilot

Norfolk, Va., May 6, 1979

The amount of radiation spewed from the Three Mile Island nuclear power plant following the accident of March 28 was twice as much as officials thought at the time, sufficient to increase the cancer death toll around Middletown, Pennsylvania, by from one to 10 persons.

This was not only the worst nuclear accident in this country, but the first one in which federal officials acknowledged that some citizens will likely pay for it with their lives.

That sobering, even shocking, thought should underline safety regulations laid down by the Nuclear Regulatory Commission for all nuclear power stations. It should eliminate any argument that the NRC is needlessly restricting utilities in their efforts to generate more and more energy from fissionable materials. It should fortify the NRC against pressure from utilities to continue their nuclear-construction and operation timetables without delaying for safety inspections or redesigning doubtful equipment.

It should caution the administration, Congress, and the utility industry that if the nation is to expand its nuclear-power generating capacity, as President Carter has urged, it can do so only if safety first is the rule. Mr. Carter and Energy Secretary James Schlesinger, in seeking means of shortening the lengthy permitting process for new nuclear plants, must now take care not to short-circuit safety measures when it appears that greater precautions are in order.

But safety is not the obligation of the federal government alone. It is a utility responsibility, too. The best method of fixing that responsibility is to make the power companies liable for damages in accidents such as that in Pennsylvania.

Utilities and Wall Street investment houses are strongly opposed to this idea. They want the consumers, not the stockholders, to pay the costs of accidents.

The head of General Public Utilities Corporation, parent firm of the utility that operates Three Mile Island, told a Senate committee last week that Wall Street will hit the power companies with a 20 percent surcharge for new capital invested in nuclear facilities unless it is understood that utility customers will pay for accidents.

"If the costs go to the customer, what lever is there to guarantee (company) performance?" asked one senator.

The utility head suggested passing the buck to the consumers this time but holding management liable next time. "It would do no good to bankrupt GPU and destroy 177,000 stockholders," he asserted.

That's debatable. The death of a careless power company might be more shocking to the utility industry than the cancer death of an innocent mother or child who happened to live too close to the accident.

If the lives of corporations are on the line no less than the lives of power users, the utilities will have to observe the safety-first rule. Self-preservation is as strong a motivator of corporations as of individuals in time of danger.

Roanoke Times & World-News

Roanoke, Va., April 12, 1979

All is quiet at Three Mile Island; the crisis is past, says Pennsylvania Gov. Richard Thornburgh. But from a financier's viewpoint, doomsday is just over the horizon. The worst-case scenario is that the huge nuclear generating plant — whose cost has been put at anywhere from $700 million to a billion dollars — may never run again.

On the other hand, nuclear apologists say we have had enough of worst-case scenarios: Some people got irradiated but nobody is dead from Three Mile Island, so don't go off the deep end about this.

All right; let us put a more optimistic face on the economic situation at the crippled plant. Let us assume that Metropolitan Edison will not lose its nuclear operating license because of infractions at Three Mile Island, and that one day the plant will be back in service, supplying electricity to Met-Ed's 340,000 customers in central Pennsylvania.

A lot must happen before that day.

First, some 80,000 gallons of radioactive water must be removed from the reactor's primary cooling system; gas and contaminated material must be cleared from the two-million-cubic-foot reactor containment building; and at least 200,000 gallons of radioactive water must be drained from the structure's floor.

Then the building must be allowed to "cool" — i.e., xenon, krypton and iodine inside the containment building must lose most of their radioactivity. Even after the reactor was brought under control, measuring devices registered contamination levels inside the building's dome at about 30,000 rems — enough to kill a person instantly. At least a month of "cooling" will be needed before anyone can enter, even in protective clothing.

After that begins the real cleanup and scrubdown. Some experts believe that as much as 350,000 cubic feet of radioactive materials must be removed: 10 times the amount normally generated by a "nuke" in a year's time. Damage to fuel rods, etc., must be repaired and the reactor itself decontaminated. Some have said this could take up to four years; a more sanguine view is two years, with the cleanup costing $40 million.

That expense is merely for bringing back to life a $700 million-plus facility that had operated only three months before the accident. Meantime, Metropolitan Edison must buy from other utilities electricity to replace Three Mile Island's output.

That replacement cost, a Department of Energy official told a congressional committee recently, will range from $500,000 to $900,000 a day; over a two-year period, at least $365 million. Who'll pay? Met-Ed's customers. The DOE official foresaw a 20 percent boost in their monthly bills. These same people were hit, three days after the Three Mile Island accident, with a $49 million-a-year rate increase, most of it based on the reactor's start-up costs of last December.

Well, tough isotopes. Who else should pay but the Pennsylvanians? They demand the electricity; they use it. But any utility may now hesitate before it asks financiers or its customers to underwrite an investment so big that could go bust in so short a time. No, there was no genuine disaster at Three Mile Island: just an accident that was contained, but at a prospective cost, at minimum, of $400 million.

CHARLESTON EVENING POST

Charleston, S. C., April 24, 1979

Who should pay the estimated $1 million-a-day cost of replacing the electric power normally generated by the stricken Three Mile Island nuclear plant in Pennsylvania? The power company's stockholders, or its customers?

The company, naturally enough, wants the customers to pay. It requested an immediate 20 percent rate increase, a move complicated by the fact that it was granted a $49.2 million rate hike only last month. Consumers, however, feel otherwise. "We will fight like hell to prevent our people from paying for this disaster," a spokesman for the Pennsylvania Association of Older People said last week. "Stockholders are the ones who took the gamble, now they want a guaranteed return."

Ordinarily, we tend to side with the stockholder, viewing him as an endangered species in socialist America, not unlike the Whooping Crane. Someone is always trying to clip his wings, to keep his little nest egg from hatching. In this case, however, our hearts are with the consumer. The purchaser of stock must assume a measure of risk with his entitlement to profit.

Footnote: The Pennsylvania Public Utility Commission is still studying this matter. It acted last week, though, to hold in abeyance the rate increase it granted last month to the operator of the Three Mile Island plant, the Metropolitan Edison Company.

The Des Moines Register

Des Moines, Iowa, April 26, 1979

The Price-Anderson Act of 1957 limits utilities' liability in the event of a nuclear power accident to a small fraction of potential damages. It exempts the manufacturers of nuclear-plant components from liability even if a manufacturer's carelessness caused the accident.

During Senate debate on the most recent extension of the act, Senator Mike Gravel (Dem., Alaska), an opponent of the liability limit, said: "If a company made, let us say, cylinders for the landing gear of 747 aircraft, it has tort liability for those cylinders. But if it were to make valves for a nuclear reactor . . . it would not have similar tort liability."

If manufacturers were liable in such situations, Gravel said, "when they are making reactor valves they would be just as careful as when they are making cylinders for 747s. . . ."

Gravel quoted William Kriegsman, a former Atomic Energy commissioner: "Do away with [the Price-Anderson Act] and you'd probably see nuclear valves coming off the assembly line in better shape."

This week, some officials of the U.S. Nuclear Regulatory Commission suggested a shutdown of all nuclear power plants made by the manufacturer of the Three Mile Island reactor. They contend that such plants have at least five features that make them liable to failure.

It seems fair to ask: If manufacturers were forced to stand behind their product, would such a warning be necessary? Would the Three Mile Island incident have occurred? A more basic question is: If the nuclear industry won't assume the risk involved, why must the public?

THE INDIANAPOLIS NEWS

Indianapolis, Ind., May 3, 1979

Recent events at the Three Mile Island nuclear power plant weren't "catastrophic," as has been suggested.

No one was killed. No one was injured. The reactor is now under control. But in at least one sense, the accident in Pennsylvania was a disaster. A financial disaster.

Cleanup: The reactor itself was insured for $300 million, which won't begin to cover cleanup costs. Design changes are not covered. The reactor's manufacturer, Babcock & Wilcox, may incur some liability. But if human error was the primary cause of the accident, that liability will be small.

Damages: A pool of nuclear insurers has already paid more than $1 million to families evacuated from surrounding areas. By law, a utility's maximum liability for off-site damages is $560 million. Two suits asking that amount have already been filed.

Replacement: It is costing $800,000 a day to replace the power that Three Mile Island can no longer generate.

General Public Utilities Corp., a three-company consortium which owns Three Mile Island, has reduced the dividends it pays to stockholders and suspended an employe stock ownership plan. GPU has halted construction on all new generating and transmission facilities and cut back on maintenance at existing plants in an attempt to conserve cash. Still, GPU says Metropolitan Edison, part-owner of Three Mile Island, faces bankruptcy "within days" unless help comes.

On April 19, the Pennsylvania Public Utility Commission temporarily barred Metropolitan Edison from collecting a previously awarded $49.2 million rate increase and voted to review an increase granted to another part-owner of the plant. Commissioners have already said that customers should not be forced to pay for the accident itself. They have, however, left unresolved whether stockholders or customers should pay the cost of replacement power.

Who should pay? The consumers? The investors? Or how about the government — just this once? One thing is certain: No more nuclear power plants will be ordered until the question of financial risk is resolved.

Asking consumers to bear the entire burden seems a poor solution. They were assured the plant was absolutely safe. They shouldn't have to shoulder all the costs. But they are still in need of power.

Forcing investors to swallow the enormous costs would be pointless, too. If Three Mile Island investors have to shoulder the replacement tab, they will lose their shirts. That lesson won't be missed by potential backers of nuclear plants. There will be no support for nuclear power if the industry cannot attact investors.

Investors and consumers, together then, must share the burden. The investors took the risk; the consumers need the power.

The financial fallout of Three Mile Island cannot be shared by the entire nation until a consensus — in the form of strong government program supporting nuclear energy — is reached. There can be no "just this once." Setting such a precedent would give utilities little incentive to preserve the extraordinary safety standards that are essential in the nuclear industry.

We are far from agreement on nuclear power. Whatever its hazards, it is almost impossible to imagine the next 30 years without nuclear energy. Until we make up our minds, we cannot end its attractiveness to investors without fully comprehending what that will mean — the end to the nuclear option.

The Washington Star

Washington, D.C., April 25, 1979

The financial fallout from the Three Mile Island nuclear accident could be more serious than the limited radioactivity that wafted over the banks of the Susquehanna. The end of the after-effects on the economics of electricity is nowhere in sight. And there seems to be considerable disagreement among some consumerists, politicians and regulators about who, in the end, must bear the costs.

Consider, first, the impact on the utility companies directly involved and their customers. Metropolitan Edison and its parent company, General Public Untilities, face additional costs of $24 million a month for electricity to replace what would have come from the closed plant. Construction costs and interest on the unproductive damaged plant run at $8 million a month, and it may not reopen for two or three years. The total bill for the reactor mishap is anyone's guess, but it obviously will mount soon into scores of millions.

The Pennsylvania Public Utilities Commission demonstrated its grasp of the situation last week by refusing to reinstate a previously granted $49.2 million rate increase for Metropolitan Edison on the ground it was premised on operation of the nuclear plant. Utility executives say there is a danger of corporate bankruptcy in the absence of rate relief. During a hearing before his subcommittee on nuclear regulation, the reaction of Senator Gary Hart was that this is the problem for utility stockholders, not consumers.

There is a strain of unreality here. Wiping out utility shareholders is not going to meet the increasing costs of electric power generation. In an industry financed ultimately by user charges, the customers bear the freight. Yet there is a vogue for populist-style regulation that would attempt to deny this fact. A member of the Oklahoma Utility Commission used his record of never approving a rate increase to get elected attorney general of that state. A new member of the Georgia Public Service Commission won election to the post with a pledge never to vote for a utility increase.

But the costs go up inexorably, and if people are going to have electricity they will have to pay for it. A case could be made for a system of national sharing of such extraordinary costs as those falling directly on the Three Mile River utility and its customers. This might relieve some of the public worry about further nuclear power investment around the country. Nuclear plants produce additional energy from sources other than imported petroleum, and in our opinion this can be done with due regard for the public safety — enhanced by the lessons of Three Mile River.

The costs of new nuclear power plants already were increasing rapidly before the Pennsylvania accident, and are expected to increase faster as a result. This eats into — and may eliminate — the present price advantage of nuclear power as compared with electricity from coal. But coal-generated power is going up in cost as well, with air pollution control a big expense. So there will be no relief from rising electric rates no matter which way the country turns. Both sources nevertheless should be cultivated.

The Three Mile Island mishap, by scaring investors about the risks of nuclear power generation, apparently is pushing up the financing costs of other electric companies. Vepco, with a big commitment to nuclear power, has just incurred millions in additional debt-retirement expenses because of the nervous climate. Investors want to be paid higher interest for risk-taking. Another indirect cost of the accident may result from more shutdowns of other plants because of safety fears, lessening the overall efficiency of nuclear facilities.

As the campaign against nuclear power continues under the impetus of the Harrisburg incident, the legal delays and resulting cost increases for new nuclear plants will further add to future electric rates. A 1,100-megawatt nuclear plant now costs $1 billion. A three-year delay is calculated to add $450 million to the cost of such a plant planned to operate in 1988.

Three Mile Island, in any case, is a burden that we all will share to some extent, both as consumers and taxpayers. We need to see that the sharing is as fair as it should be.

The Chattanooga Times

Chattanooga, Tenn., May 15, 1979

When an accident befalls a utility, like the mishap at the Three Mile Island nuclear plant in Pennsylvania last month, who pays for the cost of settling claims and repairing the facility? That's the question facing General Public Utilities Corp., parent company of Metropolitan Edison, the subsidiary that operated the Three Mile Island plant. A lot depends on how the question is answered.

William G. Kuhns, GPU chairman, laid it on the line last week to the corporation's stockholders, telling them that unless the company is granted rate increases by the Pennsylvania and New Jersey state commissions, it could run out of money in just a few months. In the meantime, GPU has imposed a hiring freeze, planned a tentative layoff of about 5 percent of its employees (some through attrition and early retirements) and reduced executive salaries.

At first glance, GPU's argument would seem ludicrous: Why should a corporation be allowed to "pass along" the costs of an accident to consumers and taxpayers? After all, wasn't the accident its fault?

A group called "Mobilization for Survival" summed up at least one popular response this way: "We feel it is an outrage that GPU has first of all endangered the lives and welfare of hundreds of thousands of people and now has the nerve to make the same people pay for it."

But the problem is more complicated than the rhetorical shorthand implicit in that statement. To obtain the short-term bank credit it needs to repair the No. 2 plant at Three Mile Island and to make other capital expenditures so that its subsidiaries can continue to supply electrical power to the region, GPU needs rate increases so it can maintain its solvency. If it goes bankrupt, everybody loses. Accordingly, it has asked the Pennsylvania Public Utilities Commission for a 9 percent rate increase for six months.

It would be foolish to deny the request. If GPU's financial underpinnings are knocked out from under it by a denial of fresh capital at this critical time, thus forcing the company into bankruptcy, it's virtually certain that a court-appointed trustee would immediately petition for rate increases to protect the interests of GPU investors.

Denying GPU's petition for rate increases might temporarily stroke the fiery outrage of some anti-nuclear protesters upset over the accident. But it would also unnecessarily burden thousands of other consumers by jeopardizing the financial health of a vital utility corporation.

Rocky Mountain News

Denver, Colo., June 23, 1979

Guess what? Somebody's got to pick up the tab for damages to the crippled nuclear power plant at Three Mile Island. The plant supplies power to customers of Metropolitan Edison Co. and Pennsylvania Electric Co., partial owners of the facility.

The Pennsylvania Public Utility Commission conducted a month long investigation into who should pay the costs of the accident.

Guess who? That's like lobbing a soft easy one up to the plate underhanded to Reggie Jackson when he's in a hitting mood and has his 40-ounce bat cocked to deliver.

The customers will pay, of course. Residential customers will be hit with roughly an 11 percent increase in monthly bills.

Outrageous.

The Pittsburgh Press

Pittsburgh, Pa., August 15, 1979

Unless new evidence comes in to change its mind, the U. S. Nuclear Regulatory Commission is prepared to declare that the accident at Three Mile Island was not an "extraordinary nuclear occurrence."

Before your jaw hits the floor, it should be explained that "extraordinary" in this usage has a precise meaning, as spelled out by the Price-Anderson Act which was designed to speed legal action in the event of a major catastrophe at a nuclear installation.

The criteria include the amount of radiation exposure to nearby residents, the degree of property contamination, and the number of people killed or hospitalized.

A tentative NRC decision is that the Three Mile Island incident did not meet any of these criteria.

There was no massive release of radiation. Nobody was killed or hospitalized. And no property outside the plant itself was rendered unusable.

What this means, if the finding stands after a 30-day public-comment period, is that people claiming damages as a result of the accident will have a tougher time collecting because they will have to prove negligence on the part of the plant's operator or the builder of the reactor and their associated companies.

On the other hand, under the Price-Anderson Act's definition of "extraordinary," they would have to show only that they suffered loss or damage and that it was caused by Three Mile Island — regardless of whether any negligence was involved.

Simple justice, of course, demands that anyone injured by Three Mile Island be compensated as fully as possible. But justice also demands that the injury be patent and provable.

As of June 1, the two syndicates which insure the nuclear industry had paid out $1.2 million as reimbursement for living expenses and lost wages to 3,000 families and individuals who evacuated the area near the plant.

This comes out to a mere $400 per claimant. So it surely represents only a fraction of the final legal cost of Three Mile Island.

But in the absence of any obvious harm to anyone's health or property because of what happened at Three Mile Island, those who are claiming hundreds of millions of dollars in compensation will have to prove their cases in court.

Lincoln Journal

Lincoln, Neb., May 9, 1979

President Carter's spot decision to meet leaders of the weekend anti-nuclear demonstration and listen to their grievances and warnings represented a triumph for organizers of the protest.

The triumph was one of recognition by the White House — and hence instant legitimacy — even though what the president said in the 20-25 minute session surely was not acceptable to those wanting to turn off the nation's 72 operating nuclear-powered electric generating plants right now.

"Out of the question," Carter said.

He's right.

Whether the nuclear plants represent 13 percent of the country's generating capacity, or something more or less, the abruptness of what was demanded at the steps of the U.S. Capitol Sunday could not be tolerated, even if most Americans tilted toward the protesters' views. And they almost certainly don't.

The rage one sees being played out currently in California because of gasoline supply shortages would be replicated, if not rendered far more ferocious, if a shutdown of the 72 current plants — including two in Nebraska — brought electric power blackouts.

Once again, this newspaper would like to comment that the health dangers and risks associated with nuclear plants are no greater, nor any less, today, after the Three Mile Island plant accident, than they were before. All that has changed — it is a welcomed change in the direction of greater maturity — is the level of public appreciation of those dangers and risks.

The most serious challenge to expanded reliance on nuclear fission as a source of electric power continues to be what to do with the wastes and byproducts which will be lethal for tens of thousands of years.

Not only have we not figured out a permanent and safe depository for this perpetually deadly cargo, we have no real short-term storage strategy, either, other than keeping an ever-increasing number of spent fuel rods in water baths at the plant sites. Year after year.

As scientists and political leaders struggle with this enormous, maybe insoluble problem, it is reasonable to think that there will be constructive changes because of the Three Mile Island accident.

One advocated by Edward Teller, a fixed proponent of nuclear power, is a much better corps of technicians operating these plants. Perhaps all such technicians may have to be federally licensed. Significant human error reportedly was involved in the near-catastrophe in Pennslyvania.

Pressure will develop — it certainly should — for a revamping of the Price-Anderson Act, which has financially shielded the nuclear industry from the consequences of plant accidents since the Eisenhower era. Price-Anderson excuses utilities with nuclear plants from costs and liability in excess of $525 million because of a single accident. Simple inflation since the 1950s would justify that figure being hoisted to at least $2 billion, and utility insurance premiums raised accordingly.

In the wake of Three Mile Island, more safety features surely will be incorporated into existing nuclear plants. Let there be new requirements, too, for utilities to begin regularly setting aside important dollar sums annually for multi-million-dollar expenses associated with ultimate plant decommissioning costs.

These things, in sum, can't help but increase the net overall cost of a kilowatt produced by a nuclear plant. As they should.

Electricity produced by nuclear power — or fossil fuels — should lay on to those who consume it and benefit today all the real costs associated with it. Under those conditions, cold economics and not hot emotions may well turn people away from nuclear power.

The Houston Post

Houston, Texas, October 1, 1979

The beleaguered nuclear power industry is developing a plan that could have far-reaching positive effects. As described to a House interior subcommittee on energy and the environment, the utilities would create an insurance pool to cover losses from shutdowns of their nuclear plants. The plan was outlined by an industry spokesman during testimony on proposed legislation to revise the nation's nuclear power policies. Details of the self-insurance pool are being worked out by the Nuclear Policy Committee of the industry's Edison Electric Institute and are expected to be ready for industry-wide consideration soon. The pool would be in addition to property and liability insurance now covering the atomic plants.

Much of the innovation being considered by the industry and Congress is motivated by the accident last March at the Three Mile Island plant in Pennsylvania. That accident has caused wide public alarm about the nuclear power that this country needs to help meet energy demands. In addition to claims of potential health hazards, critics argue that it is not fair for consumers to have to bear the high costs of plant shutdowns resulting from accidents or structural failures. The company that owns the Three Mile Island plant is reportedly paying $20 million a month for power to make up for that lost by the idling of the nuclear generator.

And there is fear within the industry that public antipathy over Three Mile Island may prompt a long moratorium on construction of new atomic power plants. The public and governmental concerns cannot be ignored by the industry. Hence the action of the nuclear utilities to adopt a program that could not only result in reduced costs to consumers but bring about some form of self-policing is encouraging. With the utilities involved in mutual financing, they will have a stronger mutual interest in what each other is doing.

The Oregonian

Portland, Ore., May 21, 1981

The word is that Congress has "no stomach" to pay cleanup costs to the owners of Three Mile Island, the badly damaged nuclear reactor. We hope the word is right, because there should be no federal bailout.

The way to prevent future such accidents is to impress upon operators, the industry and moneylenders, that careless actions will prove financially costly, even when part is paid with federal insurance, as was the case with the March 1979 incident.

While ratepayers are not to blame, they are already being made to pay for the higher costs of substitute power bought when two reactors were shut down after the accident. Forcing cleanup costs on ratepayers would be an excessive burden.

The industry needs a fair and equitable formula for sharing cleanup costs, expected to top $1 billion in the case of Three Mile Island. Such a formula, based on regional sharing of the costs, needs to be worked out before accidents occur. The federal taxpayers should not have to share in the costs of a utility's mismanagement.

Rockford Register Star

Rockford, Ill., August 13, 1980

Those people who gave you the nation's worst nuclear plant scare at Three Mile Island seem to have as weird a way of reasoning as they do of speaking.

When the plant was in trouble and trying to close down because of malfunctions, candor was the last thing to come out of the Three Mile Island hierarchy.

Now, having endured huge losses, the owners think the deficits should be passed on to electric users across the nation.

What they are specifically looking for, they will tell you — and already have in a press conference convened for this special purpose — is something like an extra $100 million yearly for the plant cleanup!

These owners reason that it is only "reasonable and appropriate" that all utility customers across the nation come up with this money, which they figure would only cost the "average" user 10 cents a month.

We have watched, recently, as some strange theories have been advanced about private enterprise.

The theory underlying Chrysler's rescue mission is that a company hugely mismanaged can be helped only with the kind of huge loan that only the federal government could, and finally did, float on Chrysler's behalf.

Now, in the case of Three Mile Island where profits came before safety, the theory seems to be that this cliff-hanger was a national entertainment and that a charge for admission is only fair.

We have news for Three Mile Island.

It was, and remains, a private enterprise. If it can't bounce back on its own, shut the place up. Let the corporation go bankrupt. That's a far more honest and direct approach than seeking access to the pockets of electric consumers across the nation who have never benefited from Three Mile Island, never will and are content not to.

THE SACRAMENTO BEE

Sacramento, Calif., January 13, 1981

It's now nearly two years since that so-called "transient" left between 600,000 and 750,000 gallons of radioactive water in the sump and basement of the crippled Three Mile Island nuclear power plant in Pennsylvania, yet with every passing month the cost estimates for the cleanup get higher and the completion date more remote.

Officials of the company that operates TMI now expect that it will cost $1.5 billion to clean up and repair the facility; other nuclear experts believe the final cost will be several times that high. Similarly, after original declarations that it will take months to do the work and that the damaged core could be removed by 1983, company officials have now set back that date to 1985 and, according to a recent report in the New York Times, "say it may be delayed even further."

The essence of the TMI problem is that the difficulties of cleanup are so complex that there is no agreement on what to do. As a consequence, the process has come to a virtual standstill. No one knows how badly the reactor's core has been damaged or whether it can be repaired, but until the water has been removed, that's regarded as a secondary problem. There is hope that by filtering out the radioactive minerals in the water, it can eventually be pumped into the nearby Susquehanna River, but company officials say they are not planning to make a formal proposal on that matter until next year. At that point the Nuclear Regulatory Commission will have to decide whether or not to approve the plan that is submitted.

The certainty is that TMI has already become the most expensive man-made mess in history. And while it's true, as people like Edward Teller often remind us, that no one was killed at TMI, it is not true that the system worked. The accident is over, but the costs continue to mount. At the rate things are going now, it may be a generation before the last bill is rendered to the taxpayers and ratepayers and the price fully paid.

Newsday

Long Island, N.Y., April 1, 1981

Two years ago today, Newsday was reporting that the danger of an explosion at the Three Mile Island nuclear power plant seemed to have increased and that evacuation plans were being expanded.

But ultimately the plant was brought to a cold shutdown; no one outside was hurt, and the risk that anyone will develop cancer as a result of exposure to radiation during the country's worst commercial nuclear accident is considered minuscule.

Even so, the danger isn't over. The Nuclear Regulatory Commission estimates that completing the cleanup and decommissioning the reactor near Harrisburg, Pa., may take anywhere from five to eight years.

What's more, it's likely to cost $1 billion. That's a quarter-billion more than the cost of building the Three Mile Island plant in the first place, and now the question is who should pay the bill.

The candidates are Metropolitan Edison's customers, the utility and its parent company, nuclear utilities in general, the State of Pennsylvania and the federal government.

Although the Pennsylvania Public Utilities Commission has prohibited the use of utility customers' money for the cleanup, it seems to us that they'll eventually have to bear some of the cost. But the major financial responsibility clearly rests with Met Ed, and meeting it could take the company into bankruptcy.

The least acceptable alternative is to have the taxpayers foot the bill. The nuclear industry is already heavily subsidized, and its critics often argue that the figures on nuclear power look better than they should because the costs of decommissioning a reactor are underestimated even under normal circumstances.

Bailing out Met Ed would hardly encourage the industry to be more accountable for nuclear safety. It should be up to the industry, not the taxpayers, to help Met Ed pay for the damage at Three Mile Island as part of the cost of being in the nuclear business.

THE WALL STREET JOURNAL.

New York, N.Y., April 10, 1981

A covey of bankers gathers today to decide whether to roll over loans to Metropolitan Edison, the unhappy owner of the Three Mile Island power plants, or whether to let Met Ed plunge into bankruptcy next week. The ongoing confusion at Three Mile Island speaks not only to the problems of nuclear power, but to the way we regulate our electric utilities.

While everyone has been watching for the last millicurie to settle from the nuclear accident, Pennsylvania has been inundated by financial fallout. Laying aside the issues of how to pay for cleaning up the mess, Met Ed, its bankers, its customers and the Pennsylvania Public Utility Commission have been engulfed with the issue of how to pay for the power Met Ed has had to buy to serve its customers while its reactors are down.

TMI-2 is still awash in 700,000 gallons of radioactive water and TMI-1 can't resume operation until the Nuclear Regulatory Commission gives its OK. Thus Met Ed has been plugged into various other power grids. Although this imported energy is a lot more pricey than the old nuclear-generated kind, Met Ed customers have been paying for it at the old rates. In June 1979, two months after the accident, General Public Utilities Corp.—the holding company for the TMI utilities—borrowed some short-term bucks from a group of banks to help bridge the shortfall.

It wasn't until a year later, however, that the Pennsylvania Public Utility Commission granted GPU some relief by allowing a fuel adjustment to cover the increased cost of the power replacement. This adjustment was not large enough to cover the interest costs of the loans—a shortsighted decision since the ratepayers will have to cover those payments eventually at a greater cost as they are rolled over. PaPUC also removed TMI-1 and -2 from GPU's rate base, further slashing the company's revenues. It's no wonder, then, that the utility continues to be embarrassed by its cash position.

Met Ed is now asking the banks to roll over the loans and lend it an extra $20 million to $25 million so it can pay a state tax bill due April 15. Without such action Met Ed will be in default.

The banks have said no dice on both accounts unless PaPUC starts behaving reasonably and allows a permanent rate increase—an utterly sensible posture since Met Ed customers were still paying, even after the fuel adjustment, among the lowest rates in the Pennsylvania area. After stalling a bit, PaPUC voted last Thursday to grant Met Ed a $51 million increase, some $25 million short of what it needed. The banks must now decide if that increase is sufficient to justify further aid.

There is a certain irony in Met Ed's coming to grief over its state tax bill. This tax is a straightforward percentage levied on gross receipts; it is applied to customers' monthly bills, collected by the utility and then paid to the state. It sounds honest enough, except that as the fuel adjustment bumped up monthly bills, the state's tax take increased accordingly. The state is thus receiving a windfall gain from the accident's unfortunate aftermath. Last year the windfall gain amounted to almost $12 million; this year it'll be over $20 million. Such profiteering is rather unseemly, we think, and the state legislature ought to waive its collection of the windfall portion.

But reasonable behavior has been in short supply over the past two years. Even small demonstrations of reasonableness could have prevented Met Ed's present predicament—PaPUC could have swiftly granted the full necessary rate increase and the NRC could have made the startup of TMI-1 its highest priority. But both PaPUC and the NRC shuffled and delayed, thus tightening Met Ed's cash bind.

Even if Met Ed goes bankrupt and vanishes from the face of the earth, its customers will still be hooked up to the same electric wires. Their electricity will still come from the same grids, and will still cost just as much. There will be no one but them to pay its price. This reality could have been faced from the first without forcing the utility to borrow and incur interest costs, which are likely to be passed along to consumers either sooner through higher rates or later through higher costs in the capital markets. If the Public Utility Commission had been willing to face it, the whole Three Mile Island problem would be much closer to a solution.

The Wichita Eagle-Beacon

Wichita, Kans., May 30, 1981

Paying for the clean-up of Three Mile Island is not the responsibility of utility users across the United States. Proposed legislation to require such nationwide assessments is neither prudent nor fair. A precedent would be established for burdening uninvolved taxpayers with the mistakes of local utilities, especially taxpayers who will in no way benefit from the power generated from these facilities.

Admittedly, General Public Utilities Corp., the holding company for the Three Mile Island nuclear reactor site, is faced with some serious problems. The estimated cost of the clean-up has reached $1 billion. The $300 million of accident insurance is almost gone. Banks have tightened credit to the company, dividends have not been paid to stockholders, and GPU is in danger of becoming the first U.S. utility to go bankrupt since the Great Depression.

None of these factors, however, justify saddling all U.S. users of nuclear energy with the clean-up costs, as the proposed legislation would require. Officials of GPU cite federal bailouts of Chrysler and Lockheed as their rationale for public assistance. The greatest concerns expressed about those two examples were that they would cause other businesses to look for similar aid. It's apparent those fears were valid, but a line must be drawn.

The American taxpayer can't be told on the one hand that he or she is living in a free enterprise system, while being forced to provide tax dollars each time a corporation makes faulty foreign investments, fails to meet the needs of its customers, or does not apply adequate safety precautions.

The 700,000 gallons of highly-contaminated water in the Three Mile Island reactor must be removed. All damage and danger of radiation must be addressed responsibly. It is not responsible, however, to expect non-involved private citizens to foot the bill.

THE COMMERCIAL APPEAL

Memphis, Tenn., May 12, 1981

IT IS TWO YEARS since a relief valve atop the pressurizer of Three Mile Island Unit 2 near Harrisburg, Pa., failed to close, causing the most serious nuclear power plant accident the nation has known.

That plant remains closed and there is still no estimate of when technicians will be able to remove all the 600,000 gallons of radiologically "hot" water that accumulated in the basement.

The GPU Nuclear Group,, which operated the TMI power plant, admitted frankly in a newsletter issued by its own communications officer in March that the spilled water presents a major risk of harmful releases to the environment. "So long as the water remains in the reactor building, there's a chance that it could seep into groundwater outside and, eventually, reach the Susquehanna River," the company's public relations flyer stated.

Several possible methods of disposing of the processed water are available, GPU officials say, including recycling, evaporation, solidification into concrete, tank storage on site and discharge into the river. But GPU admits "all have implications for releases of tritium to the environment."

Some of the waste now has started moving in truckloads to the the Washington State nuclear dump, but there is no assurance that will solve the TMI problems. Given such dimensions of the clean-up problem, it is not surprising that the cost already has been estimated at $1.3 billion. Nor is it surprising that Rep. Allen Ertel (D-Pa.), whose district includes the disabled plant, has introduced legislation in Congress which would require all utilities in the nation which operate nuclear plants to contribute to a nuclear clean-up insurance fund. Under Ertel's bill, customers of those utilities would be required to contribute an average of $3 million annually to the fund, the first $750 million of it to go to TMI and the rest to a pool to cover the cost of cleaning up future accidents.

The proposal is opposed by the utility regulators in Pennsylvania, and representatives of the Union of Concerned Scientists and six environmentalist groups testified this week in opposition to the consumer-funded insurance concept. They recommended that the utility's financial problems be dealt with by stretching out private loan repayments if GPU actually faces bankruptcy due to the clean-up costs, rather than dunning consumers nationwide.

"As long as Congress clearly signals its unwillingness to bail out GPU directly or indirectly, the normal financial processes will work to produce an acceptable private solution to GPU's cash flow shortfall," Michael E. Faden, legislative counsel to the Concerned Scientists, told the House subcommittee on energy and environment. There was a strange irony in the fact that a representative of the scientists felt compelled to give the traditional defenders of free enterprise such a lesson in business economics.

The people of the nation already are being called upon on to help companies such as GPU with the problems they created by their own failure to maintain proper safeguards in the operation of their plants. The federal government now insures against a maximum of $300 million damage at any nuclear plant. That is enough.

NUCLEAR PLANT plant operators should be required to increase their safety standards so that there will be less likelihood of future accidents of the dimensions of TMI or, perhaps, even worse. They should not be allowed the comfort of assurance that they will be reimbursed regardless of what happens in the future.

WORCESTER TELEGRAM.

Worcester, Mass., October 26, 1981

General Public Utilities down in Pennsylvania is over its head in radioactive water.

It is looking at a $1 billion price tag for cleaning up what's left of the mess at its Three Mile Island nuclear power plant, the one that leaked and steamed and was damaged two years ago in the nation's first major nuclear power industry accident.

GPU can't afford the work. It has been looking for help since the accident occurred.

Now the Reagan admininstration is planning to come to the troubled utility's aid. Energy Secretary James Edwards says the administration will ask Congress for $123 million to help clean up the crippled plant. The utility can't afford to do the job because it is paying the carrying costs of two nuclear plants, the damaged one and another, identical plant on the same site that was ordered shut for fear of a repeat of the nuclear accident. It is also paying for and passing along, the cost of expensive electricity to replace the generating capacity lost when the accident occurred.

Secretary Edwards says, if Congress approves, the federal commitment will be limited to removal, disposal and study of the reactor's damaged fuel core and accumulated atomic wastes. But he cannot put a price tag on the job. The federal share, under the administration's plan, will depend on how badly the reactor core is damaged. Edwards told a congressional committee earlier this week that Congress has already appropriated $37 million for research and development at the plant.

The logic for federal involvement in the cleanup is clear enough, if not very palatable. Utility companies provide a necessary service. They operate as regulated monopolies because of the nature of their role. The government already has a major hand in regulating the nuclear power industry and so should perhaps carry some of the responsibility for any accident. Allowing a utility company to fail because of the financial burden of the $1 billion cleanup could, just possibly, jeopardize the supply of power for two states, an unthinkable consequence.

But is a federal direct cash gift to GPU the appropriate step at this time? Some say no. U.S. Rep. Allen Ertel, D-Pa. is the author of a rival bailout proposal calling for mandatory industry-wide mutual accident insurance. He says the Edwards plan was just "a mirror trick." Alternatives include having the nuclear power industry come up with $190 million to help pay the cleanup cost. GPU, the holding company that owns the damaged reactor, would pay $245 million under the plan. New Jersey and Pennsylvania would kick in the amount of the gross receipts tax windfall they have received from higher electric bills since the 1979 nuclear accident.

Ertel says the primary financial burden of the accident rests properly with the utility and with its customers. He objects to setting a precedent that would have research and development money used to pay what should be private clean-up costs.

Pennsylvania Gov. Dick Thornburgh wants the cost shared. Sen. John Heinz, R-Pa., has proposed creating a federal nuclear insurance program for TMI and future nuclear mishaps. It would be run by the Energy Department. No federal funds would be involved.

With a variety of plans to choose from, and with at least one designed to avoid a federal share of the cost, the Congress ought to consider carefully before diverting R&D funds to the clean-up — before saddling all the taxpayers with local costs incurred by a private company as a risk of doing business. There are other models to follow.

The Reagan administration, with its insistence on attempting to balance the federal budget, cut deficits and federal spending should be wary of diverting hundreds of millions of badly needed federal dollars to the aid of a single utility company.

The Chattanooga Times

Chattanooga, Tenn., October 28, 1981

The Reagan administration has proposed that the federal government come up with $123 million over three years to help with the $1 billion tab for cleaning up Three Mile Island, site of a nuclear plant accident in 1979. Ideally, the taxpayers shouldn't have to shell out a dime to rescue General Public Utilities, but the alternative — closing the plant for good and forcing GPU into bankruptcy — is unacceptable.

Besides, the Reagan proposal is relatively modest, slightly more than 10 percent of the total cost; the rest will be borne by Pennsylvania and New Jersey, GPU, insurance companies and GPU's stockholders. In considering the Reagan request, however, Congress should not make an open-ended commitment.

Obviously GPU can't handle the cleanup cost alone — the accident has battered its finances, dropping the price of its stock from $19 to $5 — so it makes sense to spread the cost of the cleanup among those who would benefit from it. Federal participation, says Pennsylvania Gov. Richard Thornburgh, would help the utility put together a cost-sharing package to do the job. He has suggested contributions of $190 million each from the utility industry and the federal government, $245 million from GPU, $30 million and $15 million from Pennsylvania and New Jersey, respectively, and the balance from insurance.

The administration's willingness to help fund the cleanup indicates a federal interest is involved; the $123 million it proposes to spend is a rough calculation of the value of information to be gained from removing and studying the damaged reactor core. The additional information could enable the Nuclear Regulatory Commission to improve its oversight, for example, or the government's ability to maintain the stability of the nation's energy supplies. Private utilities also have a stake in GPU's financial health. If GPU went belly-up, depressed prices for other nuclear utilities' stock would make it harder for them to raise money, thus forcing higher rates for consumers.

Even those who disagree with the administration's commitment to nuclear power would have to agree that its responsibility to protect the public can be helped by learning from TMI. The relatively modest federal share of the TMI cleanup is defensible.

The State

Columbia, S.C., October 6, 1981

THE BOARD of the trade association representing the nation's power utilities has taken a responsible first step toward resolving the problem of paying for the cleanup of the Three Mile Island nuclear plant in Pennsylvania.

TMI, one will recall, had a serious but non-lethal accident in March of 1979. It is estimated that decontaminating the damaged reactor will cost $760 million. This is beyond the financial capacity of its owner, General Public Utilities Corp.

The directors of the Edison Electric Institute, the trade association, have proposed that the nation's utilities pick up 25 percent of the cost, which comes to $192 million.

Although participation by individual utilities is not binding, most of the power companies in the nation are expected to kick in their share. Those with nuclear operating licenses or construction permits are being asked for twice the contribution that will be requested of companies that don't have such plants.

The trade association's proposal is close to that put forth in July by Pennsylvania Gov. Dick Thornburgh. His cost-sharing plan suggests $192 million from the electric power industry, the same amount from the federal government, $245 million from TMI's owner, and $45 million from Pennsylvania and New Jersey, the states which drew power from TMI.

The trade association's board said that all utilities should participate to help restore the public's and Wall Street's confidence in the utility industry as a whole. Governor Thornburgh called the industry's decision "a major breakthrough in the long and frustrating impasse over TMI cleanup funding."

Pennsylvania's chief executive recently appeared before a congressional committee in an effort to persuade the federal government to follow the industry's lead.

The TMI incident was by far the most serious ever suffered by the nuclear power industry. One hopes that the lessons learned will help prevent repeats. But it is appropriate that utilities which may need similar assistance in the future contribute to the staggering cost of this cleanup.

The Washington Post

Washington, D.C., August 7, 1981

MORE THAN TWO years after the nuclear accident at Three Mile Island, the cleanup of Unit 2, the damaged reactor, is still in its earliest stages. Nearly $300 million has been spent. General Public Utilities Corp., the plant's owner, says it will soon run out of funds. Unit 1, the undamaged companion plant on the same site, has not been turned on, costing GPU $14 million per month in replacement power. Bankruptcy of the utility because of monumental cleanup costs is still a real possibility.

Though he has no direct jurisdiction over the cleanup, Gov. Richard Thornburgh of Pennsylvania has responded to the lack of urgency displayed by industry and the federal government by proposing his own plan for raising the necessary funds. The details of the plan—involving voluntary contributions by GPU, the federal government, the nuclear industry and the states of Pennsylvania and New Jersey—are less important than the sum thought to be needed: $760 million. And that breathtaking estimate is probably too low.

Yet, even if the money were available, cleanup would by no means be assured. Cleansing the nearly one million gallons of radioactive water in the plant before leaks occur is no small task, but is the easiest part of the process and should be under way soon. Removing and finding a safe way to store the damaged reactor fuel is more difficult. Though no one yet knows for sure what condition the fuel is in, it is likely to be more damaged than anything dealt with before. The biggest job, one that nobody has even begun to think about, will be finding an acceptable resting place for the huge reactor vessel, and devising a safe way to get it there.

Thornburgh's plan should be taken as a reminder that the cleanup of Three Mile Island is in danger of being forgotten by just about everyone outside of Pennsylvania. It is natural for the industry and its federal regulators to want to focus on the future, on positive efforts to ensure that such an accident never happens again. The damaged plant is, nonetheless, a national responsibility, one that Congress, the executive branch and the industry must cooperate in solving. Turning on the undamaged Unit 1, as soon as the Nuclear Regulatory Commission completes the hearings now in progress, will help. But so long as cleanup continues to inch along with no prospect of the necessary funding, and with major technical problems unresolved, the nuclear industry cannot expect a financially or politically stable future.

After nearly 30 years, the nation still has no policy for dealing with radioactive waste. Nuclear waste has always been an afterthought, something that someone else would deal with sometime in the future. Three Mile Island stands as a reminder that the future cannot be put off indefinitely.

The Philadelphia Inquirer

Philadelphia, Pa., October 24, 1981

The pieces of the Three Mile Island cleanup package seemingly are falling into place. The Reagan administration late Monday promised Gov. Thornburgh it would back his cost-sharing plan by supporting a $123 million federal appropriation to be paid over three or four years and, significantly, would consider additional assistance if warranted.

The Reagan commitment is gratifying for two reasons. The most obvious is that the decision makes it clear that cleaning up the crippled nuclear plant is viewed as a matter of national interest. Also, federal participation is the key to the success of the governor's program, which involves the federal government, Pennsylvania and New Jersey, the nuclear utility industry and General Public Utilities, owner of TMI, sharing costs jointly.

The simple fact that the Reagan administration has chosen to take an active, ongoing role in the $1 billion cleanup task at the TMI nuclear plant is enormously important to assuring that the dangerously damaged reactor won't be allowed to sit in its present state, filled with 600,000 gallons of radioactive water and deadly debris. Every month of delay in ridding the reactor of its life-threatening wastes increases the likelihood of another major accident, one which carries consequences to the public health and safety every bit as great as the events that occurred on March 28, 1979. Work there already has slowed to a snail's pace because of inadequate funds.

Under terms of the financial commitment made by the Reagan administration, the money will be allocated for "research and development." That designation was carefully and properly chosen. Certainly, there is much to be learned at TMI which now and in the future will serve as a giant laboratory for scientists and technicians. But more important, it connotes a one-time expenditure of federal dollars, not an outright grant to a utility and industry in serious financial trouble.

It was federal research and development dollars that led to the expansion of nuclear power in the United States, and it is proper that federal research and development dollars should aid in correcting the serious problems that expansion has created. There is one other fact that cannot be ignored. The cleanup task will not be completed without the federal government paying a share. The Reagan administration's decision to foot part of the bill is a responsible one.

Santa Ana, Calif., September 30, 1981

Two months ago, Gov. Dick Thornburgh of Pennsylvania outlined a $760 million proposal to clean up radioactive debris at the severely damaged Three Mile Island (TMI) nuclear reactor No. 2 near Middletown, Pa.

Thornburgh envisioned that various "contributions" toward cleanup would include $190 million from the federal government, $30 million from the State of Pennsylvania, $15 million from neighboring New Jersey, $90 million left over from TMI insurance coverage, $245 million from the utility consortium that owns TMI, and — here's the rub — $190 million from electrical users nationally via rate hikes.

The probability of that happening grew dramatically two weeks ago when directors of the Edison Electric Institute gave its blessings to Thornburgh's cost-sharing plan with the proviso that state regulatory bodies grant special rate hikes to cover the $190 million industry share plus (apparently) the $245 million that TMI's owners cannot afford.

The Institute endorsement is only advisory. But given that it represents 198-member companies which produce 80 percent of the nation's electricity, it can be expected that its endorsement will carry substantial weight.

Whether cleanup costs at TMI are rightfully a national responsibility — falling as heavily on Californians and Minnesotans as Pennsylvanians — can be argued from now to eternity.

The problem in arriving at a clear-cut yes or no is that nuclear power facilities, while operated by private utility companies, have been virtually nationalized by wholesale governmental intervention and regulations. Additionally, the power companies themselves have monopoly status from government and cannot be neatly categorized as fish or fowl.

In the normal situation when a private company goes belly up, the problem lies solely with the company itself and its stockholders. If the company's facility burns down, blows up or becomes liable for $1 billion in damages (the overall expected cost of the TMI cleanup between now and 1987), it is not a burden that properly can be saddled on every American.

However, the government is so obviously a full partner in nuclear plan operations everywhere that it becomes difficult not to see some logic in Thornburgh's proposal. Like it or not, we expect rate hikes will be involved nationally to offset the TMI cleanup operation simply because there is no other obvious solution.

Even if it is argued that TMI's No. 2 reactor be paved over and forgotten for costs dramatically less than $760 million, the problem is that TMI's owners have waiting in the wings Unit No. 1, another multi-billion dollar operation undamaged by the 1979 blowout. And unless Unit 1 is permitted to begin operations, the owners of TMI cannot pay back their $245 million share of cleanup costs yet to come. (The company has already spent $240 million in cleanup costs to date.)

The problems that began 2½ years ago with a stuck valve are a long ways from going away. One way or another it seems likely all of us will have a share of the bill.

The Evening Bulletin

Philadelphia, Pa., December 8, 1981

The owners of the Three Mile Island nuclear plant make a point in their $4-billion suit against the government. But we wouldn't want to bet that it will hold up in court.

The point they make is that 18 months before the TMI accident, a similar incident occurred at the Davis-Besse nuclear plant at Port Clinton, Ohio. As happened at TMI, a pilot-operated relief valve opened and didn't shut as it should have. Davis-Besse was being operated at only 9 percent of capacity. Otherwise, the outcome might have been the same as at TMI, which was operating at 93 percent of capacity at the time of the accident.

If the federal Nuclear Regulatory Commission had passed along information about the Ohio incident, TMI's operators would have been alert to the possibility of trouble and could have closed the valve in time.

There's no doubt the NRC should have spread the word. But keep in mind that:

☐ Primary responsibility for plant safety belongs to the operators. NRC inspections and information are important but the owners can't duck their burden.

☐ The nuclear power industry itself could have been circulating such information. In fact, it now does via the Nuclear Safety Analysis Center, set up for that purpose after the TMI accident.

The $4-billion claim is not new. It was made a year ago by General Public Utilities Corp., the owner, and was rejected by the NRC. Now GPU is trying again in the federal courts.

The NRC and the nuclear industry have as much to gain from cleaning up TMI as GPU and its customers. If the suit does nothing else, it should wake up the NRC.

●

OTHER REFRESHING warnings are being voiced by Nunzio Palladino, the former Penn State engineering dean who leads the NRC. Last week he told the nuclear power industry that it still allowed too many lapses such as poor construction practices, falsified documents and harassment of quality control personnel.

"Quality cannot be inspected into a plant," he told the Atomic Industrial Forum, "it must be built into the plant." Amen. He proposed quality control audits of nuke plants, and added that if the utilities don't do them, "we may have to require them."

Good idea. Palladino can also apply the principle with equal force to the NRC itself.

THE DAILY HERALD

Biloxi, Miss., December 9, 1981

Two months ago, we complained softly about the administration's apparent willingness to pay out about $192 million to help pay the cleanup cost at the Three Mile Island nuclear plant.

Last week, the plant's owners did something that ought to turn the administration away from volunteering any federal funds for the cleanup costs.

After months of meetings, Pennsylvania Gov. Richard Thornburg and the Edison Electric Institute devised a cost-sharing plan that called for participation by the electric-utility industry, General Public Utilities Corp., which owns TMI, and the states of Pennsylvania and New Jersey, the states which received power from TMI. Insurance would pay the remainder of cleanup costs. The high costs of the cleanup, estimated to be near $1 billion, was given as justification for calling on so many sources.

At that time, we questioned the propriety of using federal funds to undo the damage from TMI's celebrated near-disaster; after all, General Public Utilities Corp. is a profit-making organization, not a governmental operation. Why should government be responsible for helping put the plant back on line? Many businesses are struck every year by disabling accidents, some so severe as to put the businesses out of operation, and the federal government does not involve itself in restoring them. Does being nuclear put GPU into some special category?

Before these questions were answered in Congress, where any cleanup contributions would have to gain approval, good old GPUC last week filed suit against the federal government, seeking $4 billion in damages. The arrogance of GPU suing the very taxpayers it is petitioning for assistance should be sufficient to prompt the Reagan administration to pull out of any voluntary sharing of the cleanup costs.

Should GPU's lawsuit be successful, that firm will have extracted from the national treasury a sum four times greater than the estimated cleanup costs. With that sort of possibility looming, administration officials ought to begin concerning themselves with protecting taxpayers' dollars rather than donating them.

Remember the much-debated federal involvement with the Crysler Corporation? The debate then was over government loan guarantees, not outright government grants. Had Crysler been asking Congress for grants, there is no doubt that Congress would have advised Crysler, in no uncertain terms, to take the lumps resulting from failure to make a profit in the marketplace. And even though Congress finally did approve helping Crysler, many American taxpayers still disapprove that merger of government and industry interests.

If Congress studies the TMI situation carefully, it will postpone allowing government money to be used in the cleanup scheme until GPU's lawsuit is completed.

The Burlington Free Press

Burlington, Vt., November 29, 1981

The nation's first major nuclear accident in 1979 sent a wave of fear rippling through the Harrisburg, Pa., area and the Northeast, touching off public speculation about the consequences of a meltdown of the reactor core.

Thousands fled or were evacuated the vicinity of the plant because they were worried about the hazards they and their families might be exposed to. Even though the accident was contained, an imp of doubt about safety was planted in the minds of many others who lived near nuclear power plants.

Two years and millions of dollars later, Three Mile Island has come back to haunt the country's electrical consumers in another way. The Edison Electric Institute, the utility industry's trade association, is seeking legislation in Congress that would require the nation's utilities to share $32 million in cleanup costs on the basis of the percentage of their power sales and their ownership of nuclear capacity. Vermont utilities would have to bear about $400,000 of the cost, according to Public Service Commissioner Richard Saudek who opposes the measure. That would mean that state ratepayers would pay about 30 cents a year in additional bills over a six-year period to cover the cleanup costs.

While the amount may be infinitesimal for most ratepayers, several congressmen fear that it will set a precedent for imposing the burdens of other accidents on the shoulders of power consumers. And they point out that the government ultimately could spend $131 million on the cleanup. Radiation research at the crippled plant already has cost $12 million and another $19 million is budgeted for this year. The Reagan administration is seeking congressional endorsement of a $100 million expenditure for cleanup. Thus the responsibility for that portion of the cost could fall on taxpayers' shoulders. To ask them to pay another $32 million as utilities' customers might be to stretch their patience to the breaking point.

At the same time, companies like Westinghouse and Babcock & Wilcox who manufacture reactors "haven't been asked to come up with any part of the funding, according to Sen. Gary Hart, D-Colo.

The industry trade association has pledged $192 million for the cleanup and has asked for legislation to make utility contributions toward that sum mandatory so that it will be easier for its members to pass the cost along to ratepayers.

General Public Utilities Corp., owner of the disabled plant, has warned the government that it will go bankrupt if it does not receive additional help. The utility has exhausted $300 million in nuclear insurance to pay some of the $1 billion in cleanup costs.

Saudek said that the state's electricity consumers should not have to contribute to the cleanup cost "if someone tries to put it on their backs."

"I am not convinced the benefits to Vermont are sufficient to warrant Vermont ratepayers picking up Three Mile Island cleanup costs," he said.

What might be troublesome to some who may support the idea of sharing the cleanup costs is the magnitude of financial responsibility that Vermont ratepayers might have to assume should there be a similar accident at the Vermont Yankee nuclear plant in Vernon. A billion-dollar cleanup bill certainly would overtax consumers' ability to pay.

Even so, the government and the public should not be expected to pick up the cleanup costs of a private firm whose stockholders and suppliers should bear the major burden of the cost.

To ask the nation's ratepayers to pay for the cleanup is to set a precedent that will most certainly be followed in covering the costs of future nuclear accidents.

THE BISMARCK TRIBUNE

Bismarck, N.D., November 19, 1981

We sometimes wonder about the lip service paid to the "free market" system in our country.

Here we are, governed by the conservative Reagan administration that is applauded for "getting government off our backs" and "unleashing" the free enterprise system.

And then along comes the issue of who will pay for cleaning up the crippled Three Mile Island nuclear power plant in Pennsylvania.

The prevailing business philosophy would seem to have General Public Utilities — the plant operator's parent company — paying its own cleanup costs. But that's not the case, as it turns out.

Instead, the power industry, the administration, and the states of Pennsylvania and New Jersey have embraced a plan whereby much of the $1 billion cleanup cost is being footed by taxpayers and by ratepayers of virtually every privately owned electric utility in the country.

The taxpayers pay, because the administration is kicking in federal funds. And ratepayers pay because the Edison Electric Institute, a trade group that represents the nation's investor-owned utilities, has assessed its members to cover $192 million of the cleanup bill over the next six years.

Yes, if you receive a monthly bill from Montana-Dakota Utilities Co., you, also, will be paying for the Three Mile Island cleanup.

MDU has agreed to pay the assessment beginning in 1982, although it cannot count the cost as an expense recoverable through its electric bills until its next rate filing with the Public Service Commission.

Admittedly, the MDU annual share — $11,847 — is miniscule and figures out to less than a penny a month per consumer.

But the amount isn't the point. The principle is.

Why should MDU — or, more properly, MDU customers — pay at all?

MDU serves a region rich in coal and, therefore, has no intention of putting up a nuclear plant, even in the distant future. Thus, neither MDU nor its customers has any immediate interest one way or the other in whether an Eastern utility, with its nuclear power plants, is successful.

On the larger question, is it right that a utility be able to make and retain a profit when things are going well, but then seek a handout when things don't go well?

Granted, new ideas — such as nuclear power in its infancy and some forms of alternative energy in these times — often need government and other outside support, in the form of research money or tax breaks, to give them an initial chance to prove whether they're worth it.

But after a time, those strings should be cut, and the endeavor, whatever it is, should be forced to succeed or fail on its own.

It would be in the best interests of the nation and of the power industry as a whole to make the nuclear power industry sink or swim on its own, providing us all with a clearer idea of the potential real costs of nuclear power.

We hasten to add that we are not opposed to nuclear power generation, if carried out properly Nor are we opposed to the cleanup at Three Mile Island; certainly, it needs to be done for reasons of public safety.

And, we agree with Gov. Dick Thornburgh of Pennsylvania, who proposed the cost-sharing plan for the cleanup, that the lessons to be learned from the cleanup will be of "enormous benefit to Americans of all regions."

But the immediate issue is: Who pays?

Perhaps General Public Utilities should be made to repay the financial aid over the long haul, and, if that puts a financial overload on the company, then so be it.

For, nuclear power is competing with other forms of electrical generation, and the time has come to put it to the free market test.

FORT WORTH STAR-TELEGRAM

Fort Worth, Texas, April 1, 1982

The third anniversary of the nuclear accident at Three Mile Island passed last weekend with a reminder from the plant's owner, General Public Utilites Corp. of Parsippany, N.J., of the necessity for extreme caution in the making and care of nuclear power facilities.

The company said it could be forced into bankruptcy if it does not get help in cleaning up the contamination at the damaged Three Mile Island plant. Congress is considering aid for the cleanup project that it is estimated will take at least another year to complete at a total cost of about $1 billion.

Since the accident, restrictions on the construction and operation of nuclear facilities have been tightened, and that is good. But in the process of being more careful, we have added to the cost of construction of nuclear power facilities.

The power industry says that nuclear power is still a sound investment in spite of the rising cost of producing it. Some critics say the opposite and that it is unsafe.

However, there have been no major accidents at nuclear facilities since the Three Mile Island incident of March 28, 1979. That record could be due in part to the lesson we learned at Three Mile Island and to the pressure brought on the industry by critics.

The slowness of getting Three Mile Island's plant repaired and the $1 billion cost of the cleanup should make it all the more clear to the power industry that extreme caution must be taken with nuclear power. Cutting corners and careless maintenance are far too costly and dangerous.

THE CHRISTIAN SCIENCE MONITOR

Boston, Mass., March 30, 1982

What is the balance sheet for nuclear safety three years after Three Mile Island? Constructive results from America's worst nuclear power accident have to be weighed along with warnings of new risks from aging nuclear plants. The plus side will prevail if industry and government build on the increased attention to safety into which they were shocked by TMI.

This momentum, so conspicuous during the first year after TMI, seems to have been flagging. It must not be lost in the midst of the economic preoccupations that cloud the future of the nuclear industry despite all-out support from the present administration.

Protesters marked Sunday's third anniversary of TMI with reminders of continued public concern about safety. This week congressional committees are working on the still unresolved matter of how TMI's massive cleanup costs should be paid for.

The latter question is significant, to be sure, because its outcome could set a precedent for handling the costs of any future accident. A central issue is whether TMI should be considered a national problem, with federal and state taxpayers contributing along with industry to cleaning up the contaminated site. One proposal would also require electricity users all over the country to contribute through their electric bills.

The Reagan administration, with its market principles, is understandably reluctant to promote a federal bailout, though it talks about some $120 million for "research" on the problem. Surely the utmost legislative scrutiny should be provided before establishing a system by which everybody pays for nuclear accidents anywhere. More promising are proposals for an expanded property insurance system, perhaps with government participation, to take care of nuclear utility needs.

But this cleanup-financing question should not overshadow the reverberations from Three Mile Island in the more fundamental realm of preventing or limiting the extent of accidents in the first place. These gain importance in the light of this week's comments by a Nuclear Regulatory Commission safety engineer that the wrong metal was used in a number of nuclear plants, causing the high likelihood of a core meltdown before long.

To minimize accidents and their effects, the improvements spurred by TMI must be furthered: better training of operators, tighter regulatory enforcement procedures, greater focus on emergency planning. These must not be sacrificed in the accompanying effort to make licensing procedures more efficient as the nuclear industry tries to snap back – not only from Three Mile Island but from the economic vicissitudes that are causing predictions of cancellation for many plants already under construction.

The Evening Gazette

Worcester, Mass., April 3, 1982

In inventive testimony to the creativeness of those in Washington, D.C., the Senate Energy Committee has whisked through a bill designed to help pay the cost of cleaning up the Three Mile Island nuclear power plant accident.

The bill for the damages stands around $1.5 billion for the nuclear reactor accident at the plant on March 28, 1979. That's more than the utility that owns the place can afford to pay. So the pleaders have gone to Washington. Already they have succeeded in prying loose some federal money, earmarked for research, to help foot the bill.

Now the notion is to tap the owners and operators of all other nuclear power plants in the country for a share of the cleanup cost.

The Senate Energy and Natural Resources Committee, by its 12-7 vote on the measure, is apparently willing to tap the pockets of all the utility companies to help pay for Three Mile Island and to create a $750 million insurance pool to help pay for any future accidents.

Supporters of this measure believe that mandatory contributions from the nation's utilities are the only way to raise money for the project. Other proposals in the past would have required direct federal contributions to a decontamination fund, but those proposals were rejected.

Some senators who oppose the legislation are promising a fight on the Senate floor. Their reason? The proposal amounts to a national consumer bailout of the nuclear industry.

They are right. Utility companies get their money from their customers. The mandatory contribution proposal is simply an indirect way of raising money to sop up the mess left by the operators of TMI. The money would come right out of the pockets of New Englanders and others who have had the foresight to shift from reliance on oil to cheaper nuclear power. What kind of reward is that? It's no reward at all.

The pockets of all Americans were not among the "natural resources" that were being considered when the Energy and Natural Resources Committee got its name. The Senate should slap down this new proposal for what it is — an outright grab — and at a minimum, delete the Three Mile Island section from the insurance fund proposal.

The Kansas City Times

Kansas City, Mo., April 5, 1982

Kansans and Missourians unfortunately may yet pay the cleanup costs of Three Mile Island, that crippled nuclear plant far away in Harrisburg, Pa. The Senate Energy Committee has approved a bill that requires utility companies operating atomic plants to help pay $170 million toward resurrecting that white elephant, which has been out of service since March 28, 1979. Opponents complain that amounts to a national consumer bailout of the nuclear industry.

Other plans to have taxpayers and utility users foot the bill for the rest of the $1.5-billion cleanup costs still are floating around. Edison Electric Institute may seek financing from its member utility companies nationwide. Other proposals include taxpayer assistance from the federal Treasury and from the states of New Jersey and Pennsylvania.

From our standpoint, the financial burden should not be on the shoulders of customers of other utilities who reaped none of the benefits of the plant when it was operating. More rightfully it should be borne by the owners of the plant, who made the decision to build it and to assume the risk.

Utility companies contend that General Public Utilities, the chief owner of the TMI plant, will go bankrupt without assumption of costs by other utilities and their customers, or the government. A bankruptcy would lead to loss of confidence in nuclear power by investors everywhere, driving up the cost of borrowing money for utilities which eventually would pass them on to their ratepayers, or so the argument goes. But that bankruptcy seems a small price to pay in the face of the potential public backlash that could result over this and any future bailouts through customer bills.

Nuclear power still is a debatable energy source in this country, but it does exist. As long as it exists, no one can say accidents like Three Mile Island might not happen again. If they do, the same question will be raised. Who is to pay? As in the debate over the dumping of nuclear wastes, no one wants the responsibility.

An alternative is to use the national Treasury for that purpose, considering the fact that nuclear power is permitted by national policy. Another possibility advanced is a national contingency fund financed by assessments on public utilities.

The financing decisions for TMI may set a bad precedent for those who don't think utility customers across the country should come to the rescue of every plant that goes under. One thing is certain: Congress, the states and the companies building nuclear plants cannot wait until another disaster comes along and then try to address the problem again. One white elephant is enough.

The Seattle Times

Seattle, Wash., April 4, 1982

IT'S A long way between this state and Three Mile Island in Pennsylvania, but in the distorted thinking among members of the Senate Energy and Natural Resources Committee, we might as well be neighbors.

Legislation approved by the committee last week would have the already reeling Washington Public Power Supply System's utility members pay part of the tab for cleaning up the mess left after the reactor accident at the privately owned Three Mile Island nuclear plant in March, 1979.

As sponsors of the bill see it, the entire nuclear-energy establishment across the country ought to help share the cost of finishing the cleanup. For WPPSS, the bill might mean a "surcharge" of $3.9 million, payable by the local utilities that make up the system.

Among other utilities in the Pacific Northwest that stand to be hit is Portland General Electric, which would be tagged for $1.4 million because it operates the Trojan nuclear plant at Rainier, Ore.

Critics of the measure pointed out rightly that a Three Mile Island bailout would only aggravate the angry feelings entertained by thousands of irate ratepayers over the WPPSS imbroglio.

"When ratepayers find out across the country that they're being forced to participate in a bailout of a privately held utility," said Sen. Gordon Humphrey of New Hampshire, "their reaction is going to be very unfortunate." What an understatement!

A spokesman for the American Public Power Association added: "At a time when people in the Pacific Northwest are mad about their electric rates, it is kind of weird to suggest they help bail out a utility in Pennsylvania."

Wisely, this state's Sen. Henry Jackson and Oregon's Mark Hatfield had themselves recorded as against the bill, which surely ranks as one of the worst pieces of legislation to be considered in the 1982 Congress.

Oregon Journal

Portland, Ore., May 22, 1982

Congress unwisely is considering a bill to foist the costs of cleaning up the Three Mile Island nuclear plant on other utilities across the country. It's a bad precedent that Congress will regret.

The Senate Energy Committee approved a bill that would raise about $170 million over six years toward the estimated $1 billion cost of cleaning up the mess that happened in Pennsylvania in 1979.

This government imposed bailout, as amended in committee, would require utilities with nuclear plants under operation to help with cleanup costs. Those utilities with plants under construction would be spared sharing the costs. Another amendment would permit state utility commissions or commissioners to decide if the utilities could pass the costs along to ratepayers.

Sens. Henry M. Jackson, D-Wash., Dale Bumpers, D-Ark., and Mark Hatfield, R-Ore., protested the bill. They say General Public Utilities, the owner of the TMI plants, should bear the responsibility for its own damaged plant. In fact GPU has spent nearly all of the $300 million in insurance money that became available after the 1979 accident. For almost three years the Pennsylvania ratemaking authorities have refused to permit the GPU ratepayers to suffer the full burden.

The General Accounting Office says that GPU could raise $150 million a year with rates at most 15.5 percent above 1980 levels.

"The costs of the poor luck or negligence of the owners of TMI plants should not be charged to any other utility," reported Jackson, Bumpers and Hatfield. They are right. If Pennsylvania's congressional delegation persists in trying to spread TMI cleanup costs nationwide, an amendment should be added to require some nationwide assistance in paying for the dry holes that the Washington Public Power Supply System (WPPSS) is creating in this region.

OKLAHOMA CITY TIMES

Oklahoma City, Okla., November 26, 1982

NOT much is heard of Three Mile Island these days, and that's a pity because the public perception of the nuclear power plant in Pennsylvania remains a negative one, and that belies the facts. Indeed, TMI has become a sort of symbol for zealots crusading against nuclear energy.

Despite the continuing efforts of TMI's operator, General Public Utilities (GPU) to set the record straight about the accident that occurred there 3½ years ago, hair-raising stories still circulate about the purported effects of radiation on people and animals in the area.

The tenor of the stories and of much of the news coverage of TMI has elevated the malfunction to the status of a disaster. Yet, contrary to popular belief, there was no meltdown and no explosion. And, by all accounts the back-up gear worked as it should.

As one frustrated Pennsylvanian wrote to The Wall Street Journal: "Not a hair of anyone's head was injured. There was and has been no irradiation problem, and within 25 miles of the plant the following year there was a smaller number of substandard babies than in the previous year. There also has been no increase in malignancies."

He complained further that the No. 1 plant at TMI, which was not harmed, has not yet been allowed to function. The correspondent added: "Recently it was about to be given a clear signal when a judge ordered it could not go on stream until a psychological study of the surrounding inhabitants be made. He did not dictate how that was to be accomplished."

These problems, of course, pale beside the monumental financial burden of removing and "cleaning" the million gallons of contaminated water, removal of the damaged core and decontaminating the interior of the containment building. Deciding who was going to pay for all this — the company, the government, the electric industry, the nuclear business or the consumers — has delayed cleanup operations.

The bottom line for not only the Pennsylvania letter writer but for the nation is that the furor over TMI has just about stalled a vitally needed energy industry.

EVENING EXPRESS

Portland, Maine, July 30, 1983

Although the U.S. Senate's Environment and Public Works Committee has sent to the floor a bill requiring utility customers throughout the nation to pay part of the clean-up costs for the Three Mile Island nuclear plant accident, there's a good reason the panel didn't recommend approval: It's simply not a good bill.

The measure represents the first time ever that the committee has forwarded a bill to the full Senate without any recommendation. Some committee members say they didn't like the bill but thought it was too important to be killed in committee rather than on the floor.

Perhaps. It would have been better, however, if the panel had just voted the measure down. With even a slight chance for passage, there is grave danger a terrible precedent may be set.

If the Congress decides that utility customers in Maine and elsewhere must pay part of the cost of the Pennsylvania accident, would the cost of future accidents in any single area also be passed along to all the nation's ratepayers?

It would be grossly unfair for utility customers in states not served by TMI to pay higher electric bills because of the accident. Did Pennsylvania ratepayers pick up any of the $20 million tab for alternate fuel purchased during the 1979 shutdown of Maine Yankee? Of course not. Those costs were paid by the customers who used the power produced at the Wiscasset plant.

The fact that the cleanup costs for the TMI accident are unusually high—estimated at $1.3 billion—should not influence the decision on who must pay. The responsibility must be borne by the owners and customers of TMI. To legislate otherwise would be to virtually absolve all nuclear plant operators of responsibility for safe, efficient operation of their facilities by spreading the cost of unsafe, inefficient operation among all the nation's utilities and ratepayers.

WORCESTER TELEGRAM

Worcester, Mass., February 9, 1983

The nuclear-power industry, a child of public financing as the industry grew up, has taken a step toward independence.

After 25 years, the federal role in insuring the public against losses from a nuclear power-plant accident has ended.

The nation's electric utilities and private insurers now are underwriting all the protection required by law.

The federal Price-Anderson Act requires that nuclear-plant operators have $560 million in liability insurance to protect the public financially in the event of a nuclear accident.

Some $160 million in liability is provided by insurance pools. A second layer of insurance is provided by an assessment of $5 million from each plant owner, if the need arises.

Until now, the federal government has made up the the rest of the coverage needed to meet the Price-Anderson requirement of $560 million.

However, with recent federal licensing of the nation's 80th nuclear power plant — the San Onofre-3 plant in southern California — the owners' liability fund has grown to $400 million, eliminating the need for government backup.

That's a welcome sign of progress for the nuclear industry, which has had more bad news than good in recent years.

THE DAILY HERALD

Biloxi, Miss., July 29, 1983

Surprisingly, the Nuclear Regulatory commission staff has recommended Congress raise or remove the government ceiling on the nuclear industry's liability in an accident. We favor removing the limit.

The limit came about during the nuclear utility industry's infancy, in a 1957 law known as the Price-Anderson Act. Federal policy at the time was to encourage commercial development of peaceful uses for atomic energy. Congress intended to assure victims of an atomic plant accident would get more benefits than they would under normal tort law. The thinking then was that a catastrophe would bankrupt most utility companies and some victims would be left without reimbursement for their damages.

Times have changed and the law no longer serves the purpose for which it was intended.

Under Price-Anderson, the limit of liability is now $570 million — $160 million from insurance coverage utilities purchase on each reactor they operate and $410 million from a pool obtained by a one-time assessment of $5 million levied against each of the 82 commercial plants now licensed.

Since 1957, the cost of living for Americans has more than tripled. The size of nuclear power plants has incrased. Many of them are located in areas of dense population. No one knows with any degree of certainty the amount of damages a catastrophic nuclear accident would cause, but the controversial Rasmussen Report of 1975 estimated it could cause $14 billion in property damage and several thousand immediate deaths. At today's prices, $570 million wouldn't be anywhere near adequate.

Another relevant development since passage of Price-Anderson is that consumers cannot buy insurance to protect themselves from nuclear risks. Policies exclude losses by nuclear reaction or radioactive contamination. The insurance industry points to the Price-Anderson Act as justification for the exclusions.

Originally, the liability limit was $560, $50 million of which was put up by the federal government. When the number of plants reached 80, the federal subsidy was phased out. Critics of the act contend, with some merit, that there are still two substantial subsidies. The taxpayer or the accident victim, not the responsible party, will pay damages above the limit. An indirect subsidy exists because the act relieves the nuclear industry of the full cost of insuring itself against risks.

Harnessing nuclear power for peaceful uses is a worthy goal, but it ought to be accomplished without requiring American taxpayers to accept involuntarily the risks nuclear plants impose.

The Courier-Journal

Louisville, Ky., August 7, 1983

THE THOUGHT is almost 30 years late, but it's welcome all the same. The staff of the Nuclear Regulatory Commission has suggested in a draft report that Congress eliminate the $570 million limit on the liability of utilities for damage resulting from a nuclear power accident.

The limit is roughly 500 times smaller than possible actual damages from such an accident. A recent government study calculates that the worst possible accident could cause more than $300 *billion* in property damage, as well as 100,000 deaths from radiation-induced cancer and other causes.

As matters stand, victims of such an accident, or even much smaller ones, are supposed to whistle (presumably at Congress) for compensation after the $570 million limit is exhausted. In such a disaster, public funds surely would be voted to cover at least part of the cost. But both victims and taxpayers would be certain losers.

The limit has remained substantially unchanged since congressional passage of the Price-Anderson Act in 1957. It was ridiculously inadequate even before 26 ensuing years of inflation. Just as important, the liability limit, along with other government subsidies, has made nuclear power seem cheaper than it really is.

Providing bona fide liability coverage normally is part of almost any industry's routine expense. An exemption that shifts the risk to someone else is an immensely valuable subsidy. The nuclear industry has relied heavily on this and other subsidies and omissions of real costs to support its argument that nuclear power is competitive with other energy sources.

The nuclear industry, of course, argues that the chances of a catastrophic accident are small. But so are most of the other risks against which prudent people and businesses insure themselves. The odds against a tornado striking a given home in the lifetime of its owner, for example, are tiny. But insurance is considered essential nonetheless. And, if actuaries really consider the risks low, the insurance is correspondingly cheap.

Therefore, if the recommendations in the draft report are adopted, the implication is that utility ratepayers will stand behind liability costs. The report proposes that if a major accident occurred, premiums of all nuclear insurers would rise to $5 million or perhaps $10 million per reactor for as long as it took to pay off the accident. The premiums, if state public service commissions approved, could be charged to the customers.

That still would be paying after the fact, rather than before, as most industries do for their insurance. The industry still could argue that major accidents are impossible and therefore shouldn't be considered when the economics of nuclear power are discussed.

But that argument, loaded onto nuclear power's other troubles, will be hard to sustain. Only four months ago, the Supreme Court ruled that states can block new plants until somebody figures out a safe way to dispose of waste fuel. And a month later, the Nuclear Regulatory Commission fined a New Jersey nuclear power plant $850,000 for the "most significant" safety violations since the 1979 Three Mile Island disaster supposedly shocked the industry into "never again" preparedness programs. Now comes new evidence that nuclear power may not have the cost advantage so often claimed for it.

This conclusion was drawn the other day in *The Wall Street Journal* by two economists expert in the field. They reported that nuclear power, even in New England where coal costs are high and even counting the additional cost of cleaning up flue gases from high-sulfur coal, can be expected to be 70 percent more expensive than coal generation. And that's a worse-case assumption for coal, since there still are ample supplies of *low-sulfur* Eastern and Midwestern coal and since better and maybe cheaper sulfur-removal technologies are widely assumed to be on the horizon.

Assumption of nuclear power's real liability costs, even on an after-the-fact basis, would add one more heavy load to an economic scale that already is crumbling under the weight of cost escalations. John Ahearn, an outgoing Nuclear Regulatory Commission member, commented recently that he couldn't predict whether another new nuclear plant would be ordered in this country during his lifetime. Even in Europe, where energy sources are scarcer than in the U. S., nations are backing away from nuclear power.

The implications for such companies as Public Service Indiana, which already have invested heavily in nuclear plants, are increasingly grim. One result is the sort of campaign being launched by PSI to begin charging customers now for the huge rate increases that will be necessary when its Marble Hill plant goes into service in 1986 or later.

As the *Wall Street Journal* article noted, utilities, in the heady days when they were rushing into nuclear power, thought the adventure risk-free because the customers would have to pay whatever it cost, plus a reasonable profit. But almost every week now, there's new evidence that the bills may be higher than even the long-suffering utility customer can bear.

The Providence Journal

Providence, R.I., January 5, 1984

When the nuclear power industry was young, Congress placed a ceiling of $570 million on the liability of any electric utility for damages that might be caused by an accident at a nuclear plant. But that limit has long since become unrealistic. The Nuclear Regulatory Commission now proposes to replace it with a new insurance system.

The accident at the Three Mile Island plant in Pennsylvania brought the issue into focus. As it turned out, damages to property or persons outside the plant were minor, compared with the catastrophe that could have ensued if the damaged reactor had gone completely out of control, breached the containment building and loosed deadly radiation into the atmosphere.

Now the NRC is urging Congress to set no absolute limit on the right to sue for damages from an accident. Instead, it recommends establishment of an insurance fund, for which all nuclear plants would be assessed. But the assessment would be made only in the event of an accident; and then it would be limited to $10 million per plant each year for each incident. The assessments would continue until all claims had been settled.

Under that arrangement, the insurance pool would eventually be able to collect about $1 billion a year, from the 100 reactors that will be in operation. It could expand as the number of reactors increased.

The old limit contained in the Price-Anderson Act was set 26 years ago. The old limit has been dwarfed by the huge inflationary rise in costs and the value of property and damage awards by courts in other types of cases. Had it been called into play, it might have, in effect, penalized the victims of an accident through its inadequacy.

So the time has come to bring the liability provisions up to date and to place the risk where it belongs, on the owners of the utility plants, for damages from nuclear reactors as from conventional power sources.

The Three Mile Island accident resulted in payment of $28 million in claims to people who were exposed to the small amount of radiation that was released. All concerned can hope that larger payments will never be necessary. But, in view of the incidents that still occur at reactors around the country, and the fines the NRC is imposing for negligence and other infractions of rules, the industry and the government have to be prepared for future accidents. The insurance provisions are only a small part of the structure needed to deal with all contingencies.

Arkansas Gazette.

Little Rock, Ark., January 19, 1984

When Reactor Unit 2 of the Three Mile Island nuclear plant overheated and broke down, people as far away as California or Arkansas took note of the tragedy. Being simple souls who do not understand the art of creative management, most of us assumed General Public Utilities Corporation was responsible for its own error and would have to clean up the mess.

Now it turns out that citizens in the remote regions, or at least those who pay taxes, will have to pick up part of the tab. In the end, they may pay all of it — and then some.

GPU will not send out bills to everyone with the explanation that each person's share of the cleanup cost is $6 or some other amount. The collection system is a little more complicated — and certainly more effective. In the end, the cost could be considerably more than it would have been if the toll took the form of a supplement in the monthly electric bill.

The estimated cost of the cleanup is $1 billion, but GPU does not have that amount of money in its bank account. If the company borrowed the money with the promise that repayments would be made from the cash flow, interest costs would be high and GPU might not be able to show a profit for several years. Consider what that would do to the stockholders, not to mention the electric bills of customers in Pennsylvania and New Jersey.

The logical solution was to spread the cost around in a thin layer.

The federal Energy Department agreed to kick in $159 million toward the cost of dismantling and analyzing the reactor core. No one should need to be reminded that the money will have to come from revenues and borrowing. As things now stand, that will be $3 from current revenues and $1 from credit, since the government borrows one-fourth of its spending.

Pennsylvania and New Jersey will make modest contributions. Then there is the insurance that will pay a small part of the bill.

Edison Electric Institute, a utility trade group, hit on a marvelous idea for bailing out the stockholders of GPU. It proposed to "take up a collection" among other electric utilities with a goal of raising $150 million. The pitch for the donations was that electric utilities already have been hurt by the high cost of borrowing and by a general negative image. So long as the cleanup of the TMI disaster drags on, the industry will encounter public relations problems that could even add a few points to the cost of credit.

The argument was a bit vague, even for sophisticated executives in the utility business. They understand money so EEI began to talk their language.

The first step would be to convince the Internal Revenue Service to grant a "tax deduction" covering any donation made to the cleanup fund. No problem.

The IRS has ruled that any utility's donation toward cleaning up the damage "would appear to be appropriate and helpful in reducing its own financing costs and in improving the public's perception of the industry."

The ruling reportedly was in response to requests for tax deductions from Duke Power Company and Iowa Public Service Company but, for the sake of convenience, the letter was sent to all other electric utilities.

The ruling should improve the prospects for donations, since the money given to GPU now can be counted as a legitimate expense of doing business and subtracted from taxable income.

If an electric utility customer in Arkansas experiences some difficulty in seeing how a donation by his company could "appear to be appropriate and helpful," he simply does not understand intricacies involved in operating a monopoly business in a free enterprise economy.

Even the "favorable" IRS ruling may not be enough to assure collection of the whole bundle. Some utilities already have enough "expense of doing business" to take care of most of their income tax needs. The Economic Recovery Tax Act of 1981 accelerated their depreciation schedules, invited safe-harbor leasing, and made other provisions that helped reduce the tax bill — without, it should be mentioned, curbing the cash flow. Given the provisions of the law and other factors, quite a few utilities can "prove" that their rate of return is inadequate and their tax obligations, while still too high, are modest.

The next act in the little drama — the decisions that would clinch the whole deal and rescue GPU stockholders — will be played before the various state Public Service Commissions. Utilities that are eager to help fellow members of the Fraternity of Benign Monopolies can be expected to go before their appropriate Public Service Commissions and seek permission to pass the cost of the donations along to their customers.

That sort of a double-whammy would be the sweetest deal in the industry since the Dixon-Yates scheme was concocted some 30 years ago.

Even a cynic would not believe, at this point, that any PSC would go for that sort of ripoff, but some companies will give it a fling.

The IRS ruling, already in the books, is as bald as anyone has a right to expect. Now an application for a "pass through" of the donation would confirm that the utility had reached a new plateau of greed.

State Laws on Plant Safety Upheld in Silkwood Case, 5–4

In 1979, a federal jury in Oklahoma awarded $10.5 million in damages to the estate of Karen Silkwood, a former laboratory technician at the Kerr-McGee Corp. Cimarron plant who had become contaminated by plutonium. Kerr-McGee, a miner and processor of uranium, manufactured nuclear reactor fuel rods at the Cimarron plant. The major issue in the trial was the safety of the operation of the plant. Silkwood had been active in labor organizing efforts at the plant and was investigating allegedly unsafe conditions there; Silkwood family lawyers suggested that someone had tried to intimidate her through plutonium exposure. Kerr-McGee argued that Silkwood had tried to prove the Cimarron facility dangerous by deliberately exposing herself to plutonium. The jury's verdict was seen as a rejection of industry safety claims.

The award was overturned in 1981 by the U.S. 10th Circuit Court of Appeals, which accepted Kerr-McGee's argument that the award infringed on the federal government's exclusive authority in the regulation of nuclear safety.

In 1984, however, the Supreme Court ruled that federal laws did not shield the nuclear power industry from state laws protecting the safety of the workers. Justice Byron R. White, writing for the majority in the 5-4 decision, concluded that there was "ample evidence" that Congress intended to allow "persons injured by nuclear accidents . . . to utilize existing state tort [damage] law." White noted that the Price-Anderson Act of 1957, an amendment to the Atomic Energy Act, severely restricted the damages liability of the nuclear power industry in the federal courts. "It is difficult to believe," White wrote, "that Congress would, without comment, remove all [other] means of judicial recourse for those injured by illegal conduct."

In one dissent, Justice Lewis F. Powell Jr. stated that the Silkwood decision contradicted the court's stand in a recent case, Pacific Gas & Electric v. California. That ruling had maintained that states had a role in the regulation of nuclear power, but only on economic matters. (See pp. 118-127.)

BUFFALO EVENING NEWS

Buffalo, N.Y., May 26, 1979

The old saying, "it isn't the money, it's the principle" could apply with a vengeance to the nuclear power industry if the startling legal principle of the celebrated Karen Silkwood case is upheld on appeal — even if the jury's huge money award of $10 million in punitive damages is sharply scaled down.

The case itself had been made into a cause celebre among anti-nuclear protest groups. Miss Silkwood, who had suffered radiation contamination in 1974 as a lab technician for a Kerr-McGee Corp. plutonium plant in Oklahoma, was killed in an auto accident while on her way to meet a reporter to whom she had promised to document some complaints about inadequate plant safety.

The suit against Kerr-McGee was filed a year later by her family, and the case had several sensational features. One was a union claim, which Justice Department investigators rejected, that Miss Silkwood's death was not accidental but that someone had forced her car off the road. Another was a company claim that she had deliberately stolen plutonium from the plant and taken it to her apartment to try to fake a contamination complaint.

But the jury, in the end, not only found that she had not intentionally removed the plutonium from the plant, but held the company grossly negligent in allowing the plutonium to leave its plant. They awarded $500,000 for the care of her three children and $10 million in punitive damages.

What leaves the nuclear industry gasping in dismay more than the size of the verdict itself, however, is the principle enunciated by the federal trial judge, Frank G. Theis, that nuclear companies bear an absolute liability for any radiation contamination. In effect, he instructed the jurors, they need not find any negligence by the company, nor any failure to meet full government safety standards. Rather, the only way they could absolve the company of liability for any radiation harm to Miss Silkwood would be to find that she had deliberately removed the plutonium from the plant, as the company charged.

What this "absolute liability" doctrine would presumably mean, if it prevails as a precedent, is that merely living up to government safety standards would no longer shield any nuclear company from liability for any harm done a litigant by the escape of even low-level radiation from its plant. As one of the Silkwood lawyers picturesquely argued, maintaining a nuclear power plant is like keeping a caged lion and "if the lion gets away," for whatever reason, the company has to pay.

But if that is to be the new rule, then how many prospective nuclear investors — especially in the wake of so many other recent setbacks to public confidence in the future of nuclear power — are going to conclude that the game is no longer worth the risk?

ST. LOUIS POST-DISPATCH

St. Louis, Mo., May 22, 1979

It was the misfortune of the Kerr-McGee Corp. to be on trial for negligently operating its nuclear fuel processing plant in Crescent, Okla., at the same time the nation was transfixed by the events at Three Mile Island. It was Karen Silkwood's misfortune to work for Kerr-McGee at a time when quality control at the plant was lax and safety was worse. She tried to do something about that through her union, only to discover that she had been contaminated with cancer-causing plutonium. She died in a still controversial auto accident as she was on her way to give information about problems at the plant to a reporter and a union official.

Her survivors sued Kerr-McGee for negligence. The other day, six jurors found against the multi-billion dollar energy corporation and awarded Ms. Silkwood's three children $10.5 million in damages. In doing so, the jurors rejected the company's contention that Ms. Silkwood contaminated herself in an effort to embarrass the company, and they set what may be a new standard of responsibility for the nuclear industry. The award will be appealed. If it is sustained, it will mean quite simply that the price of secrecy, lying and negligence about safety in the nuclear industry will have just gone up dramatically.

The Star-Ledger

Newark, N.J., August 25, 1979

The $10.5 million award by a jury in a case involving radiation contamination of a laboratory technician could have major implications for the future of the nuclear power industry.

It raises broader public and legal concern, coming in the wake of the nuclear reactor accident in the Three Mile Island plant in Pennsylvania.

Of far reaching consequence is the precedent introduced in the contamination proceeding by the trial judge which holds the nuclear industry absolutely liable for the escape of low-level radiation.

Under this wide-ranging ruling, negligence does not have to be shown. If low-level radiation gets into the atmosphere and people outside a plant are affected, the responsible company must pay damages even if it complied with government safety standards and was diligent in trying to prevent the escape of radiation.

There is little doubt that the magnitude of the ruling would have an adverse effect on future nuclear development if it is sustained on appeal. The jury award of $10 million in punitive damages (the balance of the judgment was for actual damages) would be the highest if it is allowed to stand.

Of more transcendent ramification than the award itself is the precedent established in the contamination. For the first time, the doctrine of "strict" or "absolute" liability was applied to an incident less than a major nuclear castrophe, such as nuclear core meltdown or an explosion.

In the future — again depending on the outcome of appeals or a request for a new trial — it would mean that the "absolute" doctrine could be invoked for less serious but more frequent occurrences of less nuclear accidents.

Legal precedents and jury awards likely will have a dampening effect on nuclear development. More important, they will evoke an overdue cautionary management prudence on the side of strictest safety standards, moving toward a failsafe capability. The extremely volatile implications of nuclear generating facilities make this a matter of utmost concern for the public safety and welfare.

Oregon Journal

Portland, Ore., May 21, 1979

The Oklahoma City verdict last week which awarded $10.5 million to the family of Karen Silkwood for her plutonium contamination arose from a case which has carried with it a variety of issues.

Miss Silkwood died from an automobile crash in 1974, when she was driving to an appointment with a newspaper reporter who had agreed to look at what she said was evidence she had collected proving negligence and hazards at the Kerr-McGee nuclear facility where she worked.

The $10.5 million in damages her family was awarded stemmed not from the automobile crash, but from the fact Miss Silkwood was suffering from severe radioactive poisoning.

The case from the beginning spawned a variety of issues, including nuclear plant safety, union fights with Kerr-McGee over employee health, the company's political influence, whether government inspections and standards were adequate and the allegation by Miss Silkwood's friends that her car crash was murder.

Horror stories told about the plant where the young woman worked would be devastating if only half of them were true. The plant is now closed, but while it operated, there were widespread reports of deadly plutonium dust contaminating workers and the town where it was located.

Miss Silkwood also claimed when she was building a case against her employees that fuel rods manufactured at the plant and destined for a test facility at Hanford, Wash., were sloppily made and dangerous. Also, several pounds of plutonium were reported unaccounted for.

Kerr-McGee, which will certainly appeal the case, contended Miss Silkwood poisoned herself in attempts to build a case against the firm.

Long before the trial verdict, supporters of Miss Silkwood who also happen to be skeptics about the nuclear industry, made her into what amounts to the country's first nuclear martyr.

Oklahoma authorities should now renew criminal investigations into the case. They have generally treated the Silkwood case with doubt.

No matter what the eventual outcome in the courts, above everything else looms the specter of poor government supervision over a manufacturing process which involves the deadliest compound man has yet discovered.

The Honolulu Advertiser

Honolulu, Ha., May 22, 1979

In ordering the Kerr-McGee Corporation to pay $10,505,000 to the children of the late Karen Silkwood, the federal jury in Oklahoma City warns the nuclear industry to consider itself responsible for the safety of its employees.

The judge specifically instructed the jury that adherence to government standards alone does not necessarily absolve a company when one of its workers becomes contaminated.

The defense will challenge that point as part of its appeal. So the case no doubt will continue to be debated.

Kerr-McGee officials claim that Silkwood contaminated herself with plutonium to embarrasss the company. With its decision, the jury rejected that argument.

What is not established, however, is exactly how Silkwood came to be exposed to a dangerous level of radiation. She was active in the Oil, Chemical and Atomic Workers Union, which was about to begin contract negotiations with Kerr-McGee.

Silkwood was trying to compile evidence showing that safety measures at the plutonium processing plant were lax. She was en route to meet a New York Times reporter presumably to turn over such material when her car somehow veered and struck a culvert, killing her instantly.

The documents she was supposed to have been carrying weren't recovered.

Her death remains a mystery, along with the evidence of plant radioactive dangers she claimed to have. The verdict rendered last week does not touch on these elements.

The case still needs to be pursued vigorously.

The Washington Star

Washington, D.C., May 22, 1979

Under our legal system, solicitude for a single individual can override the multi-billion-dollar interests of a huge and vital industry. That is exactly the point of a federal jury's stunning verdict last Friday in the Karen Silkwood case.

Six federal jurors found that Miss Silkwood's contamination by plutonium shortly before her accidental death in November, 1974, was the result of negligence by her employers: Kerr-McGee's now defunct Cimmaron plant near Crescent, Okla. They awarded her estate and heirs $10 million in punitive damages, in addition to half a million in actual damages. If in the inevitable appeals the award and the judge's instructions to the jury are upheld, the U.S. nuclear industry could soon be crippled by uninsurability.

There remain more puzzles than certainties in the Silkwood case — including the major puzzle of how Miss Silkwood actually came to be contaminated. She had been a highly contentious organizer for the Oil, Chemical and Atomic Workers International Union and had accused the managers of the Cimmaron plant of neglecting the safety of its employees. As evidence of that neglect she offered her own contamination, on three occasions, by trace elements of plutonium. The plant was manufacturing plutonium fuel rods. But no one really knows how a quantity of plutonium, possibly the source of that contamination, came to be in her apartment refrigerator; nor how plutonium traces came to be *added* to urine samples she furnished to AEC lab analysts.

Shortly after the contamination episodes, Miss Silkwood died in the wreckage of her small car on an Oklahoma highway. She was on her way to a meeting with a *New York Times* reporter with, it was said, evidence of company negligence. Investigators from her union advanced the lurid theory that her death was no accident, that she had been driven from the road; but that was not the finding of the Oklahoma state police.

Even after the verdict, conclusive answers to some of the emotional questions her partisans raise about Miss Silkwood's cause, and her death, seem as far away as ever. But the verdict itself, an immediate certainty, creates great legal and financial hazards for the nuclear industry.

A critical role in the case was taken by the trial judge, who accepted a quite sweeping definition of "strict" liability. He instructed the jurors to discount evidence that the Cimmaron plant had "substantially" observed federal safety regulations. The jurors were, he said, free to determine independently the adequacy of those regulations and the company's safety practices. The jurors, said the judge, could find against Miss Silkwood only if the company proved that she herself had removed plutonium from the plant to her apartment, as lawyers for Kerr-McGee tried to prove but could not.

The Silkwood verdict put the broad issues of nuclear liability to six jurors who were not sequestered during the trial and who were expected — and instructed by the judge — to ignore the huge excitement simultaneously generated by the Three Mile Island accident and the Lemmon-Fonda movie, "China Syndrome."

In resting the economic stability of a whole industry on a narrow base of judgment, the Silkwood case fits the mold of other developments in the field of liability. Some recent product liability cases, for instance, seem to impose a virtually unconditional obligation on manufacturers of various products — drugs, cars, even ski equipment — to guarantee the public against personal injury, regardless of who is to blame. It seems to be the public view, these days, that every product, and every manufacturing process, must be risk-free or pay the price.

Congress can limit liability by statute. But the drift of congressional opinion, especially about nuclear safety, seems to be in the contrary direction today.

The judge may very possibly reduce the $10 million punitive damages award. But even if he does so, the Silkwood judgment will cast a long shadow over the economics of nuclear power. For what it says to the nuclear industry is that even diligent adherence to federal regulations, as written, cannot protect a company from negligence claims; and that those claims may be subjectively assessed by six jurors who may not be immune to the influence of general public anxiety over nuclear safety.

The Topeka Daily Capital

Topeka, Kans., May 23, 1979

Without arguing the merits of the lawsuit, the jury's award in the Karen Silkwood trial at Oklahoma City was absurd.

The jury awarded the family a total of $10.5 million, finding the Kerr-McGee Corporation guilty of negligence in the plutonium contamination of the 28-year-old woman who worked at its nuclear fuel plant.

The jury awarded $505,000 in actual damages, which would appear to be generous enough for the family and its attorneys to share. The additional $10 million was awarded for punitive damages.

Loud applause broke out in the courtroom following the verdict. And, according to the Associated Press, Silkwood attorney Gerry Spence blew kisses to the jury. One can easily guess why.

The $10.5 million award was excessive, exceeds reason and sets a dangerous example for future cases that undoubtedly will spring up.

Kerr-McGee and others in the nuclear industry have already learned — as we all have — that the best possible safeguards must be taken to protect employees and the public from the dangers of nuclear contamination. We have no doubt that the public outcry for safety in the nuclear energy field, heightened by the Three-Mile Island saga, has already spurred the industry into retooling of its systems and concern for safety. To do otherwise would be irresponsible to the industry's obligation to the public and to each company's stockholders.

Cases of extreme negligence, such as apparently was proved in the Silkwood case, should be brought before the courts on behalf of those who suffered.

But $10 million punitive awards can only tend to seriously damage, inhibit and perhaps even destroy, companies or the entire industry.

The Virginian-Pilot

Norfolk, Va., May 22, 1979

It is diverting if hardly useful to speculate on the outcome of the Karen Silkwood case had the conclusiom arrived before the Three Mile Island nuclear accident.

The near-meltdown in Pennsylvania, occurring in the early stages of the Oklahoma civil trial, may have influenced—if subtly and unintendedly—the jury's gaudy verdict. The $10 million in punitive damages loaded onto $505,000 in actual damages could be read as a broadside warning to the nuclear industry, which was slipshod in both events.

The Silkwood verdict was largely foretold before the federal district court jury in Oklahoma City reported Friday. The trial judge had ruled earlier that a nuclear industry is absolutely liable for the escape of low-level radiation, agreeing with lawyers for Miss Silkwood's estate that the employer was responsible for harm even if a plant met federal safety standards and no negligence was involved.

There is no certainty that the size of the award—or the verdict itself—will survive appeals by the defendants, the Kerr-McGee energy conglomerate and its nuclear subsidiary. But while it stands it poses problems of insurability, safety and security enforcement, and criticism for a sensitive and troubled industry.

Miss Silkwood, a 28-year-old divorcee with three children, was a union activist who worked in Kerr-McGee's Cimarron nuclear plant north of Oklahoma City. She died in the crash of her small car in November 1974. At the time she was on the way to meet a union representative and a newsman, ostensibly to show them documentation of inadequate safety at the plant where plutonium fuel rods were made. Plutonium contamination later was detected on her and in her apartment refrigerator.

How she was contaminated was unexplained. The jury in any event seemed less concerned with how than with the evidence that radiation can go public—through misadventure, through accidental transfer by workers, or even through malicious intent, which was a possibility in the Silkwood case. When the Cimarron plant closed in December 1975 its inventory was shy 40 pounds of plutonium.

Whatever her point was, Miss Silkwood didn't live to make it. She was, however, taken up as a cause—even a martyr—not only by her lawyers but by women's rights and antinuclear activists. There is a broader constituency as well for keeping the nuclear industry's jinn stoutly bottled and corked, a viewpoint evidently represented by the jury. The lesson bears learning.

TULSA WORLD

Tulsa, Okla., May 19, 1979

A SIX-MEMBER Federal Court jury in Oklahoma City Friday delivered a judgment which, if upheld, could be as big a setback for the nuclear power industry (and industry in general, for that matter) as the Three Mile Island reactor accident.

The jury found Kerr-McGee Nuclear Corp. negligent in a nuclear contamination incident involving the late Karen Silkwood, a Kerr-McGee employee, antinuclear activist and union organizer. The jury awarded the Silkwood estate $10 million in punitive damages, $500,000 in actual damages and $5,000 for loss of property which was removed after her apartment was found to be contaminated with plutonium.

Within minutes after the jury's judgment was announced, the plaintiff's lawyer was talking about the possibility of "class action" suits on behalf of other nuclear workers who may allege exposure to radiation.

The Silkwood case will be appealed, of course. The defendant's lawyers will almost certainly challenge rulings by the trial Judge which barred jurors from hearing evidence about Ms. Silkwood's personal behavior. The defense contended Ms. Silkwood contaminated herself while trying to "spike" a urine sample with plutonium in an effort to embarrass the company.

But even if it should eventually be reversed, the Oklahoma City judgment stands as another stunning public relations blow to the nuclear power industry. And it gives a new emotional boost to the anti-nuclear, anti-technology, anti-industry movement.

The jury's decision is not the final answer to all questions about legal liability in radiation exposure cases. It may raise more questions for lawyers than it answers. And it may take years before it goes through the full appeal process. But for the moment at least, it is bad news not just for Kerr-McGee, but for the whole U.S. energy industry.

Tulsa, Okla., January 13, 1984

BY NOW, any mention of Karen Silkwood's death raises all the misnformation, prejudice and hysteria that has plagued the case for years.

Tuesday, the Silkwood case was back on the front page when the U.S. Supreme Court reinstated the $10 million punitive damage award Ms. Silkwood's estate won from Kerr-McGee Corp. The estate claimed the company was liable for Ms. Silkwood's plutomium contamination. The tribunal reversed a circuit court ruling on narrow legal grounds of jurisdiction, but told the lower court it could reconsider the case on grounds of evidence and the fairness of the award.

Conspiracy buffs who see some vague but terrible plot in Ms. Silkwood's dispute with Kerr-McGee and her subsequent death will read the Court's decision as an affirmation of that belief. In fact, the decison had nothing to do with that claim. Evidence, carefully examined by the Oklahoma Highway Patrol, the FBI and a Congressional committee, indicates she died in an ordinary automobile accident. But nothing will change the view of those determined to make the Silkwood story a case history of corporate evil. Never mind the facts.

In reinstating the $10 million award by a 5-4 vote, the Court held only that the Oklahoma tort law, under which the Silkwood estate sued, was not "pre-empted" by federal laws regulating the nuclear industry.

Strong arguments can be made on either side of this complex legal issue, but that issue has nothing to do with whether Ms. Silkwood was contaminated through negligence by Kerr-McGee or contaminated herself to discredit the company. Nor is it related to the cause of her death.

The heart of Tuesday's decision is that the industry is now vulnerable to endless litigation and huge damage awards in state courts.

The threat casts another ominous shadow over the future of atomic power as a source of energy for America.

Los Angeles Times

Los Angeles, Calif., January 15, 1984

The Supreme Court decision to reinstate a negligence suit related to the nuclear industry is reassuring at a moment of growing controversy over the use of nuclear energy. In effect, the court has ruled that the industry is not beyond the reach of damage suits in state courts although an earlier decision had reaffirmed the exclusive claim of the federal government to regulate the use of nuclear energy. Both rulings are correct.

What matters here is that people now know they can take action against nuclear companies if people get hurt. The high court has reinstated a damage suit brought by the heirs of Karen Silkwood, who, according to the findings of a jury in Oklahoma, suffered radioactive contamination in 1974 because of the negligence of the Kerr-McGee Corp. Silkwood died nine days after the contamination under mysterious circumstances in an automobile wreck while on her way to meet a New York Times reporter. That tragic event has made it impossible to clarify the details of the case although it has provided the basis for a popular motion picture named for the victim.

The $10-million award of the state court was thrown out by the appeals court on grounds that it represented an intrusion by a state into the exclusive domain of the federal government. The Reagan Administration supported the appeal of the corporation from the judgment. But, as Justice Byron R. White wrote in his majority opinion, "there is ample evidence that Congress had no intention of forbidding the states from providing such remedies."

Now the case goes back to the 10th Circuit Court for review, including a careful look at the imposition of punitive damages and the amount set by the jury.

The Philadelphia Inquirer

Philadelphia, Pa., January 23, 1984

Keith Prescott drilled tunnels in Nevada and was directed by his employer to re-enter a mine to retrieve equipment. As a result, he believes, he contracted bone cancer. He sued.

If Mr. Prescott had worked for a road builder or a coal company, he also could have sought punitive damages, arguing that the order to re-enter the tunnel constituted negligent conduct and should be penalized so that his employer would think twice about doing it again.

But Mr. Prescott worked at the Nevada Test Site drilling tunnels for atomic weapons tests. The tunnel was filled with radioactive iodine following a bomb detonation. Until this month, he was denied the opportunity to sue for punitive damages because his employer was the private contractor who manages the test site for the U.S. government. The contractor only had to argue that he was complying with existing federal standards and not liable.

About 500,000 people work at U.S. nuclear facilities, civilian and military. Hundreds of thousands of other Americans live near these installations, or, like Keith Prescott, once worked there.

On Jan. 11, in a 5-4 decision, the U.S. Supreme Court granted them the same rights to seek punitive damages from private nuclear firms that any other person has. It ended one of the many double standards of liability under which the nuclear industry has operated since its inception.

The court ruled that compliance with federal standards no longer was an acceptable defense to bar punitive claims. A jury, the high court ruled, could consider whether those standards were sufficient and whether they were being enforced.

The court rejected the industry's argument that allowing a jury to award punitive damages, in effect, was allowing it indirectly to set regulations. That authority, the industry maintained, was reserved exclusively for the federal government under the Atomic Energy Act.

Such exclusivity was never the intention of the Congress, the majority ruled in upholding a $10.5 million punitive damage award to the family of nuclear-plant employee Karen Silkwood. Citing an earlier opinion, the court noted that a primary purpose of the act was, and continues to be, the promotion of nuclear power. However, "the promotion of nuclear power is not to be accomplished at all costs." One of those costs is safety.

That is the significance of the Silkwood decision. It affirms the rights of states to become active in certain matters of nuclear operations, such as plant siting and waste disposal practices. Above all, it clearly says to the nuclear industry: You shall be held to the same standards of safety and care as other industries. If you fail, you can be held financially accountable.

It is interesting to note the position of the Atomic Industrial Forum, which argued to the court that the "specter" of lawsuits seeking millions of dollars in damages "would make private industry reluctant to remain involved in nuclear energy projects."

The nuclear industry has evolved under a large, protective umbrella erected by the federal government. It has been given special protections and concessions. Similar limits have not been offered other industries, in large measure because few of them face the risk of such horrendous liabilities. That in and of itself should be reason enough to hold nuclear facilities to the strictest safety standards imaginable. But the existing regulatory scheme has failed repeatedly.

The Supreme Court took note of testimony that showed that Ms. Silkwood's employer, the Kerr-McGee Corp., had been cited for 75 minor violations at its Oklahoma plant but that none had ever resulted in a fine. Various federal investigations before and after the Silkwood incident did not turn up evidence of wrong-doing by the company. Perhaps there was none. Or perhaps those enforcing the regulations were complacent.

A jury, hearing much of the same evidence, cited Kerr-McGee for $10.5 million in punitive damages in addition to compensatory damages awarded Ms. Silkwood's survivors. That discrepancy cannot be dismissed, even considering the vagaries of jury verdicts.

The nuclear industry and federal energy officials long have argued that states lack the expertise to regulate atomic technology. That responsibility properly belongs to federal experts, they claim.

The federal government must never divest itself of final decisions concerning things nuclear, for there are too many national security and welfare issues involved. However, increased state involvement in some areas of oversight is vital.

Those who work in nuclear facilities and those who live near them will be the immediate beneficiaries, but in the long term the decision will serve the nation. No industry — especially one with such enormous safety burdens — should be allowed to function without the strictest accountability.

The Wichita Eagle-Beacon

Wichita, Kans., January 13, 1984

The U.S. Supreme Court's decision in the Karen Silkwood case opens an important avenue of redress for atomic industry workers who are contaminated or otherwise harmed while on the job. The decision, which essentially upholds an original ruling by U.S. District Judge Frank Theis of Wichita that Ms. Silkwood's heirs could collect for punitive damages, should encourage the atomic industry to place high priority on its workers' safety.

By reinstating Wednesday the $10 million punitive damages award won by the children of Karen Silkwood, the court effectively has ended the federal government's heretofore exclusive regulation of hazardous radiation in atomic plants and factories. Atomic workers who believe their employers' negligence responsible for such injuries as radiation poisoning now are free to seek punitive damages — which in effect are fines — in addition to actual damages in lawsuits brought in local courts. The worries about workers' safety expressed by Ms. Silkwood and her fellow employees at a Kerr-McGee Corp. plutonium processing plant should be alleviated as a result.

To encourage rapid development of the atomic industry, Congress in the 1950s gave the federal government, through the Nuclear Regulatory Commission, the exclusive right to monitor atomic plant safety and to punish errant employers. But, as Ms. Silkwood's contamination with plutonium prior to her death in an auto accident in 1974 illustrates, such regulation tended to favor employers' economic interests over the health of workers.

To the degree that no longer will be the case, will the benefits of the high court's decision be measured by the public.

Protests Rife After TMI, But Voters Reject Closing Plants

Following the Three Mile Island accident in March 1979, there were many headline-making antinuclear protests and demonstrations. Public opinion polls completed in April 1979 showed an increased level of fear among Americans about nuclear power plants. A New York Times/CBS poll showed only 46% of Americans favored further development of nuclear power, compared with 69% who were asked the same question in a July 1977 poll. And a Gallup survey showed an increase in the percentage of Americans who favored a cutback in nuclear plant operations until stricter safety regulations were put into effect, to 60% from 40% in a similar 1976 poll. Only 25% of those polled by Gallup, however, were in favor of shutting down all nuclear plants, and this seemed to reflect a pattern that emerged nationwide. While antinuclear demonstrations continued across the country, voters in many districts rejected measures to close functioning nuclear plants. In the first such plebiscite after the 1979 accident, Maine voters defeated a referendum proposal in September, 1980 to shut down the state's only nuclear power plant, the Maine Yankee facility near Wiscasset. The ballot issue also proposed a permanent ban on the building of other nuclear plants in the state. The Maine Yankee facility supplied about one-third of the state's electricity; opponents of the referendum argued that consumers would pay an increase of $140 million a year in electric rates if the plant was shut down. During the November, 1980 elections, Missouri and South Dakota voters rejected restrictions on nuclear power plants and Montanans turned down a measure to severely limit uranium mining. There appeared to be a greater level of public apprehension, however, concerning the radioactive waste generated by power plants. In Washington, voters approved strong limits on nuclear waste storage, and in Oregon, voters barred construction of nuclear plants without federally licensed dump sites for radioactive waste. (See pp. 70-77, 118-127.)

Public opposition to nuclear energy has remained strong in the 1980's, in part because of continuing doubts about the standards of safety regulation at nuclear plants. A Business Week-Harris Poll taken in 1983 showed that 51% of those polled opposed the construction of new nuclear plants. There is no question that this opposition has contributed to the problems of an industry still feeling the effects of the economic and political fallout from Three Mile Island.

THE CHRISTIAN SCIENCE MONITOR

Boston, Mass., September 25, 1979

The man who coordinated America's largest anti-nuclear rally so far may not exactly be a detached observer. But, with some 200,000 protesters gathered in New York on Sunday, Donald Ross indicated the growth of nuclear power as a national political issue: "No politician who favors nuclear power will be elected president in 1980." More realistically, no candidate is likely to escape having to take some position on the matter. They all ought to be doing a lot of homework.

Here are some of the questions that candidates will be or should be asked:

So you oppose nuclear energy. Does this mean you want to shut down the 71 plants that had operating licenses as of last year? To stop construction of the 90 more plants underway or permitted to go ahead? To deny permits to the 40 more that are on order but do not have the permits?

So you favor nuclear energy. Does this mean you support going ahead with all of the above? Proceeding with an indeterminate future number? Developing the breeder reactor with its new promises and risks?

If you oppose expansion of nuclear energy, what alternatives do you propose to save or supply the electricity it would have provided? Do these alternatives have safety or environmental problems? If so, how would you meet them?

If you favor expansion of nuclear energy, how would you approach the problem of accumulating hazardous nuclear waste? Would you require states which have legislated against storing it to accept it? Would you make agreements with the Indians whose lands have some of the more likely places for repositories? Would you handle the interim problem by promoting the idea of transporting spent fuel to central temporary "away from the reactor" storage pools known as AFRs?

The list could go on, including the basic questions of moral values raised in discussions of dealing with a source of both benefits and dangers in relation to the present world and future generations.

Not only New York saw a Sunday "no nukes" rally. There were others around the country. There will be more. Six months after the Three Mile Island accident in Pennsylvania there remain heightened concerns about nuclear energy. A number of investigations are seeking to find the kind of facts that have been so often concealed or distorted during the nuclear years. We await particularly the report of the President's own commission next month. It will be important for the public to feel assured that it is candid and without fear or favor.

Meanwhile, a kind of de facto moratorium continues on buying more nuclear generating machinery. Critics and proponents are at something of a stalemate. Previous projections of numbers of future plants have been reduced. We don't like the clouds of rhetoric and marijuana smoke over some of the rallies. But the new nuclear activism cannot be ignored. It could help make the country and its politicians face the crucial issue of how to live with nuclear energy or without it.

The Washington Post

Washington, D.C., September 28, 1979

SOME 200,000 PEOPLE gathered last weekend for the largest anti-nuclear rally ever held. The speakers called for a variety of different measures, but most demanded a ban on further construction of nuclear power plants and the closing of plants already in operation. Nuclear power became a symbol, or symptom, of various ills ranging from an overcentralized economy to a failure to pursue genuine arms control.

Unhappily, such rhetoric merely tends to confirm the views of the most fervent nuclear advocates that nothing is wrong with nuclear power except, in the words of one of them, "the critics, the courts, the bureaucracy, the press and the politicians." Neither view is right, and neither helps resolve the real and pressing problems associated with nuclear power.

Nuclear energy now provides about 4 percent of the nation's total energy, and 12 percent of its electricity. That doesn't sound like very much, but closing the country's 71 existing nuclear plants—which represent investments in the tens of billions of dollars—would have severe economic repercussions. In some regions the results would be devastating. Therefore, closing plants that are operating or partially constructed is both unlikely and unreasonable without clear evidence that they are unsafe and that their deficiencies cannot be corrected. The various commissions and investigations analyzing every aspect of the Three Mile Island accident may come up with such evidence; but from what has come out so far, that seems remote.

The longer range future of nuclear power is a different matter. The United States has had a de facto moratorium on new nuclear plants for several years—utilities have simply not ordered any new reactors—and this will continue at least until all the questions raised by Three Mile Island have been resolved. Nuclear power will therefore *not* experience the tremendous growth its advocates have predicted and its critics have feared. Nor will it substantially lessen the country's dependence on imported oil. Nevertheless, it can make an important—perhaps vital—contribution to a balanced energy budget during the hoped-for transition to reliance on renewable energy resources.

The real issues that have to be faced, if nuclear power is to play this modest but significant role, are reactor safety and the management of nuclear waste. It is here that nuclear skeptics could most usefully focus their attention. In particular, the government's continuing failure to come up with a waste-management plan that is technically sound and politically acceptable casts a long shadow over nuclear power's future. Nearly two years ago the president launched a major interagency study of this problem. The group's draft report, which was circulated for public comment months ago, amply testified to the magnitude of both the technical and political uncertainties. Though generally well received, the report disappeared back into the maw of the bureaucracy and hasn't been heard of again. Whatever the outcome of that particular effort, the country needs a widely acceptable nuclear waste policy, and soon. If one cannot be developed, the future of nuclear power will take care of itself.

The Hartford Courant

Hartford, Conn., October 25, 1979

Officials at Northeast Utilities are rightfully concerned that the accident at Three Mile Island last March has eroded public confidence in nuclear power.

But assurances given by the company this week that a similar accident could not occur in Connecticut will not be enough. Every company engaged in nuclear power has claimed that its facilities are virtually foolproof.

The exact sequence of events and errors of Three Mile Island are unlikely to be repeated in Connecticut, or anywhere else. Three Mile Island was a unique event. That's what most of the experts say.

If, however, the accident at the plant near Harrisburg, Pa., was unique, there would have been no need for massive investment in safety. Mistakes made at Three Mile Island have spurred the Nuclear Regulatory Commission to demand new safety precautions at plants. Studies by reactor manufactuers and utilities have also resulted in recommendations for improvements.

Northeast is complying with the regulatory commission's suggestions. Safety-related changes will cost more than $40 million in the next five years, company officials say. The plans are impressive. However, as long as errors are possible, which they always will be, people will remain uneasy.

The accident at Three Mile Island resulted from a combination of equipment failures, design deficiencies, inadequate operator training and human error. The emphasis has been on the mistakes in judgment, what officials at Northeast call the "human factor." If anything, Three Mile Island proved that the technology, the reactor's safety systems, were themselves "safer than even we believed," the Northeast engineers say.

But the technology cannot be separated from its human operators. Dangerous conditions emerged at Three Mile Island when operators did not realize that a valve had been stuck open for 2½ hours. They also apparently relied on one safety gauge, a water level indicator, instead of analyzing data from all indicators.

Safety in nuclear plants ultimately depends on decisions made by people, not machines. It depends on the operators, and on top company executives and on federal and state regulators when a crisis arises. Their performance must be evaluated, not by their ability to handle the routine, but by their reaction to freak occurrences.

The operators at Three Mile Island were said to be better trained than many. Yet, the President's Commission on the Accident at Three Mile Island has given strong hints that its final report will show the operators were not prepared to handle the job.

There may never be total public confidence in nuclear power after Three Mile Island. That does not mean the regulators and the utilities should not try their best.

Above all, companies and regulators should report to the people immediately and forthrightly when mistakes and breakdowns occur.

But to ask the public to put blind faith in technology is to ask for the impossible.

The News Journal

Wilmington, Del., July 7, 1979

Figuring out a recent Washington Post Service poll about how Americans look at nuclear energy is complicated.

A spokesman for the nuclear industry suggested that the opposing respondents were, to a large extent, opposed to the mere presence of what they regard as unpleasant things in their communities. That would be the case, he claimed, whether the question was offshore oil rigs visible from the beaches, or an airport in the neighborhood, or whatever. "It's just anti-facility" feeling, he said, "not anti-nuclear." In other words, he was claiming, it was the traditional "not in my neighborhood" feeling.

To which a member of the Union of Concerned Scientists, which is fervently critical of nuclear energy, replied, in effect, "Oh, yeah?" If the feeling is merely anti-facility, he argued, why has opposition increased so markedly since the accident at the Three Mile Island plant in Pennsylvania?

That's a point well made in reply to the anti-facility argument, but it is not what seems so interesting about the poll results.

The poll showed that, indeed, most people do not want nuclear plants built in their communities.

On the whole, respondents broke down this way: 26 percent were actually opposed to nuclear energy, 36 percent supported it and the largest of the three groups was of those who had not yet made up their minds — 38 percent.

People who are now served, or people who at least *think that they are being served*, by nuclear power, are generally more supportive of it. That may be traceable to several factors. They may feel that they already are enjoying the benefits of this form of energy. They may feel that as long as the plant is already present in the community, there is not as much need to be vehement in opposition to it. They may also perceive some benefits to their local communities from the plants — such as the plant's tax payments.

It is that huge percentage of undecided people, however, who are most interesting at this point.

This indecision, while interesting, may be fairly easy to explain.

How can the layman in the ranks of the public respond in such a poll when the scientists, the technicians, the experts are divided?

Debate in the National Academy of Science, reports a New York Times Service account of the studies of radiation effects, is on how these effects really operate. For instance, there is the question of whether some "threshold" really exists, some level at which radiation does no harm at all.

An academy committee report assumes, at this point, that any *additional* exposure, however small, leads to some increase, however small, in cancer risk. On this basis, it suggests that in addition to 325,000 cancer deaths that normally could be expected in the Three Mile area's population, one fatality and one nonfatal case might result from the accident. Or, if the effect is weaker than this committee supposes, nobody at all might contract an additional case of cancer as a result of the accident.

One fatality and one nonfatal case?

The figures seem small, although statistically they might be important and they certainly would be important to the persons affected. Of course, whether or not nuclear energy had ever been dreamed of, those two persons would not have been immune to other forms of disease or other causes of death. They'd likely suffer from *something* and certainly would die from *something* such as an automobile accident or the effects of pollution caused by coal-burning power plants.

The net effect, however, of these figures is simply to make the public uncertain about how it feels in relation to nuclear energy.

When the scientific and technical community gets things sorted out better, perhaps more of the people polled will be able to make their minds up also.

The Chattanooga Times

Chattanooga, Tenn., September 26, 1979

Public policymakers had best pay close attention to the staggering response to New York City's anti-nuclear power rally on Sunday. An estimated 200,000 persons gathered in a peaceful, almost festive, protest in response more to fears of a theoretical calamity than demonstrable affronts to the individual's rights or liberties.

Their real understanding of the complex process by which controlled reaction of nuclear fuels is directed into the production of electric energy may be minimal, but their lack of knowledge neither whets their curiosity nor lessens their fears; it merely strengthens their opposition to a program on the eve of its third decade without attributable loss of life.

To them, their concerns are real, shared by growing numbers of Americans. The New York turnout was only one of many similar gatherings across the country on that same day.

No one denies that problems still exist for nuclear power. The Department of Energy in its report last May to the Congress said the aim of the administration was to ensure that the industry remained "as a significant source of energy for the rest of this century . . . (and) as a potential backup technology for the next century."

But, it went on, this kind of development assumed the resolution of several issues, among them reactor safety and the management of nuclear wastes. The government is intensifying its work for solutions in these areas, the department said.

Meanwhile, competent scientists who have the knowledge to back their conclusions assure us that the risks involved in production of energy from nuclear fuels are being held to an acceptable level. We see no reason to discredit their judgment.

Participants in the "anti" rallies, however, would much prefer to listen to, and certainly are more deeply influenced by, the sponsors who play upon their emotions and arouse their fears of what conceivably could happen in the future.

The difficulty for the policymakers lies in the simple truth that emotional responses are easier to come by than rational reactions. Out of such rallies as that which occurred in New York, there come more fears by the uninformed, the partly informed and the misinformed, than there are challenges for proof or demands for alternatives.

The result can be pressures for referenda in which numbers alone count, and by which defensible courses of public action are defeated. Therein lies a cause for alarm. Can anyone imagine 200,000 at a rally supporting a nuclear plant?

Newsday

Long Island, N.Y., October 9, 1979

Demonstrators against nuclear power have every reason to be peaceful. Reducing the risk of nuclear proliferation is, after all, the ultimate aim of their protest, and protection of health and public safety are worthy goals along the way.

So it's disturbing to see the switch in tactics at the Seabrook nuclear plant in New Hampshire over the weekend. Armed with wire cutters and grappling hooks, protesters attempted to get through the fence and occupy the plant. They were rebuffed by club-wielding, Mace-spraying police.

The demonstrators' tactics were endorsed by the SHAD alliance, which sponsored an anti-nuclear rally that attracted some 1,500 opponents of the Shoreham nuclear plant to a rally Sunday in Hauppauge. Members said they'd support destruction of property and forceful occupation of the Shoreham site if other attempts to keep the plant from opening fail.

That might well be one way to get more media coverage, but it strikes us as counterproductive. The anti-nuclear movement comprises a large number of diverse organizations and, we suspect, an even larger number of unorganized sympathizers who are in no mood to be radicalized.

The whole nation is in fact embroiled in a debate over the future of nuclear power. The presidential commission investigating the accident at Three Mile Island has yet to issue its report, and Congress is in the process of assessing the performance of the Nuclear Regulatory Commission. It's too early to tell what conclusions may be reached.

Meanwhile, it's in the opponents' interest to keep their base of support as broad as possible. The best way to accomplish that is by education and peaceful protest, not violent confrontation.

Rockford Register Star

Rockford, Ill., October 23, 1979

What passes for "dialogue" regarding nuclear power plants is a set scenario hurled back and forth between two intransigent sides — the owners who intend to prevail and the protesters who doubt and defy this form of energy.

Recently, at Seabrook, N.H., the struggle turned ugly as police wielded night sticks and riot gas to turn back 1,500 demonstrators intent on occupying the Seabrook nuclear plant site. Thus a new issue, that of law and order, begins to assert itself.

What is being ignored are the very real concerns of Americans who read horror stories about radioactive pollutants seeping into the atmosphere, of illness and genetic disorders afflicting those living near radioactive plants, of contaminants so lethal that waste products cannot be safely stored during our lifetime.

Surely, more is at stake here than merely a dispute over rates and the right of free enterprise to pursue its profit motives.

At stake here are issues of immediate concern for every citizen.

It is not enough that someone, anyone, says there is no alternative to nuclear power. It is not enough that someone, anyone, assures that there is no danger. It is not enough that someone, anyone, makes the decisions for all of us.

The American public deserves proof — complete proof — that nuclear power either is or is not the only solution to our energy problems. The public deserves proof — complete proof — that nuclear plants are and will continue to be safe for mankind (or that they will not). The public deserves proof — complete proof — that nuclear wastes can or cannot be disposed of safely to protect our children and grandchildren.

And, on the basis of hard facts, the American public deserves the right to makes its own majority decision on the future of nuclear energy and the role we want it to play in our lives.

The producers of nuclear power are the beneficiaries of a discovery which the American people paid for with their lives and dollars during World War II.

It is the nature of this trust that it be publicly monitored and publicly restrained if peril looms.

It is not sufficient that the utility firms already in existence make the decisions for all of us.

It is not sufficient that utility owners argue they have yet to cause a mass loss of life.

The risk is sufficient that utilities, through their lobbies, already have achieved dollar limits on their liabilities if and when those deaths occur. This recognizes that the very nature of nuclear power is violence through controlled means.

Consumers asked to live with this controlled violence should have a choice. And they should know what is the bottom line of their preference, its cost not only in financial terms but in terms of community safety.

That is the point which keeps getting lost in confrontations.

It is a point that reasserts itself with each new discovery of hazard in the nuclear production of energy.

It can hardly be resolved within the industry, with a major stake in attempting to save its own investment in energy distributions systems.

It can hardly be resolved at the state level, where no state is above reaping the benefits of nuclear power while passing its risks on to its neighbors.

It rests with the federal government to provide the hard facts — the honest answers. Then a way must be found to allow the public to decide the role it wants nuclear power to play in our future.

"OK, who turned off the power. . .?"

St. Louis Globe-Democrat

St. Louis, Mo., September 27, 1980

Watching protests and demonstrations organized to entice TV cameras, an unsuspecting viewer might easily get the mistaken impression that the entire country is full of anti-nukes.

It just isn't so. The resounding vote against the proposed closing of Maine's only nuclear power plant proved a staged demonstration is not a good setting to get an accurate reading of the public's pulse.

The unprecedented referendum (it was the first balloting in the U.S. aimed at closing an existing nuclear plant) brought out 56 percent of the state's voters. The turnout exceeded Maine's general election in 1978 and was below only three presidential elections.

There was ample reason for the outpouring at the polls: The Maine Yankee plant at Wiscasset produces one-third of the state's electricity requirement. Furthermore, there is no alternative that could replace this vast amount of power at the click of a switch as anti-nukes like to believe. Most of the state is heated with expensive-imported oil that could be cut off sharply by a foreign embargo.

Nuclear power also **has become an issue in Missouri. An attempt to prevent the operation of the Union Electric nuclear plant No. 1, under construction in Callaway County, has been ordered off the Nov. 4 ballot by Cole County Circuit Judge Byron L. Kinder.**

The court ruled the proposal's title on the petitions was not explained clearly as required by law. The proposition could be restored to the ballot by the Missouri Supreme Court.

Should the nuclear foes ultimately succeed in placing the proposition on the ballot, Missourians should follow the lead of Maine voters and throng the polls to voice their support for nuclear power. Plant No. 1 will play a major role in the state's economy. It is expected to produce 12 percent of the state's electrical needs when it begins operation in 1982.

The anti-nukes attempted to pass the proposition off on the ballot as a step "to regulate the operation of nuclear power generating facilities," although it actually seeks to prohibit such generation. Missourians should read the fine print and not allow themselves to be fooled by such tactics.

The State

Columbia, S.C., September 27, 1980

ONCE AGAIN common sense has prevailed in the often emotional and distorted issue of nuclear energy. By a 3-2 vote, the people of Maine killed a proposal to shut down the state's only nuclear power plant.

That the question ever got on the ballot in the first place is testimony to the militancy of the anti-nuke organizations. They will not relent in spite of this defeat, they say.

Isn't it unfair that a loud minority can repeatedly force an issue which the majority has soundly determined? At almost every turn where the nuclear energy question has been put to public vote, the anti-nukes have lost.

It is easier to carry on causes which are essentially "against" developing nuclear power than it is to whip up enthusiasm to defend the status quo. But wouldn't it be interesting if some day there was finally a vote somewhere on the issue proposed by the proponents, like, resolved: Nuclear power is here to stay?

The controversy won't be settled in that fashion, of course, and maybe never will be unless the anti-nukes wear themselves out on the majority — or find another cause.

THE DENVER POST

Denver, Colo., October 2, 1980

THE WISDOM of holding an anti-nuclear power referendum in Maine has now been tested. It seems to have been the wrong issue at the wrong place at the wrong time.

By a vote of 230,000 to 159,000, Maine voters last week rejected a proposal to shut down the big Maine Yankee plant and ban all nuclear stations in the future. The anti-nuclear lobby evidently ran into credibility problems for several reasons:

- Maine Yankee has been reliable. It has functioned well in its eight-year life, with only one minor water spill. Two years ago it set a U.S. record by running continuously for 14 months, producing 7 billion kilowatt-hours of electricity.
- It has good economics. With a capacity of 841 megawatts, it has three times the current rating of Colorado Public Service Company's St. Vrain nuclear station. It supplies one-third of Maine's power, and its closure would deal customers an immediate setback of $140 million.
- The timing was bad in that the Arab oil cartel has just raised its average prices above $36 a barrel. Maine consumers are sensitive to this fact because the chief alternative to nuclear power is extremely costly oil from the Mideast. The other likely fuel option is coal which, in New England, raises fears of acid rain.

So Maine was not a good place to offer so stark a choice on nuclear power. Sweden, which rejected a similar referendum last spring, did the job differently. Swedes were offered three different scenarios and chose to keep present nuclear power plants — but not start new ones. Maine residents had no choice. Their anti-nuclear lobbyists insisted on a "winner take all" decision — and lost.

In the long run, however, Maine residents may have chosen the best route anyway. The more experience the world gets with nuclear power, the more comfortable we become with it. Carefully handled, it produces heat at a competitive cost and with little damage to air and water. It requires complicated and lengthy waste-storage procedures. But weighed in the balance, these may be acceptable trade-offs in securing the power we need. What happened in Maine would indicate support for this view in a traditionally conservative state.

The Oregonian

Portland, Ore., September 25, 1980

The Tuesday message of Maine voters — to keep their Maine Yankee nuclear power plant in operation — was a timely and intelligent response to an emotional issue. Moreover, the Maine mandate to keep the nuclear power plant on line is good advice for citizens everywhere.

It is doubtful that many of the 231,000 Maine voters who opposed the nuclear ban did so solely because of the current Iraqi-Iranian fighting that threatens world oil supply. Yet, that issue — the struggle between two oil-producing nations in an already politically unstable Middle East — is the best reason for protecting, as well as expanding, this nation's power supply resources, including the nuclear option.

Political analysts in Maine explained Tuesday's surprisingly large voter turnout for a single-issue referendum and the pro-nuclear victory as a "pocketbook response" of citizens to a campaign that emphasized consumer costs in replacing the power produced by Maine Yankee. While personal economics certainly should be an issue in the nuclear debate, the overriding statement from Maine to the rest of the nation should be: "Keep the nuclear power option open as one way to combat the nation's dependency on foreign oil."

The referendum question sought to close Maine Yankee and ban any future nuclear generation of electric power in Maine. Had the measure been approved (it lost by a 3-to-2 ratio), Maine's future power options would have included oil-fired turbines or thermal power plants using oil or gas. Similar initiative petitions, attempting to shut down the Trojan nuclear power plant, have been circulated in Oregon but have not drawn required citizen support.

The Maine vote, which marks the first time the issue of closing existing nuclear plants was put to voters in the United States, was a needed test of 1980s public sentiment on this complex and controversial issue. Anti-nuclear initiatives in prior years, including those defeated in California, Oregon, Washington and five other states in 1976, were too confusing and multi-faceted to gain a clear impression of public attitudes toward embracing nuclear energy as a future electric supply source.

A national referendum on the nuclear issue may be overdue, especially if nuclear power's role in this nation's struggle for energy independence from unstable foreign oil producers is explained.

Until then, the nation should view the Maine vote the way politicians used to look at Maine voters in the 19th century. The American political maxim, "As Maine goes, so goes the country," stated by political observers in 1888 who recognized Maine as a bellwether state in presidential elections, should be applied to Tuesday's nuclear referendum — if not as a prediction, certainly as a recommendation

EVENING EXPRESS

Portland, Maine,
September 24, 1980

Tuesday's referendum vote was an emphatic statement by the people of Maine on one of the most controversial issues of the times.

The anti-nuclear forces in Maine undoubtedly spelled defeat for the referendum when they included the closing of the Maine Yankee atomic plant at Wiscasset in the question submitted to the people.

Had the issue been simply "should any more nuclear power plants be built in Maine?" the outcome probably would have been a "no" as decisively expressed as the "no" in Tuesday's balloting.

Undoubtedly among the almost 231,000 who voted to keep the Wiscasset plant in operation there are thousands who have serious reservations about the nuclear generation of electricity. We're not sure that economic factors took precedence over safety. What we suspect happened was that the majority of voters made a judgment call.

They weighed the recognized hazards of nuclear power against the probability of a serious accident. Having accepted that element they then applied the economic factors which weighed heavily in favor of retention of the plant.

In view of the apathetic vote all too common in ordinary elections, there is satisfaction in a turnout of well over 390,000 voters Tuesday.

On the other hand, the vital nature of the issue, its obvious direct bearing on the lives of every citizen in the state, the intensive campaigning by both sides, might have suggested an even larger turnout.

However, a representative number did make the decision and they made it in an emphatic manner. It would have come as no surprise had the final tally been much closer.

The people of Maine may be better informed on this important issue than the residents of any other state thanks to the weeks of campaigning. They have had an opportunity to express their views in an orderly and democratic process. We think they made the right choice. We're positive they did it in the proper manner.

Des Moines Tribune

Des Moines, Iowa, October 10, 1980

A shift in the nuclear power industry's tactics could be detected as the Atomic Industrial Forum Inc. scanned the future recently. Industry leaders apparently will be reminding the public that progress does not come without risks.

In the past, the industry has stressed that the chance of dying or suffering serious harm in a nuclear accident is far less than the risks of driving a car or stepping out of a bathtub.

That reflected a defensive posture by the industry. Now it appears ready to take a more aggressive approach, going so far as to chide Americans for wanting a "risk-free society."

Several speakers hit hard on that theme at the Chicago meeting. An energy technology company executive said that a risk-free society is "a myth that can never exist." President John Silber of Boston University contended that "no-fault living" is a retreat from responsibility.

Silber said this attitude is evident in moves for no-fault auto insurance, no-fault divorce laws and the no-fault surgery shielded by malpractice insurance.

"Cost-plus contracts are no-fault business," he went on, "and federally guaranteed loans no-fault banking. Abstract art is no-fault draftsmanship. Tenure for college professors is a form of no-fault teaching, and canned laughter on television shows allows no-fault comedy. Golf carts allow no-fault exercise, and, finally, the development of the morning-after pill ensures no-fault sex.

"All this has encouraged non-consequential thinking, and non-consequential thinking has undermined realism and given rise to the belief that we can actually engage in practices that have no consequences. Among these is the belief that we can have energy without risks, that is, no-fault energy."

Most people can see the wide difference between a corporate decision that could affect the lives of millions of people and a personal decision to lessen the risk of pregnancy. To suggest that a woman who uses contraceptive methods is retreating from responsibility is asinine.

Beneath the subtle attacks on a society supposedly lacking the courage to face risks is the industry's nagging frustration with government's efforts to protect people from dangers over which they have no control.

The rhetorical flailing in Silber's comments dodges the moral principle that the recognizable risks of any action should not cause greater harm than the perceived benefits. What bothers many Americans is that corporate executives are ready to take risks that millions still think outweigh the potential benefits.

THE SAGINAW NEWS

Saginaw, Mich., September 29, 1980

It is no surprise that the voters of Maine decided to hold on to one-third of their electrical supply by voting against the shutdown of the Maine Yankee reactor. Turning off a reactor and replacing it with coal or oil in these waning years of the fossil fuel era would have been close to madness.

When it came down to the nitty-gritty last week — jobs and energy versus a utopian ideal of a risk-free society — the people of Maine demonstrated that common sense will prevail if a large enough segment of the population has something to say about it. When people see their livelihood or their pocketbooks threatened in such a graphic way, they react, speak their minds. Maine voters showed up at the polls in record numbers. The 56 six percent turnout was the biggest ever in Maine.

While the news of the Maine referendum surfaced on newspaper pages and television screens, it was interesting to observe the composition of Maine's anti-nuclear contingent. Like so many nuclear power opponents everywhere, they seemed invariably white, middle class and college educated in the liberal arts. One of the most widely quoted leaders of the shut-em-downers was a sculptor.

Sculptors have as much right as foundry workers to advance their viewpoint in a democratic society. But it is a fact that the white, middle class nuclear power opponents have little to fear from the loss of a few thousand blue-collar jobs. Those whose education and economic status have won them a place of privilege in this society are well insulated against higher energy costs. The college-educated elite and leisure-time intervenors who seem to make up the vast majority in the anti-nuclear movement could learn a lesson from the Maine Yankee referendum: most people really do care about money.

Still, though the Maine vote was an overwhelming endorsement of nuclear power there, the Maine anti-nuclear forces vowed to force the shutdown proposal to the ballot again, and again, no matter what the majority has proclaimed — loudly and clearly through the polls — it thinks is best for it.

Perhaps the best solution to the whole atomic-power issue would be a nationwide referendum to clear the air once and for all.

Sentinel Star

Orlando, Fla., September 26, 1980

MAINE voters did the nation a favor when they voted overwhelmingly to keep open the 840-megawatt Maine Yankee nuclear power plant. The vote dealt the national anti-nuke forces a stunning defeat.

The Maine referendum was the first in the nation aimed at shutting down an operating nuclear power plant. It was the culmination of a long campaign by the anti-nuclear coalition, which had hoped to use the accident at Three Mile Island to win its point. Voters in Maine, which derives a third of its electricity from the nuclear plant, proved that Yankee common sense is still a New England virtue.

The national campaign against nuclear power, which has gathered the remnants of the anti-war activists of the 1960s, is based more on ideology and emotion than on reason and facts. There is present in the movement the same disregard for reasoned debate, the same refusal to accept the will of the majority in the democratic process, and the same reliance on non-democratic tactics that characterized the worst excesses of the anti-war movement.

Foreign policy should not be decided by street demonstrators; even less should they decide so technical a question that is vital to dealing with an issue as important as energy. The future of the nuclear power industry should be decided by careful and deliberate thought, not songs and chants, and the wise people of Maine have done a lot to ensure that the deliberative process can continue.

The Philadelphia Inquirer

Philadelphia, Pa., October 4, 1981

Night after night, television screens around the nation and the world were filled with the disruptions. Masses of protesters, reminiscent of the anti-war years a decade and more ago, defied local authorities, trying to block work on the Pacific Gas and Electric Co.'s Diablo Canyon nuclear power plant near San Luis Obispo, Calif.

Pop music stars and aging hippies, lots of splendidly suntanned California enthusiasts for protesting anything that people who wear neckties support — and a lot of others — rallied and massed and interfered. Ultimately, 1,901 of them were arrested. The attempt to prevent operation of the reactor — based on claims that its proximity to an earthquake fault was dangerous and based also on simple, categorical opposition to all nuclear power — failed. The plant's operators were ready to begin loading uranium fuel rods for low-level testing that had been approved by the federal Nuclear Regulatory Commission.

End of story? The enormous engineering and power industries that have devoted billions to research and to perfecting nuclear generating techniques, supported by a huge and expert federal bureaucratic and regulatory establishment, had prevailed over the irresponsibility of a lot of wild-eyed extremists who are mindlessly opposed to progress, right?

Wrong.

•

No sooner had the last of the protesters been arrested or had left in frustration, than officials of Pacific Gas and Electric announced they had found a little problem. The intricately designed structural engineering required to prevent earthquake damage, which could lead to a nuclear accident, was amiss. It would be a few weeks before work could proceed.

That was the first day. As more time passed, more details transpired. The "first-rate screw-up," in the words of NRC commissioner Peter Bradford, had been caused by plant engineers using the wrong blueprints. To do that, Mr. Bradford, who is hardly an opponent of nuclear power, was "almost analogous to a student copying down the wrong homework assignement: No matter how brilliant the work from then on, he's just not going to get the right answers."

Except, of course, the stake is not a passing or failing grade for a brilliant student — it is whether, under the prevailing seismic conditions in the plant's location, there would be a likelihood of an accident that could have caused heavy damage to human, animal and sea life over a vast area.

And, as the days passed, a reawakened NRC found problems with four other systems of the plant — the safety injection system, the component cooling water system, the steam generator blowdown system and the hydrogen recombiners. Estimates of a delay of a few weeks began to stretch to six months.

All that may seem defiantly technical to all but the most informed fans or opponents of nuclear power. Americans in the area of Pennsylvania, New Jersey and Delaware, of course, have a certain advantage in that, since all of those arcane technical elements became part of the daily news — and anxiety — of millions of people two and a half years ago when history's worst nuclear power accident occurred at Three Mile Island.

Nobody died at Three Mile Island, of course, as the defenders of nuclear power are fond of putting on their bumpers. But today there are 600,000 gallons of radioactive waste left in the reactor building there, of a level of deadliness that even the operators cannot estimate accurately.

One thing about that is certain: If there were a rupture, which everyone involved admits is possible, and the residue escaped, it could very well wipe out all life in the Susquehanna River and well out into the Chesapeake Bay. And thus far, neither the nuclear construction industry nor the public utility industry nor the federal government has begun really to find a satisfactory way to raise the billion or so dollars necessary for cleaning up TMI.

•

Nothing blew up, or fell apart, or began to melt down at Diablo Canyon last week. And since the reactor had not been fueled there is no massive — and massively expensive — cleanup to do.

Thus Diablo Canyon stands today as the result of an important experiment.

In part because of the TMI experience, supporters of nuclear power generation have doubled and redoubled the intensity and expense and levels of expertness applied to the practicalities of nuclear plant safety.

Diablo Canyon, certainly the most controversy-burdened plant since TMI, was to be the proof of the nuclear pudding. It was to demonstrate that nuclear power, given the best of technical attention, was safe, practical and economical. It would put TMI in its proper perspective — a "learning experience," as one federal nuclear official called it not long ago. It would show that all those masses of sun-drenched protesters were just a mob of ill-informed malcontents.

Perhaps many of them are. But in the last few days that has become irrelevant to the argument. What has been made obvious by the Diablo Canyon experience is that *even* with redoubled attention to safety and design the industry was not able to put together a safe plant. The fact that the construction error was found before the plant went into operation may be considered a bit of fortuity. The other systemic weaknesses that regained NRC attention after that one was discovered may be repairable — or they may not.

No prudent person, however, could have any confidence today that every weakness in that plant has been found. And if it ever does go into operation, there will be anxiety about how many "screw-ups" have been missed.

If tens of billions of dollars tied up in nuclear power plants and plans were not at stake, if threats to millions of lives — directly and through potential genetic damage — were not at risk, if the entire nuclear power controversy were not so intractably rooted in the political and economic fabric of modern America, the recent events at Diablo Canyon might be an entertaining burlesque. Another set of gags about Murphy's Law.

It is no light entertainment, however. The Diablo Canyon saga is, rather, a vital demonstration of the limits of modern America to deal convincingly with the safety of nuclear power. And as such, it deepens the doubts that TMI cast about the prudence of the United States going further with it.

San Francisco Chronicle

San Francisco, Calif., September 23, 1981

THE LONG WAIT for a grant of permission to begin testing in their nuclear plant at Diablo Canyon, which has been tied up in regulatory red tape for far too long, ends in satisfaction for the engineers and managers of the Pacific Gas & Electric Co. We are happy for them and happy, too, for the people of Northern California, who should soon be getting a good share of their electric power from this plant. When it gets going full blast, it will produce, more cheaply, the amount of power that, in a conventional plant, would use up 20,000 barrels a year of imported oil.

All the same, we are concerned to learn that the Nuclear Regulatory Commission, though it voted unanimously to give permission for low-level testing of the new reactor, has expressed its lack of faith and confidence in the procedures followed by its subordinate regulating agency, the Atomic Safety and Licensing Board, in considering the PG&E's license application.

★ ★ ★

ONE COMMISSIONER, Victor Galinsky, criticized the licensing board for "shoddiness" and lack of fairness in the conduct of its hearings. We don't know what merit this criticism has; certainly the licensing board took its time in arriving at its favorable recommendations. Equally certainly, the commissioner's criticisms will fortify the hopes of those — Governor Brown's lawyers, for instance — still trying to obstruct the Diablo Canyon plant's coming into full operation.

It would be most unfair if the utility, which has invested $2 billion in Diablo to bring an adequate power supply to Northern California, should suffer a further time penalty in getting a full operating license because of the shoddy functioning of the Atomic Safety and Licensing Board. We would challenge the commissioners who are so critical of the performance of people in their own shop to see that this doesn't happen.

THE RICHMOND NEWS LEADER

Richmond, Va., October 9, 1981

The anti-nuclear protest at California's Diablo Canyon may have been a bust, but it still will cost California taxpayers a fortune.

According to police estimates, the daily cost of providing extra security for the Diablo area came to a cool $80,000. The total tab for handling the protest could exceed $1 million.

That the Diablo nuclear facility soon will go on line is not in doubt.

The only remaining question is whether the protesters will pitch in to reimburse California taxpayers, or whether they expect working families to pony up for their seaside vacations.

The Union Leader

Manchester, N.H., October 6, 1981

Any reader who breezed through with complete detachment the progress report in The Union Leader on construction of the Seabrook Nuclear Power Plant is not living in the world of reality.

To be sure, the placing of that 214-ton dome atop the containment building of reactor unit one, making that unit now 52 percent complete, was symbolic in a very positive way: It set the Clamshell Alliance anti-nuclear protesters on notice that they have failed in their basic goal of "shutting down" Seabrook Station, just as — across the nation in California — their Abalone Alliance soul brothers failed, despite their illegal antics of a recent weekend, to deter the Nuclear Regulatory Commission from issuing a license authorizing Pacific Gas and Electric Company to activate the Diablo Canyon nuclear reactor.

On the other hand, every consumer of electricity should know that he or she is paying the fare for these demonstrations. As a recent news account by Union Leader Staff Reporter A. S. Kneeland pointed out, the pseudo-environmentalists' interventions, combined with the vacillation of federal bureaucrats and a few labor strikes, have resulted in delays at Seabrook that have more than ***TRIPLED*** the cost estimates, from $1 billion to $3.56 billion, and delayed completion of Unit 1 from November, 1979, to February, 1984.

To say nothing of the heavy costs borne by taxpayers as the result of law enforcement efforts required to control the illegal mob actions at Seabrook.

Any taxpayer-consumer who takes an I-don't-want-to-get-involved attitude toward the Seabrook controversy has to be purblind to the fact that, like it or not, he is already ***heavily*** involved.

San Francisco, Calif., October 8, 1981

California and perhaps the entire nation owes a debt of gratitude to those persons who personally made a protest against the opening of the Diablo Canyon nuclear reactor which was built by Pacific Gas and Electric Company.

The protestors knew that they were facing derision from some quarters by some who alluded to them as just another group of hippies.

Then there were other opponents, in the persons of the news media, which blindly closed their eyes to the potential health hazards of nuclear reactors, particularly one like Diablo Canyon, which was erected on a known earthquake fault.

California and the nation owes a debt of gratitude to the protestors because their activities prodded an investigation of the plant by the federal regulatory agency which must give any nuclear reactors built by private industry the green light to operate and produce energy from radioactive materials.

Fortunately the protest seems to have awakened the regulatory agency from their long-practiced ennui of not examining requests made to them by multibillion dollar industrial giants like PG&E.

The regulatory agency has always performed to tunes played by the industrial giants, inasmuch as no members seemed disposed to lower the profits or thereby the dividends paid out to stockholders.

But the investigation brought on by the performance of the protestors brought a frightening discovery when information was released to the public that structural defects had been found in the construction of the installation revealing that the builders had used the wrong blueprints, greatly compounding the hazards if an earthquake should occur in the future.

Ever since the mishap at Three Mile Island, Pennsylvania, where a nuclear reactor experienced operational problems in which some leakage developed and large numbers of residents in the immediate area had to be evacuated from their homes, the American people have held misgivings about the erection of any more nuclear reactors in their neighborhoods.

There are also grave apprehensions about the storage of nuclear waste material, inasmuch as the industries which use such fissionable material seem to have no idea where to dump the waste materials.

The Pacific Gas and Electric Company has displayed no real concern so far since admitting to the buidling blunder, and their lack of concern causes many to wonder about the future of the Diablo Canyon nuclear reactor.

There is an added threat in the proliferation of nuclear installations, in view of an announcement by experts that the steel housing of the installations have deteriorated because of intensive radiation and could develop leaks of contaminated elements dangerous to all animal life as well as to the environment itself.

We are hard headed enough to view the use of nuclear reactors as a real danger to mankind and other forms of life because we believe that we cannot control the material for any long period, nor can we give assurance that there are no present or future hazards to the enjoyment of a normal life.

The protestors against nuclear energy installations are good, decent people who have studied well what happened to Hiroshima when that city in Japan was the first target of the awesome destructive power of nuclear weaponry.

They are people who wish to lead their lives free of any contamination created by mankind that might create health problems or shorten their life span for the benefit of people insensitive enough to hold a h igher regard for profits than they do for the health of mankind.

PG&E will no doubt have to rebuild the installation at Diablo Canyon. They have made it publicly known that the installation has already had a $2 billion price tag in construction costs. Consumers will have to wonder just how much their electricity bills will climb because of what seems a very careless move made by the construction department of the huge utility.

Platonium, or whatever the element used to produce nuclear power, is needed in only a small amount to operate either ships or nuclear power plants, in comparison to the huge amounts of petroleum or coal, making that source of energy very alluring to profit-mot[illegible]ed utilities.

But t[illegible]eroic protestors at Diablo Canyon have shown the American people just what dangers they face. Again, the American people owe them a debt of gratitude.

The Birmingham News

Birmingham, Ala., September 28, 1981

Nuclear power plants are natural catalysts for protests and news, it seems. And so it was as predictable as sunrise that several hundred demonstrators and several dozen newsmen would gather in the vicinity of San Luis Obispo, Calif., while waiting for the Nuclear Regulatory Commission in Washington to grant a license for low power testing at the Diablo Canyon power plant.

Later the numbers of demonstrators swelled to a few thousand and the reporters, camera crews and technicians to scores. A media event was well underway. It was the proclaimed purpose of the demonstrators to prevent the implementation of the NRC's decision. The demonstrators admitted openly they fully intended to break the law in order to achieve their goal.

The fact that a number of them went to jail then should not be surprising and should elicit no groundswell of sympathy.

The commission's ruling was only the latest and perhaps not the final decision in a long series dating back into the '60s. The decision was part of a long, laborious, painstaking process by which hundreds of duly elected and appointed representatives of the public at local, state and national levels — with advice and assistance from almost every available authority on the subject — have studied and deliberated this single issue: Whether, how and when a nuclear power plant should be allowed to operate in Diablo Canyon.

The public should have no doubt about what was being demonstrated at Diablo Canyon. The protestors were demonstrating that they would pay no heed to the lawfully arrived at and lawfully enforceable decisions of representatives of the people of the United States.

The public does not have to give a hoot about nuclear power or any other issue to ask: Who the heck do they think they are?

It's safe to say that the great mass of Americans will never know very much about nuclear power generation. It is, however, one of the geniuses of our kind of democracy that while the sovereign people are not expected to possess competence to decide all those questions which touch the quality of their common life, they are recognized as capable and enlightened enough to select men and women competent to decide such questions on their behalf.

Apparently the demonstrators at Diablo Canyon did not admit this capacity of the self-governed. The arrogance they displayed is more than an insult. It is a deliberate challenge to the form of government and the Constitution through which Americans exercise and have exercised their rightful authority for more than 200 years.

In flouting the sovereign, constitutional authority of the American people, what do they offer in its place? The implacable will of a lawless mob. And that is the real challenge, the real issue at Diablo Canyon — not nuclear energy at all.

ALBUQUERQUE JOURNAL

Albuquerque, N.M., November 28, 1981

Caught between the misfortunes and miscues of the industry and its government regulators, and the vociferous activity of well-organized opponents, nuclear power generation is now out of favor with a majority of Americans.

The AP-NBC news poll reports that 56 percent of respondents said no new nuclear power plants should be built, against 32 percent in favor and 12 percent not sure.

In 1977, the last time the poll asked the question, 63 percent of respondents were for additional nuclear power construction.

The national attention focused on the 1979 reactor mishap at the Three Mile Island nuclear plant was probably the biggest factor in the change in public opinion.

But ineptitude in the industry and its government watchdogs have received much publicity in subsequent years, eroding public confidence in the nuclear generating industry's ability to build and operate acceptable-risk stations — even if the technology exists.

A recent example was the Diablo Canyon nuclear power plant in California. The plant survived mass demonstrations only to have its license lifted when it was determined the plant was built with substantial variations from its original plans.

Respondents to the AP-NBC poll said they favored increased conservation and expansion of other energy sources to increased dependence on nuclear plants.

We in the coal-rich energy areas of the West must realize that dependence on other sources — such as coal — will increase our role in future energy production.

Increased exploitation of New Mexico's coal resources will have a corresponding increased impact on our environment.

Consequently, the reluctance of other parts of the country to face the real or imagined environmental risks of nuclear generation in their back yards could result in the export of the environmental risks — in the form of coal mining and transportation — to New Mexico.

Man has not yet devised a way to mass-produce energy without degradation of somebody's environment.

TULSA WORLD

Tulsa, Okla., November 24, 1981

AN ABC-Associated Press nationwide poll shows 62 percent of the people now think nuclear power should be avoided in favor of conservation and other power sources.

That is a dramatic turnaround from 1977, before Three Mile Island and resultant soaring prices for nuclear-powered generators.

Although public opinion can shift dramatically on such issues (as it has on this one) and there is still considerable evidence to show that nuclear power ought to be a source of electricity for the future, the strong consensus indicates the public has at least made a decision.

Fortunately, the United States possesses such vast reserves of coal that the country does indeed have a choice in the power generation matter.

If the public indeed is willing to "conserve" and direct its utilities and industry to use coal, it can be done.

But "conservation" is going to mean more than turning down the thermostat and turning off the lights. It's going to mean higher prices for every source of energy, because we have been shown clearly that only price will bring about true conservation.

Beyond that, the public must be willing to accept some very drastic environmental tradeoffs if we are to forego the nuclear option. Few realize how great the environmental price will be when the nation converts to coal.

It will mean almost endless freight trains of coal, a deterioration of air quality even with the best technology in use, drastic change in our life styles as we conserve at the required levels, and great physical damage to the coalfields of the West.

The odds are good that once these sacrifices begin, the public will want to look at the nuclear option again.

In the meantime, however, some direction in the energy picture is needed. It appears the public has offered it.

THE SACRAMENTO BEE

Sacramento, Calif., October 24, 1981

We've not supported demonstrations, such as those at Diablo Canyon, aimed at shutting down or otherwise impeding operations at nuclear plants which have met all formal government licensing requirements. Such protests seek to impose the moral, scientific and political attitudes of a minority on orderly licensing procedures based on the best available scientific and technical information. One can argue about those procedures, but one cannot question the proposition that, in the final analysis, reasonable judgments about plant safety can be reached only through a formal technical process and not through the acts of crowds trying to block the gate of a plant.

Despite all that, however, there is something dangerous in attempts by utility companies and others to file damage suits against demonstrators in an effort to discourage such demonstrations in the future. At least two such actions have now been filed, one in New York where the Long Island Lighting Co. (LILO) and several construction unions are seeking $2 million in damages from those who demonstrated at a LILO plant construction site, the other in California where the Pacific Legal Foundation of Sacramento (PLF) has filed suit against the Abalone Alliance and others involved in September's attempt to block the opening of the PG&E plant at Diablo Canyon.

Among the plaintiffs in the California case are the California Association of Utility Shareholders, Assembly Minority Leader Carol Hallett in whose district Diablo Canyon is situated, and a variety of other groups and individuals. In their suit they do not claim any specific damage to property or to business prospects. They claim only that in trespassing on PG&E property, illegally blocking roads and engaging in a number of other illegal acts, the demonstrators force public utility ratepayers and taxpayers to pay extraordinary expenses.

PLF, a conservative public-interest law group, denies the suit was designed to intimidate protesters. "This is not a 'Get the Left' suit," a PLF attorney said. Yet clearly its effect — should it survive in the courts — will be precisely that. Although PLF says it is only going after people who broke the law — not those who engage in informational picketing or other public education activities — there is hardly a way to limit the impact of such suits or to shield people wishing to exercise constitutionally protected rights from their chilling effect. If PLF can collect damages from those who were arrested for blocking a gate or refusing to obtain an order to move on there is no way that anyone with a picket sign or a handful of leaflets can ever again feel totally secure.

There are obvious risks in First Amendment protections of free speech and free assembly, just as there is likely to be considerable inconvenience and public expense in certain demonstrations. But that's a price that this country, as a free society, has always been willing to pay, and that it should continue to pay.

THE SUN

Baltimore, Md., January 27, 1982

When fear of nuclear disaster gripped the nation in late March of 1979, millions of Americans became frightfully aware of their ignorance of nuclear energy. What's a rem? How does a meltdown occur? Is the China Syndrome a matter of fact or fiction? Most people hadn't the foggiest idea. They were at the mercy of nuclear physicists and government specialists who seemed to speak their own foreign language.

That is a dangerous situation, concludes a study by the Carnegie Foundation. It leads to alienation and distrust of government and public figures. It also places enormous responsiblity for making highly complex decisions in the hands of technocrats who understand the complicated issues. The Carnegie report warns of "drifting unwittingly into a new kind of Dark Age—a time when small cadres of specialists will control knowledge and thus control the decision-making process."

This is not confined to the nuclear issue. What about the debate over the "window of vulnerability" and the MX missile? What about those technical decisions that must be made as communities select among competing cable television companies? Environmental disputes, zoning disputes, even school and Social Security disputes, have taken on a complexity that baffles the general public.

"Schools and colleges simply must help students understand the process by which public policy is shaped and prepare them to make informed, discriminating judgments on questions that will affect the nation's future," the report says. It is time to resurrect that old curriculum workhorse, "civics."

What the authors of the report have in mind is a broad system of courses and degrees aimed at educating adults on pressing social issues. The focus is on adults, not teenagers, because for the rest of this century most students served by higher education will be over 21 years of age. They must be offered programs that do more than simply pass the time or repeat undergraduate courses.

Such steps will require imagination and vision on the part of educators. It will call for new interdisciplinary programs to strip away the mystery surrounding complicated public issues. A new goal will have to be established in higher education: preventing this country from becoming what the Carnegie authors fear would be a nation of civic illiterates. That's an admirable focus for American colleges and universities, which have been searching without much success in recent years for an identity and relevance that fits into the America of the 1980s.

RAPID CITY JOURNAL

Rapid City, S.D., November 15, 1982

Advocates of nuclear power generation won important victories in two states in the November 2 elections.

By a 55 percent to 45 percent margin, Maine voters turned down a referendum calling for the shutdown of the Yankee nuclear power plant which supplies about 30 percent of the state's power. In voting as they did, Maine residents opted to retain the plant rather than spend $2 billion to $4 billion over the next five years for an alternative source of energy.

Idaho voters also approved a pro-nuclear initiative that would require the legislature to get voter approval before moving to impede nuclear power development in that state.

Since the Three Mile Island accident, every nuclear power plant incident has been widely publicized. But as the Wall Street Journal points out editorially, nuclear power is by no means dead even though no nuclear power projects have been started in the U.S. in over a year because of regulatory delays and legal actions. There are 82 licensed plants in operation and 76 under construction or on order. Nuclear plants now furnish 13 percent of the nation's power — second only to coal.

The Wall Street Journal also pointed out that a study of a "worst case" nuclear plant accident, "leaked" just prior to the election, was misintepreted by the Washington Post. The Post said the study showed there was a two percent chance of an accident that would cause more than 100,000 deaths and over $300 billion in damages occurring in the United States before the year 2000. The Nuclear Regulatory Commission, which had commissioned the study, challenged that interpretation and said the study calculated the probability of such a "worst case scenario" at one in one billion per year of reactor operation.

The Wall Street Journal concluded that a reading of the study conducted by the Sandia National Laboratories leaves the impression that what it really was saying is that the chances of a nuclear power plant disaster are extremely remote.

The Journal also pointed out the advantages of nuclear power. Unlike coal, it's clean. It's not subject to the uncertainties which characterize oil supplies from the Persian Gulf. If regulated intelligently, it's relatively cheap.

One of nuclear power's biggest problem is the bad press it has received despite the fact that there's a high level of support for nuclear power among scientists who know the most about the subject.

The votes in Maine and Idaho indicate the tide of public opinion may be turning against the anti-nuclear power lobby.

Arkansas Gazette.

Little Rock, Ark., January 11, 1982

The American Association for the Advancement of Science has been meeting at Washington in recent days and the reports pouring in have given the scientific community plenty to think about, especially in the area of nuclear energy and radiation.

First off, a study financed by the Department of Energy that reviewed results of 228 surveys conducted by various organizations found public support for building nuclear power plants only if they are placed in someone else's neighborhood. Opinion against local construction of nuclear power plants, the study found, runs two to one. The opposition had been even stronger in the immediate aftermath of the accident in March 1979 at the Three Mile Island plant in Pennsylvania.

Proponents of nuclear power have been busy ever since pooh-poohing the idea that any harm was done to humans, claiming that the amount of radiation released during the accident was not possibly high enough to do any damage, short term or long term. Additional reports presented at the AAAS meeting, however, tend to raise new and greater concerns not only about Three Mile Island but also about "normal" radioactive contamination around nuclear power plants and atomic weapons plants.

Biologist Bernd Franke of Heidelberg's Institute for Energy and Environmental Research reports that the Nuclear Regulatory Commission's method of calculating radiation doses may seriously underestimate the human health risk. Even normal releases from nuclear facilities, he says, can lead to overexposure because current dose limits are based on unrealistic data. The limits, according to Franke, are based on the health risk to a single segment of the population instead of many sensitive groups. In one case involving the introduction of radioactivity into the food chain, for example, the NRC, says Franke, based its calculations on data for American cows, English plants and Russian soil.

A health physicist from Georgia Tech, Karl Morgan, reports, among other things, that the cancer risk posed by low radiation doses must be revised upward by several orders of magnitude in view of new information about the Hiroshima bombing aftermath. He opposes the breeder reactor because it generates deadly, long-lived radioactive substances and poses a risk of nuclear proliferation. And Carl Johnson of the University of Colorado compared radioactive contamination near atomic weapons plants with studies suggesting abnormally high cancer rates there.

With reports of this nature brought before a distinguished scientific group, is it any wonder that the American people have such grave reservations about having a nuclear plant situated in their own backyards?

Newsday

Long Island, N.Y., March 4, 1983

"If you took away all the emotionalism surrounding this issue," Islip Supervisor Michael LoGrande remarked recently about the Shoreham nuclear power plant controversy, "I don't think the public has one iota of real information."

LoGrande probably was indulging in a bit of hyperbole. This newspaper alone has published hundreds of thousands of words about the dispute over Shoreham. But his remark reflects justifiable concern for something that has been troubling us also: the enormous amount of unreasoning fear that has colored the Shoreham debate — and the uses to which that fear has been put.

The origin of the fear is easy to trace. It comes from a vast public pool of myth, misinformation and just plain ignorance about the production of energy in general and nuclear power plants in particular.

Do you believe, for example, that nuclear power plants are capable of exploding like atomic bombs? They're not, but many Americans are convinced they are. Or do you think that during normal operation, a nuclear power plant emits more radiation than a coal-fired plant or that nuclear plants cause more air pollution than conventional plants? Neither is true, although many Americans remain unconvinced of that.

As a Newsday poll recently found, most people on Long Island would prefer to see Shoreham abandoned. And 46 per cent of those polled gave as their reason the conviction that nuclear power plants are unsafe. Other studies have demonstrated that much of this belief in the intrinsic danger of nuclear plants is based on misinformation and fear.

But instead of calmly dealing with the relative risks and advantages of nuclear plants as compared with conventional methods of producing electricity, too much of the rhetoric about Shoreham has exploited anxieties about nuclear power generation. And Suffolk County Executive Peter Cohalan has contributed to that exploitation with his arbitrary statements and positions.

Lou Howard, the presiding officer of the Suffolk County Legislature, put his finger on this point in an article that appeared in Newsday's Viewpoints section last Wednesday. Howard is the lone legislator who voted in favor of a county emergency plan for Shoreham. Cohalan and the rest of the county government said no such a plan could be devised.

In the article, Howard noted that this conclusion was predicated on eventualities that might occur, according to the county's consultants, no more than once in 50,000 years and conceivably only once in 500 million years.

In other words, the county's fundamental attack on Shoreham is rooted in an essentially irrational fear of a cataclysmic nuclear accident that — as Howard noted — is far less probable than a tidal wave.

Nuclear plants certainly carry risk, as we've observed repeatedly. If discussions of Shoreham were being conducted rationally, they would include careful and precise examination of all the risks involved in the plant's operations and they would attempt to determine what protective measures were responsive to each level of risk.

The best way of making those determinations and of resolving the Shoreham dispute would be through cooperative consultation between Suffolk County and the Long Island Lighting Co. But such consultation is unlikely. The county has dug itself into so intransigent a position on this issue that any accommodation without outside assistance seems impossible. The logical person to provide that outside aid, as many Long Islanders have perceived, is Gov. Mario Cuomo.

Cuomo proved himself an accomplished mediator long before he became governor. Now, with the power and prestige of his current office added to his mediatory skills, he might be able to help the county and LILCO find a reasonable way out of this costly and dangerous standoff.

Up to this point, Cuomo has carefully avoided involvement in the controversy. That has been a mistake. This community constitutes an important segment of his responsibility — and right now, the future of Shoreham is the number one concern of this community. The governor should intervene and try to break the Shoreham deadlock.

The Louisville Times

Louisville, Ky., September 13, 1983

The nuclear industry's new public relations campaign pushes the argument that scare stories and unreasoning hysteria over safety have contributed to the troubles facing commercial atomic power. The truth, say industry spokesman, is that reactors have operated safely most of the time and are worthy of public confidence.

One response might be that accidents in nuclear plants are financially and psychologically catastrophic, even if no injuries can be traced directly to them. Even some members of the nuclear establishment complain, moreover, that American-designed plants leave much to be desired when it comes to safety. A nervous citizen might further cite the Nuclear Regulatory Commission's discovery of a potentially fatal flaw, cracked water pipes, in a number of power plants.

But even people who accept the basic soundness of nuclear technology have reason to be distressed by accounts of construction and management failures. Shoddy workmanship can raise doubts about the reliability of any machine. Much more is at stake — the lives of thousands of people — when mistakes are made in a building that is supposed to contain lethal radioactive materials.

The defects, and particularly the substandard welding at the nearly completed Zimmer nuclear plant near Cincinnati, for instance, will do far more to hurt public confidence than a dozen anti-nuke shindigs. As recounted in *The Wall Street Journal* last week, investigators believe that 70 per cent of Zimmer's welds may not meet code requirements, hardly an endorsement of the plant's capacity for safe operation.

One explanation, according to allegations made to the NRC and congressional committees, is that many welders hired to do the work were not qualified. This, moreover, appears to have been the case at a number of construction sites. Cheating on tests, falsified records and the sale of union cards to inexperienced welders enabled many workers to get high-paying jobs they weren't prepared to do. Supervision and quality control, needless to say, were seriously deficient.

The other reason much of the public has soured on nuclear energy was most recently illustrated by the controversial Seabrook plant in New Hampshire. The New England utilities that own the plant have decided, because of rampaging costs and less-than-expected electricity demand, to all but stop work on one of two reactors.

Bad luck and unexpected delays may be partly to blame for similar dilemmas at the Marble Hill plant in Indiana and Shoreham on Long Island, where ratepayers are bracing themselves for massive "rate shock" when, and if, electricity is actually produced. But sharper management analysis of economic trends, and realistic cost comparisons with coal-fired power, might have spared utility ratepayers these monumental blunders. Some experts argue that nuclear power was never even in the same economic ballpark with coal.

Nuclear enthusiasts are also vague about the inevitable additional cost of waste disposal and plant closings. The industry, moreover, has been betrayed by its own success in persuading Congress to place a limit on its liability in case of an accident. If nuclear utilities had had to figure in the full cost of enough insurance to cover accident claims, as other businesses do, many would have chosen another energy source, and would be celebrating their good fortune today. Those that went the nuclear route would have had a powerful incentive to hire competent and honest welders.

Nuclear promoters, of course, have every right to make their case that public misperceptions are a big part of the problem. But the story will be far from complete until they also acknowledge that the public has real reasons to be alarmed.

Mental Stress Not a Factor in Startup of TMI Unit 1 Reactor

The Supreme Court April 19, 1983 ruled unanimously that the Nuclear Regulatory Commission did not have to evaluate the psychological stress on residents of communities near nuclear power plants before approving a change in the operation of those plants. The decision in the case, which arose because the NRC was considering a restart of the undamaged Unit 1 reactor at the Three Mile Island plant in Pennsylvania, was a victory for the Reagan Administration. The Unit 1 reactor had been shut down since the 1979 accident involving its sister reactor. In 1982, an appeals court mandated hearings to weigh the potential effect of a restart on the mental health of people living near the plant, as part of an environmental impact study the commission was required to conduct before approving a restart. The Supreme Court decision overturned that ruling. Justice William H. Rehnquist noted in the court's opinion that "a risk of an accident is not an effect on the physical environment." As a stress factor dealt only with such a risk, he continued, it need not be included in a study that pertained to actual effects. "Risk," Rehnquist wrote, was "a pervasive element of modern life." Emotional harm would have to "have a sufficiently close connection to the physical environment" to be included in a mandatory environmental impact study, he held.

ST. LOUIS POST-DISPATCH

St. Louis, Mo., September 1, 1979

It may be at least a year before the reactor at Three Mile Island produces electricity again. No, not the one that had the accident in March — that may never be repaired. The one that will be shut down for another year is the *other* nuclear power plant on the island near Harrisburg, Pa., the older brother of the one that went awry. It was shut down for routine maintenance at the time of the accident and has stayed that way. Earlier this month the Nuclear Regulatory Commission decided that there will have to be changes in the reactor's design, proof of fiscal and managerial stability, and improved operating procedures before the plant can be turned on again. The NRC also wants to be assured that the expected efforts to clean up the radioactive mess of the nearby TMI reactor that went out of control last spring will not interfere with the operation of the undamaged plant.

The utility holding company that owns both reactors had hoped to get the undamaged one in operation no later than Jan. 1 to ease the costly burden of importing power from other suppliers. The NRC decision is expected to raise the specter of a "financial crisis" for the utility and generate another round of rate increase requests. Once again, we predict, bankers and utility executives who normally rally around the free enterprise flag will come forth to argue that the public should pay for their mistakes.

The Philadelphia Inquirer

Philadelphia, Pa., December 16, 1980

The Nuclear Regulatory Commission recently decided that psychological stress cannot be taken into account in hearings about restarting the undamaged nuclear reactor at the Three Mile Island facility. The NRC had been requested to consider public concerns about the resumption of service at TMI's Unit One by a number of sources, chief among them the Commonwealth of Pennsylvania.

By its action, the NRC is ignoring the most crucial question that confronts nuclear power in America today: public acceptance. Reactors can be modified; redundant safety systems installed, and alarms, gauges, lights and valves added. But nuclear power will not survive in the United States if the public questions its safety.

Nuclear power's future does not rest solely with improved technology. It also rests with public confidence.

The NRC, in its 2-2 vote on the psychological stress factor (which procedurally represents a defeat on the question), has effectively brushed aside this very real issue, saying that resolving public fears is not the NRC's responsibility.

To the contrary, this must be the chief priority of the NRC, for it is an issue that cannot be severed from the responsible operation of nuclear plants in the United States. And that duty — overseeing the safe management of nuclear reactors — rests solely and entirely with the NRC.

NRC Commissioner John F. Ahearn, who voted not to take psychological stress into consideration, noted in his opinion: "The best way to decrease such stress is to insure the plant is safe if it is approved for operation." That kind of neat logic fails to adequately address the issue. The public no longer will buy official assurances of safety. The public is skeptical, and rightfully so, for TMI is the accident that never was supposed to occur.

Assurances that the clean-up operation at the damaged reactor is being supervised by "the best minds in the industry" ring hollow when the public learns, as it did the other day, that radiation may be seeping from the plant through contaminated pipe seals.

At all of the numerous public hearings that have been held following the TMI accident, testimony from individuals living near the plant has been dominated by their often poignant pleas for consideration of the stress they have been living under since March 28, 1979 when Unit Two malfunctioned. That stress will be magnified by having Unit One, undamaged in the accident, restored to service, they testify.

The NRC has failed to hear what these people are saying: Take our concerns seriously, for they are as important as the question of whether or not the reactor has newly designed valves and cooling equipment.

The NRC has agreed to reconsider the psychological stress question in its review of TMI Unit One when a fifth member of the commission is appointed by the Reagan administration. It can only be hoped that the appointee will understand that public confidence is as vital to the development of nuclear power as enhanced technology.

The Evening Bulletin

Philadelphia, Pa., January 11, 1982

Talk about legal thickets, we think the U.S. Court of Appeals in Washington has us headed for a beaut. It just ruled 2-1 that Three Mile Island's Unit 1 reactor can't be restarted until the government finds out how it might affect its neighbors' psychological health.

Until the opinions with the decision are revealed, we won't know the judges' reasoning. But if the ruling is upheld, the potential is scary. A psychological testing rule might be applied not just to atomic energy plants, but almost anything from pizza parlors to lumberyards.

No doubt some people around TMI are afraid of nuclear plants. There are those who have built on that stress whole platforms of ideological opposition to nuclear power. How does a survey distinguish between the two?

What are the criteria in the survey? How much stress is too much? And how do you measure it?

If communities near nuclear plants — maybe coal-burning plants, too — have to be stress-free, we may have seen our last plant start-up.

Rigorous guarantees of the safety of the plant ought to be enough. A Nuclear Regulatory Commission board last fall ended almost 10 months of hearings on Unit 1's restart and approved it for the first phase of "going critical."

Lately, two leaks in the unit's steam generators were discovered and delayed the start-up date until late March. We are as emphatic as ever that neither Unit 1 nor the accident-crippled Unit 2 should go back on line until they are proven safe. But even concrete proof will not allay all the stress around Middletown.

Only dismantling the plants would do that.

THE KANSAS CITY STAR

Kansas City, Mo., January 11, 1982

Never before, said Judge Malcolm R. Wilkey in his dissenting opinion, has the National Environmental Policy Act been used to require an environmental assessment of the psychological effect on area residents of a given project. But it has now, in a case involving whether or when to restart the undamaged nuclear reactor at the Three Mile Island power station near Harrisburg, Pa. It is the scene of the nation's most serious nuclear accident, in March 1979, when a near meltdown occurred in Unit No. 2. At that time Unit No. 1 was down for refueling. It has never been restarted.

In a suit brought by a group called People Against Nuclear Energy, the Nuclear Regulatory Commission had refused to consider any psychological impact in deciding when to start up TMI No. 1 again. But Judge J. Skelly Wright and Judge Carl McGowan of the appeals court ordered the NRC to study such effects before giving any start-up order. It is a most unusual ruling. Under the NEPA law the physical dislocation resulting from the construction of a project can be determined and the likely effect on the natural ecology of the area can be estimated. But the *psychological* impact? This is an effect impossible to measure meaningfully because in any typical impacted area a project will encounter reactions ranging from ardent advocacy through apathy to alarmed opposition.

TMI No. 1, moreover, is not a blueprint proposal but an existing, operating power station which has had no problems except the accident to its sister unit. In fact the operating safety lessons learned from that unhappy event should make No. 1 even more trustworthy. But it is, nonetheless, a hostage to the pathological fear that enveloped the Harrisburg area when residents thought that some sort of nuclear holocaust had been narrowly averted. As Judge Wilkey observed, it will be a colder winter around Harrisburg for lack of the electricity TMI No. 1 stands ready to produce. There is no reason to anticipate trouble when it is restored to operation. But fear bred of even a near-disaster is unreasoning, particularly when unseen radioactivity is its source. And the appeals court majority felt obliged to take that sad phenomenon into account.

The Cincinnati Post

Cincinnati, Ohio, January 25, 1982

Many people may not know it, but there were two nuclear reactors on Pennsylvania's Three Mile Island on March 22, 1979, when the nation's worst commercial nuclear accident occurred.

Only one, called Unit 2, was involved in the accident. It remains radioactively "hot" inside and may never go into operation again.

The other reactor, Unit 1, was shut down for refueling at the time. Although it was entirely undamaged, we're beginning to believe that it, too, may never produce another watt of power.

TMI-1 could have been returned to operation months ago, after it was brought into compliance with more stringent safety requirements imposed by the Nuclear Regulatory Commission on all reactors.

The NRC's excessive caution has been bad enough. But now a federal court has dealt a potentially fatal blow to TMI-1.

The U.S. Court of Appeals for the District of Columbia has ruled that the NRC "shall not make a decision to restart" the reactor until it conducts an assessment of how this might affect the psychological health and well-being of nearby residents and communities.

The suit which led to the ruling was brought by a group called "People Against Nuclear Power." Like other such groups, it wants not safe nuclear power but no nuclear power at all.

We're sure "People Against" will have no trouble finding any number of psychiatrists to testify to the adverse psychological effects of restarting TMI-1, given the irrational fears so many Americans seem to have about nuclear power. At best, it will be more years before the assessment is concluded.

In the meantime, four million customers of General Public Utilities Corp., operator of Three Mile Island, will continue to have to pay for the higher-cost electricity the company has been forced to buy from other utilities.

Someone should do a study of the psychological effects of that unnecessary burden on the public.

For ourselves, we find much in the antinuclear movement extremely depressing.

The Washington Post

Washington, D.C., January 11, 1982

TWO YEARS AGO, people living near the reactor at Three Mile Island asked the Nuclear Regulatory Commission to consider the psychological stress they suffered as a result of the nuclear accident there, in deciding whether to allow the facility's undamaged reactor to resume operations. The request was seconded by the commission's own Atomic Safety and Licensing Board. Though the board believed that neither the Atomic Energy Act nor the National Environmental Policy Act (NEPA) *required* the NRC to take the community's mental health into account, it concluded that NEPA *allowed* this to be done, and it recommended that this unusual step be taken in this special case.

The NRC, in a split vote, disagreed. With visions of adding the testimony of contending psychiatrists and psychologists to its already tortuous licensing process, and with anxiety about creating a difficult precedent, the commission forbade its licensing board to consider psychological stress. TMI's neighbors responded with a lawsuit that last week produced a court order requiring the NRC to conduct an environmental assessment, and perhaps a full-fledged environmental impact statement, on the psychological effects of starting up the undamaged reactor.

It is too early to assess the legal impact of the decision, since the two-judge majority has not yet provided its opinions. In his dissent, however, Judge Malcolm Wilkey asserts that the decision adds "an 'impact' which has never before been considered as covered" by NEPA. If he is correct, and if the ruling stands, it is likely to do serious damage to a good law. NEPA is an important tool for improving the quality of federal decisions that affect what the law calls the "human environment." But there are limits to what it should control—and, in our view, psychological stress is beyond them. If NEPA is forced to do too much, it will become impossible to administer and will inevitably be swept away.

The effects of the decision on the Three Mile Island reactor and its near-bankrupt owner are much clearer: the hoped-for March start-up date will most probably be delayed at a cost of millions per month. With the advantage of hindsight it is easy to see that NRC Commissioners Peter Bradford and Victor Gilinsky were correct two years ago in urging their agency to go beyond its strict legal requirements. Had the plant's worried neighbors been given the opportunity to vent their anxieties—reasonable or not—the lawsuit with its unfortunate precedent could have been avoided. Steps could have been taken—though admittedly unnecessary in the strict technical sense—to mitigate local fears for a fraction of the cost of the delay the utility now faces.

The lesson is clear enough—though this is by no means the first time it has been forced on the nuclear industry and its friends at the NRC. For a combination of good reasons and bad, a large number of Americans have lost confidence in this industry and in the capacity of its government regulators to ensure public safety. Winning back that confidence will be a long, slow process requiring some extraordinary steps. Obviously, to people who understand and have faith in this technology, many of these steps will seem wasteful, unneccessary or just plain silly. But the longer the industry resists acknowledging the reality of public fears, the harder the task is eventually going to be.

BUFFALO EVENING NEWS

Buffalo, N.Y., May 19, 1982

The nuclear accident at the Three Mile Island plant in Pennsylvania in 1979 damaged only one reactor in the power complex. The other unit, after studies of possible design changes and safety procedures, had been expected to resume operations soon, but now a dubious federal court ruling has delayed the startup indefinitely.

In a 2-1 decision, the U.S. Court of Appeals for the District of Columbia ruled that the Nuclear Regulatory Commission must, before the startup, study and assess the "psychological stress" that might result from a resumption of operations at the power-generating plant. If such stress were found to be significant, then a full-scale environmental-impact study would have to be ordered.

The dissenting judge, Malcolm Wilkey, expressed concern that the ruling might bring about a "court-imposed paralysis" on nuclear-power developments generally. Anytime an anti-nuclear group could "whip up sufficient hysteria," he pointed out, the psychological stress could be used as a means of blocking national nuclear policy and energy goals.

A nuclear power official called this "a truly novel situation." For the first time, the licensing of a nuclear plant has been linked to factors beyond safety or ordinary environmental considerations.

The Three Mile Island accident was, of course, the nation's worst nuclear accident, and the inevitable tension and anxiety were heightened by exaggerated scare talk about a nuclear "meltdown" — something that never came close to happening. Thus, it would appear reasonable to take special safeguards in the resumption of operations and special efforts to reassure area residents about the safety of operations.

But to hold up nuclear operations until every last psychological concern has been eliminated from every person in the area would be an unrealistic standard. Anti-nuclear advocates are using the same psychological argument in an attempt to stop the licensing of the Indian Point plant in Buchanan, N.Y. and of plants in three other states.

Nuclear power can and should play a role in providing an alternative source of power and helping to increase the nation's energy independence. Without question, everything necessary must be done to guarantee safe and secure operation of nuclear plants. But this new federal court ruling sets an excessive standard that would make the nuclear regulatory burden costlier and more time-consuming.

THE WALL STREET JOURNAL.

New York, N.Y., January 12, 1982

Just when it looked as if some headway was being made in clearing up the physical and financial wreckage left by Three Mile Island, the whole effort has been dealt a below-the-belt punch. It came, not surprisingly, from Judge J. Skelly Wright and his District of Columbia Circuit Court of Appeals.

Judges Wright and Carl McGowan outvoted Judge Malcolm Wilkey in a three-judge panel opinion that TMI proceedings, which have been dragging along for nearly three years, have not given sufficient attention to the "psychological health" of people living in the vicinity of the Three Mile plant. Hence there will be a further delay in starting up TMI-1, the undamaged power reactor that has been shut down since the accident at TMI-2 in March 1979.

The 4 million residents who buy power from General Public Utilities Corp. in the affected service area will be privileged to pay still more for unnecessary costs brought about by court and regulatory delays.

There was never any good reason for keeping TMI-1 shut down this long. It was down for fuel replacement at the time of the accident and could have been restarted after the Nuclear Regulatory Commission review of the procedures that caused the trouble at TMI-2, or more precisely the failure to follow proper procedures. The utility's management got a clean bill of health months ago, after it was ascertained that it was in compliance with more stringent NRC requirements.

If the utility had been allowed to restart TMI-1 promptly after the necessary review, it would have been relieved of some of the cost burden of cleaning up TMI-2. With both reactors down, the utility has had to buy high-cost power from other utilities in its power grid. This has forced it into costly borrowing at a time when its resources already were strained. The Pennsylvania PUC has had little choice but to give it rate increases. The only redeeming grace was that customers were paying rates lower than the Pennsylvania average to begin with.

The court suit was brought by an anti-nuke group called "People Against Nuclear Power." Lawyers for this group contend that the NRC must consider how "psychological health" in the area would be affected by a start-up, a demand the NRC had rejected as not being required under environmental law. In upholding the PANP suit, Judges Wright and McGowan gave assent to the idea that the NRC, having consulted everyone else these last three years, must now call in the psychiatrists.

There might be some purpose to be served there. It would be interesting to know what psychiatrists think about the role official misinformation about the danger during the accident played in generating hysteria. Their opinion also would be useful on why an event in which no one was hurt is persistently described as a "disaster." But none of this is worth what it would cost the GPU customers.

We, however, have suffered our own "psychological damage" from the court decision. We had thought we were returning to a political environment in which regulatory bodies and federal courts once again were capable of rational judgments. We were far too optimistic.

The Birmingham News

Birmingham, Ala., January 18, 1982

To those who oppose any use of nuclear energy, Judges J. Skelly Wright and Carl McGowan of the District of Columbia Circuit Court of Appeals may be heroes of a sort. But to many of the 4 million customers of General Public Utilities Corp., the judges may be considered villains of a piece that is going to cost them many millions of dollars for electric power.

The court voted 2 to 1 that TMI proceedings have not given sufficient attention to the "psychological health" of people living in the vicinity of the plant. Hence, forthcoming are further delays in starting up TMI-1, the undamaged ractor that has been shut down since the accident at TMI-2.

The ruling came just when it appeared that headway was being made in clearing up the physical and financial wreckage left by the accident nearly four years ago at the Three Mile Island nuclear generating facility.

No good reason since the accident has ever been given for keeping TMI-1 inoperative. At the time of the accident, it was closed down for refueling. It could have been started up after the Nuclear Regulatory Commission review of the procedures — or more accurately, the failure to follow procedures — that caused the problems at TMI-2. The TMI-1 managers were found to be in full compliance with even more stringent NRC requirements.

If the utility had been permitted to start up TMI-1 promptly after the necessary examination by the NRC had been completed, it would have helped provide a part of the cost of cleaning up the damaged reactor. But with both reactors down, the company has had to buy expensive power from other utilities in its power grid. This heavy cost has forced it to borrow costly money at a time when its resources already were strained.

In turn, the Pennsylvania Utility Commission has had no alternative but to grant rate increases which brought the lower rates for TMI customers in line with rates paid by other Pennsylvania customers not receiving nuclear generated power.

The suit was brought by a group called "People Against Nuclear Power," which contended that the NRC must consider how "psychological health" in the Three Mile Island area could be damaged by starting up TMI-1. The NRC earlier had rejected the demand on the basis that psychological determinations were not required under environmental law.

In deciding for the anti-nuclear group, the judges approved the idea that the NRC, having consulted everyone else these past 36 months, must now call in the psychiatrists, professionals not noted for consensus on many psychological questions.

One would hope that while the psychiatrists deal with the question of health, they confront the role that misinformation about the danger during the accident played in creating hysteria. They might also be asked for an opinion on why the accident in which no one was injured is obdurately referred to by most in the media a disaster.

The Washington Times

Washington, D.C., June 30, 1983

Do you worry about Three Mile Island? You should.

Not because of what happened there four years ago — that didn't amount to much — but because of what could happen there now, with another reactor being worked on by union welders who don't know the difference between a blowtorch and a hammer.

Yesterday's hearing before Orrin Hatch's Senate Labor Committee confirmed our Whitt Flora's revelation that some members of the boilermakers' union gained journeyman status by buying it. Without passing through apprenticeship programs or taking thousands of hours of on-the-job training as union rules require.

Flora's Wednesday story reported that as little as $300 will buy the full union membership needed for assignment to jobs paying as much as $60,000 a year, three times what greenhorn welders earn. Witness Gary Boring told committee staffers of union members at Three Mile Island pounding studs into the containment vessel walls with hammers, and of workers with no experience who'd *bought* the union "books" they needed to get work at the nuclear plant.

The reactor has never been put into operation because of lengthy licensing procedures and litigation. Earlier this year the Supreme Court rightly threw out a suit demanding consideration of nearby residents' "psychological stress" before operation could begin. If the nailbiters had prevailed, hardly any government action would be immune from harassing litigation charging "emotional distress."

Vague worries about grandly improbable accidents aren't reason enough to keep a nuclear plant on ice. But the possibility of defective welds on the reactor's core and the walls that are supposed to confine radioactive leaks is exactly the sort of potentially disastrous mistake the Nuclear Regulatory Commission's watchdogs are supposed to prevent.

The commission's inspectors ought to be working overtime to check every inch of welding at Three Mile Island and every other job worked on by members of the International Brotherhood of Boilermakers, Iron Shipbuilders, Blacksmiths, Forgers [!], and Helpers. The reactor mustn't be licensed until NRC *knows* it's been put together right.

The Department of Labor should investigate the sale of union credentials. Anybody involved should be stripped of union office. And Justice should follow up with criminal prosecutions — not only of corrupt unionists, but of everyone who looked the other way knowing the plant's physical integrity was being sacrificed for a few hundred dollars worth of bribes.

THE INDIANAPOLIS NEWS

Indianapolis, Ind., January 16, 1982

It is not enough that experts were brought to Three Mile Island to survey the damage and the cause of the accident in March 1979 at one nuclear reactor (TMI-2). Nor that the management of the utility has improved operations. Nor that TMI-1, shut down for refueling before the accident and undamaged, could be restarted safely.

No, the District of Columbia Circuit Court of Appeals agreed with the suit filed by the "People Against Nuclear Power" and ruled the "psychological health" of the people living near the reactor had not been considered and TMI-1 could not be restarted until it was.

Send in the shrinks, then, and keep fueling this hysteria based on false information. This is no time to be rational about nuclear energy and all the good it can do.

CHARLESTON EVENING POST

Charleston, S.C., November 8, 1982

The U.S. Supreme Court's agreement to decide if the Nuclear Regulatory Commission must consider the psychological stress on neighbors before approving the start-up of a nuclear reactor at the Three Mile Island plant in Pennsylvania suggests at first glance that the justices are about to enter a swamp where the vapors of foggy thinking, unproved theories and contagious fears obscure the paths to the higher ground of reality.

Second glance, however, suggests the high court might walk with sufficient caution to reach a decision of import without getting in over its head. Fear, or psychological stress, or mental anguish, is not exactly new to the courtroom. It bobs up every now and again in the trial of murder cases and of damage suits. It figures in insanity pleas, as the nation was forcefully reminded by the trial of the man accused of trying to assassinate President Reagan.

Because the expert witnesses tend to testify in what to laymen are confusing and often conflicting terms, some people pooh-pooh psychological stress. Yet psychological stress is very real, especially (we should think) among people who live within a stone's throw of a nuclear power plant that was closed down after an accident that could have had grave consequences.

The case the Supreme Court will consider involves an appeals court ruling that environmental law requires consideration of the psychological effect of restarting a reactor at a plant where an accident had resulted in the emission of radioactive material. The NRC argues that the ruling will adversely affect all nuclear licensing decisions and force federal agencies to consider the psychological effect of any project for which federal law requires an environmental impact statement. Such an outcome would be a matter of major concern, for the effect would be widespread and possibly inhibiting insofar as nuclear power development goes.

On the other hand, Congress wrote the environmental impact law as an "action-forcing" mechanism that would compel administrators to examine closely all possible consequences of their actions. Given the fact that man can't be too careful in avoiding the dangers of radiation, a procedure that ensures that nuclear watchdogs cover all the corners — including the anxiety of those within fallout range — does not sound too outlandish after all.

Charleston, S.C., May 2, 1983

The U.S. Supreme Court ruling that the government can allow the Three Mile Island nuclear power plant to reopen without considering the potential psychological affect on neighbors does more than help clear the way for a plant startup. It also keeps the lid on what might be a rather large can of worms.

In its opinion the high court went straight to the legal heart of the matter. It said neither individual nor community anxiety is addressed in the federal law that requires the government to examine environmental impacts when it licenses nuclear reactors. Because possible mental stress is not spelled out as a factor to be considered, the government, the state of Pennsylvania and the operators of TMI were, in effect, home free, no matter what the neighbors thought.

The Supreme Court didn't say having a nuclear power plant nearby isn't stressful — at least for some people. It said only that the law doesn't require environmental impact studies to include consideration of psychological effects on residents. The framers of the law had a reason to exclude such a requirement. Measuring psychological impact is difficult, at the very least. While "mental anguish" is a familiar factor in damage suits, it usually is related to specific physical actions someone is alleged to have taken.

Accurately predicting physical impact is one thing; predicting community anxiety levels is quite another. Requiring environmental impact statements to include mental stress considerations would lead into a swamp of theory and conjecture, of "ifs" and "coulds" and "possiblys." How do you separate the realistic fears from the unrealistic fears of those living in proximity to a nuclear power plant? Who's to define realistic?

All reasonable precautions ought to be — have to be — taken in connection with nuclear power plant construction and operations. Public safety is at stake. Environmental impact studies should be thorough enough to ensure such precautions, but reason has to proscribe the limits of such studies.

Chicago Tribune

Chicago, Ill., April 21, 1983

The U.S. Supreme Court has unanimously ruled that the government does not have to consider the public's mental health in deciding whether to license a nuclear plant. Whatever that ruling may do for the public, it will help keep the law sane.

The justices overturned an appellate court ruling that grew out of the Three Mile Island nuclear accident in 1979. The lower court upheld pleas by a local citizens' group, People Against Nuclear Energy, that residents would suffer "severe psychological stress" if the TMI plant were restarted; it ordered the Nuclear Regulatory Commission to take such emotional states into account before letting a nuclear plant operate.

But the Supreme Court held that this was asking too much. The law, it said, requires federal officials to weigh effects only on physical environment, not on people's feelings.

That shuts off a strange, mutant outgrowth—a hybrid of law and psychology—that seemed to be taking shape in the courts. Under the appeals court ruling, psychological stress tests would become a measure of legality. If a decision about a nuclear plant caused too many people to breathe faster, secrete more adrenalin or see strange things in inkblots, that decision would be illegal.

But this wouldn't necessarily settle it. Presumably those who wanted a given plant to operate could produce test results of their own showing that X percent of its neighbors were breathing normally. The court would then have to decide which set of tests was more reliable; it might have to hire psychologists to test the tests. And the only predictable result is that a lot of nuclear plants would not get built for reasons having nothing to do with their safety or usefulness.

The Supreme Court's ruling does not prevent opponents of nuclear plants from doing all they can to stop them. It just keeps federal energy policies subject to law, not to statistics on blood pressure.

THE TENNESSEAN

Nashville, Tenn., May 1, 1983

THE nuclear power industry has received some good news and some bad news from the Supreme Court of the U.S. in two separate but significant decisions.

The good news was a ruling that said psychological stress suffered by surrounding neighborhoods is not a factor in authorizing nuclear power plants. The bad news for the industry was that Congress has given the states the right to say no to new nuclear plants if the objection is based on economic grounds.

The first case arose from community objections to the restart of Three Mile Island's undamaged Unit 1 reactor. Last year a federal appeals court ordered the Nuclear Regulatory Commission to evaluate stress among local residents who might fear another accident.

The justices held unanimously that "risk of an accident is not an effect on the physical environment," and therefore is outside the impact assessment required by federal law. The ruling held implications not only for nuclear plant construction but all major federal projects where adverse community reaction might occur.

In the second case the court surprisingly upheld a 1976 California moritorium on future reactor construction. Previously it has been thought that the federal government has had exclusive powers in regulating nuclear energy. That has now been weakened to an extent.

Critical to the court's decision was its view that the California law was based on concern about the economic viability of new nuclear plants built in the absence of a permanent solution to waste disposal. Justice Byron White noted that without a permanent means of disposal the waste problem could become critical, leading to unpredictably high costs to contain the problem or, worse, shut down the reactor.

Since there have been no new nuclear plants ordered since 1978, the only practical impact of the ruling may be on plants of the future and it will put on them the added problem of state and local interests. The warning to the industry seems clear: It may be tougher in the years ahead for nuclear power.

The Chattanooga Times

Chattanooga, Tenn., April 23, 1983

In two decisions this week the U.S. Supreme Court handed the nuclear power industry a boost and a setback, which is to say the court ruled, generally speaking, in the public interest.

The first decision came Tuesday when the court held unanimously that the Nuclear Regulatory Commission need not consider psychological stress among residents in deciding whether to permit a resumption of operations at Three Mile Island I. That is the companion to Three Mile Island's Unit 2 which was damaged four years ago last month. A lower court had ruled in a lawsuit brought by residents of the Harrisburg, Pa., area that the NRC had to evaluate the psychological stress that restarting Unit 1 might cause among residents fearful of another nuclear accident. The Supreme Court disagreed, but despite the court's ruling, Unit 1 is not likely to be restarted soon.

The ruling is sensible, even though it will be no comfort to the residents who brought suit to halt the restarting. Justice William Rehnquist wrote that the National Environmental Policy Act of 1969 requires federal agencies to assess in advance the impact of actions "significantly affecting the quality of the human envionment;" the question was whether psychological stress caused by the operation of a nuclear power plant was the kind of "environmental impact" that triggers the mandatory assessment requirement.

The justices rightly said no. "A risk of an accident is not an effect on the physical environment," Justice Rehnquist wrote. "A risk is, by definition, unrealized in the physical world." Modern technology generates many risks, providing both the possibility of accidents and opportunities for achievements. But, the court said, while balancing technological gains and psychological risks may be an important public policy issue, that is not something Congress wanted federal agencies to do as part of an environmental review. The agencies would be forced to develop psychiatric expertise at the expense of the truly relevant functions they are required to perform. If they were required to assess the psychological effects of every potentially risky venture in society today, progress might well grind to a halt.

In a far-reaching decision Wednesday touching on states' rights and federal responsibility, the court ruled that until the federal government can come up with a safe way to dispose of radioactive nuclear waste, states can ban construction of new nuclear power plants — provided the ban is motivated by doubts about the economics of nuclear power rather than its safety. California asserted that the motive of its moratorium was economic, in that the lack of a national program for storing nuclear waste meant that the power plants faced unpredictable costs in the future.

Statements by nuclear power industry spokesmen notwithstanding, the court's ruling is a setback for the industry, made all the more important by its scope and unanimity. It effectively prevents the construction of new plants in at least five states — and perhaps all states if they adopt legislation similar to the California law on which the court ruled. Just as clearly, the ruling should prod the federal government to speed efforts to come up with workable technology for permanently disposing of the radioactive spent fuel that a nuclear plant creates. Legislation passed last year commits the federal government to finding such technology.

It may be that in the future, the United States will need much more energy than it is now producing, and that need may have to be met by nuclear plants. But absent the protection of a workable waste disposal plant, we'd better not count on building new plants.

Pittsburgh Post-Gazette

Pittsburgh, Pa., April 22, 1983

The nuclear-power industry won one and lost one — a big one — in the U.S. Supreme Court this week.

On Tuesday the high court, in a 9-0 decision, ruled that the government can allow a restart of the undamaged Unit 1 reactor at Three Mile Island without first weighing whether it would cause psychological damage to nearby residents. People Against Nuclear Energy, a group of about 2,000 people, had filed a lawsuit in the aftermath of the March 28, 1979, accident at the adjacent Unit 2, the nation's worst commercial nuclear accident.

In a victory for the nuclear-power industry, Justice William H. Rehnquist wrote: "We think the context of the statute shows that Congress was talking about the physical environment — the world around us, so to speak. If a harm does not have a sufficiently close connection to the physical environment, [the law] does not apply."

Thus the nuclear-power industry was saved from the never-never land of psychology where it might be almost impossible to prove that opening *any* nuclear plant would not mentally injure nearby residents fearing a potential accident.

But, like the Lord, the Supreme Court giveth and the Supreme Court taketh away. A day after the TMI decision, the court, in another unanimous decision, ruled that states have the right to ban new atomic-power plants for economic — though not safety — reasons. The justices upheld a California moratorium on future nuclear-power-plant construction, dismissing federal-government and utility challenges that the state's action thwarted federal policy of promoting nuclear power. They held that California had a proper *economic* concern over the fact that the lack of a national program for handling nuclear waste threatens nuclear plants with unpredictable future costs or even shutdowns.

Here was clearly a pro-state's rights decision. The court held that while the federal Atomic Energy Act of 1954 prohibited the states from involvement in the safety aspects of nuclear-power plants — keeping that as the exclusive domain of the federal government — they are left free to regulate in the economic area.

Ordinarily, industry likes to have power in the hands of states, figuring it is easier to win battles at the local level than in Washington. But here, clearly, is a decision that will dismay the nuclear industry. That is because the federal government has been the staunchest promoter of nuclear power, and, more pertinently, because the usual situation is reversed — legislatures and local officials are likelier to prove more vulnerable than Washington to pressures, in this case, from anti-nuclear activists.

Of course, opponents of nuclear power will have to reverse their field and, instead of stressing safety and radiation, concentrate upon economic factors. But the nuclear industry rightly can fear that the route will make little difference; the result likely will be the same: making it more difficult than ever to construct new nuclear-power plants. (No utility has sought a license to build a nuclear plant since 1978, mostly for economic reasons, but with the TMI accident no help.)

Three results now can be expected. One is that there will be pressure on other states to emulate California, Connecticut, Maine, Maryland, Oregon and Massachusetts in passing a moratorium.

Second, consumers may face higher utility rates, with the nuclear option crippled, with utilities revving up now-idle plants that use expensive oil and gas and with the ability of utilities to raise money for plant construction of any kind hurt.

Third, the federal government will be under greater pressure than ever to devise a national plan for the permanent disposal of spent fuel from every nuclear plant. That, of course, will run into the NIMBY ("not in my backyard") problem. But not until there is such a solution will the nuclear-power industry have hopes of overcoming the *economic* barriers posed by this week's Supreme Court decision.

Appendix

Is There a Future for Commercial Nuclear Energy?

Because of all the problems, discussed in previous pages of this book, that have beset the nuclear power industry—bloated electricity demand forecasts, soaring interest rates and construction costs, shutdowns, mishaps, public opposition, etc.—the future of commercial nuclear energy has been called into question. Some analysts say it looks very bleak indeed, that it may already be too late to make nuclear power economically viable. No new orders for nuclear plants have been placed since 1978, and in the current climate of uncertainty none are expected through 1990. Many utilities are turning back to coal, as the price of financing nuclear plant construction rises higher and higher. One unknown factor in the equation for the future is electrical demand; after increasing at an average rate of 7% a year for over a decade, demand for electricity increased by only 1.7% in 1980, by .3% in 1981 and actually shrank in 1982. But many analysts warn that eventually, as the economy recovers, more electricity will be needed, and that if the U.S. gives up on its nuclear energy program now, it may rue that decision in the future. In the long run, they argue, nuclear fission is still the most cost-efficient energy source.

CHARLESTON EVENING POST

Charleston, S.C., June 26, 1980

At the Venice economic summit, one of the conclusions drawn concerning the West's over-reliance on imported oil was that greater use and development of alternative sources of energy had to be pursued with vigor. One of the alternative sources specifically mentioned was nuclear power. "We underline the vital contribution of nuclear power to a more secure energy supply," the heads of state declared. "The role of nuclear energy has to be increased if world energy needs are to be met. We shall therefore have to expand our nuclear generating capacity."

How, then, must President Jimmy Carter's fellow summiteers view the Democratic Party's platform plank on nuclear energy, unanimously approved the very next day? It calls for the virtual elimination of nuclear power plants in the United States.

Some of the new boys at the summit might think this is a classic case of the Carter right hand not knowing what the Carter left hand is doing. Wiser heads, better schooled in American politics, will know the worth of party platform promises — which is exactly zero.

THE INDIANAPOLIS NEWS

Indianapolis, Ind., October 2, 1981

An international conference in Vienna last week broadcast some danger signals. It was a conference of the world's leading nuclear scientists, and the signals they sent were not "Beware of nuclear power."

On the contrary, their warning was: "Last chance to develop cheap and safe power."

While the anti-nuclear protestors were milling around the Diablo Canyon nuclear installation in California, the experts in Vienna were condemning them — and others — for their prejudice and shortsightedness.

The assembled experts were the scientists of the International Atomic Energy Agency (IAEA). Their reaction to anti-nuclear demonstrations in California and Western Europe was one of dismay. They said the world — and especially the Third World of developing nations — is on the verge of passing up the opportunity of the century of having an unlimited supply of cheap power, which could be used to replace dwindling supplies of fossil fuels. They also lamented the increasing reliance upon coal — a fuel which may be environmentally unsafe if it is used in sufficient quantities.

They commented negatively that the essential lead time needed to build a nuclear plant in the United States is now more than 10 years.

The Director General of the IAEA is Sigvard Eklund of Sweden, who in his report to the group pointed out that as late as 10 years ago it was believed that 50 percent of the world's electricity would be generated by nuclear power by the year 2000. At present rates of construction about 22 percent will be nuclear-generated. About 10 percent is the discouraging figure today.

"When man's ingenuity has enabled him to produce almost unlimited amounts of cheap energy," said Dr. Eklund, "it is a pity not to make full use of that achievement in order to improve his living conditions."

Dr. Eklund is only one of hundreds of scientists who are saddened — even angered — by the propaganda generated by a small group of protestors who are ignorant of the safety record of nuclear power facilities around the world. The scientists believe that the key to development of the backward nations is cheap power. They also believe that the continued economic well-being of industrialized nations is cheap nuclear power. Reliance upon alternative energy sources, including solar power, is extremely uncertain, they believe.

These scientists believe — and we agree — it is tragic that the free world's best hope for cheap power has been dashed by a handful of ignorant marchers and shouters.

Rockford Register Star

Rockford, Ill., October 21, 1981

When President Reagan virtually eliminated funding for research in such potential new energy sources as the sun, there were complaints from those who had hoped for alternatives to both dependence on foreign oil and the controversial nuclear program.

Then he endorsed, without reservation, an all-out commitment to nuclear power, including development of its most controversial form, the breeder reactor. That brought objections from environmentalists, from anti-nuclear forces and from those who feared the consequences of producing large amounts of bomb-grade plutonium by the breeder plants.

But none of these objections carried much weight with Congress or the business/industrial community which so badly needs major sources of energy to fuel this nation's economy.

Now, however, comes a protest from another source — one based on economic factors. This protest, a House of Representatives Government Operations Committee report, claims nuclear energy is too expensive to produce.

To the president's claim that nuclear energy costs will drop if unneeded government regulations are eased, the report says nuclear design and construction problems, not government regulations, are responsible for unforseen nuclear costs. The report also claims that any easing of government safety regulations will further undermine shaky public confidence in nuclear plants.

Now, it must be noted that this report comes from a committee dominated by Democrats, so the report has to be, at least in part, influenced by political considerations. But it must be recognized also that the nuclear industry has been plagued by huge construction cost overruns and problems with lack of public confidence.

There clearly are legitimate questions about whether nuclear energy will solve the long-range problems of providing cheap, trustworthy energy. And as long as these questions exist, there hardly can be any justification for putting, as Reagan has done, all our future energy hopes in this one source. Sabotaging research into alternative sources, as the present administration has done, simply leaves us with too few options for the future.

THE BISMARCK TRIBUNE
Bismarck, N.D., October 24, 1981

Nuclear power.

If you want to start a lively discussion at a party, just mention those two words.

If you're among like-minded people, the discussion turns into a "Can you top this?" exchange. And if you're among people with differing views, the gloves come off and the discussion can turn downright nasty.

The nastiness you're likely to encounter points up a major problem within the industry: Nuclear power advocates have not gotten their act together.

And, as a result, they're likely to encounter further public resistence. Furthermore, that resistence is likely to mean the industry will have a harder and harder time raising capital to finance construction. That, in turn, translates into higher consumer utility rates to meet the inflated costs of capital, and, in the end, nobody wins.

The industry's public image was somewhat tarnished before Three Mile Island, but there are additional problems now, and the industry seems at times to be its own worst enemy. Consider:

• The industry, and, in particular, Pacific Gas & Electric Co., continued to push the Diablo Canyon atomic power plant licensing in the face of a widespread public protests. As it turned out, nearly 2,000 people were arrested during demonstrations at the plant site last month. Even if Pacific Gas & Eletric Co. is right — even if there is no danger from the Hosgri earthquake fault three miles away — officials should have backed off from a confrontation. The confrontation has only hardened the attitudes of the most outspoken opponents to nuclear power development.

• Pacific Gas & Electric learned near the end of the two-week confrontation that there was a diagram error, and engineers had to stop fueling the reactor. A Nuclear Regulatory Commission official said, "The problem with the diagrams led to a construction error. As a result, the pipe supports may be inadequte to withstand an earthquake measuring 7.5 on the Richter scale." The pipe system is used to remove residual heat from the generator when the reactor is closed down for refueling, maintenance or an emergency. A company spokesman asserted that the problem has "nothing to do with radioactivity." That may be, but little wonder that some people are not confident the plant could withstand an earthquake.

• Three Mile Island continues to pop up in the news, and just recently, the Reagan administration decided to help pay the estimated $1 billion decontamination bill at the plant. The government may pay up to 20 percent of the cost, in addition to the $37 million previously committed for removal of the damaged reactor core in Unit 2.

In case anyone has to be reminded, that money is taxpayer money, and it is being used to help bail out a private enterprise.

• Even some within the courts system are giving the public signals that, in essense, oppose the development of nuclear power generation. The U.S. Court of Appeals in San Francisco just last week upheld California laws restricting reactors, and the unanimous decision said, "Until a method of waste disposal is approved by the federal government, California has reason to believe that uncertainties in the nuclear fuel cycle make nuclear power an uneconomical and uncertain source of energy." The ruling, by the way, does not affect the Diablo Canyon nuclear plant.

• Waste disposal is in a state of chaos. The Edison Electric Institute, a trade group of privately owned utilities, says nuclear plant owners are spending at least "tens of millions of dollars" to "re-rack" spent fuel rods in on-site storage facilities. Those costs, in turn, are being passed on to electricity users, the institute noted.

Each plant had been designed to store a certain number of spent but highly radioactive fuel rods until the government could dispose of them. But the government's plans for handling the rods became bogged down during the Carter years, and, as a result, the plants have been installing new aluminum racks with boron insulators. The boron acts as a "poison," in the jargon of the industry, and prevents neutrons from hitting fissionable uranium atoms, thus preventing a chain reaction.

A total of 50 of the 71 licensed operating plants in the United States have received NRC permission to re-rack, and 10 more applications are pending.

President Reagan has announced his intention to permit the commercial reprocessing of spent nuclear fuel rods, but there have been indications that such a plan will encounter stiff opposition in Congress.

Those are just a few of the recent problems to beset the nuclear industry, and, as we say, its image has been tarnished.

But there's an even more basic problem involved in the controversy over nuclear power. It is a problem of risk perception, and that is related to the extent of damage possible if something awful did occur.

There are worst-case scenarios that suggest how many people might die and how many more might be hurt if a melt-down occurred at a plant. Casualties are one part of the equation.

But industry sources tell us that the chance of a melt-down is extremely slight. That is another part of the equation.

The industry sources give greater weight to the slim chance of an accident, while the opponents give greater weight to the extent of damage if there were an accident.

Who is right? Neither side and both sides; for, the answer depends on the individual's perception of acceptable risk.

Add to that the continuing confrontations, the continuing design flaws and the mounting waste disposal problems, and it seems evident that there will be no end to the controversy any time soon.

The government and the industry are in a position to change all of that, however, if they will take into account the fears and build a solid track record based on candor and good experience.

THE MILWAUKEE JOURNAL
Milwaukee, Wisc., December 7, 1981

It boggles the mind to hear the continued claim of the nuclear power industry that its method of generating electricity is less costly than other major power sources. American taxpayers have already spent $12 million on the cleanup of the wrecked Three Mile Island nuclear power plant; $19 million more is budgeted for this year, and the Reagan administration has proposed spending another $100 million.

And that's just the possible federal share. TMI's owner has already used up its $300 million in nuclear-accident insurance. The utility's rate payers and/or stockholders will ante up plenty more toward the billion-dollar TMI decontamination task.

Can you imagine any accident at a coal-fired power plant costing so much?

We can't either. Yet, the nuclear power people still claim their method of generating electricity is cheaper than using coal — even as they demand more government subsidy of nuclear power. As Rep. Richard Ottinger (D-N.Y.), chairman of the House energy conservation and power subcommittee, put it:

"The Reagan administration continues to spout its rhetoric about the magic of the marketplace while it quietly spends millions of dollars to save the nuclear industry from the verdict of the marketplace."

Actually, we don't believe the relative costs of nuclear vs. coal power are sufficiently clear. Left out of the cost equation are such items as nuclear-accident cleanup, waste disposal and the eventual decommissioning of old nuclear plants, as well as such coal-power costs as mining accidents and air pollution.

We suspect that nuclear power would not turn out to be the cheapest. But, in any case, we think a thorough study would be a shocker. An unvarnished look at the full cost of electricity might help drive home the common-sense lesson that it is in everyone's interest to use all forms of energy as efficiently as possible.

The Salt Lake Tribune

Salt Lake City, Utah, December 23, 1981

Thirty years after its "accidental" birth, the American nuclear power industry is suffering the ill effects of another accident that emphasized the danger inherent in this once promising energy source.

On Dec. 20, 1951, less than 20 persons watched as the Experimental Breeder Reactor at the National Reactor Testing Station near Idaho Falls, Idaho, produced sufficient electricity to light four 110 watt bulbs. The event was the offshoot of a breeder reactor experiment. Today, some 12 percent of the nation's electricity is supplied by nuclear reactors. But the once-rosy outlook for nuclear power was dimmed in 1979 by a frightening accident at the Three Mile Island reactor in Pennsylvania which showed that despite elaborate safeguards, the danger potential remains high.

The Three Mile Island accident increased public demand for even greater regulation of an industry which was already showing financial strains caused by bureaucratic delay and direction. Since then, high costs, rising concern about nuclear waste disposal, expensive design changes and a host of other problems have virtually stopped nuclear power plant construction and imposed new burdens on existing facilities.

There is reason to believe the setback is temporary. The Reagan administration has declared a federal policy of promoting the development of nuclear power. In addition, the price of uranium fuel is low now and should remain so for several years, a direct result of government stockpiling and the slowdown in demand from the moribund nuclear industry.

In the interim, the Nuclear Regulatory Commission (NRC) is taking an unusually critical view of past safety measures and proposing stronger new ones. Though initially hard on the industry, the NRC toughness could provide a level of public confidence vital to achieving economic comeback.

In theory, nuclear power remains an admirable source of energy. The challenge is still the same as in 1951: to translate the bright promise into safe and practical reality. Thirty years after its inception, the nuclear power establishment seems to accept this premise without reservation. It is less certain that it will survive long enough to put the new-found dedication into practice.

ST. LOUIS POST-DISPATCH

St. Louis, Mo., December 28, 1981

Nuclear Power Continues To Edge Coal In 1980 Costs, Performance

Or so says the headline on a recent press release from the Atomic Industrial Forum, an industry trade organization. The text of the release was somewhat more precise. It said the edge nuclear plants enjoyed — in overall plant performance and total cost of producing electricity — was "modest."

"Nearly nonexistent" would have been a better choice of words.

By limiting its "study" to plants that were in commercial service during *all* of 1980, the AIF was conveniently able to ignore the disabled unit at Three Mile Island — out of service since the accident in March 1979 — and a second TMI unit that also sat idle in 1980.

That omission — and others — prevented the AIF from burdening its survey with facts that did not fit the message it wanted to deliver. Had the TMI units been included in the "capacity factor" calculation, for example, the edge claimed for nuclear plants would have gone to coal-fired plants and nuclear's "forced outage" edge over coal would have also been lessened. As for "availability," coal plants had the edge — 75 percent to 64 percent — the AIF reports. The gap would have been larger had TMI and other disabled nuclear plants been considered.

That brings us to the bottom line. The AIF reports that nuclear plants produced a kilowatt of power in 1980 for an average of 2.3 cents compared to 2.5 cents for coal. But its figures also indicate that coal plants that have come on line since 1976 are more likely to be available, less likely to experience forced outages and capable of producing a kilowatt of power more cheaply than their nuclear counterparts.

As a result, the nuclear-coal difference has narrowed dramatically since a 1979 AIF survey that found a nuclear-coal spread of 1.9 cents to 2.3 cents. And therein lies the challenge to those creative statisticians at the AIF who must find a way to give the edge to nuclear power in 1981 and beyond. Between 1979 and 1980, the price of a kilowatt of electricity produced from the atom went up 21 percent. The increase for coal plants was 8.7 percent, and there is little to indicate that that trend will be significantly lessened, let alone reversed.

In spite of the statistical contortions the AIF uses to claim a "modest" edge for nuclear power, utilities, utility regulators and investors are beginning to catch on. They are seeing that the bottom line advantage claimed for atomic power has an increasingly Alice in Wonderland quality about it. In fact, like the Cheshire Cat, it is disappearing into thin air.

The Evening Gazette

Worcester, Mass., February 12, 1982

According to the newspapers, a group of very important people met privately with Vice President George Bush recently to see what can be done to save the sputtering nuclear power industry from bankruptcy.

If the reports are correct, this group of utility executives, reactor manufacturers, investment specialists and state rate commissioners urged the government to pour even more money into the industry. Some even talked about nationalizing nuclear power.

That would be about the worst idea since the chain letter. A government bailout of the nuclear power industry would make Chrysler look like a good investment.

Given the relationship of the atom to military force, the government had to become involved in nuclear research and nuclear power. As a result, the whole nuclear industry is built on enormous federal subsidies.

But beyond those subsidies, mainly involving the processing and regulation of uranium, the federal government permits the free market in nuclear power to operate as far as possible. If the nuclear industry cannot survive even with that kind of help, then something is wrong.

The nuclear power industry is sick — no doubt about that. In the past two years, 23 nuclear power plants have been canceled in the United States. Not one has been started. Several under construction — including those at Seabrook, N.H., and Millstone, Conn., are in deep trouble. In Washington State last week, the Washington Public Power Supply System canceled two partly-built nuclear plants. According to Fortune, the electricity users of the Northwest will pay $7.2 billion in principal and interest before the bonds are paid off in 2018. And they will not get one kilowatt of power for that money. In Maine, a second referendum will be held on whether to close down the state's single nuclear plant at Wiscasset. In California, the Diablo Canyon plant remains dead while regulators check out various design and engineering flaws.

This is not the sort of enterprise the U.S. government wants to jump into, especially when it is wrestling with huge deficits. The government has plenty of nuclear problems right now, including the problem of what to do with radioactive wastes.

In fact, if the government had been able to solve that problem, the nuclear industry might not be in such poor shape. It would have had an answer to one of the most persistent questions raised by its critics.

A generation ago, nuclear power looked like the answer to the nation's energy problem. Regions like New England, where one-third of the electricity is generated at nuclear plants, have been well served by nuclear power.

But it is now becoming evident that fission energy is just a transition step. Something else — probably coal and alternate fuels — will gradually replace it.

Nuclear power, even with huge subsidies, is failing the test of the free market. The government should not try to keep the dying dinosaur alive.

Roanoke Times & World-News

Roanoke, Va., February 12, 1982

IF YOU LIKED the Lockheed and Chrysler bail-outs, you'll love what the nuclear-power people are pushing nowadays. They want Uncle Sam to help the ailing industry with $50 billion in low-interest loans.

Spokesmen for electricity interests met recently with Vice President Bush, Energy Secretary James B. Edwards and Commerce Secretary Malcolm Baldrige. Among those represented were eight utilities, three nuclear-power supply firms, two Wall Street investment companies, and two utility regulators.

The meeting was behind closed doors, but some participants supplied details later. Charles H. Dean Jr., chairman of the Tennessee Valley Authority, said he proposed a national nuclear-energy pool, backed by a federal nuclear-financing bank, "to assure supply of capital needed to complete plants now started and past some specified stage of construction." The investment, he said, would be repaid in seven or eight years in savings on oil imports.

Lelan F. Sillin Jr., chief executive officer of Northeast Utilities, a group of five nuclear-owning companies, had an alternative idea. He suggested that utilities be combined into large regional power companies, to be regulated by the Federal Energy Regulatory Commission. This arrangement would allow an end run around state utility commissions, which the utilities believe are too stingy with rate increases.

What this free-enterprise administration will do with these proposals is undetermined. Its budget already includes $1.6 billion to help nuclear power, a 35 percent increase. Rep. Edward Markey, D-Mass, a critic of the industry, says, "I just can't believe that Congress will go along with some kind of a $50 billion nuclear Salvation Army when one out of every 10 American workers is unemployed." One hopes not.

There's a lot of talk about the atom's dangers, but the real weak spot of the nuclear industry is the fabulous cost of building and maintaining these generating facilities. It's common now for a sizable nuclear plant to carry a multibillion-dollar price tag.

Moreover, that's only the beginning of the cost. The industry has not yet faced up to its ultimate responsibility: planning for the care of these plants when they've served their useful life of 30 years or so.

You don't just pull the plug from a nuke, lock the gate and walk away. High levels of radioactivity build up within a plant during its operation. Either you tear the facility down and dispose of the dangerous waste (where and how has not yet been determined), or you seal up the closed plant. If you choose the latter course, the public must be protected while the levels of radioactivity decline.

This means the plant itself must be closely guarded — perhaps for a century or so. In his book "Radwaste" (Random House, 1981), reporter Fred Shapiro says:

> A comparison of the long-range costs of the three alternatives is contained in a nuclear consulting company's estimate in 1979 on expenses involved in dismantling, or entombing or mothballing two 1,-264-megawatt units Detroit Edison planned to build in St. Clair County, north of Detroit. For mothballing, surveillance for a century (at $606,300 a year), and then dismantling, the total was $123.4 million; for entombment, a lesser degree of surveillance for the century (at $242,000 a year), and then dismantling, the total was $109.8 million. Prompt dismantling and disposal was estimated to cost $110.5 million. Presumably Detroit Edison took these estimates into consideration the following year when it decided not to build the reactors.

The TVA's Dean said that for about $50 billion in low-interest loans, the proposed federal bank could acquire control over 20 million kilowatts of nuclear-generating power in about 10 years.

Acquire control, yes. And ultimate responsibility, too. What a relief it would be for a utility if it could turn over to Uncle Sam the mortgage on an obsolescent nuke, and let him mothball and guard it for the next 100 years. Plenty of bum ideas have been brought to Washington, but this has to be among the worst.

The Chattanooga Times

Chattanooga, Tenn., February 16, 1982

Scratch a member of the nuclear industry and you're likely to get a sermon on the wonders of the free enterprise system, which makes a recent meeting between industry representatives and Vice President Bush all the more interesting. At that meeting the industry requested $50 billion in low interest federal loans to bail the industry out of a jam.

Of course, the nuclear industry enjoys some of the same advantages as other businesses, such as tax credits for job development and depreciation allowances (accelerated under last year's tax bill) for plants and equipment. Perhaps realizing that resistance to a bailout might prove strong, the industry suggested to Mr. Bush that the Federal Energy Regulatory Commission assume the rate regulation responsibility now exercised by state utility commissions. Pointing out that the FERC already governs wholesale rates, the industry officials argued that the change they suggested would improve regional coordination. Perhaps so, but let's not lose sight of what they really want: a lot more money.

One reason lies in the fact that whenever nuclear plants are shut down, the operators not only have to use more expensive fuel to make up for the lost capacity, they also have to continue collecting high rates to pay off the cost of an idle plant. Who's to blame? Nunzio J. Palladino, the pro-nuclear power advocate appointed by President Carter to the Nuclear Regulatory Commission, said that during his first six months as chairman, the large number of deficiencies at some plants "showed a surprising lack of professionalism in construction and operation of nuclear facilities. The responsibility for these deficiencies rests squarely upon the shoulders of management."

The nuclear industry's problems are undoubtedly serious, but many of those problems are its own fault. Until the industry cleans up its own act, Congress should turn a deaf ear to pleas for a bailout.

THE KANSAS CITY STAR

Kansas City, Mo., November 26, 1982

A nuclear power industry group has reported that the cost of producing electricity from coal increased almost twice as much as the cost of power from nuclear reactors last year. The Atomic Industrial Forum surveyed all 44 utilities which operate both types of plants and found that total costs—both capital and operating—of nuclear generation averaged 2.7 cents a kilowatt-hour. Coal-burning plants had a cost of 3.2 cents a kwh while oil-burning utilities soared up to 6.9 cents.

A year earlier, in 1980, the AIF survey figures were 2.3 cents for nuclear, 2.5 cents for coal and 5.4 cents for oil. The industry association suggested that a reason for the more rapidly increasing coal costs could be the burden of fitting coal plants with expensive scrubbers to control air pollution. Previously coal had been closing the cost-advantage gap between it and nuclear power.

This comparison goes back to the reasons atomic energy seemed such a bonanza when it first emerged, as a clean, relatively low-cost power source. It suggests too why most of the nation's utilities still are clinging grimly to nuclear in the face of so many problems and so much doomsaying. Nuclear does have an inherent cost advantage if coal is burned with the proper air quality safeguards, and the long-playing acid rain controversy is likely to bring stronger demands that this be assured.

Part of nuclear power's woes trace to an economic recession that, along with unexpectedly successful conservation in the face of higher prices, caused projections of future electric demand to be far too high. Many nuclear reactors have been shelved, at heavy losses of investment already made, because they simply aren't needed yet.

But nuclear energy has more basic problems in plant design and safety, and the ultimate disposal of its radioactive wastes. High costs of borrowed capital, environmental opposition and a lengthy, complex licensing process have also warped the nuclear dream. The electric generating cost advantage that once looked so good has been damaged by various other factors in the nuclear equation, not all of which are readily resolvable.

ARKANSAS DEMOCRAT

Little Rock, Ark., August 11, 1982

Foes of nuclear power have made such a thing of picketing power plants as creatures of the devil that you'd think they're endangering what's only a fledgling energy industry. But the nuclear-power industry has not only arrived, it's on its way toward doubling and will become a prime career field for the 1980s and beyond.

Peterson's Guides, a Princeton concern that follows job trends, puts demand for nuclear engineers 25th on a list of 58 engineering specialties. Meanwhile, 40 major universities have nuclear engineering programs, and in 1980 – latest year for which figures are available – 500 men and women got degrees.

The Atomic Industrial Forum – a private association of 600 manufacturers, utilities, universities and unions – says the industry even now has many openings for engineers and technicians as well as for a large number of support people: radiologists, metallurgists, plumbers, nurses, technical writers, etc.

As for the industry's status, AIF says that nuclear energy is now next to coal as a source of power in this country and that the number of nuclear plants will more than double to well over 150 in the 1980s – creating 16,000 new jobs.

Dr. Paul E. Gray, president of the Massachusetts Institute of Technology, a focus of research for nuclear energy production, says that though less than 15 per cent of all the country's kilowat hours of electricity are now nuclear-produced (as against 50 per cent, for example, in France), we have enough depleted uranium in storage to provide for *all our energy needs* – not just our electric needs – for the next 300 years.

It's this astounding statistic that backs up the widespread industry demand for breeder reactors operating on reprocessed plutonium. But there remains the problem of poisonous nuclear waste – which occasions most of the opposition to nuclear-power generation. However, Gray says that MIT and other universities are working to perfect thermonuclear fusion reactors, which produce no long-lived nuclear waste, and that these will be operational after the turn of the century.

Columbia, S.C., November 20, 1982

ELECTRICITY generated by the V.C. Summer Nuclear Station finally began flowing to Midlands homes at 5:59 p.m. Tuesday after a costly decade of construction and controversy.

The nuclear power plant, which will operate for S.C. Electric and Gas Company and the S.C. Public Service Authority (Santee-Cooper), is said to have cost $1.2 billion. That is a heavy investment, indeed, but the owners believe it will be cost-effective or they wouldn't have built it.

The new reactor on the Broad River north of Columbia near Parr went on-line at a pivotal point in our country's nuclear experience. It is one of five new reactors to start up this year. Together they will add 9 percent to the nation's nuclear power production. An additional 22 reactors are scheduled for completion by the end of next year.

It is a fact, however, that no new licenses for reactors have been sought by utilities in the past two years. And given the cost of construction today, the economic conditions and the regulatory problems, there aren't likely to be any new ones sought soon.

The nuclear industry is in trouble. While its safety record is excellent, there are serious unresolved safety problems regarding the condition of active and aging reactor cores, radioactive waste disposal, spent-fuel reprocessing, licensing and regulations.

The nuclear industry problems are exacerbated by the relentless challenges of militant anti-nuclear organizations. In many cases, their efforts to delay licensing of power reactors at various stages of construction have added appreciably to the costs.

Nevertheless, the central issue is one of safety. Plant owners should be subjected to reasonable challenges to make sure the public health is well protected, and the Nuclear Regulatory Commission must also be kept on its toes. But a great deal of opposition has been specious, and often outright attempts to scuttle the nuclear program.

The State has expressed before its confidence in American technology, especially its nuclear technology. No other industry has safeguarded the public as effectively.

Earlier this year, Dixy Lee Ray, a past chairman of the Atomic Energy Commission and a former governor of the state of Washington, somberly addressed the National Coal Association:

"Do not rejoice in the demise of nuclear power, for the same thing can happen to you. The opponents of the central generation of electricity are strong; the momentum is with them; and they can use the same tactics against you which have been used against nuclear power. These are the tactics of fear and doubt, and uncertainty and delay. . . . The mistake made by the nuclear industry was to try to accommodate all the objections whether they were reasonable or not. . . . Too late they began to realize that no amount of accommodation would ever satisfy the opponents. No amount of safety is ever enough . . . ."

We hope that is not the epitaph of the nation's nuclear industry. It could be.

Nuclear energy has a very significant role in making this country independent of foreign oil.

THE WALL STREET JOURNAL.

New York, N.Y., November 9, 1982

Groups that lobby against nuclear power have had things their own way for some years now. The Three Mile Island accident fueled their drive. Since then they have taken pains to publicize every nuclear power glitch they can lay hands on, no matter how minor. Costly regulatory delays and litigation have been responsible in part for the fact that no new nuclear power projects have been started in the U.S. in over a year.

But nuclear power is by no means dead. There are 82 licensed plants in operation in the U.S. and 76 under construction or on order. Nuclear is now second only to coal as a power plant fuel. And although other nations have forged ahead of the U.S. relatively—France gets nearly 40% of its power from the atom, compared with 13% for the U.S.—nuclear power use will continue to grow for some time despite the critics.

It may be too that the public opinion tide is beginning to turn. Last week, the no-nukes lobby overreached by trying to get Maine voters to approve a referendum calling for the shutdown of the Maine Yankee nuclear power plant, which supplies about 30% of the state's power. Although "environmental" issues normally do well in Maine, voters decided, 55% to 45%, that they would rather live with the atom than spend $2 billion to $4 billion over the next five years for substitute energy.

Idaho voters approved a pro-nuclear initiative which would require the legislature to get voter approval before moving to impede nuclear power development in the state. Massachusetts voters passed a relatively mild anti-nuke initiative requiring referenda on future plant sites and low-level waste repositories.

Not the least interesting thing about these votes was the way a last-minute campaign to whip up anti-nuke hysteria backfired. Just before the election Rep. Edward J. Markey (D., Mass.), chairman of the House Interior and Insular Affairs subcommittee on oversight and investigation, leaked a report to the Washington Post saying that a "worst case" nuclear accident could cause more than 100,000 deaths and over $300 billion in damages.

The "worst-case" scenario used in the study involved the meltdown of the reactor's core, massive failure of safety systems and a huge release of radiation into the atmosphere combined with the worst possible weather and evacuation conditions. As reported by the Post, the study sounded like hot news. It had been conducted for the Nuclear Regulatory Commission by the Department of Energy's Sandia National Laboratories. The Post story said it had found approximately a "2% chance of such an accident occurring in the United States before the year 2000."

That's a pretty scary statistic all right, but the NRC—which is by no means soft on nuclear safety issues—challenged the Post's interpretation. In truth, said Robert Bernero, director of the agency's risk analysis division, the study calculates the probability of such an accident at one in one billion per year of reactor operation.

The utility industry was overjoyed at this chance to fight back against nuclear scare mongers. The Edison Electric Institute, the Atomic Industrial Forum and a utilities group took out full-page ads in the Post and other newspapers charging that "sensational and inaccurate press coverage of a valid and valuable study may have frightened and concerned millions of Americans." The headline on the ad practically chortled in large type, "Most of the Stories Were Wrong."

If the force of this counterattack makes our colleagues in the press a bit more wary of approaches from the no-nukes crowd it will probably be a good thing. A reading of the Sandia study leaves the impression that what it really was saying indeed was that the chances of a nuclear power plant disaster are extremely remote and that anyone who was overly excited by the Three Mile Island accident might want to reconsider. After all, nuclear power clearly has certain beauties: Unlike coal, it's clean; unlike oil, it's not hostage to Persian Gulf uncertainties; if regulated intelligently, it's relatively cheap.

It seems to have had more bad press than it has deserved. A recent study published by Public Opinion magazine suggests that there is a rather sharp divergence on nuclear power safety between journalists and the scientists who know the most about the subject. Samplings of opinion among scientists found a high level of support for nuclear power; 99% of the nuclear experts thought the risks of nuclear energy "acceptable" and 100% thought enough information was available to solve remaining problems. Journalists, particularly in the electronic media, were heavily of the opinion that the scientists are wrong.

"Leading journalists' attitudes toward nuclear energy also are correlated with their political ideologies," commented the Public Opinion pollsters dryly. "The more liberal the journalist, the more likely he is to oppose nuclear energy."

They may be onto something there. And the voters in Maine and Idaho may be too.

Post-Tribune

Guarding Your Interests Daily

Gary, Ind., December 16, 1982

A negative chain reaction is running through the nuclear energy industry. The future of nuclear power, in contrast to its heralded, bright promises, is dimming.

It is a reality that energy experts and utility companies should accept, because they must look for alternatives. The key word is cost.

Utilities are abandoning plants under construction, and socking customers with big rate increases to cover the losses. That is what NIPSCO did, ending work on its Bailly plant after years of controversy. Its customers will pay millions for the project that was barely begun.

We still have a problem with the justification in that, but the point now is beyond philosophy — it is a costly fact.

The North Anna plant near Mineral, Va., is another example. A few years ago, its projected cost was $2.2 billion. Now it is projected at $5.1 billion. The company has decided not to build, but wants rates increased so that customers will pay $540 million in costs.

Another blow to customers comes AFTER a plant is running. The Long Island Lighting Co. in New York increased the cost of its plant four times in the last six months. Now it wants to raise rates by 50 percent during the next three years to help pay for it. The plant is scheduled to be operational in 1983.

The story is general. Costs are deadly. Sixteen commercial nuclear plants have been canceled this year. A utility official said, "It's the demise of nuclear construction, if not the demise of nuclear power." We don't cite that with any great satisfaction, but he may be right. Syndicated columnist Robert Walters, citing regulatory militancy and the cost factors, says the future of nuclear power is indeed bleak.

Walters adds another dimension to the problem — the cost of producing nuclear energy. A plant in Port Gibson, Miss., will begin to generate power next year. It will be the most powerful commercial nuclear generating station anywhere. In 1980, utility officials said it would save customers $1.9 billion in the first seven years and more after that.

Now, that claim has been cut by more than 80 percent, and officials admit that the plant may not create any savings and could cost more than stations using traditional fuels.

The country has, and is, paying a tremendous price for what seemed to be an exciting adventure into a new world of energy. What have we learned?

Without the inexpensive nuclear energy so widely predicted and with the shocking increases in the prices of electricity and natural gas, the need for a national energy policy is painfully obvious. Potbellied stoves and kerosene lamps just do not sound inviting.

THE  SUN

Baltimore, Md., August 12, 1983

The latest blow to the credibility of the nuclear power program was the discovery of cracks in pipes in the Peach Bottom nuclear plant in Pennsylvania on the Susquehanna River near the Maryland border. The federal Nuclear Regulatory Commission says the cracks didn't pose a threat to the plant and nearby residents, but the public probably won't listen. After years of disasters and nuclear disasters, technological and financial, public faith in nuclear power is exhausted.

This is not as threatening a state of affairs as it would have been 15 years ago, when rapidly growing electric demand was seen as an essential condition for an ever-expanding gross national product and nuclear energy the bright star of the power supply future. Recent years have produced a startling upset in the GNP-electric demand equation, as conservation programs took hold in ways that couldn't have been foreseen a few years earlier. In one recent year, there was no electric demand growth at all. Now, at the end of a major recession, no one is quite certain what an expanding economy will do to electric demand. The only certainty is that the 7 percent annual growth rates of the Sixties will not be soon repeated.

So what does all of this mean for future power supply? Conservationists insist that the nation has reached, or will soon reach, a "steady state" in electric demand where it will need only to replace wornout or obsolete plants. Probably more likely is a resumption of growth, except at a much slower rate than that of the Sixties.

Some of the required new power plants will be fueled with oil, but not too many; the nation must be careful not to let OPEC put it in a vise as it did in 1973-74 and 1979. Coal is increasingly less attractive as a power-plant fuel as recent reports tighten the links between coal-fired power plants and acid rain. That leaves nuclear power, and such "exotic" sources as solar-electric ("photovoltaic") cells, geothermal power and so on. The exotic sources may be about to get a new lease on life, after slim pickings during the early years of the Reagan administration. Oil companies are buying up photovoltaic companies, including Maryland's Solarex. The U.S. Department of Energy, which indicates it may provide a higher level of support for solar energy and conservation in future budgets, may see the same handwriting on the wall.

In the meantime, we're glad Baltimore Gas and Electric Company has a relatively trouble-free nuclear power plant at Calvert Cliffs — a plant that has substantially reduced the size of electric bills paid by BG&E customers. Nuclear power, which we believe can be made safe and reliable, may get another chance some day.

St. Louis Globe-Democrat

St. Louis, Mo., January 3, 1983

The demise of the nuclear power industry claimed by opponents is premature. The industry has been hurt but it is far from dead. Opponents have cited the fact that no new reactors are being ordered and that there have been numerous cancellations of other orders as evidence of the nuclear industry's demise.

The cancellations and the lack of new reactor orders are due to a number of factors.

One of the major reasons is the enormous amount of time it takes to build a nuclear plant in this country — an average of 12 years. The tremendously time-consuming and costly procedures, paperwork and hearings required to obtain licenses and permits to build and operate a nuclear plant have put an unfair handicap on what should be a growth industry.

In Japan, nuclear power plants are built in five years, less than half the time required in this country. If U.S. licenses and regulations could be streamlined to bring construction time for an American nuclear power plant down to seven years, there would be a strong demand for new reactors.

Under today's much too long construction time and high interest rates, half the total cost of a new nuclear plant is consumed by interest payments. Many unnecessary delays impose a tremendous financial burden on utilities building nuclear plants.

The long recession and a continuing national effort to conserve power use also have removed the need for many nuclear power plants which had been scheduled.

Nevertheless, nuclear fuel continues to be the most economical source of power despite claims by opponents. Commonwealth Edison of Chicago, in its annual report, lists the following costs of fuel consumed per million BTU of power output: coal, $2.47; nuclear, $.41; oil, $6.79, and natural gas, $4.58.

These figures, however, don't take into consideration capital costs. When capital costs are considered the relative advantage of nuclear power is cut considerably but Commonwealth officials say it still is more economical to use than coal. If U.S. reactor plants could be built in seven years instead of 12, the economic advantage over coal and other fuels for electricity generation would be greatly increased.

Nuclear fuel and coal continue to be the two most promising sources of energy in the nation's future. It therefore is dangerous to throttle the nuclear industry by unnecessarily stretching out the time it takes to build a nuclear plant. If the economy should turn up sharply, the current slack in energy demand could disappear quickly and turn into a shortage. That is when the nation could suffer for its lack of foresight in providing a good climate for the nuclear industry.

BUFFALO EVENING NEWS

Buffalo, N.Y., June 2, 1983

The recent Buffalo News series, "The Fading Dream," has spotlighted the problems and uncertainties plaguing the nuclear power industry. The great hopes held out initially for nuclear power generation as a marvelously reliable and low-cost substitute for coal and oil have faded with time because of developments which were not foreseen a generation ago and over which the utilities and other energy producers had little control.

In the main, as the articles by reporter Robert J. McCarthy pointed out, the grinding to a halt of new nuclear-plant construction within recent years is not attributable as much to the fears aroused by the Three Mile Island crisis as it is to the fact that utilities cannot afford the staggering price tag.

In the face of reduced growth in electricity demand during a sluggish national economy and declining prices for foreign oil, the mammoth construction costs and sophisticated safety precautions required for nuclear plants, combined with high interest rates, provide a powerful deterrent to nuclear technology despite the significant savings in operating costs after plants come on line.

Proponents and opponents of nuclear power agree that sagging public confidence in the atom as an energy source must be restored if construction is to move ahead once an improved economy increases electrical demand. They also generally agree that, from a strictly economic standpoint, if the technology is to continue as a viable power generation source, the current regulatory climate in Washington cannot continue.

While public concerns about safety provide valid reasons for sterner federal regulations over nuclear design and engineering, the nuclear industry suffers from the lack of a "clear, consistent energy policy," in the view of utility leaders.

They cite, among other things, the "blind spot" between approval of plant construction and actual operation, leaving utilities uncertain whether costly plants will ever be allowed to start operations.

The deplorable delay in federal action to ensure adequate storage facilities for nuclear wastes has also contributed markedly to public opposition to new nuclear power plants. There is no question that this problem must be solved if nuclear power is to have a viable future.

For all of the negative factors, however, existing nuclear plants provide a vital substitute for dependence on costly foreign oil, and the nuclear option should remain part of the nation's energy "mix" along with fossil fuels, hydro power, conservation and solar and other new sources.

Nuclear sources currently provide 16 percent of New York State's electricity and 12 percent in the nation as a whole. A state master plan, adopted in 1982, cited the nuclear role in holding down a statewide reliance on oil for electrical generation that is more than double the national rate.

The problems are formidable, but the present hiatus in new plant proposals provides an opportunity for Washington to resolve questions of safety and regulation for the day when an improved economy and a restored demand for electricity may make investments in new nuclear plants necessary and worthwhile.

Los Angeles Times

Los Angeles, Calif., November 7, 1983

Victor Gilinsky, a member of the Nuclear Regulatory Commission, gave the utilities and the nuclear-power industry some good advice the other day. They should concentrate more on mastering the safe operation of existing reactors, he said, and less on trying to make unrealistic projections of nuclear-power development come true.

Gilinsky recalled that utilities all over the country rushed to place orders for nuclear-power reactors in the late 1960s and early 1970s. In retrospect it is obvious that the reactor-building boom, which involved a rather sudden change from medium-size reactors to large reactors, was based on far too little appreciation of economic realities and far too little industry experience with the construction and operation of reactors.

The boom in new reactor orders was followed by a severe retrenchment from which the industry is still suffering.

This country now has the largest nuclear-power program in the world. Existing reactors, with a generating capacity of 63,000 megawatts, account for about 12% of U.S. electric-power production. If you throw in the reactors under construction that seem likely to be completed, the nation will in a relatively short time have 125 plants with a capacity of more than 110,000 megawatts.

What has thrown the nuclear industry into a deep funk is the collapse in new orders. Of the six new reactors ordered worldwide last year, none will be built in the United States. In fact, no American utility has applied for a nuclear-reactor construction permit since 1978. About 100 reactor orders have been canceled, and Gilinsky estimates that more than a quarter of the 57 plants nominally under construction will never be completed.

A number of factors are involved, including high interest rates, massive cost overruns, construction delays, an unanticipated leveling off of energy demand, lower-than-expected prices for oil and other competitive fuels and, very important, the justifiable concern over nuclear-plant safety in the wake of the incident at Three Mile Island.

Unfortunately, as Gilinsky points out, many leaders in both the government and the nuclear industry "have continued to cling to the assumptions and expectations of the boom period."

The nuclear industry likes to imagine that its main problem is overregulation. And no doubt regulatory procedures do contribute to the fact that it takes 12 years to build a reactor in this country, compared with five or six years in Japan and South Korea. But it would be insane to soften safety regulations, as the industry and the Reagan Administration propose, at a time when nuclear safety remains so obviously in question.

By now it is plain that the Three Mile Island incident in 1979 cannot be dismissed as a one-in-a-billion aberration. Last February an automatic shutdown system failed at a nuclear-power plant in New Jersey. More recently the Nuclear Regulatory Commission heard allegations that shoddy construction work on an Ohio reactor was deliberately covered up.

Gilinsky draws the obvious conclusion that the solution to safety problems cannot simply be left to the utilities and the reactor builders. Too much is at stake, and the industry is not structured to cope with the problems.

A major problem, as the commissioner points out, is that the nuclear industry's experience—such as it is—is fragmented. Of the 44 utilities that now operate nuclear reactors, 19 operate only a single unit. Only three have five or more reactors.

"It is only a slight exaggeration," Gilinsky said, "to say that we get nearly 60 different solutions to every safety problem."

Some utilities are just not qualified to operate potentially dangerous reactors. Gilinsky tells of visiting one plant that was planning to "start up next year with a crew of 31 operators, not a single one of whom had ever held a commercial license."

It would help immensely to settle on a small number of standardized designs. This would simplify the regulatory process, the training of skilled reactor operators and the task of getting quality construction. Proposals to create operating companies with the skills and manpower to run nuclear plants for a number of utilities also are worth exploring.

But the first prerequisite is a change in attitude. As Gilinsky put it, the industry needs to "stop moaning about the fact that that there will not be hundreds of reactors in the United States by the year 2000 . . . . We need to concentrate on making sure that the nuclear-power reactors that we have are safe, economic and reliable."

Amen.

'I ONLY CLAIMED TO BE PERFECT.'

THE BLADE

Toledo, Ohio, May 22, 1983

DURING the last 10 years electric utilities have been forced into a massive abandonment of nuclear power-plant construction projects initiated in the days when demand for electricity seemed likely to rise at an ever-faster clip.

When the Arab oil embargo and the ensuing bleak economic situation sent demand plummeting, utilities began to cancel these costly construction projects en masse. In fact, nearly half of all nuclear capacity ordered since the inception of nuclear power has been canceled.

The U.S. Department of Energy has taken a careful look at this bleak situation and arrived at an encouraging conclusion. DOE has found that consumers and the nation as a whole would benefit if construction were resumed at many of the 100 canceled plants.

Resumption of construction not only could mean more favorable electric rates in the future — as recent experience at the Davis-Besse plant has shown — but it also represents the cheapest way for America to meet future increases in demand for electricity.

Those are perhaps the most optimistic words so far about this unsettling situation. But they hardly can be expected to send utility executives scrambling to dust off the blueprints and resume construction on scrubbed plants.

Building a nuclear power plant today means construction costs in the $1 billion range, painfully high financing costs, and long regulatory delays. But DOE's words and previous actions by the Reagan administration to speed licensing of nuclear plants nonetheless offer some reassurance of a brighter future for nuclear power.

As The Blade has noted in the past, it is vital for Americans to avoid a myopic outlook on nuclear power and the imperfections of the current nuclear power industry. Despite the lull in the energy crunch today, this nation's energy future clearly lies with some form of nuclear power. When nuclear-power technology is used properly, it can be safe, clean, and economical. As the technology and the industry mature, these characteristics will become even more pronounced.

Today's construction slump in the nuclear power industry would vanish quickly if the American economy returned to full vigor, thus boosting demand for electricity.

But will the nuclear industry reach that point healthy enough to make the technological leap to the next generation of power plants — breeder or fusion reactors? And will the current lull in construction foretell a national shortage of electric generating capacity in the 1990s and beyond?

Answers to such questions certainly warrant greater attention than they have been getting from DOE and other agencies.

The Courier-Journal

Louisville, Ky., January 24, 1984

IF THE FIGHT over nuclear power had been a boxing match, the referee in any reasonably humane jurisdiction would have stopped it long ago. The three knockdowns in the current round — a nearly completed Illinois plant denied a license, the scuttling of Marble Hill, and the decision to convert the 97 percent complete Zimmer plant in Ohio to coal — came after it was clear that the nuclear industry no longer could keep its gloves up.

But the boxing-match analogy is incomplete, because there is no clear-cut winner. At Marble Hill and elsewhere, some utilities that gambled on nuclear power face bankruptcy unless they persuade state authorities that the cost of their mistakes can be extracted from their customers. Even at Zimmer, it's not clear that much of the $1.6 billion investment will be useful in converting to coal.

The nearest thing there is to a winner is the coal industry. "Coal's future could rival its past," read a headline in Sunday's *New York Times*, and coal surely will gain as a result of the nuclear fiasco. A senior official at the parent company of Consolidation Coal estimates that coal sales to utilities will grow to more than a billion tons annually by the year 2000. Much of this, if acid rain controls are adopted, may be low-sulfur Eastern Kentucky coal.

But the coal industry, as always, has been sluggish in developing the technology to make burning its fuel environmentally acceptable. Rather, it has concentrated on arguing against clean air. And even if the acid-rain and other problems are solved, the "greenhouse effect" is on the horizon.

Some scientists think the climate, even at the present rate of burning fossil fuels, will be altered disastrously within the next few years by the amount of carbon dioxide being released into the atmosphere. This "greenhouse" phenomenon, unlike other pollution, isn't the result of impurities in the fuel. It's an unavoidable part of the combustion process. So other energy sources — whether renewable ones such as solar and windpower, or new forms of nuclear generation — seem certain to replace fossil fuels in the long run.

Coal people still generally believe that widespread use of renewable energy sources is far in the future. But Charles Komanoff, head of a New York energy consulting firm, notes that the capacity of solar and other renewable energy systems ordered during the last three years exceeds that for nuclear and coal plants combined.

Some argue that the need for large, centralized generating facilities is fast fading as these new sources become available, and as efficiency in generating and using electricity increases. Despite the large gains in conserving electricity during the past few years, Mr. Komanoff thinks the surface has barely been scratched.

Nevertheless, nuclear power's knockout doesn't mean its permanent retirement. Along with the increased efficiency comes a trend toward electrification of industrial processes. And while today's generating capacity far exceeds demand, obsolete plants someday must be replaced.

Some experts say the United States went down the wrong road with its light-water nuclear technology, compared with techniques developed in other parts of the world. Westinghouse and General Electric are working with Japanese companies to develop plants that are easier to build and operate. And somewhere on the horizon is nuclear fusion, which some think will be more useful and less risky than any technology now available.

Whatever comes next in the nuclear industry, one can hope that something has been learned from the billions of wasted dollars and years of wasted time. The lesson surely is that science rather than salesmanship should rule the nuclear game.

The nuclear industry was allowed to barge ahead in the face of vast uncertainties, selling hardware and services to utilities that had only the foggiest notion of the difficulties. By hindsight, it was inevitable that the utilities, without close supervision, would build themselves into messes such as Marble Hill and Zimmer.

Deregulation is the fashion today in many fields, and properly so. But the nuclear fiasco can be traced largely to authorities who were promoting when they should have been regulating. What's needed now is not a celebration or a wake, but a careful re-examination of the nation's energy future.

The Virginian-Pilot

Norfolk, Va., January 24, 1984

Reports of the death of nuclear power are greatly exaggerated.

Certainly, the spectacularly bad news over the past two weeks has the nuclear industry on the canvas, groggily taking the count on its hands and knees.

First, there was the Jan. 13 announcement that the U.S. Nuclear Regulatory Commission's licensing board would not allow a $3.5-billion plant near Rockford, Ill., to operate because of inadequate quality controls.

Next, on Jan. 16, the Public Service Co. of Indiana announced a halt to the half-finished Marble Hill plant after putting $2.5 billion into it.

And, last Saturday, the owners of the nearly complete Zimmer nuclear plant near Cincinnati announced it would be converted to burn coal. It was either that or scrap the plant, which has cost $1.6 billion so far.

This litany will have the ring of familiarity to Vepco customers who have watched their own utility, a nuclear pioneer, cancel atomic facilities over the past few years after spending hundreds of millions of dollars on construction.

But there's life in the nuclear industry yet. The pro-nuclear forces are stubborn and patient.

Evidence of this can be seen in Vepco's decision to donate $600,000 last year and $1 million this year to a subtle, skillful pronuclear public relations campaign.

Nuke advocates have learned a hard lesson. They're a lower-key bunch today than a generation ago, in the infancy of their industry, when utility and government officials were promising more than they could deliver. (Electricity from nuclear facilities would be "too cheap to meter.") Such claims did nothing to prepare the public for massive cost overruns and soaring bills to pay for facilities subsequently abandoned.

The industry blames overly zealous regulation and high interest rates for its problems. But utility officials kept building the plants even after Americans dramatically curbed their electricity appetite in response to OPEC and inflation.

Today, the score stands in favor of the anti-nuclear crowd. Utilities are reluctant in the extreme to order new nuclear power stations, given the hostile political and regulatory environment, and given investor skepticism about atomic energy. And with oil prices stable and low growth in electricity demand, power companies don't feel much pressure to build more plants, anyway.

But the world energy market is volatile; it can turn on a dime. What if Iraq launches an air strike upon Iran's oil processing facilities? What if the civil strife in Lebanon widens into a regional conflagration, resulting in yet another Arab embargo of Western nations friendly to Israel, i.e., the United States? These are obvious scenarios. There are others that can crimp our oil supply, and another run-up in fossil fuel prices will certainly embolden nuclear advocates.

This will be particularly true if acid rain legislation passes, raising the cost of coal-burning power plants.

Cost and risk are relative variables, dependent upon the price and problems of alternative fuels. Simply put, the negatives of nukes may not loom as large to the public if we're in danger of running out of energy, if, for example, a cartel declares economic war again on the United States.

Then, the public may perceive the social benefits of nuclear power to outweigh the social costs of occasional accidental releases of radiation. Politicians and regulators will feel pressure to grease the skids for the construction of more nukes if presented with a public clamor.

The nuclear power industry is in awful fettle for the moment. But Americans will grow more tolerant of nuclear plants if the alternative is, say, no air conditioning in July or hundreds more lakes and streams rendered lifeless by acid rain.

DESERET NEWS

Salt Lake City, Utah, February 8-9, 1984

Any growth in the faltering U.S. nuclear power industry is probably dead for the rest of this century.

That bleak forecast was made this week by the U.S. Office of Technology Assessment and officially confirms what this page has noted in recent editorials.

Management, regulatory, and technical problems, all leading to enormous costs, plus a growing public opposition, have brought the N-power industry to a standstill. No new projects have been started since 1978 and 100 have been cancelled since 1974.

If that situation persists for very many more years, the industry will be hard pressed to maintain its ability to produce any kind of nuclear reactors. If that happens, any needed N-power reactors would have to be imported from foreign nations.

However, the Office of Technology Assessment study does offer a suggestion that might pull the industry out of the doldrums: Adopt a basic standardized design for all N-power plants instead of designing each one separately.

The idea makes sense. In fact, it's hard to understand why it hasn't been done before. It already is being used in France and Canada, where N-power plants are steadily being built.

One of the major obstacles to construction of nuclear power plants in the U.S. has been the regulatory process, which can slow work to a crawl and take years to get all the necessary approvals.

A standardized design would seem to offer streamlined licensing, stabilized regulation, faster construction, lower costs, and better management. It undoubtedly also would contribute to easier training of workers and better safety.

If power companies could get together and agree on one basic design, it would give American consumers an important energy option for the years ahead.

The Morning News

Wilmington, Del., January 21, 1984

The Coalition for the Postponement of Nuclear Power can declare victory. Don't expect it to go home.

For there are heroes out there — and the term of course includes the many heroines. They are armed to the teeth with data and will stay on the barricades seeking further postponement after postponement. They will endure until there is a gilt-edged guarantee that there is no nuclear plant anywhere with a single leaky pipe, sticky circuit breaker or drowsy operator. That may be long coming.

The coalition has special reason to celebrate now. Its feisty holding action has played no small part in producing a situation where no additional nuclear plants are planned. Some under construction have been crushed by costs. Others — as nearby in Salem, N.J. — switch off and on as sterner regulators, the media and adamant anti-nuclear activists watch. Nuclear is shelved. What of the week?

- The U.S. Department of Energy reported that construction costs for 77 percent of nuclear plants now operating were double — in some cases quadruple — first estimates.
- A Nuclear Regulatory Commission panel refused to grant an operating license to a nearly completed, $3.35 billion Commonwealth Edison plant in Illinois.
- Public Service of Indiana abandoned its Marble Hill power plant after spending $2.5 billion on the project.
- In Ohio, a city council resolution called on Cincinnati Gas and Electric to abandon its Zimmer nuclear plant.
- On Wall Street, bond ratings of several utilities with large nuclear projects were lowered.

There is an industry reservoir of optimism. A Public Service Electric and Gas spokesman declared that construction schedules for that utility's Hope Creek station near Salem are being met and said, "It looks very promising."

That reflected a forecast by the Atomic Industrial Forum of a nuclear "renaissance." AIF President Carl Walske's assessment is that as demand for electricity rises with increased economic activity, there are but two real choices, nuclear and coal. He breezily stated, "There is no question but that nuclear will be highly competitive when the time comes to make decisions for expansion."

No one says the dog is dead yet, but Worldwatch Institute analyst Christopher Flavin, in "Nuclear Power, the Market Test," put it this way: "Nuclear power has lost substantial economic ground compared to coal-fired power . . . Only in France and a few other nations are new nuclear plants a less expensive power source than new coal plants."

It is partly the industry's fault, its slow recognition of public insistence on the highest construction and operating standards. Acceptable waste disposal technology may exist, but social, economic and political bars are high.

Minuses involved in alternatives help keep props under nuclear: Petroleum is finite and politically vulnerable; coal brings hazards to face-workers and contributes to such tremendous problems as acid rain and the warming Greenhouse Effect; solar's bright promise is slow in being realized; even wood burning is causing alarm over pollution.

Things being as they are, whale oil is out.

THE COMMERCIAL APPEAL

Memphis, Tenn., January 24, 1984

THIS YEAR began with a jolt for the nuclear power industry.

History may not remember it for a specific incident in the way that March 28, 1979, will be recalled. In the days that followed the mishap at Three Mile Island, millions of people first realized that nuclear reactors may not be safe as they had been led to believe.

Now, events across the country demonstrate that the public, the utilities and the regulators are continuing to sour on nuclear power for either safety or economic reasons.

In Washington, the Department of Energy reported that three-fourths of the nation's nuclear reactors cost consumers at least twice as much as the preconstruction estimates.

In Cincinnati, three utilities said they would try to convert their nearly completed Zimmer nuclear power plant to burn coal. They said it would have been too expensive to complete it as a nuclear facility and weren't sure they could get an operating license from the federal government.

In Illinois, the Nuclear Regulatory Commission unanimously denied an operating license for the nearly completed $3.35 billion Byron plant near Rockford. NRC officials said they were unconvinced that construction work was satisfactory. It marked the first time that such a license was flatly refused.

In Indiana, the Public Service Co., said it didn't have enough money to meet increased costs and abandoned efforts to finish its Marble Hill plant. The utility has already spent $2.5 billion on the reactor. It hopes ratepayers eventually pick up the tab.

Nuclear power has been touted as a sure source of stable energy, as an important cog in the drive for diversification of fuel sources. Using uranium as a cheap fuel source was alluring. It still is. But capital construction costs have ballooned, partially because of the required voluminous studies and safety measures. The increase in energy demand has slowed. The quality of design and workmanship in nuclear plants is undergoing more scrutiny. Questions still surround the disposal of radioactive nuclear waste.

The Harvard Business School, in its 1980 report on energy, said, "In the 1980s, the burden lies squarely on nuclear advocates and regulators to satisfy skeptical people that nuclear plants are being designed, built, and operated with the competence and skill that the country was so long, so monotonously, and, it turns out, so incorrectly assured were already commonplace." Nuclear critics, it continued, "must be sufficiently objective to recognize that proofs of design and management competence can never be absolute; some unknowns and imponderables will always remain."

IS THE NUCLEAR industry already dead? No new nuclear plant has been bought in six years. The decision to abandon the twin-reactors on Marble Hill brought to 100 the number of plants canceled since 1974. The Tennessee Valley Authority has been at the forefront of utilities that have acted to cut their loses by canceling partially-completed plants.

Conservation, load management, cogeneration and so-called "soft path" energy sources such as solar and wind power have made a dent in the nation's energy picture. But if sustained economic recovery leads to explosive growth, they may not be enough to put a cap on energy demand.

Nuclear power was expected to play an important role in the nation's energy future as we entered the 21st century. Energy stability depended on it. The industry's woes may content those who seek to reject nuclear power out-of-hand, disregarding the facts and minimal risks that are inherent throughout all of technology.

But even objective observers with high hopes for a nuclear future must be getting increasingly concerned. Sloppy workmanship, poor planning and skyrocketing costs might not let nuclear power live that long.

AKRON BEACON JOURNAL

Akron, Ohio, January 28, 1984

THE DECISION to convert the William H. Zimmer nuclear plant to a coal-burning facility is the latest in a string of bad news for the nuclear industry. Since 1978, no utility has ordered a nuclear power plant, and 50 plants have been canceled.

In the last year, the Nuclear Regulatory Commission has ordered the closing of five nuclear plants to inspect cooling pipes for cracks, and the Washington Public Supply System defaulted on the $2.25 billion it borrowed to build two nuclear plants.

And safety problems, cost overruns and construction delays threaten the completion of several plants.

Few industries were more profoundly affected by the four-fold increase in OPEC oil prices during the 1970s. Inflation sent the cost of materials and labor soaring. The estimated cost of the Zimmer plant, for instance, went from $240 million in 1969 to more than $3 billion.

Higher costs for electricity from nuclear plants have reduced, if not eliminated, nuclear's price advantage over coal-burning plants.

Inflation also took its toll in higher interest rates. And as interest rates soared in the late 1970s, it became staggeringly expensive to finance the construction of huge plants.

In addition, the energy crisis sent consumers scurrying for ways to conserve energy. Power demand forecasts were suddenly off the mark in a stagnant economy.

But economic realities are not solely responsible. Since the accident at Three Mile Island, the nuclear industry has been plagued by declining public confidence.

For many in Ohio, these concerns about safety were heightened by stories about the construction of the Zimmer plant appearing in the Beacon Journal this week. Steve Stecklow, a reporter for the Philadelphia Inquirer, cataloged a decade of faulty welds, sloppy wiring, falsified records and a chilling neglect of quality control and safety.

As Mr. Stecklow noted, building a nuclear power plant is one of the most complex jobs on earth. For instance, more than a million welds are required, and many of them are part of an elaborate safety system designed to prevent radiation leaks and other serious accidents.

It almost goes without saying that regulation of nuclear plant construction must be strict and fair. And yet the NRC, Mr. Stecklow points out, appeared disinterested in the problems at the Zimmer plant.

After a decade of warnings and as the plant neared completion, the NRC finally stopped construction for an inspection. That inspection found myriad problems, and led to the decision that Zimmer had no future as a nuclear power plant.

Plant managers often complain about cumbersome government regulations delaying construction and driving up costs. But the complaints are hardly universal. An executive with the Arizona Nuclear Power Project maintains he has no problems with NRC regulations.

And industry analysts argue that mixed reviews about the regulatory process indicate that sound management is lacking in some areas of the industry, and that may be its largest obstacle in the future.

It is safe to say that the nuclear industry does have a future in the United States. Other sources of electricity will become scarce and more expensive.

How bright that future is, however, depends on public confidence in the industry's ability to build safe plants, on better control of construction costs and, most important, on the rate of growth in the nation's future use of electricity.

The Times-Picayune
The States-Item

New Orleans, La., February 8, 1984

The congressional Office of Technology Assessment has a gloomy prognosis for the nuclear power industry. Reciting a number of industry shortcomings, the office proclaims that the "nuclear era is drawing to a close."

While that conclusion might be too drastic, the civilian nuclear power industry clearly has fallen on hard times, shattering the once rosy forecasts of its more enthusiastic backers.

Many of the industry's troubles are common knowledge. The safety issue, always worrisome, became a crucial factor in the industry's decline with the equipment failure at Pennsylvania's Three Mile Island plant in 1979.

The accident severely damaged a reactor core, setting off a national alarm and shutting down the expensive facility. That incident, says the congressional office, was "a watershed in U.S. nuclear power history because it proved that serious accidents could occur."

The nuclear power industry has also been plagued with huge cost overruns, overestimated growth in electricity demand, difficulties in obtaining rate increases from politically sensitive state regulatory agencies and increasingly tough requirements from federal regulatory agencies.

All these factors have led to a massive loss of confidence in the industry by the public, investors, safety regulators and, in some cases, managers of the utilities themselves.

It will not be easy for the industry to regain the public's confidence. Yet, as the Office of Technology Assessment points out, many utilities have built nuclear reactors within acceptable cost limits and operated them safely and reliably. Some Western European nations have been able to do so for years without notable difficulty.

Ironically, the key to restoring public confidence in the industry is not in less regulation but in stricter on-site, step-by-step evaluation of any new plants from the ground up. Nuclear power can yet be an important part in the nation's future energy mix if the safety issue can be resolved, costs controlled and public confidence restored. Meanwhile, the nation must fall back on the old fossil fuels — oil, natural gas and coal — to meet the bulk of its energy needs.

San Francisco Chronicle

San Francisco, Calif., January 20, 1984

THE NUCLEAR POWER industry is taking a fearful beating as 1984 opens. Consider just the past week's news.

On Monday, the Public Service Company of Indiana said it was abandoning a plant at Marble Hill after spending $2.5 billion to bring it halfway to completion.

On Tuesday, the Cincinnati city council began considering a resolution calling on the Cincinnati Gas & Electric Co. to abandon its Zimmer nuclear power plant because "astronomical costs are clearly unaffordable."

The previous Friday, big Commonwealth Edison lost a request to a Nuclear Regulatory Commission licensing panel for an operating permit at its nearly completed Rockford, Ill. plant for lack of quality control.

The week's fourth staggering project, Consumer Power Company's Midland, Mich. plant, began to come under the raised eyebrows of industry analysts who say it may not be able to meet its estimated $4.4 billion cost.

LONG DELAYS; doubled, tripled and quadrupled construction cost overruns; high interest rates, and recurring quality failures have accounted for these mounting disasters. It is hard for sponsors and advocates of nuclear reactor plants for generating electric power to keep a stiff upper lip. Particularly when, in the wake of the Three Mile Island accident, the Nuclear Regulatory Commission licensing reviews grow tougher and more demanding.

The Energy Department, summing up where the industry stands, suggests that final construction costs for three quarters of the plants now operating have at least doubled startup estimates. In fact, 28 percent were four times over, and in some cases costs went up more than seven times.

We imagine that those holding anti-nuclear attitudes may take cheer from this melancholy record, but we don't. Consumers are going to have to pay all the costs lost down the drain, and — make no mistake about it — the same consumers will have to pay more for their electric bills as the years run on than they would have had to pay had the shut-down plants come on line.

IT MAY BE WELL to listen to the pro-nuclear side. Stephen D. Bechtel Jr., who heads one of the world's largest nuclear power constructors, offers the following argument:

"There are those who say it no longer makes economic sense to build nuclear plants. It is true that, because of higher financing costs and a protracted regulatory process, initial costs of a nuclear plant are greater than comparable costs for a fossil-fuel plant. But 30 or 40 years down the road, when all the costs are totalled — including the high one-time costs of nuclear construction versus the high long-term costs of oil — nuclear-generated power will have cost the consumer considerably less."

In 1982, uranium-generated electricity cost 3.1 cents per kilowatt hour; coal-fired electric power 3.5 cents, but electricity from oil, 7 cents — double the cost of either coal or nuclear.

So we suggest that those willing to look into the future will find nothing to say hurray for in the past week's news of the nukes.

The Boston Globe

Boston, Mass., January 25, 1984

A major Midwest utility operating more nuclear plants than any other in the country, is denied an operating license for its newest plant, nearly ready to go on line. An Indiana utility decides to halt work on a nuclear plant after investing $2.5 billion of its own and its customers' money in a nuclear plant about half finished. An Ohio utility decides to convert a nearly completed nuclear plant to coal. A New York utility is blocked from opening a completed nuclear plant on Long Island because of uncertainty over emergency evacuation plans.

Is nuclear power dead? No.

The cases, and a number of others besides, are instead evidence of the enormous changes that have afflicted electric utilties over the past decade and of the need for better management of electric utilities of all types and for further improvements in their continuing regulation.

The three worst problems imposed on utilities have been essentially out of their control: steep rises in the price of fuels used to generate electricity, steadily rising construction costs and persistently high interest rates. These three factors have pushed up the costs of all kinds of power generating plants, not just nuclear.

Some problems represent management shortcomings. Inadequate preparation for the scale of problems represented by nuclear plants is surely the most outstanding. Electric utilities have seldom attracted the keenest management brains. Business schools that pay close attention to marketing, finance, investment banking and so on have not treated utility management as a specialty. Perhaps as a consequence, the quality of management in utilities has typically ranged all the way up to pretty good. It would be interesting to know how nuclear power would have developed had it in part been in the hands of, say, IBM.

Utility managements have exposed a few shortcomings in the fields of commercial banking and investment banking, as well. Minor league utilities have been able to coax billions of dollars in investment from houses that missed the implications of too much emphasis on just a few basic designs, on obsession with the notion of great savings in mammoth plants, on the firm but partly mistaken belief that technology would develop fast enough to keep up with the expanding demands of safety and reliability.

No new nuclear plant projects are in sight for years although most uncompleted plants will probably be finished. In one respect, that is too bad because it might force the industry to try alternatives it failed to explore earlier – for instance the smaller, gas-cooled reactors the British use with considerable success in their nuclear power program. Unless there are breakthroughs in other forms of energy, nuclear power is apt to be a significant player for years to come, probably even on a further expanded basis. The pause in its growth with these episodes should be as an opportunity to deal more successfully with some of its problems than as the foreboding of its end.

Is Nuclear Fusion an Alternative to Fission?

Our present nuclear reactors are fission reactors; they depend on the controlled fission of heavy radioactive elements, usually uranium, to provide useful energy. In the nuclear fission reaction, energy is derived from the splitting of these heavy atomic nuclei into lighter ones. In nuclear fusion, the opposite occurs; energy is derived from the combination of two lighter nuclei into a heavier one. The most common fuel, hydrogen, is available in unlimited quantities. Thus, fusion reactions have been the subject of much research as an alternative to fission for the production of power.

Fusion reactions, however, require extremely high temperatures—about 100 million degrees Celsius (or 180 million degrees Fahrenheit)—to occur. This is because the atoms must overcome a natural repulsive force due to their electric charges before they can combine. (Fusion reactions are thought to be responsible for the high temperatures of stars.) At these high temperatures, the hydrogen atoms become completely ionized, existing as separate electrons and nuclei in an electrically-neutral gas known as plasma. In order to maintain and confine the superheated plasma, the fuel must be surrounded by a magnetic field; material containers cannot withstand the heat, interfering with the reaction. At the forefront of U.S. nuclear fusion research is a magnetic containment device built at Princeton University. Called the Tokamak Fusion Test Reactor, this machine received its first test in December 1982. Scientists achieved—although for only one-twentieth of a second—the first controlled production and confinement of plasma, representing a milestone in the quarter-century history of fusion research.

Scientists at Princeton hope their reactor will be able by 1986 to reach "breakeven," or the point at which the controlled fusion reaction will produce more energy than is spent in creating the conditions for it to occur. Fusion reactors are still in the developmental stage, and there are huge practical difficulties to be overcome in sustaining the reaction and confining the plasma. Most scientists agree that it will be some time before fusion researchers can develop reactors that would be able to supply commercial power. Reactors similar to the Tokamak reactor at Princeton are under construction in Japan, the Soviet Union and Great Britain.

TULSA WORLD

Tulsa, Okla., January 10, 1982

For some time now, man has labored at recreating the energy-producing fusion process employed by the sun. Fusion energy has long been thought to hold the key to the world's most pressing need: a relatively clean, inexhaustible energy source.

Nuclear fusion plants would not be risk-free, but the environmental dangers associated with them are far smaller and more acceptable than the problems posed by the nuclear fission plants now in use.

The key to the development of fusion energy is money. In 1980, Congress approved spending $20 billion overall to fund research and development aimed at making a fusion plant commercially feasible. But the Reagan Administration's talk of cutting science funding 12 percent across the board concerns researchers.

The major stumbling block to development of fusion energy has been the extreme heat needed to make the process work. To be commercially practical, a means must be found to heat light elements such as deuterium and tritium gas to a temperature of at least 100 million degrees centigrade.

American scientists have been able to reach temperatures high enough to make nucleii fuse, but not high enough to sustain the process. But last year a major breakthrough occurred. Scientists at Princeton University managed to obtain a fusion yield several times greater than ever produced before.

The irony," says the director of Princeton's fusion research center, "is that if the Russians had done it it might have been worth at least another $100 million to the United States program. We seem to need another Sputnik."

The point is well taken. Too often America's science spending is a reaction to Soviet developments. It is true in military spending; it is true in fusion research.

But that should not be so. America must make a strong committment to sustained research in fusion. Congress made that commitment in 1980. The president should make sure that that commitment is kept.

The Providence Journal

Providence, R.I., May 27, 1981

Someday, when the definitive history of the 20th century's "energy crisis" is written, notice may be taken of an experiment last week in a laboratory at the University of Rochester. There, scientists zapped a tiny pellet of hydrogen gas with a huge bolt of energy, creating for an instant what was in effect a microscopic-sized star. The scientists cheered.

As well they might. Their experiment, using powerful laser beams to smash into the nuclei of billions of hydrogen atoms, had produced more energy than any previous non-military experiment in nuclear fusion. It did not seem to matter that the reaction was over in a tenth of a billionth of a second, or that the laser blast required an electrical jolt equal to 10 times all the generating capacity in the United States. What mattered was proving a principle: that the intense light from high-powered lasers can be persuaded to imitate the sun and annihilate hydrogen atoms to produce the vast energy contained in nuclear fusion. To this end, the Rochester experiment may have made a signal advance.

For a generation, scientists have sought ways to duplicate the fusion process by which the sun consumes its own hydrogen at a prodigious rate, releasing vast amounts of energy in the process. The dream: an energy source built around a virtually inexhaustible supply of fuel. The theory: smashing a hydrogen atom together with such force that it "fuses," creating a new atom of helium and a simultaneous burst of energy. Yet the obstacles are great: the almost incredible temperatures and pressures required to overcome the natural repulsion of hydrogen atoms and bring their nuclei together.

Until recently, most fusion research sought to create these conditions by building "magnetic bottles" that would contain the target hydrogen long enough to zap it with a mighty electrical jolt. But at Rochester, the laser enthusiasts believe they have a better way: they beamed 24 ultra-powerful laser beams onto a tiny hydrogen pellet, and it instantly became hotter than the sun. Someone in the laboratory calculated that the laser burst had produced 12 billion neutrons, well beyond what the experimenters had hoped.

Advancing from this achievement to a sustained fusion reaction still represents a vast leap, and some skeptics doubt that it ever will be economically practical. But the Rochester laser-fusion experiments give reason to think that laboratories before long may reach the long-sought "break-even" point, at which a fusion reaction gives forth at least as much energy as it consumes. So long as this field of research continues to show progress, it deserves the continued backing of government. Despite the mountainous technical obstacles still remaining, nuclear fusion holds too much potential not to be explored to the fullest.

AKRON BEACON JOURNAL

Akron, Ohio, January 9, 1982

LAST SUNDAY'S New York Times Magazine contains a thoroughgoing report bringing readers up to date on the status of work toward power generation from nuclear fushion — and raising doubt as to whether, in the squeeze of current federal belt-tightening, this work will be funded well enough to keep up its momentum.

This is disturbing. It seems as if fusion work should get No. 1 priority among government-backed energy research and development projects.

Fusion, power source of the sun and other stars, is the reaction that makes possible the terrible power of the hydrogen bomb. In it, atomic nuclei of such light elements as hydrogen isotopes combine to form heavier elements — for hydrogen, helium.

This reaction converts a much larger part of the original elements' mass into energy than does fission, the atom-splitting reaction powering the original atomic bomb and present-day nuclear electrical generating plants. Thus the power yield from a given amount of fuel is vastly greater. And the fuels are plentiful.

Further, because the elements involved are simpler, the process is "cleaner" than fission. It can be carried out without creating waste containing heavy radioactive isotopes with half-lives long enough to threaten human health for millennia, as fission of heavy elements does.

Since well before the success of the first H-bomb, physicists have been wrestling with the problems entailed in harnessing fusion to peaceful use. This requires achieving a sustained, continuous reaction at a controlled level, at fantastic heat.

Within the last year, physicist Jeremy Bernstein reports in the magazine article, a team at Princeton has come closer to this than anybody else has so far been able to do.

They achieved a reactor temperature of 70 million degrees Celsius — almost but not quite hot enough for sustained fusion at a practical level. And they obtained a yield of fusion energy "several times larger than had ever been produced before in a controlled experiment."

There is still a long, hard and to some extent uncertain way to go before many of us can switch on fusion-powered light bulbs; nobody is predicting commercial application earlier than well into the 21st century.

But the development at Princeton, plus progress in fusion work elsewhere, says Professor Bernstein, makes it appear now that "fusion scientists will, indeed, be successful."

That's the good part — encouraging to those who see in fusion a way to tame the world's energy problems for eons without generating massive ecological problems.

The bad part is the funding worry.

In 1980 the Congress authorized overall expenditure of up to $20 billion to let the United States "aggressively pursue research and development" on fusion. On this basis the Department of Energy laid out a program that would have set up by about 1984 a national center where industries, universities and the national laboratories could work together on the problems involved.

This, it was hoped, would make possible completion of a prototype generating plant by about the year 2000 — with substantial commercial application perhaps a decade beyond.

But the money squeeze in Washington now has put all that in doubt, making likely a far slower pace.

Clearly the government cannot achieve badly needed cuts in spending if every program is defended to the death. But in this instance — considering the size of the prize at the end of the line and the relatively modest government spending required; about $1 billion a year — severe cutbacks seem shortsighted.

The Oregonian

Portland, Ore., November 28, 1982

Whether produced by fission or fusion, atomic energy produces radioactive materials, most of them dangerous. But in chain reactions based on protons rather than neutrons, radioactive byproducts would not be produced, nor would the stuff of bombs become a byproduct.

Such a clean, nuclear power source, which largely fuels the stars, is being seriously pushed by a growing number of physicists. They are bucking a majority who believe proton fusion reactions would not warrant serious study until the end of this century.

A leading exponent of the proton technology is Bogdan Maglich, senior researcher for the European Organization for Nuclear Research, who is working in laboratories at Princeton, N.J. He believes particle beams are the most promising way to ignite a proton fusion reaction. Proton fusion has been thought a remote possibility because temperatures, ranging up to 10 billion degrees centigrade, or well above the heat range not yet achieved by neutron fusion experiments, were believed required to fire up the reaction.

But Maglich believes particle accelerators, those giant tunnels used for work on the basic parts of matter, can be used to provide the key to designing a proton fusion reaction using the elements boron, lithium and hydrogen. The super temperatures would not be needed.

It is Maglich's belief he can achieve such a reaction in the laboratory within five years, which would permit, he reports, the construction of fusion power plants no larger than a home furnace. What Maglich has to do is keep the positive-charged protons, used to bombard boron atoms, from scattering, or being mutually repulsed since they have the same positive electrical charges. Again, many physicists think this is impossible.

The possibilities of such an energy source are awesome. But before anyone discounts the prospects of such a scientific achievement in the short term, it is good to recall that science has a way of duplicating the wondrous things going on in the stars. Just splitting the atom stunned a lot of people.

DESERET NEWS

Salt Lake City, Utah, March 9, 1982

Once heralded as the answer to U.S. energy needs, nuclear power now faces an uncertain future.

The nation has 72 functioning nuclear power plants and another 72 are still under construction, but the dream of meeting electric power demands with atomic energy seems to be nearing a dead end.

Construction on three nuclear plants was halted this week by the Tennessee Valley Authority. That makes seven such plants cancelled or delayed this year and 30 in the past three years. No new nuclear power applications have been submitted for five years.

Yet the need for more electric power continues to rise and ordinary fossil fuels won't last forever.

The problem is at least three-fold, including costs that are soaring out of sight, concerns over safety, and an intractable problem of what to do with radioactive wastes that are a by-product of such power stations.

The latter problem could be eliminated at a stroke by a workable nuclear fusion plant — a station that could produce unlimited power from sea water and do so without any radioactive waste.

Current nuclear power plants operate on the principle of atomic fission, the splitting of atomic particles to release energy. Fusion does just the opposite, fusing atomic particles together to release vast amounts of energy — the same process that fires the furnace of the sun. The process leaves no waste.

However, the technological obstacles to fusion are enormous and not all scientists are certain a harnessed fusion process is possible. But the idea holds out enormous prospects for the future if it can be achieved.

In 1980 Congress passed, and President Carter signed, a bill committing the nation to having a demonstration fusion power plant in operation by the year 2000.

Unfortunately, that commitment has not been backed with the sense of urgency and the kind of effort and resources used to produce the atomic bomb or land men on the moon. Yet the space program is small potatoes compared with the results of a successful fusion power plant.

The Reagan administration has reduced spending on fusion research in the proposed 1982 budget to $465 million, down from an earlier figure of $525 million. Even David Stockman, the president's own budget-cutter, had called for an increase to $535 million.

The problems involved in trying to slash bloated government spending and keep deficits down are admittedly gigantic, but nuclear fusion has so much potential that it seems short-sighted to pull on the reins.

While trying to pay today's bills, let's not forget the critical need to invest in the future.

THE SAGINAW NEWS

Saginaw, Mich., January 21, 1983

The announcement at Princeton University of the first successful test of a nuclear fusion reactor seems to have hit with all the impact of a snowflake.

It deserves more of a hail than that.

Unlike the fission process by which today's nuclear power generators operate, fusion offers the potential of a limitless supply of cheap energy without the danger of radioactivity.

The problem is that scientists, despite their excitement about the Princeton test, do not expect fusion to be available for energy until the year 2025 at the earliest.

That's not much of a thrill for those chilled by this winter's fuel bills.

But the New World wasn't built in a day, either — and Harold Furth, director of Princeton's Plasma Physics Laboratory, compared the successful reactor test to Columbus's discovery of America.

Come to think of it, the news that a new world had been found beyond the horizon probably struck 15th century Europe with all the impact of that same snowflake. But the future kept its promise. Everyone weary of OPEC, suffering from China syndromes and afraid of overdoses should pull for fusion energy to do the same.

Arkansas Gazette.

Little Rock, Ark., January 2, 1983

As military research was busy developing the hydrogen bomb, it occurred to someone in the administration of President Harry Truman that the theory behind the H-bomb also could be applied to peaceful and productive uses. The theory was for nuclear fusion, and a small appropriation was included in the fiscal 1951 budget for research and development.

Relatively small amounts have been added to this initial "seed" money in the intervening years and on Christmas Eve at Princeton University the most encouraging results of the long-term $3.5-billion investment were achieved. Scientists there, in simple terms, turned on an experimental Tokamak fusion test reactor and it worked for five hundredths of a second, leading the lab director to comment: "It's like Columbus finding the New World. The question is not how big it is, but that they found land."

Scientists remain a long way from bringing this new world of fusion into practical use for humankind. Many additional billions of dollars will be necessary and a fairly general prognosis is that 40 years will pass before it is commercially feasible. The critical next step is refining the technology so that the fusion reactor produces more power than it uses, but the test at Princeton on Christmas Eve is greatly encouraging.

In some scientific circles, the controlled fusion process (the H-bomb demonstrated *un*controlled fusion of atoms) has been considered a competing technology with the breeder reactor. It was argued during the Nixon administration that more emphasis, and money, should be directed toward the breeder — the Clinch River project in Tennessee — because it had a better chance of proving feasible sooner.

Developments in the interim now place that conclusion in great question. The breeder's estimated cost has exploded from $700 million to $8.8 billion. Demand for electricity has fallen from 7 per cent growth annually to 2 per cent, and the earliest the Clinch River plant is projected to be feasible is the year 2025 — roughly the projection for a feasible fusion reactor.

Fusion has great potential advantages over the breeder process. Unlike the breeder, or the existing fission reactors in use across the world, fusion would provide a relatively safe, economical and environmentally acceptable way of producing electricity. Its fuel would be almost limitless, coming as it would from common sea water.

Congress would serve much better by abandoning the Clinch River breeder reactor and by spending the money instead on cleaner, more productive alternative modes of energy. One of these clearly is the fusion process. Fusion is still a long way from realization as a useful process, but it offers far greater dividends than does the breeder.

The Virginian-Pilot

Norfolk, Va., January 6, 1983

Some readers occasionally chide us for printing so much bad news — but some news is better than most of us realize. Take the development at Princeton University the other day. Physicists working on fusion energy successfully produced the first burst of hot, electrified gas, called plasma, in a test financed by the federal government.

The burst lasted only a 20th of a second but, like the Wright Brothers' Kitty Hawk flight, it had enormous significance. Within five years it is expected to lead to the next important step and eventually to a new, virtually limitless source of commercial energy.

Fusion is the name for energy derived from the molecular process that causes a hydrogen bomb to explode. The trick the Princeton scientists — and physicists in the Soviet Union, Japan and Great Britain — are working on is harnessing this explosive energy for peaceful use.

Last week's breakthrough occurred in the Tokamak Fusion Test Reactor, so named by the Soviets who built the first one. Tokamak is a Russian acronym meaning "current in a doughnut-shaped device." The device is 25 feet in diameter.

So far, it takes more energy to operate the Tokamak than it produces. The next step is to produce a plasma with as much heat and energy as it devours, and the target date is 1986. Confidence in achieving this objective was boosted when last week's objective was reached on the first try.

Laymen might not appreciate how big — or good — this news is, but Dr. Harold Furth, director of the Princeton lab, said, "It's like Columbus finding the New World. It's not how big it is but that they found land." Fusion power for industrial and residential use isn't expected to materialize until well after the turn of the century, but then it took a little time after 1492 for the New World to be settled.

This, incidentally, is one of the federal government's better investments — $454 million budgeted for fusion this year. It is divided among a number of research groups, including the University of California's Livermore Laboratory, Massachusetts Institute of Technology and the national laboratories at Oak Ridge, Tenn., and Los Alamos, N.M. It is one program that shouldn't be cut back in the quest to reduce red ink in Washington.

Houston Chronicle

Houston, Texas, January 12, 1983

The successful split-second test of a fusion reactor at Princeton is being compared with the invention of the wheel and the discovery of electricity, and no such event should be allowed to pass without comment.

The theory of fusion power promises reactors with an unending supply of hydrogen fuel from water and without the threat of core meltdown. As the manager of the program said, "The theory looks good and the test supports the theory."

Mankind is consuming Earth's storehouse of energy at a tremendous rate. It can't be repeated too often that there is a finite amount of petroleum and coal under the Earth's crust. Whether one believes that there is a 50-year supply or a 200-year supply, there is a limit and the energy will become increasingly expensive to obtain. Solar energy can take up some of the load. Nuclear p wer can provide enormous amounts of power but is running into technical and public relations obstacles. Breeder reactor research is continuing but won congressional support last year by only one vote.

Fusion is now being called the hope of the future, and while it may be, two points must be considered. First, the future is a distant one — perhaps 50 years before the first commercial reactor will start lighting homes. Second, the research will cost upward of $20 billion — perhaps many times that amount, and the funds will all be federal taxes.

With the increased pace of technology, the research time may be shortened. But before fusion power is a reality, a long-term commitment must be made to push ahead with that research. If fusion isn't the answer, another must be found. Otherwise, there will be some cold, dark nights in the 21st century.

The Knickerbocker News

Albany, N.Y., December 30, 1982

It was viewed as much a historical event as the Columbus landing in the New World, but the long-range implications carry even more telling implications.

In a Princeton University laboratory at exactly 3:06 a.m. last Friday, physicists successfully tested a fusion reactor. In doing so, they opened the dawn of a future age which could use safe, peaceful and limitless electric power.

Basically, fusion creates power by squeezing matter and heating it up. It is how energy is made in the sun, as opposed to the atom-splitting, waste-producing fission process used in today's nuclear power generators.

Fusion power — although experts believe it will not be available until well into the 21st century — could supply energy sources to our homes and industries after our dwindling fossil reserves are used up. In effect, it would be our survival tool.

As dramatic as the announcement was, it was somewhat dimmed by the voices of fusion advocates who say the federal government's support of the program has declined. They point out the Reagan Administration requested $444 million for magnetic fusion research in the 1983 fiscal year, down slightly from last year and a 25 percent reduction in real terms from the fiscal year 1977.

This downward spending trend is discouraging. The program needs all the financial backing it can get. Money can be found, especially in the burgeoning military budget.

Fusion research and development should be a top priority, not for the benefit of our generation — but for generations to come

The Pittsburgh PRESS

Pittsburgh, Pa., January 2, 1983

Little by little, nuclear physicists are closing in on a goal that has tantalized them for more than 30 years — ever since the hydrogen bomb was developed.

The aim is to harness the awesome power of colliding hydrogen atoms — nuclear fusion — in a controlled and sustained manner.

Such an achievement would be tantamount to bringing the sun down to earth and would open the way to a virtually limitless source of energy to light the world's homes and power its industries.

★ ★ ★

On Christmas Eve, a three-story-high, $314-million fusion test reactor at Princeton University, known as a tokamak, was turned on for the first time.

For a split second, it produced a burst of plasma — hot, electrified gas. That lasted only 50-thousandths of a second, and more energy was burned up than produced in the process.

But Dr. Harold Furth, director of Princeton's Plasma Physics Laboratory, was ignited to exclaim: "It's like Columbus finding the New World."

Well, maybe.

The first test did prove out the principle underlying the doughnut-shaped device. But other fusion reactors have had similar successes, and we are still far from any practical utilization of the process.

Dr. Furth now is confident that by 1986 the tokamak will reach the all-important breakeven point where it will generate at least as much energy as it consumes.

The next goal will be to sustain the nuclear fire long enough to justify building a prototype of a commercial fusion-power plant, perhaps early in the 21st century.

★ ★ ★

Three other countries, the Soviet Union, Britain and Japan, are building larger tokamaks, but the Princeton model is the first to go into operation.

This is one nuclear race, however, from which all nations are bound to benefit.

Immense engineering problems remain to be solved before fusion power becomes feasible. Yet each success in the laboratory is one more step away from theory and one more step toward the day when the sun may indeed be brought to earth to free mankind from dependence on dwindling supplies of fossil fuels.

The Dallas Morning News

Dallas, Texas, December 31, 1982

Solar power, the gathering of the sun's rays, may be one of the future's important energy sources. But the process that generates that power directly, on the sun itself, also may be re-created on earth, as a landmark experiment at Princeton University's Plasma Physics Laboratory has demonstrated. And that source could someday dwarf other energy sources now available.

Scientists at the lab have successfully started up the $314 million Tokamak Fusion Test Reactor for the first time, stripping the electrons from nuclei in a fuel and forming a plasma. This is the first step toward development of a commercial fusion process that could provide low-cost energy in the 21st century. For the next step, the researchers hope to attain "ignition" of the plasma — in that stage, more energy will be generated than is necessary to start the process.

Fusion is the process by which the sun generates energy, combining atoms rather than splitting them as in the fission process now used in nuclear generators.

Fusion reactors would be a definite improvement over the present fission system because less radioactivity is released. Although development of the engineering necessary to build a commercial fusion reactor is just beginning, scientists are sure that the fusion process will be much safer. Important research is being done in this field now at a fusion institute at the University of Texas at Austin.

No overnight breakthroughs are expected in the tedious process of harnessing fusion energy. But Princeton's initial success with its new Tokamak certainly should increase optimism.

The Des Moines Register

Des Moines, Iowa, January 6, 1983

The dream of nuclear fusion as a source of power for the world's future came a step closer to reality in late December when scientists at Princeton University fired up their fusion reactor for a tiny fraction of a second, and found that it works. It was cause for rejoicing.

Like nuclear fission, nuclear fusion would generate electricity by producing heat to drive turbines. But unlike fission, which splits atoms to produce the heat and creates monstrous dangers from radioactive by-products, fusion puts atoms together. Fusion reactors would be far less radioactive, far more efficient than fission. No combustion products would be released into the atmosphere, and the plants would not be subject to meltdowns.

Princeton fired up its fusion reactor and found that it works. The future need not be so distant as some say.

The successful start-up of the Princeton fusion reactor was a "magical event," in the words of Harold Furth, project director. Such terms have become commonplace since work began on the fusion concept. "Research into fusion power has often seemed more like a fairy-tale quest than scientific effort," a Christian Science Monitor writer said eight years ago. The first experimental reactor was called the "perhapsatron."

The fusion reaction is the same means by which the sun turns matter into energy, giving off heat. And the fuel supply is all but unlimited; the best fuel is deuterium, an isotope of hydrogen found in abundance in seawater.

With electromagnets holding hydrogen isotope gases within containers, the gases are heated to incredible temperatures, causing them to fuse into helium atoms with the release of an enormous amount of energy. The trick — which may not be solved for another generation — is to sustain a reaction long enough to create more energy than is consumed.

The next challenge is to reach the necessary temperature — in the neighborhood of 100 million degrees Celsius — for controlled fusion. Scientists at the University of Texas thought they had reached that peak in the early 1970s, but, Furth said, "the fog of history has clouded over that claim." He estimates that the necessary heat will be attained in the mid-1980s.

What the Princeton experiment proved, he said, is that the Princeton reactor is a workable vehicle for a fusion reaction.

The promise of fusion energy is more than worth the generation of research still needed, given fusion's potential. Fusion-energy plants located within cities could heat and cool buildings as well as generate electricity. The heat is sufficient to dissolve a city's garbage and sewage problems by reducing the muck to its chemical elements. The foulest of organic wastes and micro-organisms would become harmless carbon, oxygen and hydrogen.

Princeton's success offers another argument for putting more resources into fusion — the poor step-child of energy programs for years — and withholding money from the Clinch River Breeder Reactor. The lure of the breeder is that, like fusion, it creates more fuel than it uses, but the fuel is highly radioactive plutonium, the stuff of nuclear weapons.

In a discussion of energy alternatives eight years ago, Nobel-laureate physicist Hannes Alfven wrote: "It is often claimed that technical fusion reactors will not be developed before the year 2000 and, therefore, fusion energy should not be mentioned in the present debate. The causality chain may be the reverse: As the breeder-reactor lobby does not like the competition with the fusion alternative, this is eliminated by the claim that it belongs to a very distant future."

Princeton helped show that that future need not be so distant — unless the politically supported but scientifically discredited breeder reactor continues to be given precedence over fusion research.

THE COMMERCIAL APPEAL

Memphis, Tenn., January 2, 1983

MAN HAS MOVED another step closer to what ultimately will be his greatest scientific achievement — the fusion of hydrogen and other light elements into heavier elements in imitation of the reaction that goes on perpetually on the sun.

That step was taken on Christmas Eve in the laboratories of the same university where Albert Einstein did the work that resulted in man's last great scientific achievement — the fission of elements to produce horrendously powerful explosive weapons or to create the heat that produces electricity. The scientists at Princeton University succeeded in their attempts to heat hydrogen gas to produce what they call a plasma in which the electrons and the nuclei of the atoms are separated.

It was an achievement which Harold Furth, director of the Princeton laboratory, said was "like Columbus finding the New World." It was the same allusion that Enrico Fermi and his associates used in announcing to President Franklin D. Roosevelt that tney had successfully created the first controlled fission under the stands of Alonzo Stagg Field in Chicago 40 years ago.

The next step on that journey of scientific discovery will be the sustaining of the tremendous heat necessary for such a reaction to the point where the nuclei of atoms will be propelled and collide and join — actual fusion.

When that stage is reached, man will be in command of the greatest source of energy ever imagined.

HE WILL be free of dependence upon oil and natural gas and wood. He even will be able to abandon his thus-far cumbersome efforts to harness the sun's energy.

Man will be free to devote his energies to useful endeavors, for no single nation or group of nations any longer will control the means of production. A new day will dawn for all mankind when man finally create's his own sun. And it appears we can see that day now just over the horizon.

THE LOUISVILLE TIMES

Louisville, Ky., January 5, 1983

The successful Christmas Eve test of the new nuclear fusion reactor in Princeton, N. J., brought forth a barrage of all too familiar predictions. The machine, it was said, could usher in an era of unlimited, safe and clean energy.

A couple of decades ago, the advocates of nuclear *fission*, the technology on which the commercial nuclear power industry is based, similarly promised an abundance of cheap energy that could be produced without serious risk to people or the environment. Their forecasts proved disastrously wrong.

This time, however, the scientific community is appropriately tempering high expectations with caution.

One reason is that commercial use of the fusion process, which creates energy by combining, rather than splitting, atoms, probably lies decades in the future. Many technical problems remain to be solved. No one is sure that fusion will work on a large scale at a manageable cost.

In addition, there's reason to hope that the "nuclear priesthood" has abandoned some of its technological arrogance. After the blunders, miscalculations and deceit that have plagued our misadventures with fission power, the advocates of fusion surely know better than to offer the public another miracle cure with no unpleasant side effects.

That having been said, the success reported by the physicists who run Princeton's doughnut-shaped "tokamak," as the fusion reactor is called, offers many more reasons for celebration than concern.

If the technological complexities can be mastered, fusion does promise to produce vast amounts of energy without the problems that have made fission increasingly unacceptable. Much of the hydrogen "fuel" can be derived from seawater. While current fission reactors produce such highly radioactive and hard-to-dispose-of substances as plutonium, strontium and krypton, fused hydrogen atoms yield inert helium.

Fusion also leaves radioactive waste, but it is neither so dangerous nor so long-lived. Even though the "fire" in a fusion reactor, which operates on the same principle as the hydrogen bomb and the sun, will eventually reach 100 million degrees, scientists insist the reaction can't get out of control.

All this has yet to be demonstrated. However, the inherent advantages of fusion make the Reagan administration's determination to squander hundreds of millions of dollars on the $3.3 billion Clinch River breeder reactor all the harder to understand. Even Congress is coming to realize that the breeder gobbles up too large a share of limited federal research money even though it shows no sign of being economically feasible. The breeder, moreover, will introduce plutonium, the stuff of which bombs are made, into everyday commerce.

Fusion experiments are by no means cheap. They will get $454 million in government support this year and the eventual cost may reach $15 billion. But investments in these projects, which have begun to show results at Princeton and elsewhere, and in the grievously shortchanged solar technologies offer a better hope of an energy payoff in the 21st Century.

THE SUN

Baltimore, Md., January 10, 1983

Alas, it may not be till the next century that fusion energy, the power of the stars and the H-bomb, becomes a source of domestic energy for man. Controlled fusion has never been achieved. Even after it is, a prodigious engineering job will remain to find ways to convert the energy to usable forms. Harold Furth, director of a Princeton fusion project, said that turning on the project's fusion machine recently was "like Columbus finding the New World." The simile would have been more precise if he had said it was like launching the Nina, Pinta and Santa Maria from Spain.

Indeed, Dr. Furth's remark was suspiciously similar to Enrico Fermi's message to the White House in 1942 after the Italian scientist turned on the world's first atomic pile under a stadium at the University of Chicago: "The Italian navigator has entered the New World." But Dr. Furth should have been more restrained, for while Dr. Fermi was reporting on the initiation of the first chain reaction with nuclear fuel, Dr. Furth's fusion reactor didn't even contain real fuel.

Still, the Princeton machine is not to be sneezed at. A doughnut-shaped device of a type called the Tokamak, it is the first one designed actually to achieve "ignition"—a controlled fusion reaction that will generate enough energy to sustain itself. That may happen in three years. Turning the machine on was a major step, a demonstration that the huge magnets that compress and heat the fuel actually work. Real fuel, heavy hydrogen, wasn't used; plain old hydrogen, which can't actually "fuse," was.

Fusion, like the more familiar fission, converts matter into energy. Atoms of heavy hydrogen are fused together to make a new element, helium, while releasing energy in the process. The Princeton machine is one approach of several being tried in the U.S. The energy-hungry Japanese probably are ahead of the Americans in major ways. The Russians, who invented the Tokamak (both the machine and the name), have advanced machines on the drawing boards. Fusion promises cheap energy from endlessly abundant fuel, without pollution. That's very much worth working toward, but it's still some way off. It will be even further off for the U.S. if this nation continues to spend on fusion at only a third the rate of the Japanese.

BUFFALO EVENING NEWS

Buffalo, N.Y., January 4, 1983

With the successful test of a fusion reactor, the United States has moved one crucial step closer to the goal — long sought by scientists — of harnessing nuclear fusion as a feasible source of virtually infinite energy supplies.

Using a test vessel at Princeton University financed by the Department of Energy, physicists successfully heated hydrogen gas to form a so-called "plasma," a step necessary before atomic nuclei can be fused to release enormous bursts of energy. The more dramatic breakthrough will hopefully occur in 1986, if the Princeton project can demonstrate after extensive testing that it is possible to generate more fusion energy than the extreme energy required to produce it.

Dr. George Keyworth, President Reagan's science adviser, hailed this preliminary demonstration of the heating process as providing "a tremendous amount of hope for fusion power." He described the Princeton machine as "our major new fusion facility, our flagship and probably the most single important facility in the next generation of energy technology."

Among scientists and energy experts, fusion power has long been regarded as the potential key to meeting the world's energy needs. Unlike nuclear fission, fusion involves the combining rather than the splitting of atoms — without the radioactive hazards fission produces.

The generation of electricity by the joining of hydrogen atoms, in a process similar to that which powers the sun, would bring vast advantages over other forms of energy. Hydrogen atoms, distilled from sea water, would continue to supply energy needs long after the exhaustion of uranium and fossil fuels derived from oil and coal.

Proponents of fusion research say that the establishment of fusion power on a commercial scale would ultimately be cheaper and more efficient than other energy sources, long-lasting, relatively free from pollution and flexible in its ready availability of electricity production. Moreover, fusion would have strong advantages over both solar and geothermal power, both of which are too widely dispersed to provide the concentrated quantities of power needed for many industrial and living requirements.

In the opinion of most scientists, the achievement of fusion power on a commercial scale, if possible at all, is a long way off. It will involve the awesome technological conquest of practical ways to cope with temperatures that preclude the use of most available materials for containment of a fusion reaction. Even so, some experts are confident that the engineering problems can be overcome, and research on various approaches to achieving a sustained and controlled fusion reaction is moving ahead at various laboratories in this country and abroad.

Some fusion advocates contend that research has suffered from Washington's budget constraints, and it is indeed true that the $314 million spent on the Princeton program thus far is minuscule compared with the cost for the Apollo space project.

While the so-called Tokamak machine at Princeton has taken the lead and lends hope for eventual fusion mastery, research on alternative engineering approaches including the use of laser beams is proceeding and could yield more feasible solutions in the long run. Before the nation assumes any costly gamble on any one research effort, we should make sure we are investing in the soundest approach. There will be time enough to make such a judgment if future fusion tests prove successful.

THE INDIANAPOLIS STAR

Indianapolis, Ind., January 4, 1983

An awe-inspiring flash of the future, a split-second burst, which is expected to lead to machines that can produce limitless safe electrical power in time to come, is one reason for optimism during the new year.

Physicists said last week at Princeton University's Plasma Physics Laboratory that at 3:06 a.m. on Friday, Dec. 24, they produced the first burst of plasma — hot, electrified gas — in a $314 million test vessel.

It took more power to produce the 1/20th of a second burst than it created.

But the scientists say by 1986 the process is expected to break even, producing as much as it consumes, which will point the way to commercial nuclear fusion machines.

Fusion duplicates the energy-making processes of the hydrogen bomb and the sun. A fusion reactor would join two forms of hydrogen atoms. The main fuel would be sea water. Power would be produced with much less radioactivity and dangerous waste than is done by the fission process used in present-day reactors.

Scientists say the arrival of fusion power will take a long time. One says it may not come until "well into the 21st century." The reason is that it involves "probably the most difficult technology man has ever attempted to develop."

Physicists are generally not given to dramatizing their work. But Dr. Harold Furth, director of the Princeton laboratory, said of the first burst: "It's like Columbus finding the New World. It's not how big it is but that they found land."

Considering the source, that's saying something.

The Princeton lab gets $120 million a year of the $454-million-a-year fusion power research program being financed by the Department of Energy. Considering the potential, it may be one of the wisest expenditures the federal government ever made.

The era opened up by fusion energy may be as revolutionary as the era opened by the discoveries of Columbus.

Detroit Free Press

Detroit, Mich., January 5, 1983

IT IS possible that Dec. 24, 1982, will go down in history as a date as decisive and important as Dec. 2, 1942, when the first controlled atomic reaction occurred under the grandstand at Stagg Field at the University of Chicago.

On Christmas Eve, 1982, the giant Tokamak fusion reactor at Princeton University stripped electrons in hydrogen from their nuclei, creating a plasma, the first necessary step toward the production of fusion power. The plasma endured only a fraction of a second, but that was good enough for the researchers. They are not yet ready for a sustained fusion reaction that will create more power than it uses. Maybe in a few years, but there is much to be known and done before then.

Fusion is the opposite of fission, the splitting of atoms, the present source of atomic energy. The fusion of atoms is the process that keeps the sun going, and creating fusion means creating temperatures and pressures similar to those in the sun. The known problems are great and the unknown problems are expected to be even more difficult.

The idea behind power from fusion is so simple that when it was first advanced there seemed little doubt that science would come up with the fusion answers in a hurry. In practice, however, the behavior of atomic particles at fusion temperatures necessitated learning a completely new science and constructing fantastic reactors to operate at temperatures ranging from absolute zero to 20 million degrees Fahrenheit.

If in fact the fusion process can be made economically practical it will be worth the billions that will have to be spent to achieve it. Even ordinary sea water could be used for fusion fuel, lessening dependence upon finite supplies of oil, coal and uranium.

Fusion would not result in such disposal problems as are posed by present-day atomic power. The fuel residues would have half lives measured in terms of days, rather than thousands of years as in the case of uranium-based atomic power by-products and wastes.

A quiet race is on to make nuclei in fusion experiments behave properly. Princeton's Tokamak is a design proposed by Andrei Sakharov, the exiled Russian dissident, and an associate, Igor Tamm. The Soviets are continuing the research, as are the Japanese and the British.

Fusion power in commercial quantity is a long way off. But if it is developed, the world of the future will be profoundly affected, and the anniversary of the Princeton Tokamak's start-up long remembered.

The Oregonian

Portland, Ore., January 2, 1983

At 3:06 a.m. on Christmas Eve, physicists at Princeton University's Plasma Physics Laboratory demonstrated the feasibility of fusion power — a limitless energy source that, when combined with recent breakthroughs in solar photovoltaic cell development, could solve the world's energy problems in the 21st century.

Make no mistake, this is a scientific discovery of great magnitude — one that, taken successfully to demonstration stage in the 1990s, could punch holes in the Club of Rome's doomsday report that based much of its no-growth forecast on declining conventional energy resources such as oil and gas.

The successful fusion experiment is earth-shattering for skeptics, some of whom believed that the Tokamak Fusion Test Reactor, first developed by the Soviet Union, was the wrong avenue to harness the hydrogen bomb-making process for peaceful energy-making applications.

Magnetic containment fusion had been strictly a drawing board dream until Dec. 24, 1982, when Princeton scientists produced the first burst of a stable plasma of hot, electrified gas for one-twentieth of a second. These are not exciting numbers, as it required more energy to produce the fusion burst than the burst itself created. But the task of obtaining a break-even burst or a profitable burst in future Tokamak experiments is straightforward physics compared to the two-decade-long struggle to achieve the first successful controlled fusion reaction.

The milestone experiment at Princeton should signal President Reagan and Congress to revise fusion power's budget in fiscal 1985-86, leading to a 1986 experiment when scientists hope to produce a Tokamak reactor burst lasting several seconds, using a stable plasma (with desired pressure and density) that reaches 100 million degrees Celsius, compared to last Friday's experiment in which the plasma temperature was cool by desired fusion power standards at 100,000 degrees Celsius.

Success in the 1986 experiment would mean that the United States could build the world's first commercial nuclear fusion plant well ahead of the predicted 2012 date that most scientists believed possible for demonstrating this technology.

The cost of demonstrating and commercially developing this power source, which obtains its fuel from a virtually inexhaustible supply of hydrogen in sea water, will be significant. Thus, the role fusion might play in 21st century life largely will become an economic question. The overriding scientific question about fusion was answered at Princeton.

The Times-Picayune
The States-Item

New Orleans, La., January 1, 1983

The United States is preserving its precarious lead in the long, expensive and technologically difficult drive to develop nuclear fusion as a permanent alternative to nuclear fission and fossil fuels for generating electric power. The successful testing of Princeton University's new $314 million Tokamak research reactor was a world first for the largest version of the kind of reactor that has been in the mainstream of fusion research.

But other nations are at work on Tokamaks, and other types of reactors are in various stages of research, so the lead, to the extent that such things matter in this kind of endeavor, remains up for grabs. The potential of fusion as an energy source is so great, however, that the United States should keep research and development fully funded.

The Tokamak test may not seem like much — plasma, a superhot gas, was produced in the reactor chamber's magnetic field for 50 thousandths of a second, and it was not the kind of plasma that would produce fusion and thus energy. But it showed that the machine worked. Now work can begin to achieve "ignition" — the fusion equivalent of fission's chain reaction producing a continuous flow of energy.

Fusion developers maintain that it holds out several promises for outperforming fission and fossil fuels. Fusion fuel is a form of hydrogen that occurs in limitless supplies in sea water. The energy produced per unit of fuel far exceeds that of fission and fossil fuel plants. It is relatively clean, its radioactivity being low-level and short-lived. Plants cannot explode, fuel cannot melt down or escape into society as was threatened by the Three Mile Island accident.

But containing the process that keeps the sun burning is an engineering task of considerable magnitude. Fusion involves heating a hydrogen gas to millions of degrees, and the only "container" that can hold such a material is a strong magnetic field. Various methods of heating the material are now being researched, and Princeton's Tokamak will test an American innovation called neutral beam injection. The major research task now is to reach and pass the break-even point at which the reactor will produce more energy than is used to produce the reaction.

Cost is the major non-technical unknown. Fuel supply and efficiency would clearly reduce fuel costs, but the cost of the plants might be even more prohibitive than the cost of fission reactors are becoming today. That will depend on economic conditions too far ahead to predict, for fusion is not being thought of as a practical commercial energy producer until early in the next century.

But the current energy situation demands developing alternative sources, and if one that can use a common, limitless fuel in a process that does not damage environment or life seems a practical possibility, the search should be given high priority.

Index

Glossary

Breeder Reactor: A nuclear reactor, also called a *fast reactor* or *converter reactor,* which converts uranium to plutonium and produces more fuel than it burns. See discussion on pp. 129, 130.

"China Syndrome": Another term for meltdown, referring figuratively to the melting of a reactor's uranium core all the way through the Earth, to China. Also, the name of a 1979 film whose plot revolved around a near-meltdown at a fictional nuclear plant.

Containment Building: A structure surrounding the nuclear reactor, usually constructed of concrete reinforced with steel, to act as a barrier against the escape of radioactive material.

Coolant: The cooling fluid, usually water, that continuously flows between the fuel rods to remove energy from the nuclear reactor's core.

Cooling Tower: A chimney used at both nuclear and coal-fired power plants to cool the boiler water, which is not involved in the reaction itself. Its shape, that of a cylinder widened at both ends, has become familiar through photographs of Three Mile Island and other power plants.

Core Meltdown: Often described as the worst possible accident that could befall a nuclear reactor. It refers to the conceivable end result of a loss-of-coolant accident in which the fuel rods become so hot as to melt, releasing radiation.

Decommissioning: The process of shutting off and dismantling a nuclear reactor after its useful life of 30 to 40 years is over.

Fission: The disintegration or splitting of atomic nuclei into two or more parts, releasing energy. See discussion on p. 236.

Fuel Rods: Tubes made of zirconium, a corrosion-resistant metal with a high melting point, filled with pellets of fuel (uranium). They are a dozen feet long and very thin, and when bathed in coolant hover at a temperature of about 600° Fahrenheit.

Fusion: The forced combination of two atomic nuclei into a heavier nucleus. See discussion on p. 236.

Loss-of-Coolant Accident (LOCA): Any situation in which the coolant, which removes heat from the reactor's core, is cut off. In addition to the primary cooling system, all nuclear reactors have backup cooling systems to prevent meltdown in the event of a LOCA—these are called Emergency Core Cooling Systems (ECCS).

Millirem: A convenient measurement of radiation exposure used to assess potential health hazards. The larger *rem* (roentgen equivalent man) indicates the extent of biological damage to a human cell, and is equivalent to an average adult male absorbing one rad of radiation. The average amount of radiation received by an American from all sources in one year is typically estimated at 200 millirems. The exposure level from a standard diagnostic X-ray is between 45 and 75 millirems. The statutory limit for occupational exposure to radiation is five rem per year.

Nuclear Regulatory Commission (NRC): An independent regulatory agency established in 1974 to take over the functions previously performed by the Atomic Energy Commission. These include licensing builders and operators of nuclear plants, establishing licensing standards, and inspecting the plants.

Plutonium: See discussion on pp. 129, 130, 144.

Power Plant: Basically a water heater. The heated water creates steam, the steam drives a turbine that turns a generator, and the generator produces electricity. In nuclear plants, it is the fission process which heats the water.

Rad (radiation absorbed dose): A measure of the amount of any kind of ionizing radiation absorbed in body tissue.

Radiation: Energy released in the form of particles (alpha or beta particles) or waves (gamma rays), emitted by a radioactive material.

Radioactive or Nuclear Waste: See discussion on p. 69.

Reactor Core: Tightly stacked bundles of fuel rods, usually numbering 30,000 to 40,000 in all, and together measuring around ten to 15 feet in diameter. This is where the chain reaction occurs at a nuclear plant to create energy.

Reprocessing of Fuel: See discussion on pp. 129, 144.

Roentgen: A measure of the quantity of X-ray (or gamma ray) radiation in the air.

Uranium: A heavy radioactive element that exists naturally as a mixture of three isotopes: uranium 234, 235 and 238. It is used as the fuel for commercial nuclear power plants.